Progress in Colloid and Polymer Science
Volume 140

Series Editors
F. Kremer, Leipzig
W. Richtering, Aachen

Progress in Colloid and Polymer Science

Recently published and Forthcoming Volume

Intelligent Hydrogels
Volume Editors: Gabriele Sadowski, WalterRichtering
Vol. 140, 2013

UK Colloids 2011
Volume Editors: Victor Starov, Peter Griffiths
Vol. 139, 2012

Trends in Colloid and Interface Science XXIV
Volume Editors: Victor Starov, Karel Procházka
Vol. 138, 2011

Trends in Colloid and Interface Science XXIII
Volume Editor: Seyda Bucak
Vol. 137, 2010

Gels: Structures, Properties, and Functions
Volume Editors: Masayuki Tokita, Katsuyoshi Nishinari
Vol. 136, 2009

Colloids for Nano- and Biotechnology
Volume Editors: Hórvölgyi, Z.D., Kiss, É.
Vol. 135, 2008

Surface and Interfacial Forces From Fnndamentals to Applications
Volume Editors: Auernhammer, G.K., Butt, H.-J , Vollmer, D.
Vol. 134, 2008

Smart Colloidal Materials
Volume Editor: Richtering, W.
Vol. 133, 2006

Characterization of Polymer Surfaces and Thin Films
Volume Editors: Grundke, K., Stanun, M., Adler, H.-J.
Vol. 132, 2006

Analytical Ultracentrifugation VIII
Volume Editors: Wandrey, C., Cölfen, H.
Vol. 131, 2006

Scattering Methods and the Properties of Polymer Materials
Volume Editors: Stribeck, N., Smarsly, B.
Vol. 130, 2005

Mesophases, Polymers, and Particles
Volume Editors: Lagaly, G., Richtering, W.
Vol. 129, 2004

Surface and Colloid Science
Volume Editor: Galembeck, F.
Vol. 128, 2004

Analytical Ultracentrifugation VII
Volume Editors: Lechner, M.D., Börger, L.
Vol. 127, 2004

Trends in Colloid and Interface Science XVII
Volume Editors: Cabuil, V., Levitz, P., Treiner, C.
Vol. 126, 2004

From Colloids to Nanotechnology
Volume Editors: Zrinyi, M., Horvolgyi, Z.D.
Vol. 125, 2004

Aqueous Polymer Dispersions
Volume Editor: Tauer, K.
Vol. 124, 2004

Trends in Colloid and Interface Science XVI
Volume Editors: Miguel, M., Burrows, H.D.
Vol. 123, 2004

Aqueous Polymer - Cosolute Systems
Volume Editor: Anghel, D.F.
Vol. 122, 2002

Gabriele Sadowski • Walter Richtering
Editors

Intelligent Hydrogels

Springer

Editors

Gabriele Sadowski
TU Dortmund
Thermodynamik
Dortmund
Germany

Walter Richtering
RWTH Aachen
LS für Physikalische Chemie
Aachen
Germany

ISSN 0340-255X ISSN 1437-8027 (eBook)
ISBN 978-3-319-34768-4 ISBN 978-3-319-01683-2 (eBook)
DOI 10.1007/978-3-319-01683-2
Springer Cham Heidelberg New York Dordrecht London

Printed on acid-free paper

Springer is part of Springer Science+Business Media (www.springer.com)

Preface

Hydrogels are polymer networks built from hydrophilic polymers, copolymers or even polyelectrolytes. Their network structure and their hydrophilic nature enable them to respond to external stimuli such as changes in temperature, pH, additive concentration or electrical fields. Due to their unique properties they can act as smart materials and offer a whole series of interesting applications.

This book summarizes the work of about 25 research groups in the field of responsive hydrogels. These groups worked together from 2006 until 2012 in the priority research program "Intelligent Hydrogels" funded by the German Science Foundation DFG. The research activities were focused on three aspects: hydrogel synthesis, modeling and simulation of thermo-physical hydrogel properties as well as on new hydrogel applications.

Contributions related to the synthesis of thermo-sensitive hydrogels mainly included the work on new building blocks for physical hydrogels which were synthesized following different synthesis strategies. One research group synthesized a homologous series of low-molecular-weight amphiphiles based on maleamic acid, another one synthesized polymers consisting of a long hydrophilic thermo-responsive middle block, which was end-capped by two small, strongly hydrophobic blocks. In both cases, the hydrogel formation upon self-assembly was found to depend on the hydrophilic/lipophilic balance of the building blocks. Thus, the length of alkyl chain and the size of the hydrophobic end blocks, respectively, could be used to trigger the thermo-responsive behavior as well as the structure of the resulting hydrogels.

In charged hydrogels, the number of physical crosslinks can also be increased by introducing counter ions into the gel. It could be shown that the responsive properties caused by ion replacement and the combined water diffusion/exchange processes could be monitored by ^{23}Na and ^{27}Al magnetic resonance imaging experiments. Moreover, ^{23}Na NMR could be also used to monitor the thermo-reversible melting and gelation processes of amphiphilic physical hydrogels as well as the effective pore size of the gels.

Besides bulk gels, also hydrogels grafted on surfaces or microgel particles are of particular interest for specific applications (e.g. for biosensors or for fast-release systems). Investigations of poly(*N*-isopropyl acrylamide) (PNiPAM) grafted network dynamics by fluorescence correlation spectroscopy revealed that the mobility of tracers in these networks is mainly influenced by crosslink density as well as by tracer-host interactions. Phase transitions of thermo-responsive microgel particles as function of crosslinker density were investigated by means of DLS and SANS.

A particular focus of the priority program was to improve the mechanical properties of hydrogels and to extend the range of hydrogel applications by the synthesis of composite materials. A whole variety of nano composite hydrogels with improved mechanical properties has been developed. One way was to link star-shaped hydrophilic prepolymers using linear triblock copolymers, whereas the middle block serves as molecular spring for additional energy dissipation during mechanical load. Another strategy is the introduction of silica nanoparticles into the gel, whereas an amino-functionalization of the nanoparticle surface acts as additional covalent cross links within the hydrogels.

Thermo-responsive PNiPAM nano- and microgels have been incorporated in hydrogel matrices leading to composite hydrogels with varying swelling properties.

Deposition of metal nanoparticles or other molecules in microgels with stimuli-responsive behavior make them promising materials which may act as nanoreactors. Gold, platinum and palladium nanoparticles were successfully generated in-situ in copolymer hydrogels. The studies indicate that the gold-containing gels exhibit the highest catalytic activity, and that the gels can be repeatedly used for catalysis.

Spherical and dumbbell-shaped core-shell microgel particles were synthesized and used for the immobilization of catalytic metal nanoparticles. The catalytic activity of the nanoparticles could be modulated both by the volume transition and the change of polarity of the thermosensitive shell of the carrier system. It could be shown that these microgel particles can also act efficiently for the immobilization of enzymes.

Using magnetic nanoparticles even opens completely new possibilities for hydrogel applications since those hydrogels are able to respond to magnetic fields. A variety of complex gel structures were designed showing interesting combinations of mechanical and thermal properties.

Another interesting application of core-shell hydrogels are hydrogel particles with hydrophobic shells. Whereas literature research usually focuses on core-shell hydrogel structures having *hydrophilic* shells, in the priority program core-shell particles were synthesized which are composed of a water-swellable core and a hydrophobic elastic shell. These composite hydrogels can be applied to design switchable surfaces which become hydrophobic in the presence of water and thus offer a lot of new applications. They could be synthesized in a stepwise approach but also in a *one-step synthesis* by a water-in-oil inverse suspension polymerization.

Another series of contributions is concerned with measuring and modeling physical and thermodynamic properties of hydrogel systems. The influence of salt and pH on the swelling equilibria of polyelectrolyte hydrogels was measured and experimental data were correlated using analytical thermodynamic models. Moreover, a modeling strategy for describing the influence of additives like organic solvents, dissolved polymer molecules and electrolytes on the swelling/deswelling behavior of hydrogels was developed.

To model also the stimulation of polyelectrolyte hydrogels by electrical fields, an analytical formulation was derived that accounts for chemical, thermal as well as electrical effects in hydrogel systems which is based on balance laws of continuum chemo-electro-mechanics. As a result, the coupling of the different effects could be clearly demonstrated and quantified.

The different modeling strategies including analytical methods as well as molecular simulations are discussed in a review article with emphasis on molecular simulations using coarse-grained models.

The influence of temperature and pH on the self-assembly of physical gels was studied by molecular-dynamics simulation leading to a qualitative description of the experimentally-found behavior. Another group applied the self-consistent field theory to polymer and copolymer blends with reversible crosslinks. It could be shown that depending on crosslink strength and crosslink asymmetry, the structure of the system changes from homogeneous to hexagonal to lamellar.

Finally, hydrogel properties are discussed and investigated towards new applications for water desalination or as chemo-mechanical sensors for a continuous monitoring of metabolites.

Thus, this special issue gives a comprehensive overview about the different aspects of hydrogel research in Germany ranging from hydrogel synthesis, polymer physics and modelling to hydrogel applications. Only the interplay of these different disciplines allows understanding the structure-property relationships of these materials and thus designing hydrogels having the physical and chemical properties required for specific applications.

Dortmund, Germany — Gabriele Sadowski
Aachen, Germany — Walter Richtering

Contents

Supramolecular Chromaticity and Thermoresponsive Hydrogels: A Self-Assembly Study on Maleamic Acid-Based Amphiphiles

Andreas Bernet, Marina Behr, Rodrigo Q. Albuquerque, Marko Schmidt, Jürgen Senker, and Hans-Werner Schmidt

Abstract

A new class of homologous low molecular weight amphiphiles based on maleamic acid was synthesized and investigated in terms of its self-assembly behavior in bulk and in solution. The unexpected yellow color as bulk material and in organic solvents was revealed by means of spectroscopic and theoretical investigations to originate from intermolecular π-π interactions yielding supramolecular chromophores. It was found that the length of the alkyl chain of the amphiphiles and the resulting hydrophilic/lipophilic balance dictates the aggregation mode in bulk. One special compound exhibiting an n-tetradecyl chain was found to form stable thermoreversible supramolecular hydrogels in aqueous sodium hydroxide solutions. The corresponding hydrogels feature a rare cellular bilayer-based morphology and can be transferred into viscoelastic solutions upon heating and vice versa.

Keywords

Bilayer • Vesicle • Wormlike micelles • Supramolecular chromophore

Introduction

Low molecular weight (LMW) hydrogelators are an interesting class of relatively small molecules with molecular weights typically below 3.000 g mol^{-1} that are able to form macroscopic gels in aqueous media [1]. In such supramolecular or "molecular" gels [2] the solvent phase is entrapped in mesoscopic, three-dimensional networks consisting of supramolecular aggregates of the corresponding gelator molecules. The intermolecular interactions are generally non-covalent and are based on forces such as hydrogen bonds, π-π-, van der Waals- and electrostatic interactions, charge-transfer complexes, hydrophilic/hydrophobic interplay and combinations thereof. This allows for the control of the build-up and disassembly process of the LMW gels by external triggers such as temperature, pH, salt content, or irradiation with light. Being able to respond to external stimuli, these macroscopic gels are often called "smart" or "responsive" LMW gels. Within the last decades, LMW gels have become an interesting new class of functional materials [3] and promising applications especially of LMW hydrogels have been already reported. For example, LMW hydrogels have been used for cell growth scaffolding [4], wound treatment [5], dye removal [6], protein filtration [7], sensoring of biomolecules [8], nanoparticle preparation [9], and especially for drug-release applications [10].

We have already demonstrated that carboxylic acid amphiphiles featuring aromatic amide units represent

The online version of this chapter (doi:10.1007/978-3-319-01683-2_8) contains supplementary material, which is available to authorized users.

A. Bernet (✉) • M. Behr • H.-W. Schmidt
Macromolecular Chemistry I, Bayreuth Institute for Macromolecular Research (BIMF), Bayreuth Center for Colloids and Interfaces (BZKG), University of Bayreuth, Universitätsstrasse 30, 95440, Bayreuth, Germany
e-mail: andreas.bernet@uni-bayreuth.de

R.Q. Albuquerque
Theoretical Physics IV, University of Bayreuth, Universitätsstrasse 30, 95440, Bayreuth, Germany

Institute of Chemistry of São Carlos, University of São Paulo (USP), Av. Trab. São-carlense 400, 13560-970 São Carlos-SP, Brazil

M. Schmidt • J. Senker
Inorganic Chemistry III, University of Bayreuth, Universitätsstrasse 30, 95440, Bayreuth, Germany

G. Sadowski and W. Richtering (eds.), *Intelligent Hydrogels*, Progress in Colloid and Polymer Science 140,
DOI 10.1007/978-3-319-01683-2_1,

a promising structural motif for the development of supramolecular hydrogels [11]. Therefore we aimed to investigate if this strategy can successfully be used for the preparation of maleamic acid-based amphiphilic LMW hydrogelators. Maleamic acid-derived compounds are of special interest due to several reasons. They exhibit high stability towards hydrolysis in acidic media [12], show bioactivity [13], are synthetic precursors, for instance for the preparation of technically important maleimides [14], form interesting mesoscopic aggregates in aqueous media [15], and they feature a polymerizable aliphatic double bond. Furthermore, the double bond of maleic acid derivatives offers the possibility for additional functionality as its geometry can be switched by external triggers from *cis* to *trans* configuration. This allowed the development of a photoswitchable LMW gelator system based on a maleic acid bisamide [16]. Furthermore, one maleamic acid-based amphiphile has been reported to act as LMW hydrogelator [17]. Additionally, this compound was utilized as polymerizable emulsifier in copolymerization reactions of styrene and n-butyl acrylate [18] and for molecular imprinting [19].

We have synthesized a homologous series of maleamic acid-based amphiphiles that only differ in the length of their n-alkyl substituent. All compounds were analyzed in bulk by means of Fourier-transform infrared (FT-IR) spectroscopy and X-Ray diffraction (XRD) techniques. The behavior of selected compounds in organic solvents was investigated by UV-vis spectroscopy and confirmed by quantum chemical methods. We then evaluated the potential of these compounds to form thermoreversible hydrogels. The discovered hydrogel system was thermally characterized and the supramolecular mesoscopic structure was elucidated.

Experimental Part

Materials and Methods

All chemicals were used as received without further purification. Organic solvents were distilled prior to usage. Water was desalted by a standard ion exchange setup. Sodium hydroxide solutions were prepared by mixing weighted amounts of NaOH (technical grade) with distilled water under stirring.

Size-Exclusion Chromatography (SEC)

For test of purity, oligomeric size exclusion chromatography (oligo SEC) was performed utilizing a Waters 515 HPLC pump and stabilized THF as eluent at a flow rate of 0.5 mL min^{-1}. 20 μL of a solution with a concentration of approximately 1 g L^{-1} was injected into a column setup comprising a guard column (Varian, 5×0.8 cm, Mesopore gel, particle size 3 μm) and two analytical columns (Varian, 30×0.8 cm, Mesopore gel, particle size 3 μm). For detection of compounds, a WATERS 486 tunable UV detector at 254 nm was used.

Proton Nuclear Magnetic Resonance (^{1}H-NMR) Spectroscopy

^{1}H-NMR experiments were carried out on a Bruker Avance AC 300 spectrometer at 298 K (300 MHz). The compounds were dissolved in DMSO-d_6, which also acted as internal reference.

Thermal Analysis

TGA/DSC measurements were performed on a NETZSCH Simultane Thermoanalysenapparatur STA 409 under nitrogen atmosphere. The crucible consisted of aluminium oxide and as reference substance kaolin was used. Nitrogen flow rate was set to 75 cm^3 min^{-1}. A heating curve from 25 °C to 650 °C (10 °C min^{-1}) was applied, and the melting peak maximum was chosen for further evaluation.

Mass Spectroscopy (MS)

Mass spectra were recorded on a Finnigan MAT 8500 apparatus (EI, 70 eV) using direct injection mode.[1]

Elemental Analysis

Elemental analysis (C, H, N) was carried out with a HEKAtech EA 3000 instrument. The theoretical amount of all elements was calculated using Isis Draw.

Preparation of Xerogels (Dried Hydrogels) by Lyophilization

Xerogels were obtained by placing a small piece of hydrogel in a glass vessel followed by shock-freezing in liquid nitrogen and subsequent drying under reduced pressure ($p < 0.1$ mbar) for 24 h.

[1]Only molecule peak ($m^{\bullet +}$) and peaks with relative intensities ≥ 10 % are listed in the corresponding characterization datasets for compounds **1–6**.

Fourier-Transform Infrared (FT-IR) Spectroscopy

FT-IR spectra of bulk material as obtained from synthesis and of lyophilized hydrogel samples were recorded with a Perkin-Elmer Spectrum 100 FT-IR spectrometer in ATR mode. All samples were powdered prior to the measurements.

X-Ray Diffraction (XRD)

XRD analysis of bulk material and of hydrogels was performed on a Huber Guinier-Diffraktometer 6000 (Huber Germany), equipped with a Huber Quarz-Monochromator 611, a Cu-anode ($CuK_{\alpha 1}$-beam, $\lambda = 154.051$ pm, X-ray generator from Seifert Company Germany), a Huber SMC 9000 stepping motor controller and a self-developed gate system, primary beam stopper and sample oven. XRD analysis of the hydrogel samples was carried out by filling preheated Mark tubes (outer diameter: 2 mm, wall thickness: 0.01 mm) with a hot gelator solution. The samples were sealed and stored at room temperature for 24 h prior to the XRD measurements. For the XRD analysis of bulk materials Mark tubes with an outer diameter of 1 mm were used. All XRD data is presented as recorded without further treatment such as baseline correction or curve smoothing.

Structure Evaluation by X-Ray Powder Diffractometry

All details for the structure evaluation of compound **1** by X-Ray powder diffractometry can be found in the SI.

Gelation Tests ("Inverted Test Tube Method")

A mixture of amphiphile and the corresponding aqueous media (1.0 mL of solvent; 10 g L^{-1} of amphiphile) was placed in a screw cap vial (inner diameter: 12 mm; filling height: 9 mm) containing a small stirring bar. After the vial was sealed, the mixture was heated to approx. 95 °C under stirring. Samples were allowed to stand at 18 °C for 24 h. Gel state was defined if upon inversion of the test tube no gravitational flow was observed.

Scanning Electron Microscopy (SEM) Studies

Bulk material as obtained from synthesis was placed on a standard sample holder equipped with a conducting adhesive pad ("C Leit-Tabs", Plano Germany), sputtered with platinum and examined with a LEO 1530 (FE-SEM with Schottky-field-emission cathode; in-lens detector) using an accelerating voltage of 2–4 kV. Xerogels were placed on a standard sample holder equipped with a conducting adhesive pad ("C Leit-Tabs", Plano Germany) and examined with a FEI Phenom desktop SEM.

Cryogenic SEM Experiments (Cryo-SEM)

Cryo-SEM experiments of a hydrogel prepared from 10 g L^{-1} of **5** in 0.1 M aq. NaOH sol. were carried out using a Gatan Alto 2500 Cryotransfer apparatus and a JEOL JSM-7500 F/FA FE-SEM at an acceleration voltage of 0.7–1.0 kV.

Cryogenic Transmission Electron Microscopy (Cryo-TEM)

For cryo-TEM studies, a 2 μl droplet of the sample was placed on a hydrophilized lacey carbon filmed copper grid (200 mesh, Science Services Germany), where most of the liquid was removed with blotting paper leaving a thin film stretched over the carbon net holes. The specimens were instantly shock frozen by rapid immersion into liquid ethane and cooled to approximately 90 K by liquid nitrogen in a temperature-controlled freezing unit (Zeiss Cryobox, Zeiss NTS GmbH Germany). The temperature was monitored and kept constant in the chamber during all the sample preparation steps. After freezing the specimens the remaining ethane was removed using blotting paper. The specimen was inserted into a cryotransfer holder (CT3500, Gatan Germany) and transferred to a Zeiss EM922 EFTEM (Zeiss NTS GmbH Germany). Examinations were carried out at temperatures around 90 K. The TEM was operated at an acceleration voltage of 200 kV. Zero-loss filtered images ($\Delta E = 0$ eV) were taken under reduced dose conditions (*ca.* 100–1,000 e nm^{-2}). All images were registered digitally by a bottom mounted CCD camera system (Ultrascan 1000, Gatan Germany) combined and processed with a digital imaging processing system (Digital Micrograph 3.10 for GMS 1.8, Gatan Germany).

Conductivity Measurements

Specific conductivity was measured using a Conductivity Meter CDM 230 equipped with a CDC861T 4-pole Pt conductivity cell (Radiometer Analytical France). The temperature of the sample vial was controlled using a Lauda RC6 cryostat. The samples were heated to *ca.* 85 °C and subsequently cooled down to *ca.* 15 °C. The resulting heating and cooling rates were *ca.* 1.2 °C min^{-1} and 3.5 °C h^{-1}, respectively. Values were recorded using a computer equipped with a RS232 interface at a time interval of 30 s during cooling and of 5 s during heating sequences.

Micro-Differential Scanning Calorimetry (μDSC)

μDSC measurements were performed on a Setaram MicroDSC III (Setaram France). A hot solution (*ca.* 85 °C) of a molten hydrogel (*ca.* 0.7 mL) was filled into a Hastelloy C276 sample cell which was immediately sealed to avoid evaporation. The reference cell was filled with an equal amount of pure solvent, and the sealed cells were allowed to stand at room temperature overnight. Prior to measurement, the sample was kept at the desired starting temperature for at least 1 h to ensure thermal equilibration.

UV-Vis Absorption Spectroscopy

UV-Vis absorption was measured at room temperature using a HITACHI U-3000 spectrometer and Hellma Suprasil 110 QS cuvettes (optical path length: 10 mm).

Computational Methods

Computational details, atom coordinates and visualization of HOMO and LUMO of **1** as monomer and as dimer can be found in the SI.

Synthesis of Compounds 1–6

General Procedure for the Synthesis of 1–6

Equivalent amounts of maleic acid anhydride and the corresponding 4-alkyl aniline derivative were dissolved separately in dichloromethane (ca. 50 mL of solvent per gram of anhydride or aniline derivative, respectively). The solutions were mixed slowly under stirring while being cooled using a water bath. The reaction mixture was stirred for additional 3 h. Subsequently, the product was fully precipitated by the addition of hexane, filtrated using a glass pore filter (pore size: 3) and washed several times with small amounts of a mixture of hexane and dichloromethane (1:1). The obtained solid was dried under reduced pressure at room temperature for 24 h. Compounds were obtained as yellow (**1**, **2**) and white (**3**–**6**) solids. Yield range: 87–95 %, 91 % average. No further purification was carried out as SEC analysis of **1**–**6** confirmed a purity of ≥99 %.

Characterization of Compounds 1–6

1: (2Z)-4-[(4-hexylphenyl)amino]-4-oxobut-2-enoic acid

Yield: 89 %. M. p.: 149.7 °C. ^{1}H-NMR: 13.26 (br s, 1 H); 10.37 (s, 1 H); 7.50 (d, J = 8.3 Hz, 2 H); 7.13 (d, J = 8.3 Hz, 2 H); 6.45 (d, J = 12.0 Hz, 1 H); 6.28 (d, J = 12.0 Hz, 1 H); 2.50[2] (2 H); 1.52 (m, 2 H); 1.25 (m, 6 H); 0.84 (t, J = 6.5 Hz, 3 H). Molecular weight: 275.35 u. MS (m/z): 275 (m^{*+}), 230, 204, 177, 107, 106. Anal. calcd. for $C_{16}H_{21}NO_3$: C, 69.79; H, 7.69; N, 5.09; found: C, 69.72; H, 7.68; N, 5.13.

2: (2Z)-4-[(4-octylphenyl)amino]-4-oxobut-2-enoic acid

Yield: 90 %. M. p.: 140.1 °C. ^{1}H-NMR: 13.26 (br s, 1 H); 10.37 (s, 1 H); 7.52 (d, J = 8.4 Hz, 2 H); 7.14 (d, J = 8.4 Hz, 2 H); 6.47 (d, J = 12.1 Hz, 1 H); 6.30 (d, J = 12.1 Hz, 1 H); 2.51 (2 H); 1.53 (m, 2 H); 1.25 (m, 10 H); 0.85 (t, J = 6.6 Hz, 3 H). Molecular weight: 303.40 u. MS (m/z): 303 (m^{*+}), 205, 106. Anal. calcd. for $C_{18}H_{25}NO_3$: C, 71.26; H, 8.31; N, 4.62; found: C, 71.05; H, 8.44; N, 4.62.

3: (2Z)-4-[(4-decylphenyl)amino]-4-oxobut-2-enoic acid

Yield: 95 %. M. p.: 144.3 °C. ^{1}H-NMR: 13.25 (br s, 1 H); 10.36 (s, 1 H); 7.50 (d, J = 8.4 Hz, 2 H); 7.12 (d, J = 8.4 Hz, 2 H); 6.45 (d, J = 12.1 Hz, 1 H); 6.28 (d, J = 12.1 Hz, 1 H); 2.54 (m, 2 H); 1.53 (m, 2 H); 1.23 (m, 14 H); 0.84 (t, J = 6.5 Hz, 3 H). Molecular weight: 331.46 u. MS (m/z): 331 (m^{*+}), 233, 106. Anal. calcd. for $C_{20}H_{29}NO_3$: C, 72.47; H, 8.82; N, 4.23; found: C, 72.52; H, 8.99; N, 4.28.

4: (2Z)-4-[(4-dodecylphenyl)amino]-4-oxobut-2-enoic acid

Yield: 87 %. M. p.: 142.2 °C. ^{1}H-NMR: 13.26 (br s, 1 H); 10.37 (s, 1 H); 7.52 (d, J = 8.4 Hz, 2 H); 7.14 (d, J = 8.4 Hz, 2 H); 6.47 (d, J = 12.1 Hz, 1 H); 6.30 (d, J = 12.1 Hz, 1 H); 2.55 (m, 2 H); 1.53 (m, 2 H); 1.23 (m, 18 H); 0.85 (t, J = 6.7 Hz, 3 H). Molecular weight: 359.51 u. MS (m/z): 359 (m^{*+}), 261, 106. Anal. calcd. for $C_{22}H_{33}NO_3$: C, 73.50; H, 9.25; N, 3.90; found: C, 73.17; H, 9.55; N, 3.93.

5: (2Z)-4-[(4-tetradecylphenyl)amino]-4-oxobut-2-enoic acid

Yield: 94 %. M. p.: 140.6 °C. ^{1}H-NMR: 13.31 (br s, 1 H); 10.38 (s, 1 H); 7.51 (d, J = 8.4 Hz, 2 H); 7.11 (d, J = 8.4 Hz, 2 H); 6.46 (d, J = 12.1 Hz, 1 H); 6.29 (d, J = 12.1 Hz, 1 H); 2.53 (m, 2 H); 1.51 (m, 2 H); 1.21 (m, 22 H); 0.83 (t, J = 6.4 Hz, 3 H). Molecular weight: 387.57 u. MS (m/z): 387 (m^{*+}), 289, 106. Anal. calcd. for $C_{24}H_{37}NO_3$: C, 74.38; H, 9.62; N, 3.61; found: C, 74.39; H, 9.60; N, 3.32.

6: (2Z)-4-[(4-hexadecylphenyl)amino]-4-oxobut-2-enoic acid

Yield: 93 %. M. p.: 136.7 °C. ^{1}H-NMR: 13.07 (br s, 1 H); 10.44 (s, 1 H); 7.51 (d, J = 8.4 Hz, 2 H); 7.14 (d, J = 8.4 Hz, 2 H); 6.45 (d, J = 12.1 Hz, 1 H); 6.30 (d, J = 12.1 Hz, 1 H);

[2]The multiplicity of this signal cannot be determined as it is covered by the signal of DMSO. However, the presence of this peak can be proven by HH-COSY NMR experiments. This is also the case for compounds **2**–**6**.

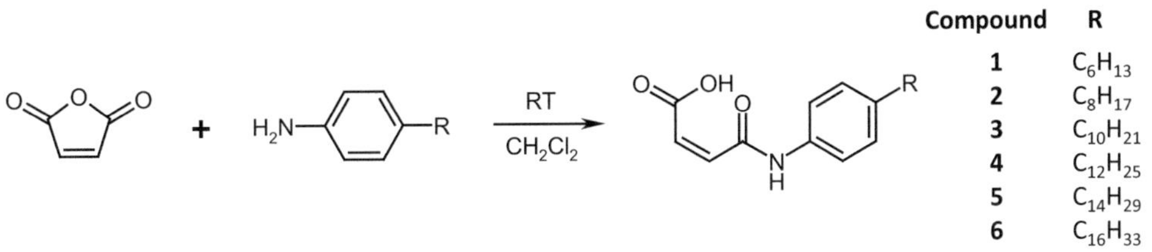

Scheme 1 Synthesis and chemical structures of the investigated maleamic acid-based amphiphiles **1–6**

Table 1 Color and characteristic FT-IR absorption peaks of compounds **1–6** as obtained from synthesis. All values are given in cm^{-1}. ν: stretching band. Original data can be found in the SI (see Fig. S10–S15)

				ν(CH_2)			
Compound	Color	ν(N-H)	ν(O-H)	*Antisymmetric* ν_{as}	*Symmetric* ν_s	ν(C = O) CO_2H	ν(C = O) *CONH*
1	Yellow	3,267	3,201	2,925	2,864	1,698	1,629
2	Yellow	3,288	3,212	2,923	2,850	1,697	1,633
3	White	3,306	3,210	2,917	2,850	1,723	1,629
4	White	3,306	3,210	2,916	2,850	1,723	1,629
5	White	3,307	3,210	2,916	2,850	1,724	1,631
6	White	3,307	3,210	2,916	2,850	1,724	1,631

2.54 (m, 2 H); 1.53 (m, 2 H); 1.23 (m, 26 H); 0.85 (t, J = 6.6 Hz, 3 H). Molecular weight: 415.62 u. MS (m/z): 318, 317, 290, 289, 107, 106. Anal. calcd. for $C_{26}H_{41}NO_3$: C, 75.14; H, 9.94; N, 3.37; found: C, 75.00; H, 10.20; N, 3.40.

Results and Discussion

Synthesis

Synthesis of the maleamic acid-based amphiphiles **1–6** was accomplished by turnover of maleic acid anhydride with equimolar amounts of the corresponding 4-alkyl substituted aniline derivative in dichloromethane at room temperature (Scheme 1). This highly effective one-step ring-opening synthesis yielded **1–6** in excellent yields (87–95 %; 91 % average). Size-exclusion chromatography (SEC) confirmed the high purity of the compounds (≥99 %). Interestingly, the shorter-tailed compounds **1** (n-hexyl) and **2** (n-octyl) were colored intensively yellow in bulk (see SI Fig. S1), whereas the compounds **3–6** were obtained as white powder. This is remarkable, as all compounds feature the same molecular chromophoric group and differ only in their alkyl chain length.

1–6 were fully characterized by proton nuclear magnetic resonance spectroscopy (^{1}H-NMR), mass spectroscopy (MS), elemental analysis (EA), differential scanning calorimetry (DSC) in combination with thermogravimetric analysis (TGA), X-ray powder diffractometry (XRD), and Fourier-transform infrared spectroscopy (FT-IR). Evaluation of the thermal properties of **1–6** showed that all compounds are thermally stable up to at least 136 °C and decompose subsequent to melting (see SI Table S1).

XRD of 1–6 as Obtained from Synthesis

XRD analysis of **1–6** as obtained from synthesis revealed that all compounds are of crystalline nature (see SI Fig. S2–S7). Furthermore, the XRD spectra of **5** and **6** feature unequivocal [*h*00] Bragg peak series (2θ ratio of 1:2:3:4:5:6), thus clearly indicating a lamellar arrangement in the bulk material. Additionally, some of the [*h*00] diffraction peaks can also be recognized in case of shorter-tailed compounds **3** and **4** (see SI Table S2 and Fig. S9). The tendency of **5** to form lamellar structures is also expressed in the microstructure of **5** as obtained from synthesis (stacks of thin platelet-like crystals; see SI Fig. S8).

FT-IR Investigations of 1–6 as Obtained from Synthesis

FT-IR investigations of **1–6** showed an interesting behavior dependent on the corresponding alkyl chain length (Table 1). Characteristic IR absorption peaks of amide, n-alkyl, and carboxylic acid groups [20, 21] can be recognized. Although all compounds show H-bonded amide units in bulk, two main groups of compounds can be distinguished: shorter-tailed compounds **1** and **2** (group A) and **3–6** featuring longer alkyl chains (group B). With increasing alkyl chain length a transition from group A to group B occurs that is accompanied by several changes in the characteristic IR absorption bands.

The ν(N-H) stretching band is blue-shifted (*ca.* 39–18 cm^{-1}) to 3,306–3,307 cm^{-1}. The red-shift of $\nu_{as}(CH_2)$ (*ca.* 7–8 cm^{-1}) points out a change in the n-alkyl chain packing [22]. The carbonyl stretching band ν(C = O) of the carboxylic acid moiety is blue-shifted (*ca.* 26 cm^{-1}). Altogether, this is a clear indication that an increase of the alkyl chain length changes the hydrophilic/lipophilic balance (HLB) and yields a significant change in the packing motif of the amphiphiles in bulk. Interestingly, this is in agreement with the change of chromaticity, as only **1** and **2** (group A) are colored yellow as bulk materials.

Crystal structures of related compounds N-phenyl maleamic acid [23], N-(4-chlorophenyl) maleamic acid [24], and N-(4-methylphenyl) maleamic acid [25] have been reported. These molecules are nearly planar in the crystalline state and exhibit intermolecular distances between the maleamic and the phenyl unit in the range of 3.3–3.4 Å, thus indicating strong π-π-interactions. The hydroxyl group of the carboxylic acid moiety forms an intramolecular hydrogen bond with the carbonyl group of the amide unit. Furthermore, the molecules are interconnected in the crystal by hydrogen bonds between the amide N-H and the carbonyl group of the carboxylic acid moiety. IR absorption data of N-(4-bromophenyl) maleamic acid, N-(4-chlorophenyl) maleamic acid and N-(4-methylphenyl) maleamic show N-H stretching frequencies in the region of 3,195–3,260 cm^{-1} and two C = O stretching frequencies of 1,690–1,706 cm^{-1} and 1,622–1,634 cm^{-1} [26]. This fits very well to the FT-IR absorption bands observed for the short-tailed compounds **1** and **2**. Therefore we can conclude that **1** and **2** feature a comparable molecular arrangement including intramolecular C = O···H-O and intermolecular N-H···O = C hydrogen bonds.

To confirm these conclusions, the structure of **1** was evaluated by X-Ray powder diffraction (for experimental details, Rietveld profile and visualization of the unit cell see SI S16–S19). Indexing of the powder pattern lead unequivocally to a monoclinic metric and the structure could subsequently be solved and refined using the Rietveld algorithm in the non-centrosymmetric spacegroup $P2_1/c$ (a = 14.36 Å(5), b = 10.90 Å(4), c = 12.59 Å(5), β = 127.84 °(1), V = 1,556 $Å^3$ (2), wR_p = 3.1 %). The results show that all amide and carboxylic acid groups are hydrogen-bonded in the crystal as expected (see SI Fig. S18). The carbonyl group of the amide moiety forms an intramolecular hydrogen bond with the hydroxyl substituent of the acid moiety exhibiting a distance (C = O···H-O) of 1.55 Å. Furthermore, infinite chains are formed by intermolecular hydrogen bonds between the N-H group of the amide and the carbonyl group of the carboxylic acid moiety featuring a distance (N-H···O = C) of 1.93 Å. Significant π-π interactions of the double bond of the maleamic acid moiety and the 1,4-phenylene unit can be assumed due to their close distances of *ca.* 3.71 Å (SI Fig. S19). These findings strongly indicate that the yellow color of **1** as bulk material originates from intermolecular π-π interactions.

Contrary to compounds **1** and **2**, the longer-tailed compounds **3–6** have to be molecularly arranged in another fashion. Unfortunately, efforts to elucidate the crystal structure of model compound **5** by powder diffractometry remained unsuccessful. However, information about the molecular arrangement can be deduced from FT-IR and XRD data. As the relevant FT-IR absorption bands of molecular units of **3–6** that are capable of hydrogen bond formation are practically identical (deviations of only 1–2 cm^{-1}), we can conclude that both type and strength of intermolecular interaction and therefore the overall molecular arrangement of **3–6** in bulk have to be the same. XRD analysis already showed that compounds **5** and **6** clearly exhibit a lamellar arrangement. Considering the XRD and the FT-IR results, we can assume lamellar order for all the longer-tailed compounds **3–6**. Furthermore, the value for the C = O stretching band at 1,723–1,724 cm^{-1} is a very strong indication for infinite chains of hydrogen-bonded carboxylic acid moieties [27] and excludes the presence of cyclic dimers of carboxylic acid units. An *all-trans* ("zig-zag") configuration of the n-alkyl chains of compounds **3–6** is proven by the values of asymmetric (2,917–2,916 cm^{-1}) and symmetric $\nu(CH_2)$ stretching bands (2,850 cm^{-1}) [28]. These findings together with the observed lamellar arrangement in the bulk material strongly indicate that van der Waals interactions among the n-alkyl chains begin to control the self-assembly process from the threshold value of 10 carbon atoms or more (compounds **3–6**).

UV-Vis Investigations of 1 and 5 in Solution

Although all compounds feature identical chromophoric units, only the short-tailed compounds **1** and **2** are obtained from synthesis as yellow bulk materials. However, all compounds form yellowish solutions in organic solvents. As short-tailed model compound that exhibits yellow color both in bulk and in solution, **1** was chosen. For comparison purposes, the long-tailed compound **5** that features white color in bulk and yellow color in solution was selected.[3] We have investigated both the influence of solvent polarity and the concentration of amphiphile on the UV-Vis absorption of **1** and **5** in solution. Figure 1 shows the normalized UV-Vis absorption spectra of **1** in different organic solvents at a concentration of 10^{-4} mol L^{-1}.

A pronounced negative solvatochromism is observed. With a decrease in the polarity of the organic solvent, a red-shift of the UV-Vis absorption maximum occurs. From DMSO to THF and MeOH to iPrOH, the corresponding peak maxima shift from 283 nm to 303 to 314 nm. Compound

[3] **5** has been favored over **6** that features the longest alkyl chain because the former shows higher solubilities in organic media.

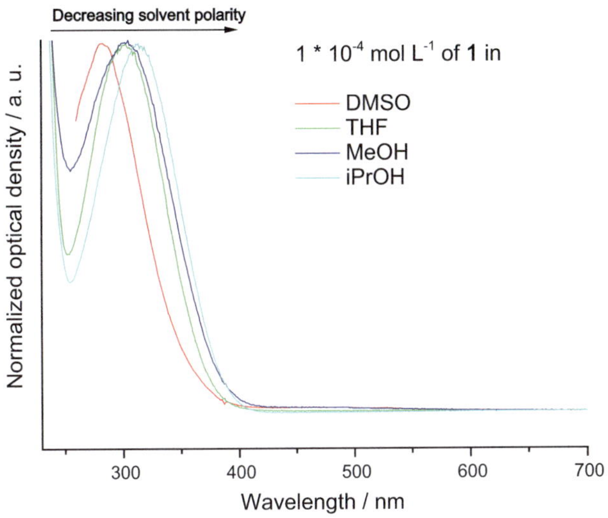

Fig. 1 Influence of the solvent polarity on the UV-Vis absorption of samples of **1** at a concentration of 10^{-4} mol L^{-1} in DMSO, THF, MeOH and iPrOH. The corresponding peak maxima are located at 283 nm, 303 nm, 303 nm, and 314 nm

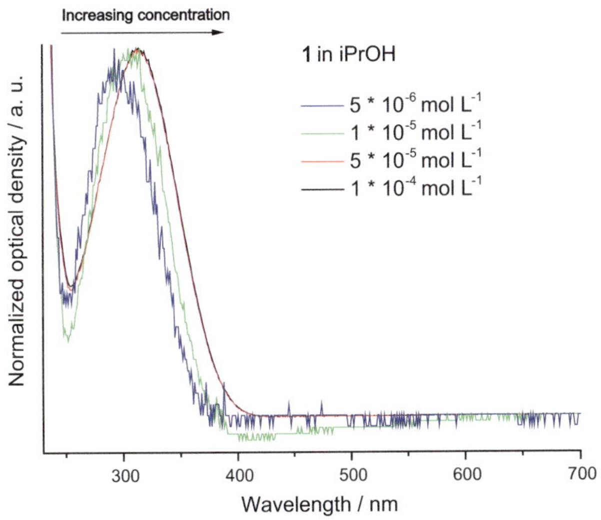

Fig. 2 Influence of the concentration on the UV-Vis absorption of **1** in iPrOH. The peak maxima are red-shifted upon increasing concentration of **1**. (For original data see SI Fig. S21)

5 shows comparable behavior in DMSO and THF (see SI Fig. S20). We assume that the yellow color of the solutions of amphiphiles originates from intermolecular aggregation, and that this aggregation is of non-covalent and therefore reversible nature. Therefore we investigated the influence of amphiphile concentration on the UV-Vis absorption. As an example, the UV-Vis spectra of **1** at different concentrations in iPrOH (where the observed effect is especially pronounced) are presented in Fig. 2.

Increasing the concentration of **1** leads to a red-shift of the absorption maxima. UV-Vis absorption spectra of a concentration series of **5** in DMSO also show similar behavior (see SI Figs. S22 and S23). Thus, we can conclude that the observed yellow color arises from intermolecular aggregates. An increase in monomer concentration shifts the equilibrium between (colorless) monomers and (colored) aggregates to the side of the aggregates and yields stronger absorption in the visible region. Furthermore, this could also explain the observed negative solvatochromism. A more polar solvent such as DMSO is able to break up intermolecular forces, and consequently the resulting absorption is blue-shifted. In a less polar solvent such as iPrOH, intermolecular aggregation is more likely to occur, and therefore the absorption is red-shifted.

Computational Investigations

To support our assumption that the yellow color of solutions and of bulk of **1** originates from intermolecular aggregation, Density Functional Theory (DFT) and Time-Dependent DFT (TDDFT) calculations of monomer and dimer of **1** were carried out (Fig. 3).

The results show that **1** tends to form intermolecular π-π interacting complexes. In this dimeric complex the molecules of **1** are arranged in an anti-parallel fashion. The intermolecular π-π distance was calculated to be *ca.* 3.2 Å. Although the dispersion correction used in the DFT is known to overestimate the van der Waals stabilization, significant π-π interaction can still be assumed. Furthermore, dimerization of **1** leads to a significant red-shift of the excited states (for exact values of calculated energy levels see SI Table S3). These findings qualitatively explain the red shift shown in Fig. 2 for higher concentrations of **1** and the resulting yellow color thereof. We also propose that the yellow color of **1** and **2** as obtained from synthesis originates from comparable π-π interactions. The alkyl chain substituent of both compounds seems to be short enough so that the polar interaction of the π-systems overrules the van der Waals interaction of the non-polar alkyl chains in bulk. The DFT results also suggest that the blue shift shown in Fig. 1 arises from differences in solubility among the solvents used, DMSO showing a monomer-like band and iPrOH the aggregate-like band.

These findings are also important in regard to the desired application of the amphiphiles as possible LMW hydrogelators where an appropriate HLB is crucial in order to avoid too high or too low solubilities in the aqueous media. Being able to vary the HLB by the choice of alkyl chain length, we expected to find a suitable amphiphile capable of hydrogel formation.

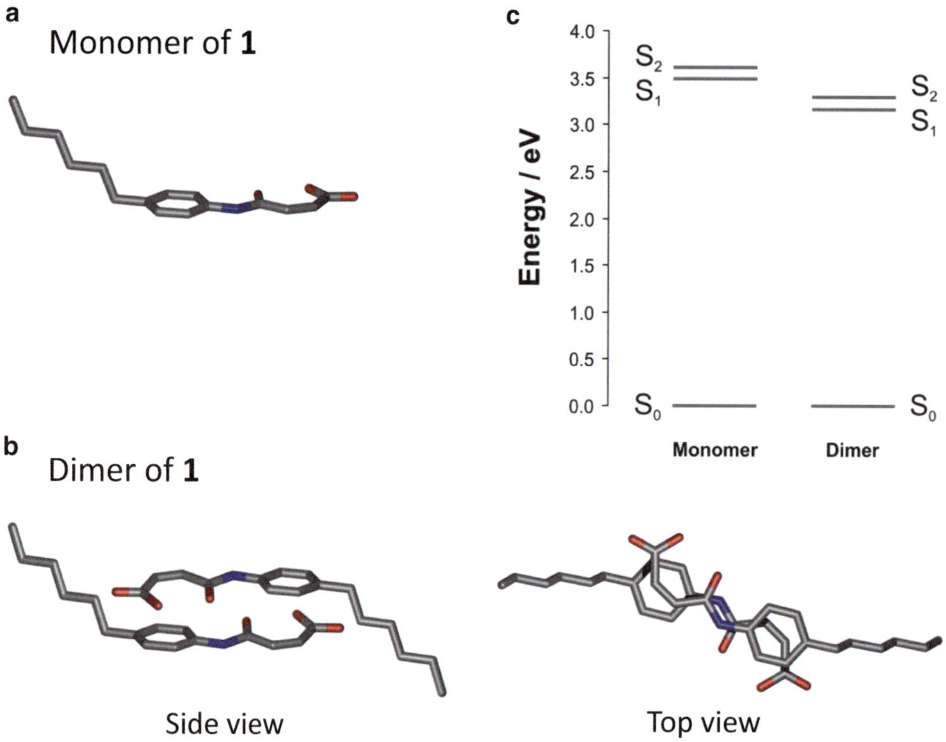

Fig. 3 DFT-optimized geometries of (**a**) monomer of **1** and (**b**) dimer of **1** and (**c**) corresponding excited states S_1 and S_2. The atom labels are: carbon (*grey*), nitrogen (*blue*), oxygen (*red*). The hydrogens were omitted for clarity. The average distance of the π-systems in the case of the dimer is *ca.* 3.2 Å. Visualizations of HOMO and LUMO orbitals of monomer and dimer of **1** can be found in the SI (Fig. S24–S27)

Examination of Hydrogel Formation Potential of 1–6

The potential of **1–6** to form supramolecular thermoreversible hydrogels in aqueous solutions exhibiting different pH values was evaluated. We found that solely compound **5** featuring an n-tetradecyl substituent was able to form hydrogels in 0.1–1.0 M aqueous NaOH solutions (aq. NaOH sol.) at initial concentrations of 9–10 g L^{-1} of **5**. Under these conditions **5** is completely transformed into its corresponding sodium salt **5Na** due to the high alkalinity of the solvent as proven by FT-IR absorption spectra of a lyophilized hydrogel sample (absence of the ν(C = O) peak at 1,724 cm^{-1}; see SI Fig. S28). Additionally, hydrogels could be obtained from **5** in aqueous potassium hydroxide solutions. However, these samples decomposed via precipitation within a few hours.[4] The macroscopic appearance of a sample comprising 10 g L^{-1} of **5** in 0.1 M aq. NaOH sol. during thermally induced hydrogel formation is shown in Fig. 4. Upon mixing of **5** with aq. NaOH sol. at room temperature, a suspension is formed (Fig. 4a). Heating under stirring at elevated temperatures yields transparent, but highly viscous solutions (Fig. 4b). These viscoelastic samples exhibit flow-birefringence upon shaking a sample between crossed polarizers. Interestingly, further increase of the temperature does not destroy the viscoelastic state (Fig. 4c). Cooling to room temperature leads to a free-flowing, translucent sample that turns more and more opaque over time (Fig. 4d). When being stored at room temperature for 24 h, a transformation into a stable translucent hydrogel occurs (Fig. 4e).[5]

The process is fully thermoreversible: Heating of a hydrogel sample restores the transparent, highly viscous sol state. Such a thermally induced transformation of a highly viscous/viscoelastic clear solution to a free-flowing solution *upon cooling* is extremely rare. Comparable behavior has been reported from a system consisting of cetyl trimethylammonium bromide and 5-methyl salicylic acid, where a transformation from vesicles to entangled wormlike micelles occurred upon heating [29].

[4]The hydrogel samples prepared from **5** in aq. NaOH sol. can be stored under ambient conditions for more than a year without any recognizable decomposition.

[5]It has to be noted that solutions of **5** in aq. NaOH sol. at elevated temperatures as well as hydrogels of **5** (where in both cases **5** is transformed in situ into the corresponding sodium carboxylate **5Na**) do not show any color. In contrast, a comparable sample of **1** in aq. NaOH sol. (**1Na**) exhibits a pale yellow color. We explain this by a strong difference of the HLB that leads to different favored aggregation modes of the corresponding sodium carboxylates **1Na** and **5Na** (π-π vs. van der Waals interactions).

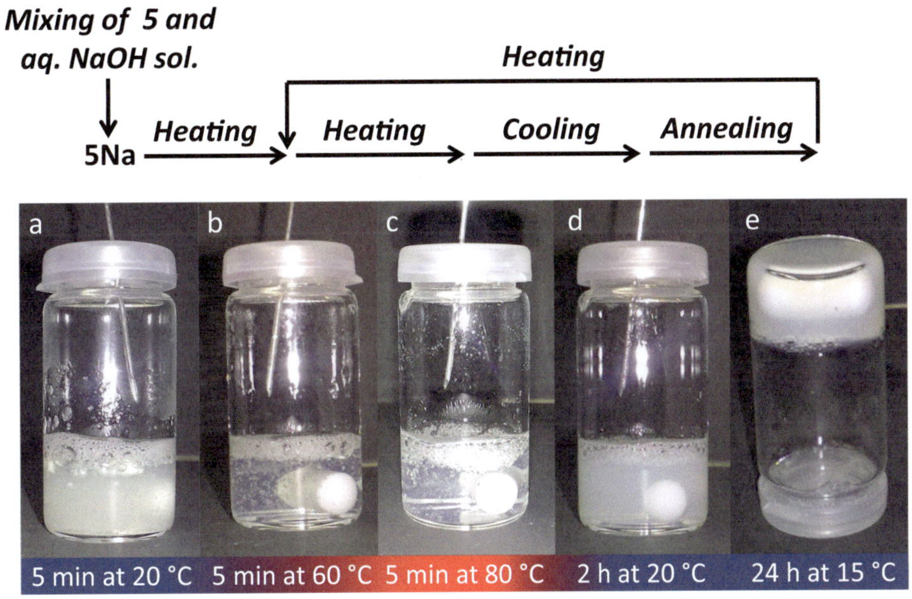

Fig. 4 Macroscopic appearance of a sample prepared from 10 g L^{-1} of **5** in 0.1 M aq. NaOH sol. during hydrogel preparation. (**a**) fluid suspension; (**b**) and (**c**) highly viscous transparent solution with entrapped air bubbles; (**d**) translucent low-viscosity sample; (**e**) stable translucent hydrogel

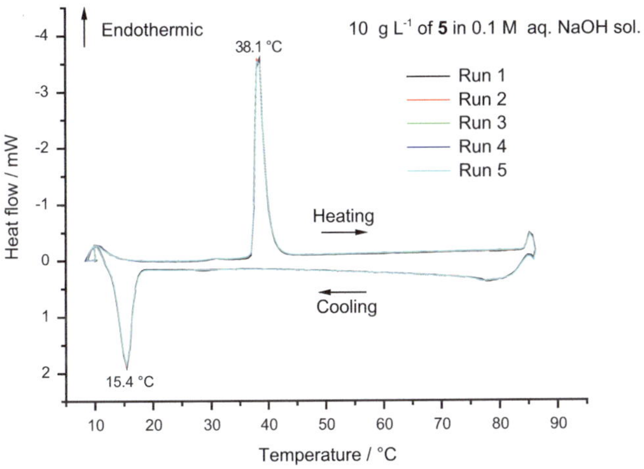

Fig. 5 μDSC traces of a hydrogel sample prepared from 10 g L^{-1} of **5** in 0.1 M aq. NaOH sol. during several subsequent heating/cooling cycles. Temperature profile: 10 °C – 85 °C – 10 °C. Heating/cooling rate: 1.0 °C min^{-1} (Note: the peaks at the beginning and at the end of each heating/cooling cycle are caused by the μDSC apparatus due to changes in the temperature profile)

μDSC Experiments of Hydrogel of 5Na

Thermoreversible gel-sol and sol-gel transitions of a hydrogel prepared from 10 g L^{-1} of **5** in 0.1 M aq. NaOH sol. were monitored by μDSC experiments. The hydrogel sample was equilibrated at 10 °C, heated to 85 °C and subsequently cooled down to 10 °C (heating/cooling rate: 1.0 °C min^{-1}). This process was repeated several times. The corresponding μDSC traces are shown in Fig. 5.

The results clearly show that the hydrogel sample exhibits excellent thermoreversibility as the superimposition of five subsequent heating/cooling cycles results in practically only one single μDSC trace. An endothermic peak corresponding to the gel-sol transition can be recognized in the heating curve at 38.1 °C.[6] In the cooling curve an exothermic peak originating from the sol-gel transition is visible at 15.4 °C. The relatively high enthalpy difference upon gel melting ($\Delta H = 23.48 \pm 0.28$ kJ mol^{-1}) indicates a transition from a highly ordered (crystalline) hydrogel state to a much less ordered sol state. Furthermore, a pronounced hysteresis exhibiting a temperature difference of more than 20 °C between gel melting and gel formation takes place under the applied conditions.

Temperature-Dependent Conductivity Measurements During Gel-Sol and Sol-Gel Transitions

The high sodium salt content of the hydrogel samples enables us to gain additional information about the gel formation and dissolution process via monitoring the conductivity. Therefore, temperature-dependent conductivity measurements of samples prepared from 10 g L^{-1} of **5** in 0.1 M aq. NaOH sol. during gel-sol and sol-gel transition were carried out. The results are presented in Fig. 6.

[6]The onset of the melting peak of the μDSC traces is in good agreement with macroscopically observed gel melting at *ca.* 35–36 °C ("falling steel ball method", *ca.* 1 °C min^{-1}).

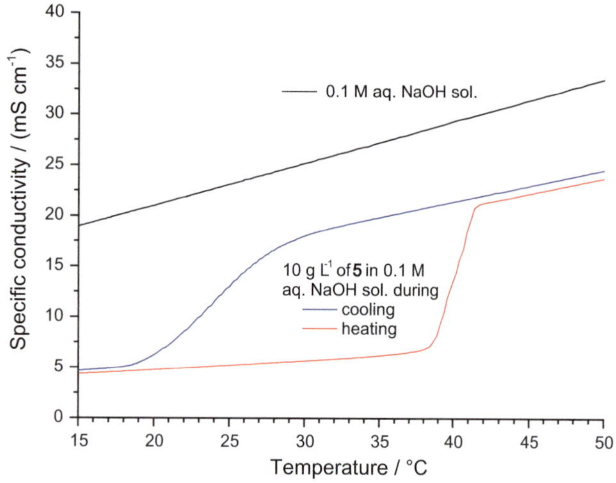

Fig. 6 Influence of the temperature on the specific conductivity of samples prepared from 10 g L^{-1} of **5** in 0.1 M aq. NaOH sol. For comparison purposes the conductivity curve of pure 0.1 M aq. NaOH sol. is also shown

The conductivity curves show a step during heating (gel-sol transition) and cooling (sol-gel transition) at *ca.* 40 °C and 24 °C, respectively. This non-linear change of specific conductivity is a clear indication that the supramolecular structures in the hydrogel of **5Na** are influencing the diffusion of ions in the sample. This can be due to the presence of a partially closed network of **5Na** in the hydrogel state. Additional diffusion test using hydrogel samples of **5Na** and solutions of phenolphthalein in 0.1 M aq. NaOH sol. also demonstrated that the hydrogel system is partially open-porous, as the dye was able to diffuse into the hydrogel (not shown).

Cryogenic Electron Microscopy (Cryo-EM) Studies

Cryogenic transmission electron microscopy (cryo-TEM) experiments were carried out in order to get an insight into the native morphology of supramolecular assemblies of **5Na**. Cryo-TEM images of a solution prepared from 1 g L^{-1} of **5** in 0.6 M aq. NaOH sol. at different magnifications are shown in Fig. 7. Bag-like, tubular and vesicular structures exhibiting dimensions of up to several hundreds of nm are visible. The thickness of the rims of these assemblies (indicated by the arrows in Fig. 7) was estimated to be *ca.* 4–6 nm. This is comparable to twice the molecular length of **5Na** and indicates a bilayer arrangement of the amphiphile. This is interesting, as single-chain amphiphiles that are able to form stable bilayer structures in aqueous solutions are rare and therefore have attracted scientific attention [30].

As samples comprising **5Na** at concentrations above the critical gelation concentration were not suitable to be subjected to cryo-TEM experiments, we decided to examine hydrogel samples of **5Na** by cryo-SEM methodology. By shock-freezing a hydrogel sample prepared from 10 g L^{-1} of **5** in 0.1 M aq. NaOH sol. followed by fracturing and partially subliming at low temperatures, the native morphology of **5Na** within the hydrogel could be visualized (Fig. 8). The visible cellular morphology features vesicular compartments with diameters of several μm that are separated by comparatively thin membranes.

XRD Investigations of Pristine and Lyophilized Hydrogel (5Na)

In order to get more insight into the molecular arrangement of **5Na** in the hydrogels, XRD experiments were carried out both on a pristine hydrogel and on a xerogel obtained by lyophilization of the corresponding hydrogel. The resulting XRD diffractograms are presented in Fig. 9.

The XRD pattern of the pristine hydrogel (Fig. 9a) features three relatively weak peaks with corresponding d-spacings of 4.20 nm, 0.43 nm, and 0.39 nm. The first peak (4.20 nm) is attributed to the bilayer thickness. As the molecular length of **5** was estimated to be *ca.* 2.6 nm, this indicates an interdigitated and/or tilted arrangement of the alkyl chains of **5Na** in the bilayers. The second (0.43 nm) and third peak (0.39 nm) are attributed to an in-plane scattering and allow for deducing a centered-rectangular alkyl chain packing. It is known that in-plane scattering within bilayer structures is likely to occur [31]. To prove that these peaks are not only artifacts, we have also examined a xerogel obtained by lyophilization of the pristine hydrogel (Fig. 9b). Due to the [*h*00] Bragg peak series (2θ ratio of 1:2:3), a lamellar arrangement of **5Na** in the xerogel can be concluded. As lyophilization allows to some extend a preservation of the initial structures, the cellular morphology could be preserved during the drying process (for a SEM image of the lyophilized sample see SI Fig. S29). Therefore it is reasonable to assume that the membranes observed during the cryo-SEM experiments also consist of a (multi-)bilayer arrangement of the amphiphile **5Na**, although the scattering intensities of the diffractogram of the pristine hydrogel sample (Fig. 9a) are too weak to clearly identify higher order scattering peaks. In the diffractogram of the xerogel, two additional Bragg peaks corresponding to d-spacings of 0.42 nm and 0.39 nm can be found (Fig. 9b). This is in good agreement with the results from the pristine hydrogel. In contrary, the d-spacing of the [100] Bragg peak of the lyophilized hydrogel is shifted to 3.45 nm. This can be explained by drying effects on the postulated hydrated (multi-)bilayers, thus reducing the effective thickness without affecting the in-plane order.

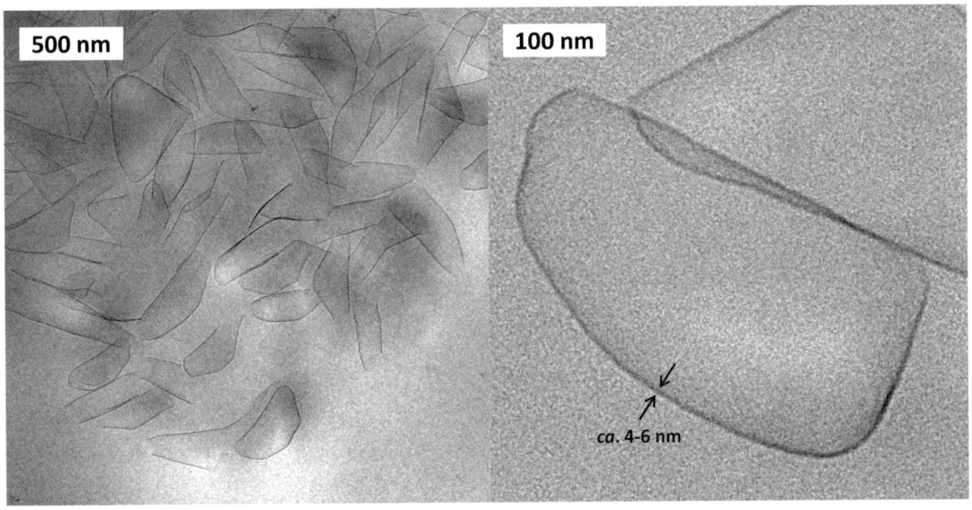

Fig. 7 Cryo-TEM images of a solution prepared from 1 g L^{-1} of **5** in 0.6 M aq. NaOH sol. at different magnifications. Bag-like, tubular and vesicular structures are visible

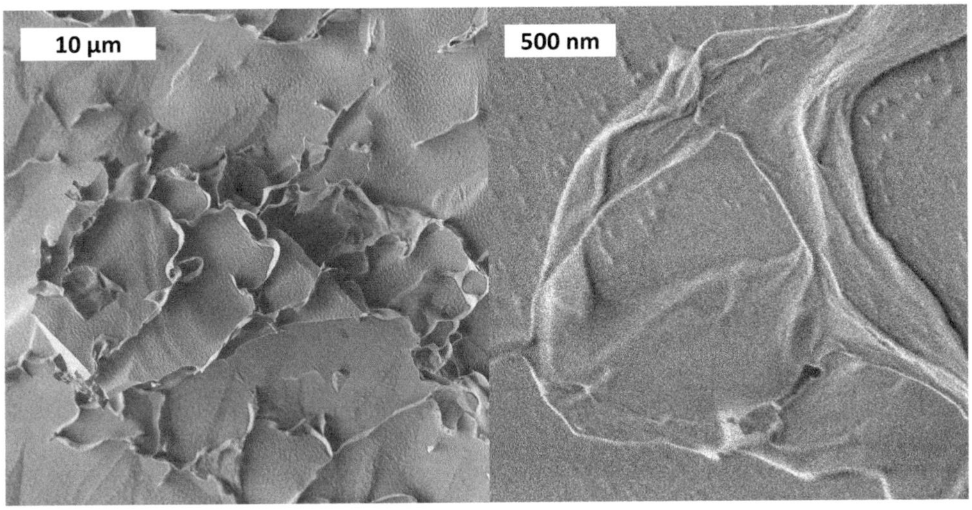

Fig. 8 Cryo-SEM image of a hydrogel prepared from 10 g L^{-1} of **5** in 0.1 M aq. NaOH sol. at different magnifications. *Left*: a cellular morphology consisting of vesicular compartments that are separated by very thin membranes can be recognized. *Right*: detail view of the surface of a collapsed vesicle (Note: The small spots are artifacts caused by the NaOH residues)

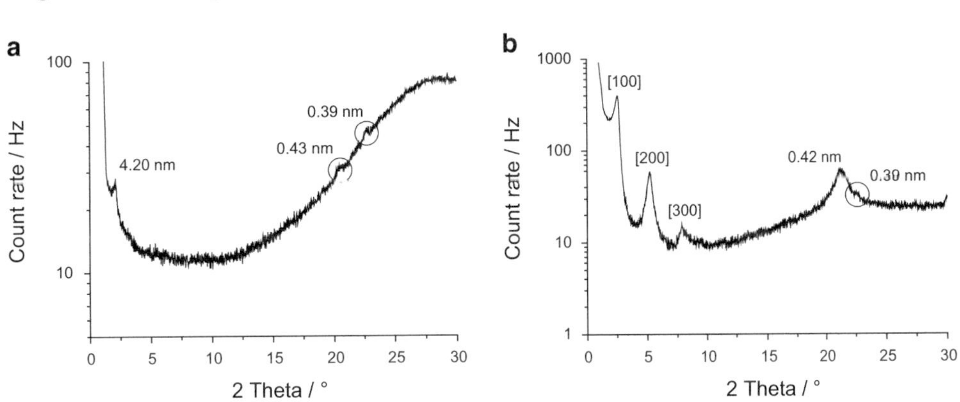

Fig. 9 Comparison of XRD patterns of **5Na** as (**a**) pristine hydrogel and (**b**) xerogel obtained by lyophilization. All hydrogels were prepared from 10 g L^{-1} of **5** in 0.1 M aq. NaOH sol.

Summarizing all the results from the cryo-EM experiments and from the XRD investigations, we can propose a packing model of **5Na** within the hydrogel: the amphiphile **5Na** is densely packed in a quasi-planar 2D-crystalline arrangement and builds up membranes consisting of (multi-) bilayers with the sodium carboxylate headgroups at the membrane surface protruding into the solvent phase. At the mesoscale a cellular morphology is formed that entraps the solvent phase, thus leading to macroscopic hydrogel formation. Although a LMW hydrogelator has been reported that formed cellular superstructures upon lyophilization [32], no comparable cellular meso-morphology formed by a LMW hydrogelator is known to us up to now.

Postulated Meso-Morphological Transformation During Hydrogel Formation and Dissolution

The experimental findings allow us to postulate a mechanism of the meso-morphological transformations of the supramolecular assemblies of **5Na** during hydrogel melting and anew formation. As shown by the cryo-SEM images, **5Na** is assembled in relatively large, flat (multi-)bilayers in the hydrogel state. The *all-trans* configuration of the alkyl chains indicated by the FT-IR results and the high in-plane order of the bilayers that was concluded from the XRD investigations lead us to assume a highly ordered state of the alkyl chains in the hydrogel. In combination with the high melting enthalpy evaluated by μDSC experiments this allows for concluding a "crystalline" state of the amphiphiles in the hydrogel. This is reasonable, as it has already been postulated that the aggregation of sodium carboxylate-bearing amphiphiles can involve a crystallization of the sodium carboxylate headgroups [33]. We assume that the hydrogel formation involves both headgroup crystallization and a significant increase alkyl chain ordering ("alkyl chain crystallization").

Upon heating, the highly ordered arrangement of amphiphiles "melts" when the gel-sol transition temperature is exceeded. The viscoelastic behavior of the resulting clear sample in combination with the observed flow-birefringence allows for deducing the presence of entangled wormlike micelles [34]. Cooling to room temperature leads to the formation of small vesicles, being expressed in significant light scattering and in the strong decrease of viscosity. Over time, these vesicles fuse and grow in size, thus leading to an increase in turbidity of the sample. Finally, the size of the vesicles is sufficient to allow contact between them, thus forming a macroscopically stable opaque to translucent hydrogel. It is reasonable to assume that the growth of the vesicles is related to the anew crystallization of the carboxylate headgroups and of the alkyl chains, and that this disorder-order transition is slow, allows significant supercooling and therefore yields a pronounced hysteresis between gel melting and anew formation. This is in agreement with the results from the μDSC experiments and the macroscopic hydrogel formation behavior.

In summary, the findings are highly interesting, as to the best of our knowledge up to now only *transformations* from fibrillar assemblies in the hydrogel to vesicles upon environmental changes are known in literature. For example, bolaamphiphiles with carboxylic acid headgroups have been reported to form vesicles *or* fibrillar hydrogels, depending on the pH value [35]. Additionally, fibrillar hydrogels produced by the self-assembly of nucleoside phosphocholine amphiphiles could be *transformed* into vesicles by heating [36]. Furthermore, an N,N-disubstituted maleamic acid derivative was reported to form stable vesicular aggregates in alkaline media, whereas the corresponding saturated succinamic acid underwent a transformation from vesicles to a "gel-like" state [16].

Conclusion

We have synthesized a new class of amphiphilic maleamic acid derivatives and investigated the self-assembly in bulk and in solution. It was demonstrated that the choice of alkyl chain length adjusts the hydrophilic/lipophilic balance (HLB), and that the HLB dictates the aggregation mode in the bulk material. The unexpected chromaticity of the maleamic acid derivatives could be attributed to intermolecular interactions of the π-systems of the amphiphiles, thus leading to the formation of supramolecular chromophores. Compound **5** (in form of its in situ formed sodium salt **5Na**) featuring an n-tetradecyl substituent was successfully applied in aq. NaOH sol. to produce thermoreversible supramolecular hydrogels. These gels exhibit a rare cellular meso-morphology that is based on highly ordered (multi-)bilayers formed by the amphiphile **5Na**. The unusual gel formation and melting behavior allowed proposing a new mechanism of hydrogel formation by the growth and fusion of vesicles induced by two-dimensional amphiphile crystallization.

Acknowledgement Financial support by the German Research Foundation (DFG) in the frame of Priority Programme SPP 1259 "Intelligente Hydrogele" is gratefully acknowledged. Computational studies were carried out in the frame of Research Training Group GRK 1640 "Photophysics of Synthetic and Biological Multichromophoric Systems". We thank M. Bieligmeyer, M. Schieder, C. Stelling, N. Al Nakeeb, S. Ganzleben and J. Failner for their support during synthesis and characterization of the compounds. B. Gossler is acknowledged for the cryo-SEM investigations. We are indebted to Dr. M. Drechsler for conducting the cryo-TEM experiments and Dr. M. Krekhova for the ff-TEM examinations. We thank B. Brunner (Prof. A. Jess, Chemical Engineering) for carrying out the elemental analysis.

Electronic Supplementary Material

Below is the link to the electronic supplementary material. NINNI MAURER et al suppl information.docx

References

1. a) de Loos M, Feringa BL, van Esch JH (2005) Eur J Org Chem 17:3615–3631; b) Estroff LA, Hamilton AD (2004) Chem Rev 104:1201–1217
2. Weiss RG, Terech P (eds) (2006) Molecular gels. Springer, Dordrecht
3. a) van Esch JH, Feringa BL (2000) Angew Chem Int Ed 39:2263–2266; b) Sangeetha NM, Maitra U (2005) Chem Soc Rev 34:821–836; c) Hirst AR, Escuder B, Miravet JF, Smith DK (2008) Angew Chem Int Ed 47:8002–8018
4. a) Cheng G, Castelletto V, Jones RR, Connon CJ, Hamley IW (2011) Soft Matter 7:1326–1333; b) Orbach R, Adler-Abramovich L, Zigerson S, Mironi-Harpaz I, Seliktar D, Gazit E (2009) Biomacromolecules 10:2646–2651; c) Wang W, Wang H, Ren C, Wang J, Tan M, Shen J, Yang Z, Wang PG, Wang L (2011) Carbohyd Res 346:1013–1017
5. a) Xu X-D, Liang L, Chen C-S, Lu B, Wang N-L, Jiang F-G, Zhang X-Z, Zhuo R-X (2010) Appl Mater Int 2:2663–2671; b) Yang Z, Xu K, Wang L, Gu H, Wie H, Zhang M, Xu B (2005) Chem Commun 35:4414–4416
6. a) Adhikari B, Palui G, Banerjee A (2009) Soft Matter 5:3452–3460; b) Song S, Feng L, Song A, Hao J (2012) J Phys Chem B 116:12850–12856
7. Gao Y, Long MJC, Shi J, Xu B (2012) Chem Commun 48:8404–8406
8. Ikeda M, Fukuda K, Tanida T, Yoshii T, Hamachi I (2012) Chem Commun 48:2716–2718
9. a) Mitra RN, Das PK (2008) J Phys Chem C 112:8159–8166; b) Chakrabarty A, Maitra U, Das AD (2012) J Mater Chem 22: 8268–18274; c) Dash J, Patil AJ, Das RN, Dowdall FL, Mann S (2011) Soft Matter 7:120–8126; d) Piepenbrock M-OM, Clarke N, Steed JW (2011) Soft Matter 7:412–2418
10. a) Huang R, Qi W, Feng L, Su R, He Z (2011) Soft Matter 7:6222–6230; b) Díaz Díaz D, Morin E, Schön EM, Budin G, Wagner A, Remy J-S (2011) J Mater Chem 21:641–644; c) Huang Y, Qiu Z, Xu Y, Shi J, Lin H, Zhang Y (2011) Org Biomol Chem 9:2149–2155; d) Rodríguez-Llansola F, Miravet JF, Escuder B (2011) Chem Commun 47:4706–4708; e) Naskar J, Palui G, Banerjee A (2009) J Phys Chem B 113:11787–11792; f) Sutton S, Campbell NL, Cooper AI, Kirkland M, Frith WJ, Adams DJ (2009) Langmuir 25:10285–10291; g) Liang G, Yang Z, Zhang R, Li L, Fan Y, Kuang Y, Gao Y, Wang T, Lu WW, Xu B (2009) Langmuir 25:8419–8422; h) Cao S, Fu X, Wang N, Wang H, Yang Y (2008) Int J Pharm 357:95–99; i) Friggeria A, Feringa BL, van Esch J (2004) J Control Release 97:241–248; j) Ikeda M, Ochi R, Wada A, Hamachi I (2010) Soft Matter 1:491–498; k) Jadhav SR, Chiou B-S, Wood DF, DeGrande-Hoffman G, Glenn GM, John G (2011) Soft Matter 7:864–867; l) Boekhoven J, Koot M, Wezendonk TA, Eelkema R, van Esch JH (2012) J Am Chem Soc 134:12908–12911; m) van Bommel KJC, Stuart MCA, Feringa BL, van Esch J (2005) Org Biomol Chem 3:2917–2920; n) Zhang J, Guo D-S, Wang L-H, Wang Z, Liu Y (2011) Soft Matter 7:1756–1762; o) Vemula PK, Cruikshank GA, Karp JM, John G (2009) Biomaterials 30:383–393
11. a) Bernet A, Behr M, Schmidt H-W (2011) Soft Matter 7:1058–1065; b) Bernet A, Behr M, Schmidt H-W (2012) Soft Matter 8:4873–4876
12. Isaev RN, Ishkov AV, Lobanova TV (2001) J Anal Chem 56:249–252
13. Arnold LA, Kosinski A, Estébanez-Perpiñá E, Guy RK (2007) J Med Chem 50:5269–5280
14. a) Lin K, Lin J, Cheng C-H (1996) Polymer 37:4729–4737; b) Zhao Y-W (2007) Huagong Jishu Yu Kaifa (Technology and Development of Medical Industry) 36:38–41; c) Xing J-J, Liu L, Qian J-H, Zhang Y-P (2009) Huaxue Shiji (Chemical Reagents) 31:352–354
15. Paleos CM, Margomenou-leonidopoulou G, Margaritis LH, Terzis A (1985) Mol Cryst Liq Cryst 129:127–135
16. Frkanec L, Jokić M, Makarević J, Wolsperger K, Žinić M (2002) J Am Chem Soc 124:9716–9717
17. Gaspar LJM, Baskar G (2005) J Mater Chem 15:5144–5150
18. a) Gaspar LJM, Baskar G (2005) Chem Commun 28:3603–3605; b) Gaspar LJM, Baskar G (2006) Langmuir 22:2795–2801
19. Zhang S, Fu X, Wang H, Yang Y (2008) J Sep Sci 31:3782–3787
20. Kar T, Debnath S, Das D, Shome A, Das PK (2009) Langmuir 25:8639–8648
21. Lim M, Hochstrasser RMJ (2001) Chem Phys 115:7629–7643
22. Nayak MK, Kim B-H, Kwon JE, Park S, Seo J, Chung JW, Park SY (2010) Chem Eur J 16:7437–7447
23. Lo KM, Ng SW (2009) Acta Cryst E65:o1101
24. Prasad SM, Sinha RBP, Mandal DK, Rani A (2002) Acta Cryst E58:o1296–o1297
25. Prasad SM, Sinha RBP, Mandal DK, Rani A (2002) Acta Cryst E58:o891–o892
26. Srivastava AK (1978) Z Kristallogr 148:21–28
27. a) Lim M, Hochstrasser RM (2001) J Chem Phys 115:7629–7643; b) Fujii Y, Yamada H, Mizuta M (1988) J Phys Chem 92:6768–6772; c) Murty TSSR (1971) J Phys Chem 75:1330–1332; d) Bulmer JT, Shurvell HF (1973) J Phys Chem 77:256–262
28. Liu Y, Wang T, Liu M (2012) Chem Eur J 18:14650–14659
29. Davies TS, Ketner AM, Raghavan SR (2006) J Am Chem Soc 128:6669–6675
30. Angayarkanny S, Vijay R, Baskar G, Mandal AB (2012) Langmuir 28:9378–9386
31. Marsh D (2012) Chem Phys Lipids 165:59–76
32. Bieser AM, Tiller JC (2007) J Phys Chem B 111:13180–13187
33. a) Wang D, Hao J (2011) Langmuir 27:1713–1717; b) Yuan Z, Lu W, Liu W, Hao J (2008) Soft Matter 4:1639–1644; c) Raue M, Bernet A, Küppers M, Stapf S, Schmidt H-W, Blümich B, Mang T (2013) Sodium NMR relaxation: a versatile non-invasive tool for the monitoring of phase transitions and the estimation of effective pore sizes of supramolecular hydrogels. In: Sadowski G, Richtering W (eds) Intelligent hydrogels. Springer, Cham/Heidelberg/New York/Dordrecht/London, pp 45–52
34. a) Cates ME, Candau SJ (1990) J Phys Condens Matter 2:6869–6892; b) Hoffmann H (1994) Statics and dynamics of worm-like surfactant micelles. In: Herb CA, Prud'homme, RK (eds) Structure and flow in surfactant solutions. American Chemical Society, Washington, DC, pp. 2–31
35. Wang T, Jiang J, Liu Y, Li Z, Liu M (2010) Langmuir 26:18694–18700
36. Moreau L, Barthélémy P, El Maataoui M, Grinstaff MW (2004) J Am Chem Soc 126:7533–7539

Thermo-responsive Amphiphilic Di- and Triblock Copolymers Based on Poly(N-isopropylacrylamide) and Poly(methoxy diethylene glycol acrylate): Aggregation and Hydrogel Formation in Bulk Solution and in Thin Films

André Laschewsky, Peter Müller-Buschbaum, and Christine M. Papadakis

Abstract

In this feature, we provide a comprehensive view and conclusions on recent investigations on the micellar aggregation of amphiphilic model polymers, the subsequent hydrogel formation, and the thermoresponsive behavior. The results obtained in bulk solution as well as in thin films are combined and compared, from the structural as well as kinetic point of view. The studies used two extensive series of diblock and symmetrical triblock copolymers, which were prepared by reversible addition-fragmentation chain transfer (RAFT). Derived from the thermo-responsive parent polymers poly(N-isopropylacrylamide) (PNIPAM) and poly(methoxy diethylene glycol acrylate) (PMDEGA), respectively, both series exhibit a lower critical solution type phase transition in aqueous media in the range of 30–40 °C. The model polymers consist of a long hydrophilic, thermo-responsive middle block, which is end-capped by two relatively small, but strongly hydrophobic blocks made from various vinyl polymers, preferentially from polystyrene. Their aggregation and hydrogel formation as well as their thermo-responsive behavior are systematically studied in dilute and concentrated aqueous solution as well as in thin films. For that, complementary methods were applied such as turbidimetry, fluorescence correlation spectroscopy (FCS), dynamic light scattering (DLS), small-angle X-ray (SAXS) and neutron scattering (SANS), rheology, white light interferometry, atomic force microscopy (AFM), optical probes, X-ray (XRR) and neutron reflectivity (XRR), grazing-incidence small-angle X-ray (GISAXS) and neutron scattering (GISANS) as well as attenuated total reflectance Fourier transform infrared spectroscopy (ATR-FTIR). All amphiphilic block copolymers self-organize at several hierarchical levels in bulk solution as well as in thin films. First, the association of the hydrophobic building blocks results in micelle-like aggregates. Then, the micelles cluster and eventually form networks, that make the systems gel. At elevated temperatures, the hydrophilic blocks undergo a collapse transition, inducing major structural changes at the molecular as well as supramolecular levels. Characteristic differences between PNIPAM and PMDEGA based solutions and thin films are worked out, concerning the self-organization, the width and hysteresis of the transition and the switching kinetics. Thin films of PNIPAM and PMDEGA based polymers differ with respect to long ranged correlations and the stability against dewetting. When probing polymer collapse, aggregation behavior, segmental dynamics and mechanical properties of the micellar solutions and the hydrogels,

A. Laschewsky (✉)
Department of Chemistry, Universität Potsdam, Karl-Liebknecht-Str. 24-25, Potsdam-Golm, D-14476 Germany
e-mail: Laschews@uni-potsdam.de

P. Müller-Buschbaum (✉) • C.M. Papadakis (✉)
Physik-Department, Technische Universität München, Lehrstuhl für Funktionelle Materialien/Fachgebiet Physik Weicher Materie, James-Franck-Str.1, Garching D-85748, Germany
e-mail: muellerb@ph.tum.de; papadakis@tum.de

G. Sadowski and W. Richtering (eds.), *Intelligent Hydrogels*, Progress in Colloid and Polymer Science 140,
DOI 10.1007/978-3-319-01683-2_2,

both the chain architecture and the chemical nature of the thermo-responsive block are found to play an important role for the detailed phase behavior.

Keywords
Triblock copolymer • Amphiphile • Polymer micelle • Hydrogel • Lower critical solution temperature • Aggregation behavior • Phase transition kinetics • Thin film

Introduction

Polymer hydrogels can be produced by a number of strategies. Common to all is the need of cross-links that keep the polymer network together, either by chemical or by physical bonds [1, 2]. The latter strategy has the advantage to be reversible, and thus to enable the reshaping and the assembly (or disassembly) of hydrogels by adapting physical parameters. A class of polymers particularly suited to form hydrogels, are amphiphilic block copolymers, which aggregate into micelle-like structures at low concentration, and eventually gel due to the formation of permanent or temporary networks at high concentrations [3–5] Characteristically, the hydrophobic blocks in such copolymers are responsible for the primary aggregation, while the hydrophilic blocks ensure the compatibility of the aggregates with the aqueous matrix and control the extent of swelling of the gels [4–7]. For long, hydrogel formation of simple AB diblock copolymers consisting of a hydrophilic block “A” and a hydrophobic block “B” has been studied. In recent years, the interest has been extended to the more complex symmetrical triblock copolymers of the BAB type, i.e., to copolymers which are made of a hydrophilic center and two hydrophobic ends [8–10]. In fact, for amphiphilic triblock copolymers, the self-assembly depends not only on the chemical nature of the A and B blocks, and on their absolute as well as their relative lengths, but also on the block sequence (ABA or BAB). While ABA systems tend to form core-shell micelles in aqueous solution similar to diblock copolymers [6], copolymers of the BAB type may associate into flower-like micelles in dilute solution, with both hydrophobic blocks of the same macromolecule being part of the same micellar core [10, 11]. The flower-like micelles may eventually evolve to clusters of interconnected micelles and finally to a network, when the polymers share their hydrophobic end blocks between two micelles with increasing concentration. Such bridging conformations will favor gelation of the systems, as they provide additional physical cross-links [5, 10–12]. From this point of view, the molecular structure of amphiphilic BAB polymers reflects the one of hydrophobically end-capped associative polymers, which is known as efficient associative thickeners [13–16].

The possibilities for block copolymer based hydrogels – and inevitably also their complexity – has been increased by superposing the concept of so-called “smart” systems onto their supramolecular self-assembly. In these smart hydrogels, the swelling/deswelling behavior of the polymers (in our specific case in water) is controlled by an external stimulus, such as a specific value of pH or temperature [17–21]. In this context, the use of amphiphilic triblock copolymers that exhibit a lower critical solution temperature (LCST) for switching hydrogels, has found particular interest in recent years [22–25]. LCST behavior in aqueous media is widespread amidst polar non-ionic polymers [26–28], and the switching process is non-invasive, i.e., can be achieved in a materially closed system. Classical examples for thermo-responsive polymers exhibiting a LCST in water under ambient pressure are poly(ethylene oxide) (also often named poly(ethylene glycol), PEG), poly(vinyl alcohol), methylcellulose, elastin-like proteins, and poly(N-isopropyl acrylamide), which is arguably the best studied representative of the class [29–31]. Thermo-responsive triblock copolymers derived from PNIPAM with, for instance, polystyrene (**PS**) end blocks in aqueous solution have been shown to form flower-like core-shell micelles with a thermoresponsive shell [32, 33], and at high polymer concentration, all the classical block copolymer mesophases [32, 34].

The synthesis of amphiphilic thermo-responsive block copolymers had been hampered for long by synthetic difficulties. In fact, the highly polar and often hygroscopic thermo-responsive blocks tolerate only in exceptional cases the rather harsh classical methods to prepare block copolymers, such as living ionic polymerization [35]. The situation has much changed with the advent of the reversible deactivation radical polymerization (RDRP) methods (formerly often called “controlled free radical polymerization”) [36], which emerged in the late 1990s. Since then, the synthetic toolkit for making amphiphilic (including thermo-responsive) block copolymers has been dramatically enlarged and has increasingly enabled the possibilities to tailor specific polymer designs [37, 38]. In particular, we have profited from these advancements for preparing two series of block copolymers via the radical addition-fragmentation chain transfer (RAFT) polymerization method [39, 40]. The series are based on the use of the well-established poly(N-isopropyl

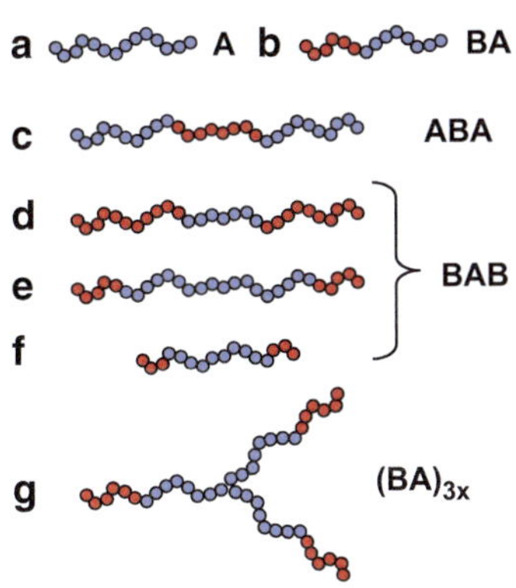

Scheme 1 Schematic architectures of the thermo-responsive polymers studied: (**a**) homopolymer references, (**b**) amphiphilic BA diblock copolymer, (**c**) amphiphilic ABA symmetrical triblock copolymer, (**d–f**) amphiphilic BAB symmetrical triblock copolymers of varying relative and absolute block sizes, (**g**) amphiphilic $(BA)_{3x}$ 3-arm star block copolymer (*light blue/light grey* shade = thermo-responsive block A, *red/dark grey* shade = hydrophobic block B)

acrylamide) PNIPAM [29], and of the – so far – little studied poly(methoxy diethylene glycol acrylate) PMDEGA [41] as hydrophilic and thermo-responsive building blocks. Both PNIPAM and PMDEGA exhibit a LCST in the physiologically particularly interesting range of 30–40 °C. Yet, as will be shown below, both polymers show characteristic differences with respect to their thermo-responsive behavior despite their very close LCST values and thus ought to be considered as complementary rather than competitive options for implementing "intelligent" hydrogel systems.

Structural Features of the Investigated Model Polymers

In order to gain more insight in the temperature-dependent self-organization of such amphiphilic thermo-responsive block copolymers, we varied the macromolecular structure of our model systems broadly (Scheme 1), by varying the molecular architecture (BA diblock *versus* BAB symmetrical triblock *versus* $(BA)_{3x}$ 3-arm star block copolymers), as well as the relative and absolute sizes of the thermo-responsive hydrophilic A and the permanently hydrophobic B blocks [24, 25, 42–48]. For comparison, these systematic variations were complemented by PNIPAM and PMDEGA homopolymer references and by some ABA symmetrical triblock copolymers, in which the permanently hydrophobic block forms the central block that is framed by the thermo-responsive hydrophilic blocks.

For the various amphiphilic block copolymers, we end-capped the thermo-responsive blocks by – mostly short – hydrophobic blocks, with a particular emphasis on **PS** as model hydrophobic polymer (see Tables 1 and 2). In order to get more insight into the structure–property relationships of such smart block copolymers, we varied systematically their most important macromolecular structural variables, namely, the absolute and relative block lengths, and the polymer architecture. Thus, we prepared amphiphilic AB diblock, BAB triblock, and $(BA)_{3x}$ triarm diblock star polymer analogs. Moreover, we have explored the role of the nature of the hydrophobic blocks, by varying the hydrophobicity of the repeat units as well as their glass transition temperatures [24], to learn about the effect of glassy or "molten" hydrophobic micellar cores. Last but not least, we have incorporated deuterated hydrophobic blocks into the polymers, too [44–48]. This facilitates the use of neutron scattering [49, 50] and spectroscopy techniques [51, 52], which are powerful tools for structural as well as dynamic investigations of mesoscopic systems.

From the pool of RDRP methods, we selected the RAFT method for synthesizing the polymers. First of all, RAFT is particularly suited to polymerize acrylic and styrenic monomers, which have been in our primary focus. Furthermore, the RAFT method confers two non-identical end groups to the polymers, which allow in a rather unique way to verify the quality of polymer intermediates and to support the – generally cumbersome – molecular characterization of amphiphilic copolymers by end-group analysis [43, 53, 54]. Detailed molecular information on the polymers studied is a keystone to interpret the results of the various investigations correctly, to maximize the reliable information that can be extracted, and to deduce general insights into such thermo-responsive supramolecular systems. Tables 1 and 2 summarize the two series of amphiphilic block copolymers synthesized, based on the thermo-responsive blocks PNIPAM and PMDEGA, respectively. The corresponding representative chemical formulas are illustrated in Figs. 1 and 2.

The variation of the hydrophobic B blocks in the copolymers of **PNIPAM** allowed us to elucidate the effects of the hydrophobicity of the various repeat units (growing with **PS** < **PtbS** and **PDBBA** < **PEHA** < **PODA**) as well as of their glass transition temperatures (growing with **PEHA** < **PDBBA** < **PODA** < **PS** < **PtbS**) on the aggregation behavior and thermo-sensitivity [24].

General Thermosensitivity of the Polymers in Aqueous Solution

While the PNIPAM and PMDEGA homopolymers dissolved easily in water at 20 °C, the dissolution/dispersion of the block copolymers required special measures, such as extended times for dissolution under shaking and/or the dispersion in non-selective organic solvents before mixing with water and removal of the cosolvent by

Table 1 Composition of the PNIPAM-containing homopolymers and amphiphilic block copolymers, and their cloud points in dilute aqueous solution

Entry	Polymer	M_n [10^3 Da]	DP_n of hydrophobic blocks	DP_n of PNIPAM block	Cloud point[a] [°C]	Ref.
	Homopolymers					
1	**$PNIPAM_{320}$**	36	–	320	31.0	[43]
2	**nbc-$PNIPAM_{340}$**[b]	39	–	340	30.0[c]	[55, 56]
3	**$PNIPAM_{430}$**	49	–	430	31.0	[43]
4	**$PNIPAM_{580}$**	66	–	580	31.0	[24]
	Diblock copolymers BA					
5	**$P(S_{50}$-$NIPAM_{160})$**	23	50	160	31.6	[42, 46]
	Triblock copolymers ABA					
6	**$P(NIPAM_{80}$-S_{27}-$NIPAM_{80})$**	21	27	160	30.5	[43]
7	**$P(NIPAM_{105}$-S_{30}-$NIPAM_{105})$**	27	30	210	30.0	[43]
	Triblock copolymers BAB					
8	**$P(tbS_9$-$NIPAM_{280}$-$tbS_9)$**	34	17	280	31.5	[24]
9	**$P(tbS_{17}$-$NIPAM_{490}$-$tbS_{17})$**	60	34	490	31.5	[24]
10	**$P(S_{11}$-$NIPAM_{185}$-$S_{11})$**	23	22	185	31.0	[43]
11	**$P(S_{11}$-$NIPAM_{280}$-$S_{11})$**	34	22	280	31.2	[44]
12	**$P(S_{11}$-$NIPAM_{370}$-$S_{11})$**	44	22	370	31.3	[44]
13	**$P(S_{15}$-$NIPAM_{185}$-$S_{15})$**	24	30	185	31.0	[43]
14	**$P(S_{24}$-$NIPAM_{550}$-$S_{24})$**	67	48	550	31.5	[24]
15	**$P(S_{30}$-$NIPAM_{530}$-$S_{30})$**	66	61	530	31.5	[24]
16	**$P(S_{40}$-$NIPAM_{790}$-$S_{40})$**	98	81	790	31.5	[24]
17	**$P(S_{45}$-$NIPAM_{650}$-$S_{45})$**	82	89	650	31.5	[24]
18	**$P(S_{60}$-$NIPAM_{700}$-$S_{60})$**	92	119	700	31.5	[24]
19	**$P(d8S_{11}$-$NIPAM_{220}$-$d8S_{11})$**	27	22	220	31.2	[47, 48]
20	**$P(d8S_{10}$-$NIPAM_{390}$-$d8S_{10})$**	27	46	390	31.5	[44, 45]
21	**$P(DBBA_8$-$NIPAM_{310}$-$DBBA_8)$**	40	16	310	31.5	[24]
22	**$P(EHA_{11}$-$NIPAM_{380}$-$EHA_{11})$**	47	21	380	31.5	[24]
23	**$P(EHA_{18}$-$NIPAM_{610}$-$EHA_{18})$**	77	35	610	31.5	[24]
24	**$P(EHA_{25}$-$NIPAM_{700}$-$EHA_{25})$**	88	50	700	31.5	[24]
25	**$P(EHA_{30}$-$NIPAM_{670}$-$EHA_{30})$**	87	60	670	31.5	[24]
26	**$P(ODA_6$-$NIPAM_{280}$-$ODA_6)$**	35	11	280	31.5	[24]
27	**$P(ODA_{18}$-$NIPAM_{800}$-$ODA_{18})$**	100	36	800	31.5	[24]
28	**$P(ODA_{27}$-$NIPAM_{1050}$-$ODA_{27})$**	140	54	1050	31.5	[24]

[a] At 1 g L^{-1} (heating run, onset of turbidity)
[b] Bearing two short hydrophobic butylsulfanylthiocarbonylsulfanyl end groups
[c] At 0.25 g L^{-1} via dynamic light scattering

evaporation or dialysis, in particular when bearing relatively long hydrophobic blocks (number average degree of polymerization $DP_n \geq 15$). Only copolymers with very short hydrophobic blocks ($DP_n < 12$) dissolved rather readily in water.

Turbidimetry of dilute aqueous solutions was used to map the general trends of thermo-responsive behavior of the various polymer structures. Already at this level, a major difference between the PNIPAM-based and the PMDEGA-based polymers becomes evident. For the PNIPAM-based polymers, a nearly constant cloud point of 30–32 °C is observed, independent of the molar mass of the polymers, the nature of the end groups (defined by the specific RAFT agent used), length and nature of the hydrophobic blocks, or the type of copolymer architecture (homopolymer, BA diblock as well as BAB and ABA triblock copolymers), as illustrated in Fig. 3 [24]. Also, variation of the polymer concentration between 0.2 and 40 wt% only marginally affected the position of the cloud points (see also Fig. 6 below) [24]. This is in good agreement with the literature, in particular also with reports on the missing effect of hydrophobic end groups or attached blocks, respectively, if the PNIPAM block outpasses a minimum molar mass (> 10^4), and if the hydrophobic end groups are equivalent to a hexadecyl chain or bigger [62, 63]. This effect has been attributed to local micro-phase separation in the aqueous environment, and to the resulting

Table 2 Composition of the PMDEGA-containing homopolymers and amphiphilic block copolymers studied and their cloud points in dilute aqueous solution

Entry	Polymer	M_n [10^3 Da]	DP_n of hydrophobic blocks	DP_n of PMDEGA block	Cloud point[a] [°C]	Ref.
	Homopolymers					
29	**$PMDEGA_{53}$**[b]	10	–	53	24.5	[41]
30	**$PMDEGA_{138}$**[c]	17	–	100	45[d]	[57, 58]
31	**$PMDEGA_{153}$**[b]	27	–	153	38.3	[41]
32	**$PMDEGA_{513}$**[b]	90	–	513	41.2	[41]
	Diblock copolymers BA					
33	**$P(S_{11}\text{-}MDEGA_{101})$**[e]	19	11	101	34.6	[41]
34	**$P(S_{11}\text{-}MDEGA_{172})$**[e]	31	11	172	38.0	[41]
35	**$P(S_{11}\text{-}MDEGA_{275})$**[e]	49	11	275	38.9	[41]
36	**$P(S_{11}\text{-}MDEGA_{331})$**[e]	59	11	331	40.0	[41]
37	**$P(S_{11}\text{-}MDEGA_{513})$**[e]	91	11	513	40.1	[41]
	Triblock copolymers BAB					
38	**$P(S_8\text{-}MDEGA_{41}\text{-}S_8)$**[b]	10	16	41	20.5	[25]
39	**$P(S_8\text{-}MDEGA_{53}\text{-}S_8)$**[b]	12	16	53	22.1	[25]
40	**$P(S_8\text{–}MDEGA_{93}\text{-}S_8)$**[b]	19	16	93	26.0	[25]
41	**$P(S_8\text{-}MDEGA_{180}\text{-}S_8)$**[b]	34	16	180	30.1	[25]
42	**$P(S_8\text{-}MDEGA_{337}\text{-}S_8)$**[b]	61	16	337	33.8	[25]
43	**$P(S_8\text{-}MDEGA_{452}\text{-}S_8)$**[b]	81	16	452	35.4	[25]
44	**$P(S_8\text{-}MDEGA_{659}\text{-}S_8)$**[b]	117	16	659	38.1	[25]
45	**$P(S_{11}\text{-}MDEGA_{126}\text{-}S_{11})$**[f]	24	22	125	36.6[d]	[58–60]
46	**$P(S_{15}\text{-}MDEGA_{550}\text{-}S_{15})$**[b]	99	30	550	37	[61]
	3-arm star block copolymers					
47	**$P(MDEGA_{78}\text{-}S_8)_3$**[g]	45	23	234	27.2	[41]
48	**$P(MDEGA_{231}\text{-}S_8)_3$**[g]	124	23	693	34.1	[41]

[a]Cloud points at 3.0 g L^{-1} (heating run, onset of turbidity)
[b]Using RAFT agent 1,2-bis[4-(*tert*-butoxycarbonyl)benzyl sulfanylthiocarbonylsulfanyl] ethane
[c]Using RAFT agent 1-[(4-methoxybenzyl)sulfanylthiocarbonylsulfanyl] propane
[d]Cloud points at 1.0 g L^{-1} (heating run, onset of turbidity)
[e]Using RAFT agent 1-[(4-carboxybenzyl)sulfanylthiocarbonylsulfanyl] butane
[f]Using RAFT agent dibenzyltrithiocarbonate
[g]Using RAFT agent 1,1,1-tris-[3-(4-(*tert*-butoxycarbonyl) benzylsulfanylthiocarbonyl sulfanyl) propanoyloxy] ethane

efficient shielding of the associated hydrophobic components from contact with water [63, 64].

In striking contrast, PMDEGA-based polymers show a very pronounced effect of the molar mass even up to high molar masses (Fig. 3). Moreover, the effect of the molar mass is counter-intuitive, as even for hydrophilic end groups, the cloud point apparently increases with increasing molar mass [41]. Furthermore, the cloud points are very sensitive to the chemical nature of low or high molar mass end groups (and thus to the specific RAFT agent used and to the length of the hydrophobic block(s) attached), as well as to the polymer architecture. Diblock, triblock and star block copolymers even of identical ratio of hydrophilic-to-hydrophobic block sizes exhibit significantly varying phase transition temperatures (Table 2, Fig. 3). In parallel, the cloud point of dilute as well as concentrated aqueous solutions of PMDEGA-based polymers display a marked dependence on the polymer concentration (Fig. 4) [25, 41]. With increasing concentration, the cloud point passes through a minimum, which shifts to lower concentrations with increasing molar mass, in agreement with predictions from simple thermodynamics.

Another striking difference is the different kinetics of the collapse or reswelling processes of the two different thermo-responsive blocks. Turbidimetry studies of the PMDEGA-based polymers show virtually no difference between the heating and cooling curves when slow rates were applied (as 0.1 °C/min), i.e., the collapse and reswelling of the polymer coils do not show a notable thermal hysteresis [41]. In contrast, PNIPAM-based polymers exhibit characteristic differences between the heating and cooling curves in turbidimetry [24, 42, 44]. The notable hysteresis is putatively attributed to a reorganization process of the PNIPAM coils above the cloud point, where part of the secondary amide moieties form hydrogen bonds between neighboring repeat units instead of with water as below the phase transition [65].

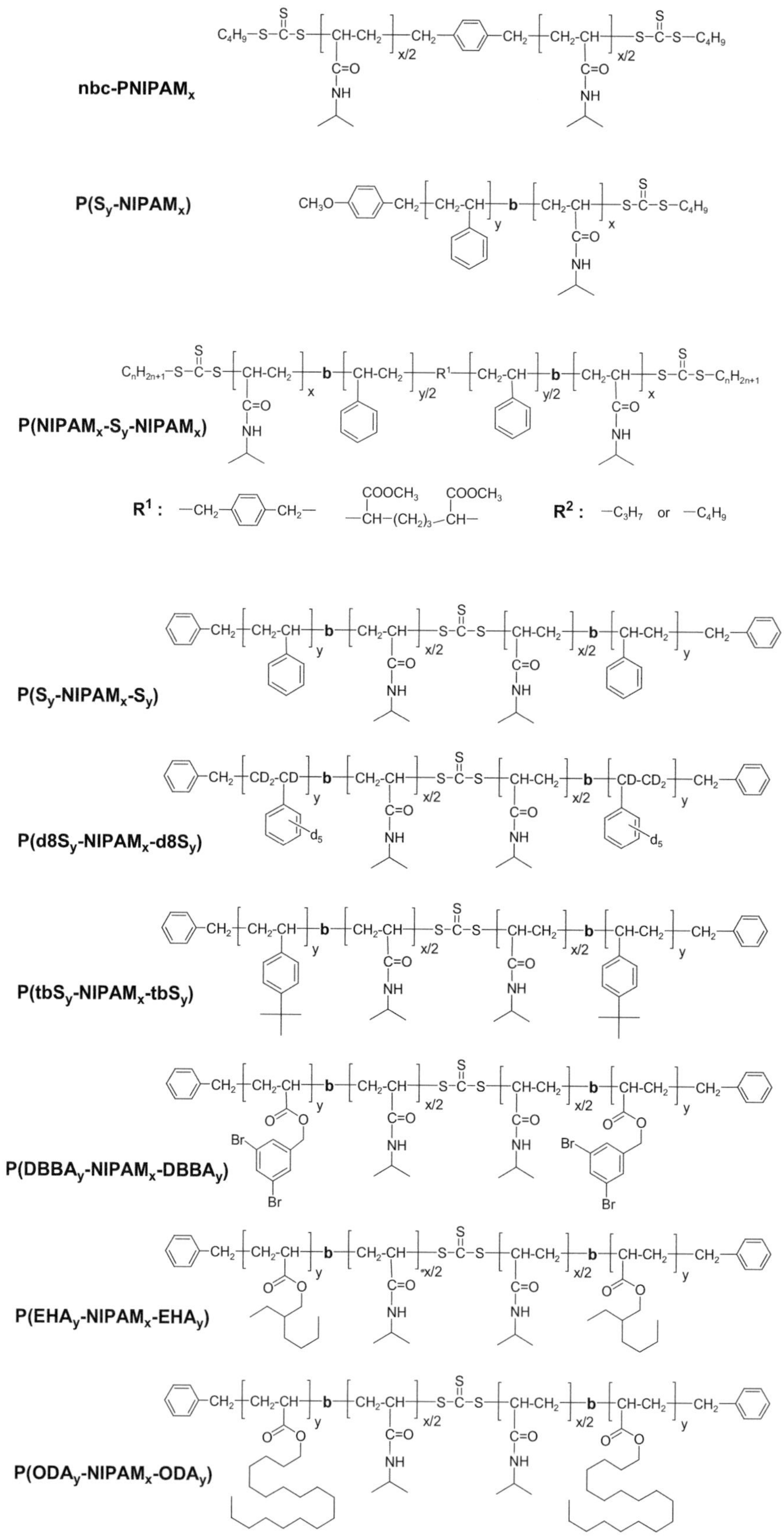

Fig. 1 General chemical structures of the synthesized polymers based on poly(N-isopropyl acrylamide) PNIPAM

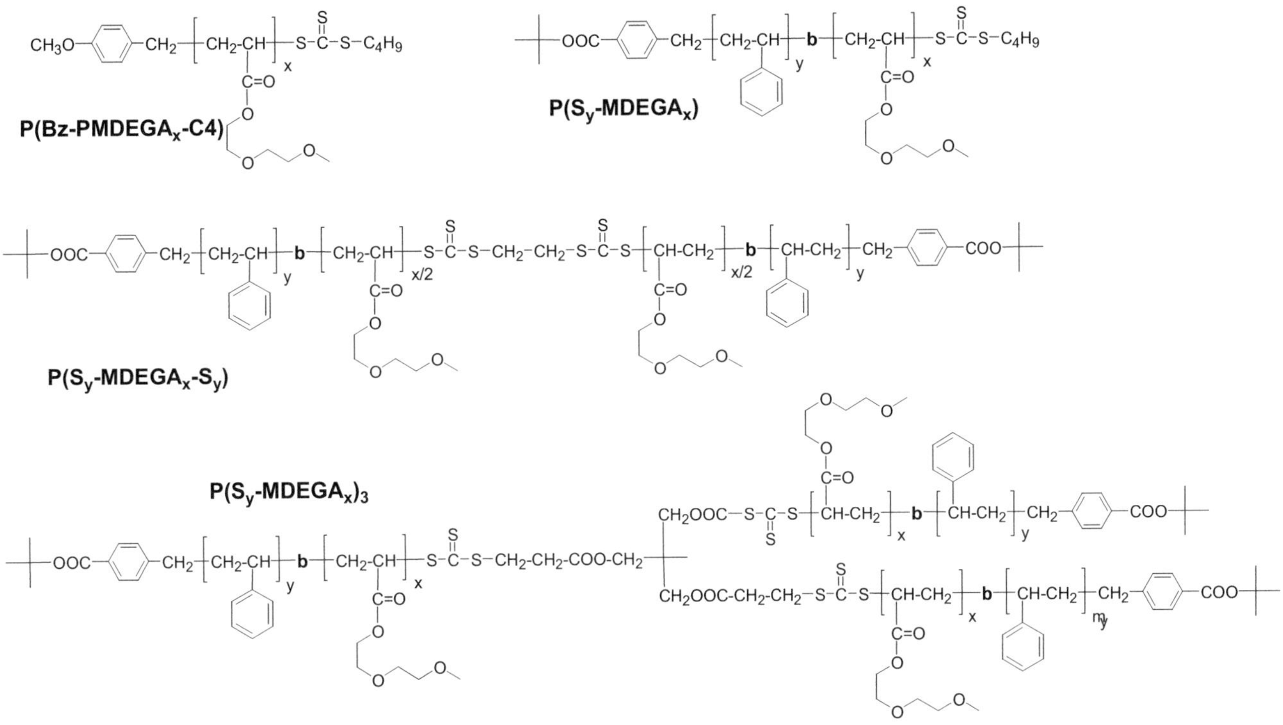

Fig. 2 General chemical structures of the synthesized polymers based on poly(methoxy diethylene glycol acrylate) **PMDEGA**

These extra H-bonds must be broken and the reorganization must be reversed when cooling the dispersions below the phase transition temperature. The PMDEGA chains, in contrast, cannot establish hydrogen bonds between themselves, and thus may react faster. These molecular differences may also contribute to the fact that PMDEGA-based polymers macroscopically phase-separate from their solutions when kept for some time above the cloud point. In contrast, PNIPAM-based polymers and micelles only aggregate to larger clusters with hydrodynamic diameters in the typical range of 100–500 nm, thus forming so-called mesoglobules [64, 66, 67].

Clearly, the detailed chemical nature of the thermo-responsive polymers is of extreme importance for their behavior in aqueous solution. Thus, the choice of a specific polymer for implementing supramolecular and responsive systems is not simply a matter of the exact value of the phase transition temperature. Moreover, the behavior of such systems cannot be ascribed and predicted by a simplified model considering just size, volume fraction and distribution of the various hydrophilic and hydrophobic polymer fragments within the macromolecules, based just on general architectural patterns as shown in Scheme 1.

The dilute and concentrated aqueous solutions/dispersions of the copolymers are clear to the naked eye at 20 °C. However, as evident from dynamic light scattering studies, the copolymers form aggregates already in highly dilute aqueous solution with hydrodynamic diameters typically in the range of 5–50 nm, as would have been expected according to their amphiphilic structures. Thus, we can assume that they form micelle-like structures, with a hydrophobic core made of the various B blocks, and a hydrophilic corona made of the thermo-responsive A blocks. A more detailed discussion of the aggregate structures is given further below.

Hydrogel Formation

Different from the homopolymers PNIPAM and PMDEGA, the aqueous solutions of the amphiphilic blocks copolymers become highly viscous with increasing concentration and eventually gel [24, 25]. The efficiency of gel formation as well as the shape of the phase diagrams may vary strongly not only with the relative and absolute sizes of the various hydrophilic and hydrophobic polymer blocks contained, but also with the architectures, i.e., whether diblock, triblock or star block copolymers are used (cf. Scheme 1). This is exemplified in Table 3 and Fig. 5. In the case of PMDEGA serving as thermo-responsive hydrophilic block, it is evident that the efficiency of hydrogel formation increases from the BA diblock via the BAB triblock to the star block structure. Further, for a given size of the hydrophobic block, hydrogel formation is favored by increasing the length of the hydrophilic block. An increase of the size of the hydrophobic PS blocks favors hydrogel formation, too, yet

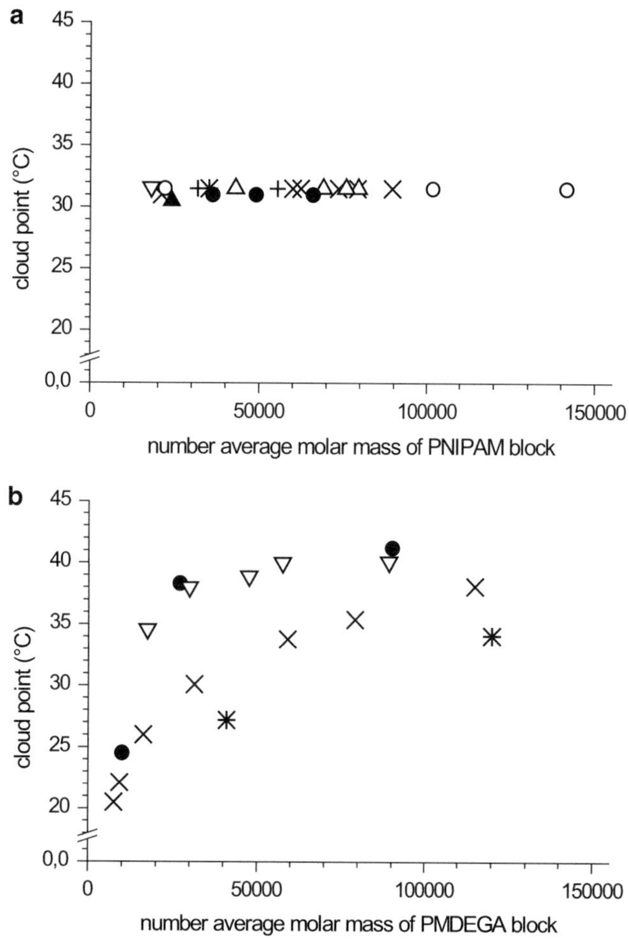

Fig. 3 Comparison of the LCST-type behavior of amphiphilic block copolymers of different structures and architectures in dilute aqueous solution, in dependence on the nature and absolute length of the thermo-responsive block (data from refs. [24, 25, 41, 42]). (**a**) **PNIPAM**-based polymers: (•) homopolymer; (∇) BA diblock copolymers with B = **PS**; (×) BAB triblock copolymers with B = **PS**; (+) BAB triblock copolymers with B = **PtbS**; (✳) BAB triblock copolymers with B = **DBBA**; (Δ) BAB triblock series with B = **PEHA**; (O) BAB triblock series with B = **PODA** (variable size of the B blocks, 1.0 g L^{-1} polymer in water, heating rate 1 °C min^{-1}). (**b**) **PMDEGA**-based polymers: (•) homopolymer (using RAFT agent 1,2-bis[4-(*tert*-butoxycarbonyl)benzyl sulfanylthiocarbonyl sulfanyl] ethane); (∇) BA diblock copolymers with B = **PS$_{11}$**; (×) BAB triblock copolymers with B = **PS$_8$**; (✳) $(BA)_3$ 3-arm star block copolymers with B = **PS$_8$** (3.0 g L^{-1} polymer in water, heating rate 1 °C min^{-1})

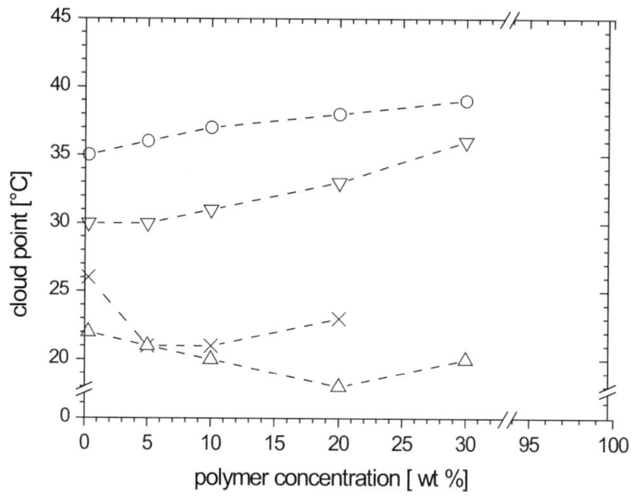

Fig. 4 Evolution of the cloud point temperatures of aqueous solutions of **P(S$_8$-MDEGA$_n$-S$_8$)** block copolymers with concentration: (O) **P(S$_8$-MDEGA$_{452}$-S$_8$)**, (∇) **P(S$_8$-MDEGA$_{180}$-S$_8$)**, (×) **P(S$_8$-MDEGA$_{93}$-S$_8$)** and (**Δ**) **P(S$_8$-MDEGA$_{53}$-S$_8$)** (Data taken from ref. [25]. Broken lines are meant as guide to the eye)

the direct solubility in water rapidly dwindles with increasing the hydrophobic blocks' size. While copolymers with **PS$_{11}$** blocks (at least according to visual inspection) can still fully dissolve in water, copolymers with **PS$_{15}$** blocks do not dissolve anymore, at least not within practical time scales. Hence, for such copolymers an indirect dispersion process must be applied, *e.g.*, dissolution in a water-miscible organic solvent, dilution by excess water, subsequent removal of the organic cosolvent, and adjustment to the desired concentration [24]. We also observe that BAB block copolymers with relatively short PMDEGA blocks, as in **P(S$_8$-MDEGA$_{53}$-S$_8$)**, form biphasic systems in water [25]. This may be correlated with the general phase diagram of hydrophobically modified telechelics [15], which are employed as associative thickeners, indicating the formation of aggregates even at high dilution. When the average polymer chain length becomes smaller than the average intermicellar distance in the dynamic network, phase separation into a viscous concentrated and a dilute low-viscous phase occurs that is driven by entropy. Characteristically for the PMDEGA-based systems, we observe that, for a given concentration, the tendency for hydrogel formation decreases markedly with increasing temperature, and, when the gel–sol borderline is reached, the hydrogels disintegrate. This can occur much below the LCST-type phase transition temperature of the thermo-responsive block, giving rise to phase diagrams of versatile shape and in a sensitive dependence on the precise molecular structure (cf. Fig. 5) [25].

While the general correlation rules between the molecular structure and the gelling behavior as discussed above for the PMDEGA-based systems might intuitively seem logical for amphiphilic block copolymers, the PNIPAM-based systems behave completely differently. It is not only in their thermo-responsive behavior, as visualized by the cloud point lines in Fig. 5, where PMDEGA- and the PNIPAM-based systems show fundamental differences. The very weak dependence of hydrogel formation on the molecular structure of PNIPAM-based block copolymers seems as characteristic for these systems as it is striking. In fact, only one partial phase diagram of a PNIPAM-based block copolymer is displayed in Fig. 5, as the differences between all the block copolymers listed in Table 1 have been shown to be small at best [24].

Table 3 Hydrogel formation as function of **PMDEGA**-based block copolymer architecture and relative sizes of the hydrophilic and hydrophobic blocks at 21 °C, according to tube inversion tests. Data taken from refs. [25, 41, 61]

Polymer	Architecture	5 wt%	10 wt%	20 wt%	30 wt%
$PMDEGA_{513}$	Homopolymer	L	L	L	L
$P(S_{11}\text{-}MDEGA_{172})$	BA diblock	L	L	L	VL
$P(S_{11}\text{-}MDEGA_{275})$	BA diblock	L	L	L	VL
$P(S_8\text{-}MDEGA_{53}\text{-}S_8)$	BAB triblock	Biphasic	Biphasic	Biphasic	Biphasic
$P(S_8\text{-}MDEGA_{93}\text{-}S_8)$	BAB triblock	L	L	VL	Soft gel
$P(S_8\text{-}MDEGA_{172}\text{-}S_8)$	BAB triblock	L	VL	Soft gel	Hard gel
$P(S_8\text{-}MDEGA_{452}\text{-}S_8)$	BAB triblock	VL	Soft gel	Hard gel	Hard gel
$P(S_{15}\text{-}MDEGA_{549}\text{-}S_{15})$	BAB triblock	Hard gel	Hard gel	Hard gel	Hard gel
$P(MDEGA_{231}\text{-}S_8)_{3x}$	$(BA)_{3x}$ star block	VL	Soft gel	Hard gel	Hard gel

L free flowing liquid, *VL* viscous liquid

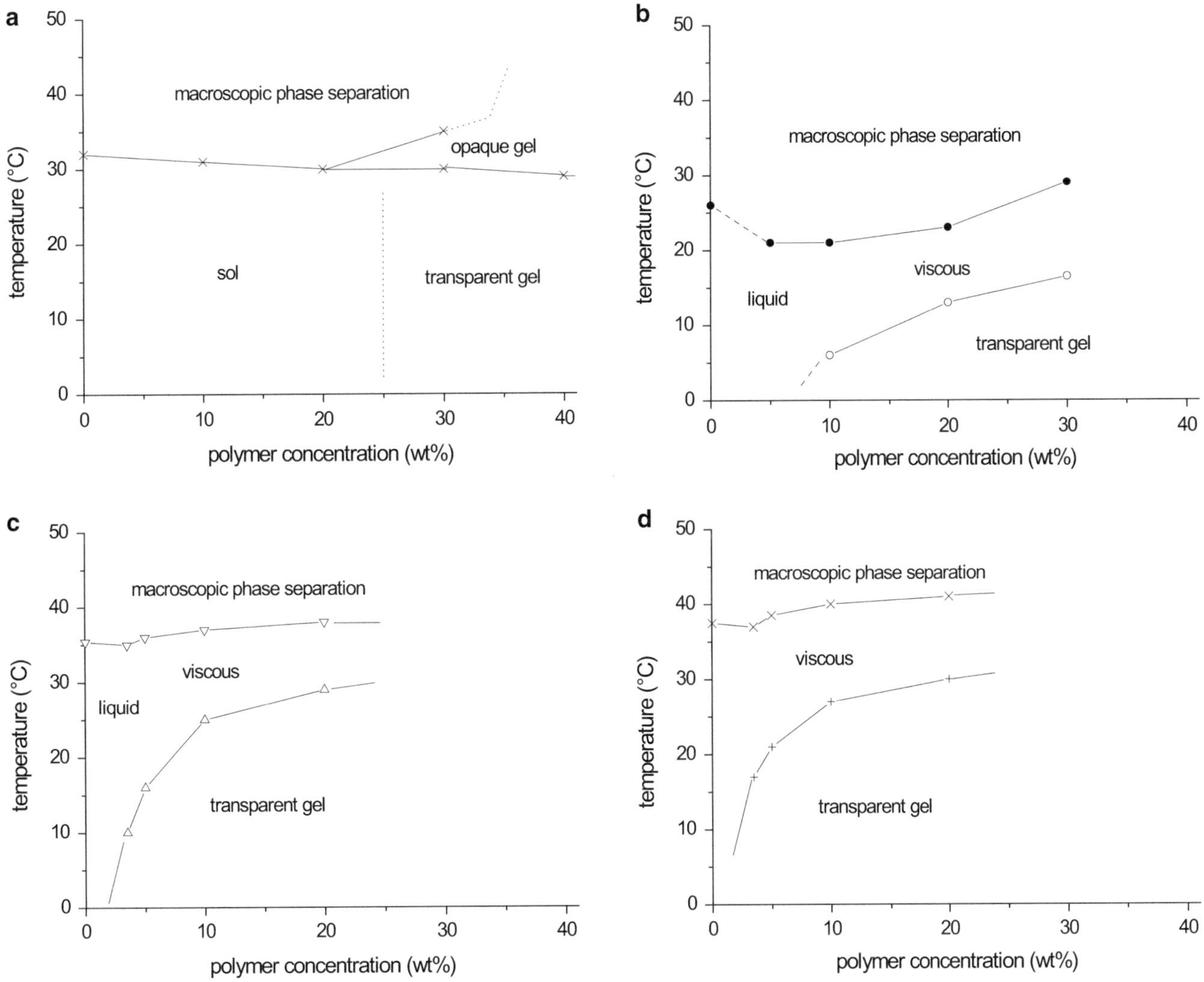

Fig. 5 Partial phase diagrams of amphiphilic BAB triblock copolymers in water. (**a**) **$P(S_{24}\text{-}NIPAM_{550}\text{-}S_{24})$**, (**b**) **$P(S_8\text{-}MDEGA_{93}\text{-}S_8)$**. (**c**) **$P(S_8\text{-}MDEGA_{452}\text{-}S_8)$**. (**d**) **$P(MDEGA_{231}\text{-}S_8)_{3x}$** (Data taken from refs. [24, 25, 61])

Depending on the detailed length of the PNIPAM-block, and on the size and chemical nature of the hydrophobic block(s), the necessary minimum concentration for gelling varies somewhat. But otherwise, the shapes of the diagrams are virtually the same for all types and lengths of the hydrophobic blocks studied, and for all studied lengths of the PNIPAM-block. Moreover, if a concentration is reached that suffices to induce hydrogel formation at low temperature, the

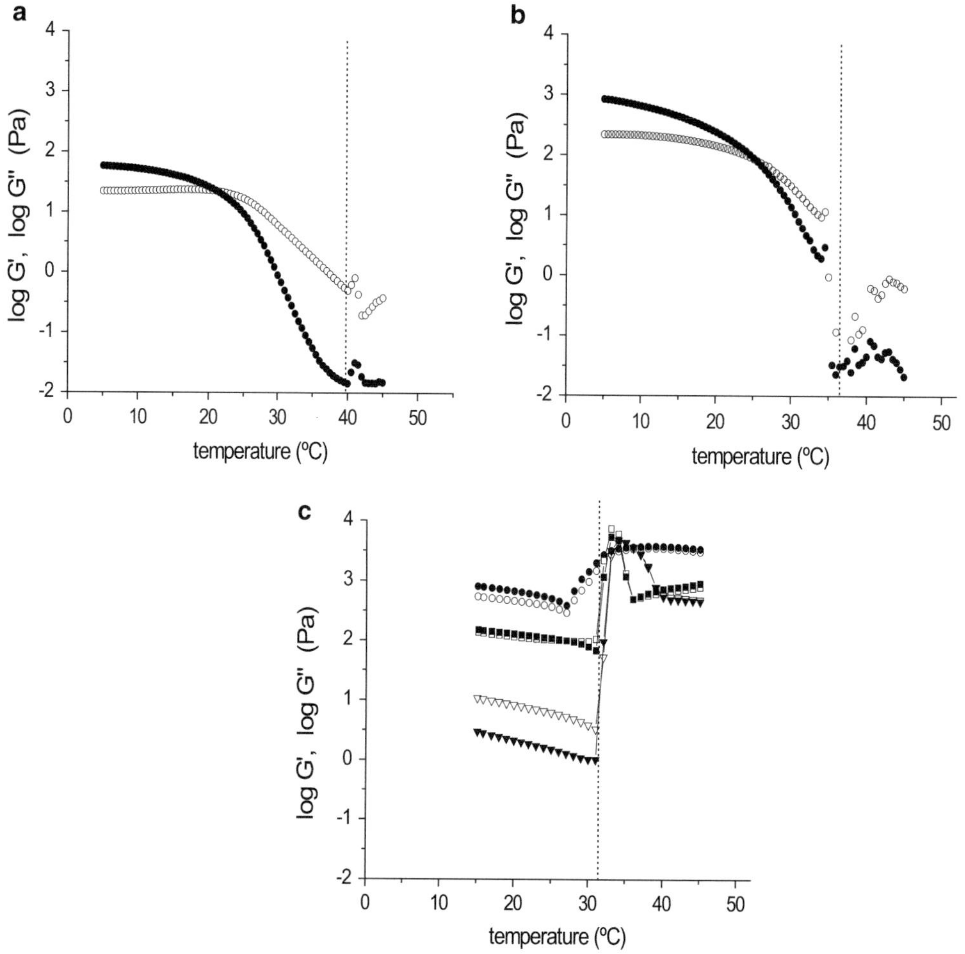

Fig. 6 Evolution of the storage (*full symbols*) and loss (*open symbols*) modulus of aqueous solutions of amphiphilic BA and BAB triblock copolymers with the temperature upon heating. (**a**) **P(S$_{10}$-MDEGA$_{450}$)** at 30 wt%; (**b**) **P(S$_8$-MDEGA$_{452}$-S$_8$)** at 10 wt%; (**c**) **P(S$_{24}$-NIPAM$_{550}$-S$_{24}$)** at 20 (▼, ▽), 30 (■, □) and 40 wt% (●, ○). *Vertical dotted lines* indicate the position of the phase transition according to turbidimetry (Data taken from refs. [24, 25, 61])

gel state persists up to the cloud point of the PNIPAM-block. Even more strikingly for intermediate polymer concentrations of 20–30 wt%, the thermal collapse of the hydrophilic block apparently enhances the ability for gelling, instead of causing a breakdown of the hydrogels (cf. Fig. 5), at least for quite some temperature window of 5–10 °C above the cloud point (for a putative explanation of this phenomenon, see discussion of Fig. 6 below).

These qualitative findings concerning hydrogel formation are corroborated and can be refined by rheological studies. The characteristic, differing behaviors of the PMDEGA- and the PNIPAM-based copolymers are illustrated in Fig. 6. Clearly, the BAB triblock copolymers are much better suited to form gels than BA diblock copolymer analogs (criterion: storage modulus $G' >$ loss modulus G''). The comparison of Fig. 6 exemplifies that even at much higher concentrations, the gel–sol transition temperature of a diblock copolymer **P(S$_y$-MDEGA$_x$)** occurs already at lower temperatures than for the analogous BAB triblock copolymer **P(S$_y$-MDEGA$_x$-S$_y$)** despite the similar lengths of the hydrophobic as well as of the hydrophilic block. Also, the hydrogels formed by the diblock copolymer are mechanically much weaker, i.e., G'(diblock) $< G'$(triblock). Furthermore, Fig. 6 illustrates that the PMDEGA-based triblock copolymers can form hydrogels already at rather low concentrations (10 wt% and less, see Fig. 6), while PNIPAM-based copolymers of comparable architecture require much higher concentrations for gelation (Fig. 6). Regarding the examples displayed in Fig. 6 for instance, gel formation occurs only from about 30 wt% solutions on, though both the sizes of the PS-blocks and of the hydrophilic block are much bigger than in the analogous PMDEGA based triblock copolymer.

Even more distinct, however, is the evolution of the moduli with temperature: The strength of the gel of the PMDEGA-based copolymer continuously weakens with temperature until the gel point is crossed and *G'* gets smaller than *G"*, and finally, the viscosifying effect virtually disappears above the cloud point (Fig. 6). In strong contrast, the thermal softening of the gels is not only much weaker for PNIPAM-based copolymers, but the hydrogels are strongly reinforced when passing through the cloud point (Fig. 6). This suggests the existence of balanced attractive interactions between the collapsing PNIPAM-shells of the block copolymer aggregates (see below), which may be at least in parts explained by the intermolecular H-bonds formed between different PNIPAM-chains following the collapse transition. This explanation is supported by the fact that similar to the turbidimetry studies in dilute solution, we observe strong hysteresis for the rheological behavior of the PNIPAM-based copolymers during heating and cooling cycles, while this is much less the case for PMDEGA-based copolymers [24, 25, 41].

Aggregation and Segmental Dynamics in Solution

The transition behavior of diblock copolymers **P(S-NIPAM)** and triblock copolymers **P(S-NIPAM-S)** and **P(S-MDEGA-S)** is mainly determined by the chain architecture: The absence or presence of bridging between micelles seems to be the most important feature. For instance, the collapse and aggregation behavior at the cloud point is shown in Fig. 7. A diblock copolymer with a short PS block and a long PNIPAM block was chosen which is expected to form spherical micelles. Diblock copolymers **P(S_{50}-NIPAM$_{160}$)** form micelles in dilute aqueous solutions which shrink slightly as temperature is increased towards the cloud point (Fig. 7). At the cloud point, their hydrodynamic radius decreases drastically, i.e., the micellar shell has collapsed. At higher temperatures, large aggregates ("clusters") are observed which, interestingly, are stable in size at 80 nm. Two triblock copolymers **P(d8S_{10}-NIPAM$_{390}$-d8S_{10})** and **P(S_8-MDEGA$_{452}$-S_8)** having very short PS block as well as thermoresponsive blocks which are approximately twice as long as the ones of the **P(S_{50}-NIPAM$_{160}$)** diblock copolymer were chosen for comparison. In contrast to the diblock copolymer, both triblock copolymers immediately form large clusters at the cloud point, and single collapsed micelles cannot be observed (Fig. 7). The cluster size depends on the nature of the thermo-responsive block: For **P(d8S_{10}-NIPAM$_{390}$-d8S_{10})**, the size is ~70 nm, i.e., similar to the diblock copolymer (Fig. 7), whereas the ones in **P(S_8-MDEGA$_{452}$-S_8)** are much bigger with ~1000 nm in radius. We assume that PNIPAM becomes glassy when water is released (dry $T_g = 133$ °C [68]), which hampers the growth of the clusters, in agreement with the literature [64]. In contrast, PMDEGA (dry $T_g < -40$ °C) stays liquid-like, enabling the growth of very large clusters. In all cases, the micellar structure is preserved in the clusters [25, 44, 46].

All copolymers studied form spherical core-shell micelles in aqueous solution: star-like micelles for **P(S-NIPAM)** diblock copolymers and flower-like micelles with a certain degree of bridging for **P(S-NIPAM-S)** and **P(S-MDEGA-S)** triblock copolymers [25, 44, 46]. Small-angle neutron scattering (SANS) on concentrated solutions revealed that diblock copolymer **P(S_{50}-NIPAM$_{160}$)** forms core-shell micelles with a peculiar behavior around the cloud point (Fig. 8) [46]: The micellar radius increases towards the cloud point and shows critical behavior, whereas the core radius is relatively constant, as expected. Thus, the micellar shell swells as the cloud point is increased; we attribute this behavior to the release of water from the shell and the transient formation of a dry brush which is thicker than a wet brush. Interestingly, the micellar radius decreases only gradually over a broad temperature region. The correlation between the micelles is liquid-like and is described by the hard-sphere radius – basically half the distance between the centers of the micellar cores – and the volume fraction η, i.e., the packing density of the micelles. For the **P(S_{50}-NIPAM$_{160}$)** diblock copolymer, η decreases as the cloud point is approached, i.e. the interaction between them becomes weaker, but above the cloud point, it increases again when the clusters of collapsed micelles form. The transition region is relatively broad, which we attribute to the absence of bridging between the micelles. The star-like micelles gradually form larger and larger clusters as the temperature is increased above the cloud point.

For the triblock copolymer **P(d8S_{10}-NIPAM$_{390}$-d8S_{10})** at 220 g L^{-1} in D_2O, the micellar and the hard-sphere radius are quite constant below the cloud point, and decrease abruptly at the cloud point (Fig. 8). The core-shell structure is preserved in the collapsed state. The volume fraction behaves similarly to the one of the diblock copolymer. These micelles made from BAB triblock copolymers are already partially bridged below the cloud point, and the collapse of the PNIPAM blocks leads to an abrupt collapse of the shell and the mutual distance. In time-resolved SANS measurements during a temperature jump across the cloud point [47, 48], pre-transitional fluctuations in a narrow region below the cloud point could be resolved: Already below the cloud point, the correlation between the micelles decreases, and the formation of aggregates is a complex multi-step process (not shown).

The triblock copolymer **P(S_8-MDEGA$_{452}$-S_8)** behaves differently: No core-shell structure could be detected. Instead, the micelles could be described as homogeneous spheres with small scale fluctuation reflecting the structure of the shell. The micellar radius increases as the temperature is raised above the cloud point and then decreases slightly,

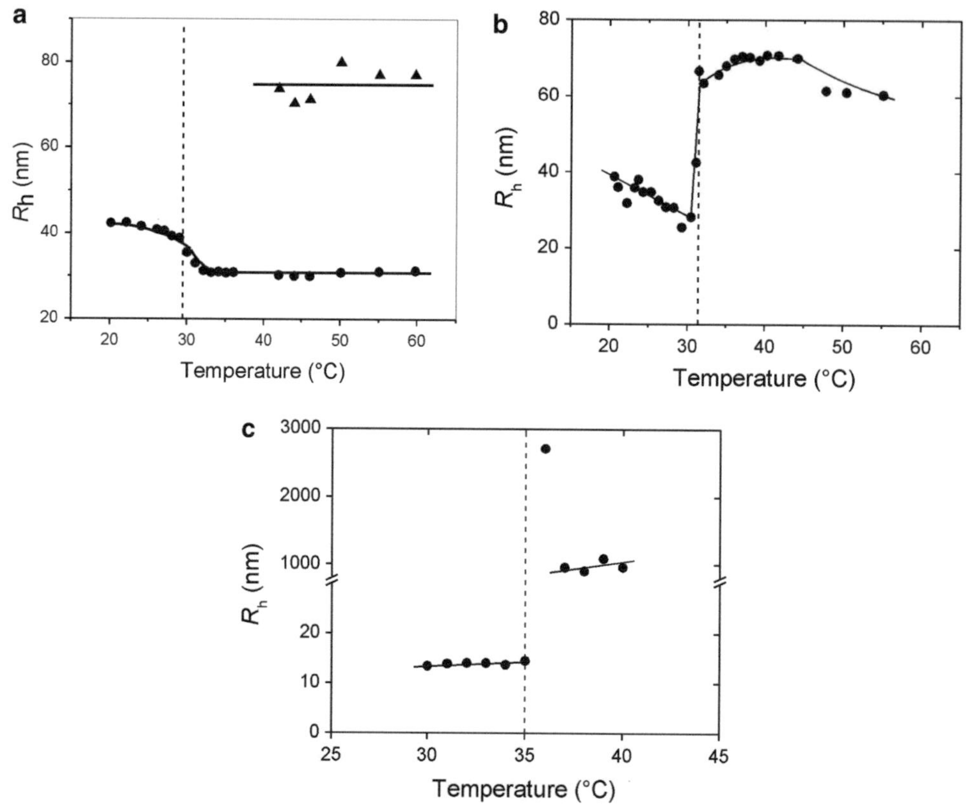

Fig. 7 Hydrodynamic radii of aqueous solutions (**a**) **P(S$_{50}$-NIPAM$_{160}$)** at 0.20 g L^{-1} [42], (**b**) **P(d8S$_{10}$-NIPAM$_{390}$-d8S$_{10}$)** at 0.20 g L^{-1} [44], and (**c**) **P(S$_8$-MDEGA$_{452}$-PS$_8$)** at 3.0 g L^{-1} [25] as determined using dynamic light scattering. Heating runs

i.e., no abrupt decrease is observed as in the PNIPAM based systems (Fig. 8). The hard-sphere radius decreases smoothly in the entire temperature range studied without any discontinuity at the cloud point. The only drastic change at the cloud point is the steep minimum of the volume fraction: Again, the correlation between micelles gets gradually lost as the cloud point is approached, but is reinstalled in the collapsed state. Clearly, there are several striking differences of the **P(S$_8$-MDEGA$_{452}$-PS$_8$)** system with respect to the **P(d8S$_{10}$-NIPAM$_{390}$-d8S$_{10}$)** system, which reflect the different chemical nature of the thermo-responsive block. Changes may also be due to the length of the PS block which possibly passes some critical limit between 8 and 10 monomers. Scheme 2 summarizes the behavior observed for the three systems.

Neutron spin-echo spectroscopy revealed a number of dynamic processes in concentrated solutions of **P(S$_{50}$-NIPAM$_{160}$)** and **P(d8S$_{10}$-NIPAM$_{390}$-d8S$_{10}$)** below the cloud point: the segmental dynamics of PS (in case of protonated PS), the so-called breathing mode of PNIPAM, i.e., the collective dynamics of the swollen PNIPAM shell, and the diffusion of entire micelles (Fig. 9) [44, 46]. The different relative block lengths together with the two architectures (star-like and flower-like micelles) allowed us to vary the grafting density of PNIPAM blocks at the surface of the PS core between 0.20 nm^{-2} and 0.61 nm^{-2}. These values were deduced from the core sizes found in SANS measurements. For both grafting densities, the breathing mode was dominant over the single-chain Zimm dynamics, which is consistent with other studies [69, 70]. With increasing polymer concentration, the segmental mode of the PNIPAM block becomes slower, as expected. Similar findings are made with **P(d8S$_{10}$-NIPAM$_{390}$-d8S$_{10}$)**, but in the swollen state, the segmental mode of the PNIPAM block is slower than in the diblock. We attribute this difference to the back folding of the PNIPAM block into the same micellar core, or to the bridging of two micelles.

Interestingly, for both systems, the segmental dynamics in the collapsed state are independent of polymer concentration and faster than in the swollen state. We attribute this finding to a "freezing in" of the segmental dynamics of the PNIPAM shell except a few strands that stick out into the solution.

The segmental dynamics thus reflects the micellar architecture: Because of the relatively high grafting density of shell blocks on the core surface, the breathing mode of the shell blocks dominates and is influenced by back folding and bridging.

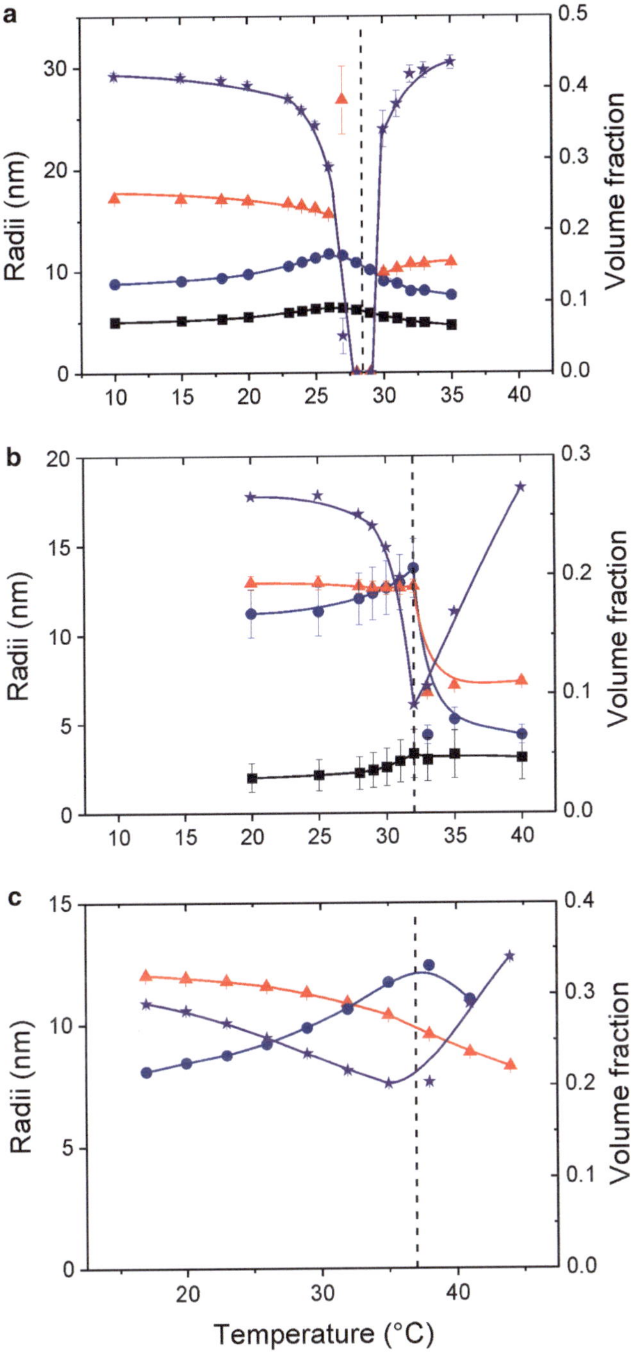

Fig. 8 Micellar dimensions as determined using SANS. (**a**) **P(S$_{50}$-NIPAM$_{160}$)** at 470 g L^{-1} in D$_2$O [46]. (**b**) **P(d8S$_{10}$-NIPAM$_{390}$-d8S$_{10}$)** at 220 g L^{-1} in D$_2$O [44], and (**c**) **P(S$_8$-MDEGA$_{452}$-S$_8$)** at 10 wt.% in H$_2$O [25]. *Filled squares*: core radius, *filled circles*: micellar radius, *filled triangles up*: hard-sphere radius (*all left axis*); *stars*: volume fraction (*right axis*). *Dashed lines*: cloud points

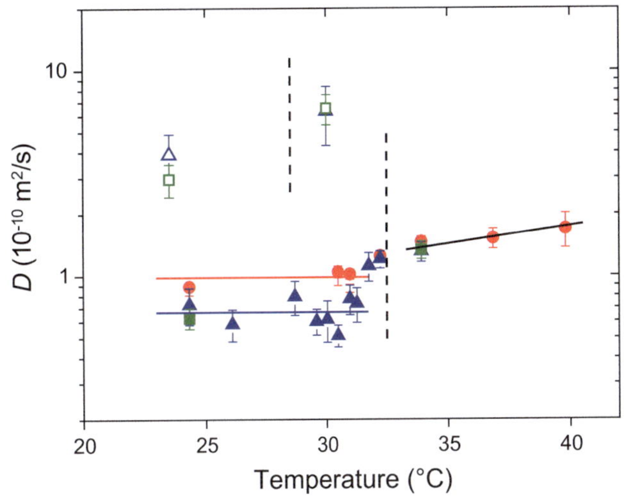

Fig. 9 Temperature-dependence of the diffusion coefficients of the segmental dynamics of the PNIPAM block. Open symbols: **P(S$_{50}$-NIPAM$_{160}$)** [46], *closed symbols*: **P(d8S$_{10}$-NIPAM$_{390}$-d8S$_{10}$)** [44], both in D$_2$O. *Circles*: 50 g L^{-1}, *triangles up*: 200 g L^{-1}, *squares*: 300 g L^{-1}. The *dashed lines* indicate the respective cloud points

The studies of the bulk behavior have thus revealed a number of differences between **P(S-NIPAM)**, **P(S-NIPAM-S)** and **P(S-MDEGA-S)**, concerning the transition temperature, the width and hysteresis of the transition, the temperature-dependent mechanical properties, the collapse and aggregation behavior of micellar solutions and the segmental dynamics. Beyond the behavior in bulk solutions, complementary investigations on thin films attracted our interest as well, because many potential applications of such "smart" hydrogels, for instance in sensors, switches or valves, require their use as (thin) films. Of fundamental interest are questions related to methods for the preparation of homogeneous films, the stability of thin films, the interaction of the thermo-responsive block copolymers with the substrate, confinement effects imposed by the thin film geometry, anisotropic response such as uniaxial vs. three-dimensional swelling, and – more general – the structural rearrangements during swelling and collapse at the cloud point. Compared to the bulk studies, the thin film geometry offers the possibility to control the degree of swelling by using water vapor atmosphere, to determine the water content and the film thickness independently, and to distinguish lateral structural changes from those along the film normal. The next sections describe accordingly the structure of thin thermo-responsive films, their swelling and collapse transition as well as switching kinetics.

Thin Films

Thin thermo-responsive films of homopolymer **PNIPAM**, **PNIPAM** end-capped with *n*-butyltrithiocarbonate at both ends (**nbc-PNIPAM$_{340}$**) [55, 56, 71], amphiphilic diblock copolymers **P(S-NIPAM)** [72, 73] as well as triblock copolymers **P(S-NIPAM-S)** were prepared and studied on flat, smooth silicon supports. Analogous experiments

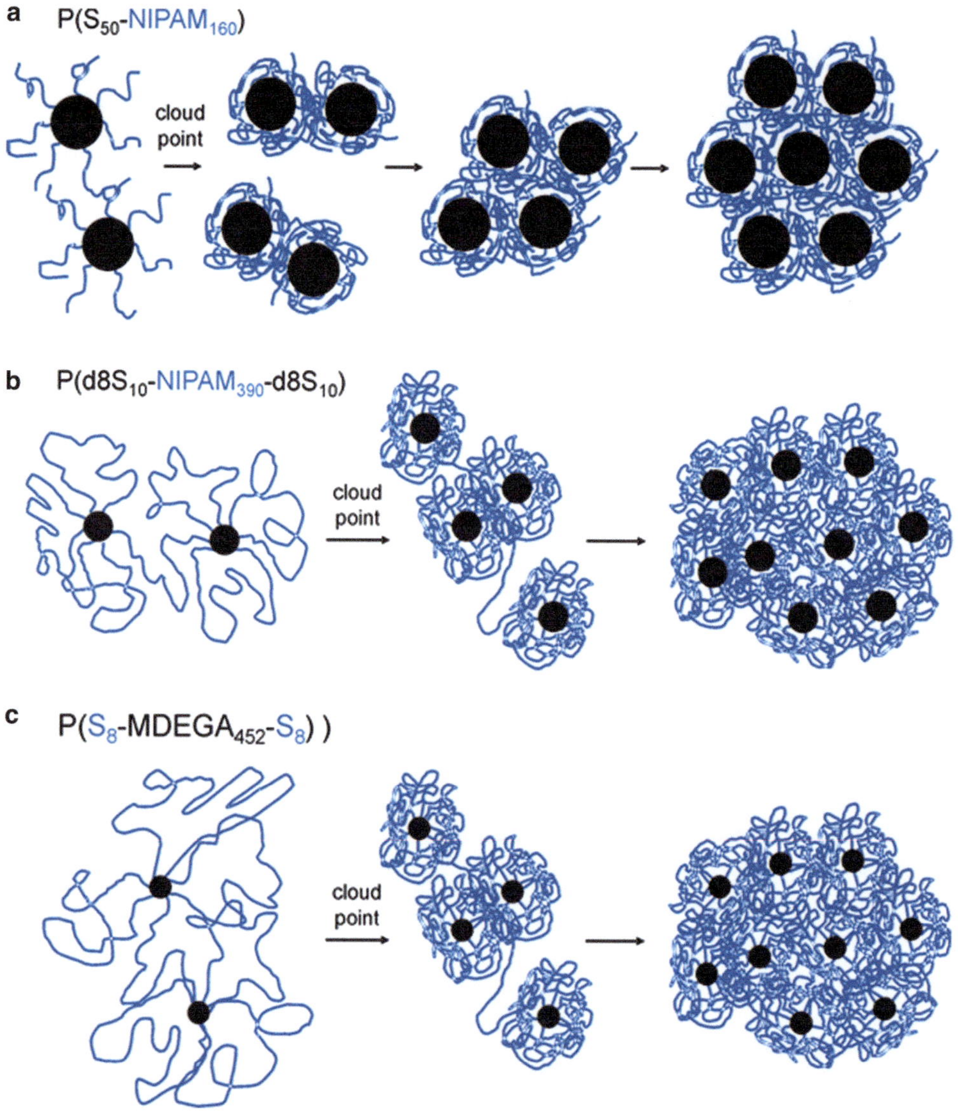

Scheme 2 Model of the phase behavior of the three systems studied in solution. For simplicity, the micellar cores of PS are sketched as homogeneous *circles*. The *blue lines* denote the thermo-responsive PNIPAM or PMDEGA blocks

were performed with homopolymer **PMDEGA** [57, 58] and the amphiphilic triblock copolymers **P(S-MDEGA-S)** [59, 60]. By spin coating on pre-cleaned substrates, smooth and homogeneous films are obtained as characterized by optical microscopy, atomic force microscopy (AFM) and X-ray reflectivity (XRR), whereas it was reported that very thick and bulk-like **PNIPAM** gel films (thicknesses on the order of 60 μm) were inhomogeneous on a micrometer scale [74]. In combination with the described substrate cleaning protocol, spin-coating from dioxane solutions turned out to result in the smoothest and most homogeneous films as compared to other solvents, *e.g.* water. No large aggregates were present in these solutions. Film thicknesses and polymer concentrations of the solution used in spin coating were found to correlate linearly for all polymers studied (see Fig. 10). This enables precise tailoring of the film thickness in a simple way for all the investigated thermo-responsive polymers, and thus, the film thickness was varied from nanometers to hundreds of nanometers. Figure 10 exemplifies XRR measurements of **nbc-PNIPAM$_{340}$** films. From the fit to the data, the density profile along the surface normal of these initially prepared films is extracted. It shows the enrichment of PNIPAM chains at the hydrophilic substrate. Thus the interaction with the substrate influences the arrangement of the polymer chains with respect to the substrate.

Using grazing-incidence small-angle X-ray scattering (GISAXS), the inner film morphology as well as long-ranged correlations in the films have been investigated. With respect to the inner film morphology, it turned out that already the very short *n*-butyltrithiocarbonate end groups of **nbc-**

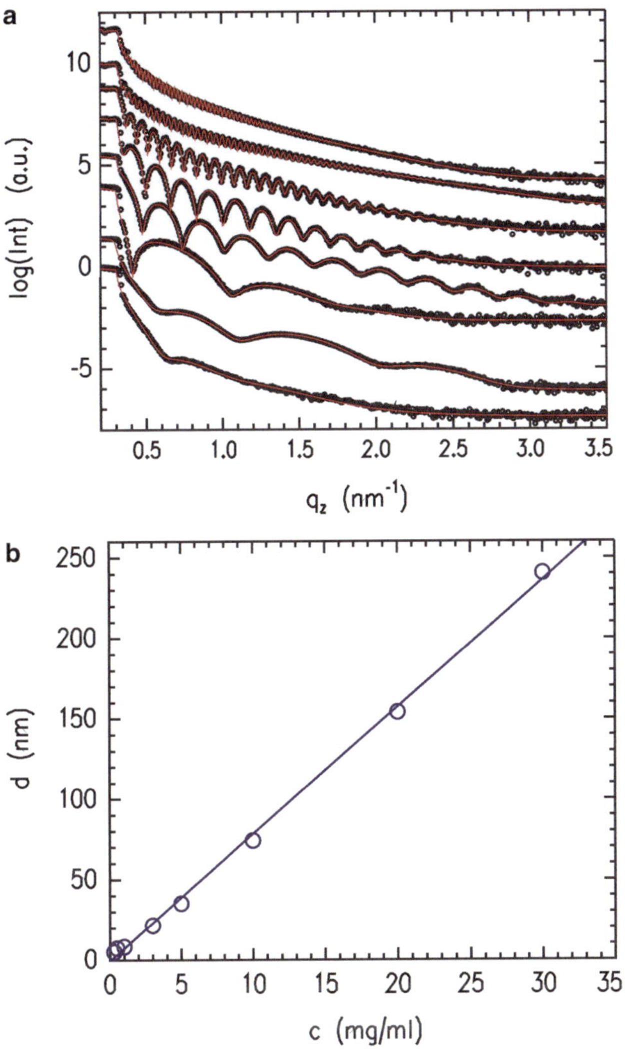

Fig. 10 Characterization of thermo-responsive hydrogel films of **nbc-PNIPAM$_{340}$** spin-coated from dioxane at room temperature and ambient conditions: (**a**) Representative XRR data (*dots*) shown together with model fits (*lines*) for the thickness regime covered in this investigation. With increasing film thickness (5, 7, 8, 22, 35, 74, 154 and 241 nm from *bottom* to *top*), the curves are shifted along the y-axis for presentation. (**b**) Film thickness *d* plotted as a function of the polymer concentration in the solution used for spin-coating. The *solid line* is a linear fit. (Reprinted with permission from Macromolecules 41, 3209–3218, 2008. Copyright (2008) American Chemical Society)

PNIPAM$_{340}$ can be regarded as miniaturized hydrophobic blocks of a BAB-type block copolymer (B denoting a hydrophobic and A a thermo-sensitive block), because an internal structure has been found [55]. Nevertheless, in case of diblock and triblock copolymers, *e.g.*, **P(S-NIPAM)**, **P(S-NIPAM-S)** and **P(S-MDEGA-S)**, the inner morphology of the analogously prepared spin-coated films is much better defined, due to an increase in long-ranged order, and a well established micro-phase separation structure is observed [57–60, 72, 73] In case of **PNIPAM** films, correlated roughness is observed [55]. Thus the surface of the thermo-responsive film is partly correlated with the surface of the underlying substrate, meaning that the roughness spectrum of the substrate is transferred through the polymer film. In contrast, **PMDEGA** based films do not have this spatial correlations. Hence, such correlations are absent and both roughness spectra are independent [57, 60]. This difference between **PNIPAM** and **PMDEGA** films is attributed to the very different glass transition temperatures T_g of both polymers. Due to the high value of T_g(PNIPAM) of about 133 °C, the surface structure installed during spin coating is frozen-in. Very similar to the behavior of other glassy homopolymers such as for example polystyrene [75, 76], the liquid surface couples to the roughness spectrum of the substrate, and the **PNIPAM** chains have no possibility to relax into an energetically more favorable state. In contrast, the low value of T_g(PMDEGA) of < -40 °C gives sufficient mobility to the **PMDEGA** chains to rearrange after spin coating and to erase the correlated roughness. In fact, due to the mobility of the PMDEGA chains, the films roughen during storage under ambient conditions as measured with optical microscopy and AFM. An onset of dewetting is visible already after 1 day, and holes of different diameter are found on the films' surfaces [57].

Thin Film Swelling and Collapse Transition

From the various polymers studied in bulk solution and hydrogels (Tables 1 and 2), we chose **nbc-PNIPAM$_{340}$**, **P(S$_{50}$-NIPAM$_{160}$)**, **PMDEGA$_{138}$** and **P(S$_{11}$-MDEGA$_{126}$-S$_{11}$)** for closer inspection in thin films. The LCST-type transition from a swollen to a collapsed film has been studied for films exposed to water vapor atmosphere, where the degree of swelling is tuned via the humidity of the atmosphere surrounding the thermoresponsive films [58].

The swelling process of PNIPAM and PMDEGA films was monitored by in-situ neutron reflectivity, revealing two regimes (see Fig. 11). In the first regime, the film thickness remains constant and the water volume fraction increases rapidly, as water occupies all holes and vacancies of the polymer film given by the free volume of the film [71]. In the second regime, water is absorbed by the hydrophilic polymer chains, causing the film to swell, so that film thickness and water volume fraction increase slowly.

This second process can be described by a modified diffusion process using a model explaining gel swelling kinetics [77–79]. In this model, swelling (or shrinking) follows first-order kinetics, and is not considered as a pure diffusion process. Information about the thin thermoresponsive film such as shear modulus and osmotic modulus as well as the relaxation time of the swelling can be extracted from the model fit [56, 58, 72, 73]. The actual values for PNIPAM

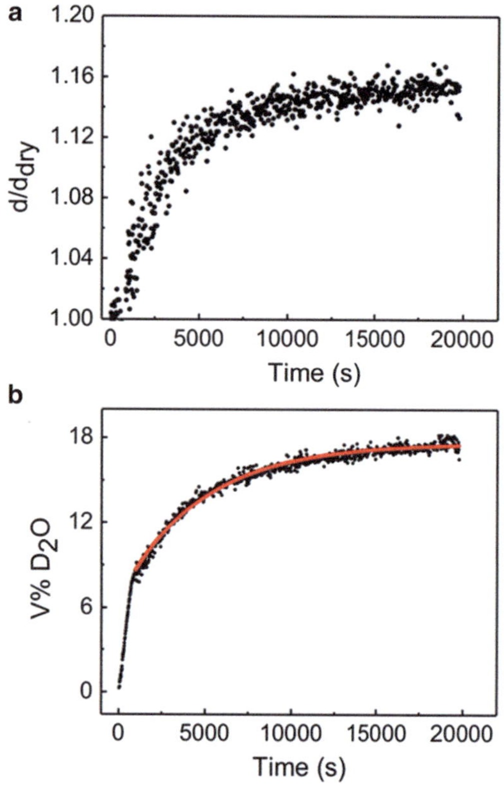

Fig. 11 Response of the $\mathbf{PMDEGA_{138}}$ film with thickness of 35.9 nm to swelling at 23 °C: (**a**) Relative film thickness and (**b**) D_2O volume fraction (V% D_2O) as a function of swelling time, and the *solid line* in (**b**) is a fit of a model as explained in the text (Soft Matter 8, 5241–5249 (2012) – Reproduced by permission of The Royal Society of Chemistry)

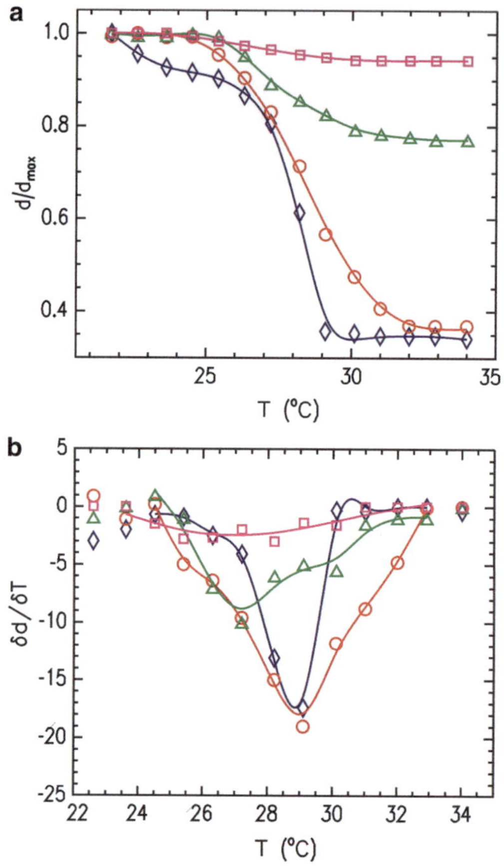

Fig. 12 (**a**) Temperature dependent changes of the film thickness measured for dry $\mathbf{nbc\text{-}PNIPAM_{340}}$ films exposed to saturated water vapor and (**b**) first derivative indicating the transition temperature and the width of the transition. The swelling curves are measured for film thickness of 10.5 (rhombs), 40 (*circles*), 105 (*triangles*) and 200 nm (*squares*). The *solid lines* are guides to the eye. (Reprinted with permission from Macromolecules 41, 3209–3218, 2008. Copyright (2008) American Chemical Society)

and PMDEGA films differ due to the different mechanical properties.

When monitoring the accompanying change in film thickness during swelling at a fixed temperature, an increasing swelling capacity is found for decreasing film thickness [55, 57]. In combination with the hydrophilic substrate surface, which favors the PNIPAM or PMDEGA segments as well as water, the swelling might be increased by a type of entropic spring change in the conformation of the polymers. With increasing film thickness, this effect gets weaker because fewer molecules are in a perturbed conformation [55]. In the literature, only a few swelling experiments for thin thermo-responsive hydrogel films have been reported, which mostly have focused on PNIPAM end-grafted on the substrates [80–83], thus a direct comparison of our results is difficult. For sure, the conformation of end-tethered PNIPAM in thin film geometry will differ from the one of non-covalently bound chains, which is our case.

In general, all investigated **PNIPAM** and **PMDEGA** films show a collapse transition upon heating. Figure 12 shows the example of the temperature dependent changes of the film thickness measured for $\mathbf{nbc\text{-}PNIPAM_{340}}$ films of different thickness [55]. As compared with the corresponding aqueous solution, the transition temperature is shifted to lower values. The transition temperature decreases with increasing hydrogel film thickness, demonstrating that even 200 nm thick films are not bulk-like and differ from solutions. In addition, the width of the transition increases with film thickness. The 10.5 nm thick $\mathbf{nbc\text{-}PNIPAM_{340}}$ film exhibits a very sharp transition within a temperature change of 2 °C, which is small as compared to bulk samples, while for 200 nm thick films, the observed transition width is equivalent to the bulk behavior [55]. Looking to the changes in more detail, it appears as if the transition at an intermediate film thickness of 105 nm occurs in two steps: one dominant step at the temperature as seen for thicker films and one less pronounced step at the temperature found for thinner films. Hence, thin films and the part of thicker films that is near to the substrate, are affected by the interaction with the underlying substrate.

PMDEGA films show the same characteristics of thermo-responsive behavior in thin films as PNIPAM-based polymers: the transition temperatures are reduced in thin films as compared with solutions, and the transition temperatures decrease with increasing film thickness. However, compared to **PNIPAM** films, the width of the transition is increased. This difference in the width of the transition may be related to the different mechanisms of water release. In the case of **PNIPAM**, two types of bound water coexist [65, 84, 85]. The first type is bound by hydrogen bonds to the amide group, while the second type forms a water cage surrounding the hydrophobic moieties. When the temperature increases above the LCST-type transition temperature, both types of water are released from the film. According to the model of cooperative hydration of the **PNIPAM** chain, a pearl-necklace like chain conformation is present in hydrated PNIPAM due to the large hydrophobic isopropyl side groups [86–88]. This consecutive sequence is dehydrated as a whole, causing a sharp transition. In contrast, **PMDEGA** does not dispose of strongly polarized H-atoms, and thus does not contain strongly or weakly acidic protons. As a consequence, **PMDEGA** can act in H-bonds only as acceptor (due to the free electron pairs of oxygen in the ether groups), but not as donor. The ester groups (anchoring the oligoethyleneglycolethers to the polymer backbone) are very weak acceptors and are commonly not considered to be able to form H-bonds, in particular with water. Thus, no cooperative effect is present in **PMDEGA,** and the transition is significantly broader than for **PNIPAM**.

Noteworthy, inert metals such as gold can be deposited homogenously on top of such thin thermo-responsive hydrogel films, as shown, for instance, for **nbc-PNIPAM$_{340}$**. This gives access to applications such as sensors or nano-switches. The deposition of a metal contact on top of the thermoresponsive films does not affect their ability to switch from an extended to a collapsed chain conformation, and thus to change the film thickness from an initial thickness to a lower value [56].

Switching Kinetics in Thin Films

In-situ neutron reflectivity was applied for investigating the switching kinetics. The swollen thermo-responsive films were exposed to an abrupt change in the temperature (called temperature jump) from a start temperature below the phase transition, to a value above the transition. Such kinetic investigations were performed analogously for both **PNIPAM** and **PMDEGA** based films. The use of deuterated water gives the necessary contrast to follow the changes in water content and in film thickness. In all cases, a complex response of the thermo-responsive film to the change in temperature is observed. Film thickness and water volume fraction show a multi-step behavior.

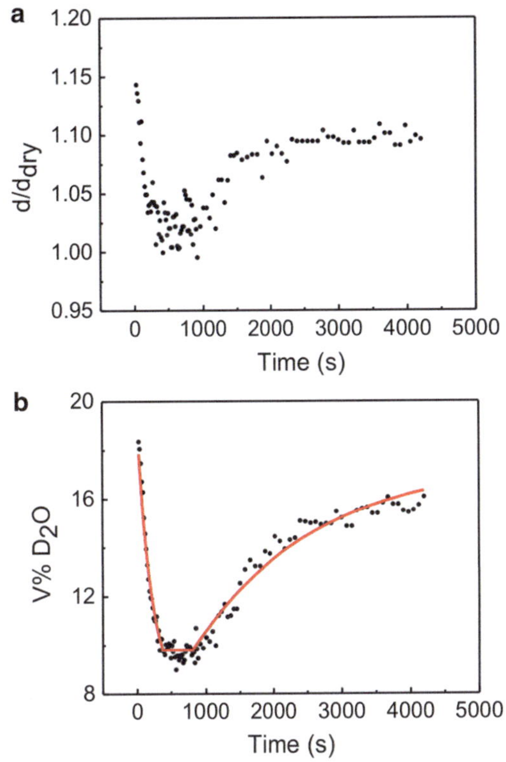

Fig. 13 Response of **PMDEGA$_{138}$** film with thickness of 35.9 nm to a temperature jump from 23 °C to 45 °C: (**a**) Relative film thickness and (**b**) D_2O volume fraction (V% D_2O) as a function of time. The *solid line* in (**b**) is a fit of a model as explained in the text (Soft Matter 8, 5241–5249 (2012) – Reproduced by permission of The Royal Society of Chemistry)

Figure 13 illustrates the example of **PMDEGA$_{138}$** hydrogel films [58]. The relative film thickness and the D_2O volume fraction (vol% D_2O) are plotted as a function of time t, with $t = 0$ being defined as the time at which the temperature is changed from 23 °C to 45 °C. The transition temperature is at 40 °C for the investigated film thickness of 35.9 nm. Three distinct regimes in the response are detected. The initial response of the **PMDEGA** film is fast (regime 1). The film reacts by decreasing the water content from 18 % to 9 %. The water release is accompanied by a shrinkage of the film back to its original dry thickness. The whole process takes only 400 s. When the film thickness is decreased to its initial value, a D_2O volume fraction of 9 % remains in the film. This residual water may occupy vacancies and holes in the porous **PMDEGA** film, which does not add to the film thickness (cf. discussion above), and is strongly bound to the polymer chains.

After the fast shrinkage event, film thickness and water volume fraction stay nearly constant for some time (regime 2). As the film is still located in the water vapor atmosphere after the collapse, the chains are assumed to slowly relax into a more favorable conformation. Due to this rearrangement, the polymers are able to bind more water, and thus the films can reabsorb a certain amount of water. This results in an increase of film thickness (regime 3). Thus, the hydrogel film reacts with a complex three-stage process to the abrupt temperature jump.

When introducing an internal micro-phase separated structure, as for example in films of the amphiphilic triblock copolymers **P(S-MDEGA-S)** [60], or by adding a metal top layer [56], this general multi-stage process is not changed. In all cases, a response which can be understood as shrinkage, reorganization and reswelling is seen. In the case of thin thermo-responsive films based on **PNIPAM,** the duration of the reorganization step is only short. This again can be explained by the different type of hydrogen bonds formed by **PNIPAM** as compared with **PMDEGA**.

Conclusions

The methods of RDRP, in particular of RAFT polymerization, enabled us to prepare two families of amphiphilic thermo-responsive block copolymers, which vary systematically the chemical building blocks (i.e., nature and size of the hydrophilic, thermo-responsive as well as of the hydrophobic blocks), as well as the architecture the amphiphiles (homopolymer vs. diblock vs. symmetrical triblock *vs* star block copolymer). In particular we successfully fabricated two analogous polymer series based on PNIPAM and PMDEGA, respectively, as hydrophilic, thermo-sensitive "smart" blocks with a transition temperature in the biologically particularly interesting range of 30–40 °C.

For the characterization of bulk solutions and gels, the combination of a number of methods allowed us to investigate a large concentration range, ranging from below the critical micelle concentration to the highly concentrated regime. The temperature-dependent structural and dynamic properties of the solutions and gels could be characterized on a large range of length and time scales and could further the understanding of the macroscopic characteristics, such as the macroscopically observed phase behavior as well as the mechanical behavior. Kinetic measurements enabled us to follow the complex structural changes around the cloud point in great detail.

Thin films were successfully prepared on solid supports and investigated concerning film morphology, long range correlations, stability, swelling behavior and thermal response. X-ray and neutron scattering in combination with real space imaging techniques enabled us to follow the water incorporation as well as the chain collapse induced water release. The interaction with the substrate as well as confinement effects introduced by the film thickness turned out to influence the film structure as well as the film swelling.

In summary, the combination of modern synthesis methods, comprehensive solution, hydrogel and thin film studies with a number of different experimental methods allowed us to get a deeper understanding of the self-assembly as well as the collapse transition of this interesting system. This way, we could especially get an insight into the role of the chain architecture and the nature of the thermo-responsive block.

Acknowledgment The results presented are the outcome of combined research activities over a period of 6 years, which crucially depended on the creativity, labor, zeal and enthusiasm of many dedicated post-doctoral, Ph.D. and master students at Potsdam University and at TU München, who are (in alphabetical order) Joseph Adelsberger, C. Adrián Benítez-Montoya, Achille M. Bivigou-Koumba, Charles Darko, Alexander Diethert, Anastasia Golosova, Abhinav Jain, Gunar Kaune, Juliane Kristen, Amit Kulkarni, Andreas Meier-Koll, David Magerl, Ezzeldin Metwalli, Anna Miasnikova, Gabriele De Paoli, Monika Rawolle, Matthias A. Ruderer, Katja Skrabania, Kordelia Troll, Weinan Wang and Qi Zhong. Moreover, using large scale facilities, the approved beamtime and support by the beamline scientists (Peter Busch, Robert Cubitt, Sergio S. Funari, Isabelle Grillo, Olaf Holderer, Jan Perlich, Vitaliy Pipich, Stephan V. Roth) are acknowledged. Financial support was provided by Deutsche Forschungsgemeinschaft (DFG) via the priority program SPP 1259 "Intelligente Hydrogele" (grants LA611/7, MU1487/8 and PA771/4), by Fonds der Chemischen Industrie (FCI), and by German Academic Exchange Service (DAAD). In the priority program, we enjoyed fruitful collaboration with Regine von Klitzing (TU Berlin), Thomas Hellweg (Universität Bielefeld), Walter Richtering (RWTH Aachen), and Norbert Stock (Universität Kiel). Also, we gladly thank our external partners Franz Faupel (Universität Kiel), Eckhard Görnitz, Michael Päch (both Fraunhofer Institute for Applied Polymer Research, Potsdam), Marianne Hanzlik (Technische Universität München), Klaus Rätzke (Universität Kiel), Ralf Röhlsberger, and Kai Schlage (both DESY Hamburg) for their valuable input and support.

References

1. Kopeček J (2009) J Polym Sci, Part A: Polym Chem 47:5929–5946
2. Vlierberghe SV, Dubruel P, Schacht E (2011) Biomacromolecules 12:1387–1408
3. Wanka G, Hoffmann H, Ulbricht W (1994) Macromolecules 27:4145–4159
4. Mortensen K (2001) Coll Surf A 183–185:277–292
5. Hamley IW (2005) Block copolymers in solution: fundamentals and applications. Wiley, Chichester
6. Booth C, Attwood D (2000) Macromol Rapid Comm 21:501–527
7. Cohen Stuart MA (2008) Colloid Polym Sci 286:855–864
8. Kimerling AS, Rochefort WE, Bhatia SR (2006) Ind Eng Chem Res 45:6885–6889
9. Madsen J, Armes SP (2012) Soft Matter 8:592–605
10. Zhulina EB, Borisov OV (2012) Macromolecules 45:4429–4440
11. Halperin A (2006) J Macromol Sci C: Polym Rev 46:173–214
12. Tsitsilianis C, Iliopoulos I (2002) Macromolecules 35:3662–3667
13. Annable T, Buscall R, Ettelaie R, Whittlestone D (1993) J Rheol 37:695–726
14. Winnik MA, Yekta A (1997) Curr Opin Colloid Interface Sci 2:424–436
15. Chassenieux C, Nicolai T, Benyahia L (2011) Curr Opin Colloid Interface Sci 16:18–26
16. de Molina PM, Herfurth C, Laschewsky A, Gradzielski M (2011) Prog Colloid Polym Sci 138:67–72

17. Chaterji S, Kwon IK, Park K (2007) Prog Polym Sci 32:1083–1122
18. Tokarev I, Minko S (2009) Soft Matter 5:511–524
19. Liu R, Fraylich M, Saunders BR (2009) Colloid Polym Sci 287:627–643
20. Kratz K, Hellweg T, Eimer W (2000) Coll Surf A 170:137–149
21. Stieger M, Richtering W, Pedersen JS, Lindner P (2004) J Chem Phys 120:6197–6206
22. Dimitrov I, Trzebicka B, Müller AHE, Dworak A, Tsvetanov CB (2007) Prog Polym Sci 32:1275–1343
23. Tsitsilianis C (2010) Soft Matter 6:2372–2388
24. Bivigou-Koumba AM, Görnitz E, Laschewsky A, Müller-Buschbaum P, Papadakis CM (2010) Colloid Polym Sci 288:499–517
25. Miasnikova A, Laschewsky A, De Paoli G, Papadakis CM, Müller-Buschbaum P, Funari SS (2012) Langmuir 28:4479–4490
26. Taylor LD, Cerankowski LD (1975) J Polym Sci Part A: Polym Chem 13:2551–2570
27. Platé NA, Lebedeva TL, Valuev LI (1999) Polymer J Jpn 31:21–27
28. Schlaad H, Diehl C, Gress A, Meyer M, Demirel AL, Nur Y, Bertin A (2010) Macromol Rapid Comm 31:511–525
29. Schild HG (1992) Prog Polym Sci 17:163–249
30. Aseyev V, Tenhu H, Winnik F (2006) Adv Polym Sci 196:1–85
31. Aseyev V, Tenhu H, Winnik F (2011) Adv Polym Sci 242:29–89
32. Nykänen A, Nuopponen M, Laukkanen A, Hirvonen S-P, Rytelä M, Turanen O, Tenhu H, Mezzenga R, Ikkala O, Ruokolainen J (2007) Macromolecules 40:5827–5834
33. Zhou XC, Ye XD, Zhang GZ (2007) J Phys Chem B 111:5111–5115
34. Nykänen A, Nuopponen M, Hiekkataipale P, Hirvonen S-P, Soininen A, Tenhu H, Ikkala O, Mezzenga R, Ruokolainen J (2008) Macromolecules 41:3243–3249
35. Riess G (2003) Prog Polym Sci 28:1107–1170
36. Jenkins AD, Jones RG, Moad G (2010) Pure Appl Chem 82:483–491
37. Moad G, Rizzardo E, Thang SH (2008) Acc Chem Res 41:1133–1142
38. Matyjaszewski K, Müller AHE (eds) (2009) Controlled and living polymerizations. From mechanisms to applications. Wiley-VCH, Weinheim
39. Barner-Kowollik C (ed) (2008) Handbook of RAFT polymerization. Wiley-VCH, Weinheim
40. Moad G, Rizzardo E, Thang SH (2012) Aust J Chem 65:985–1076
41. Miasnikova A, Laschewsky A (2012) J Polym Sci, Part A: Polym Chem 50:3313–3323
42. Troll K, Kulkarni A, Wang W, Darko C, Bivigou-Koumba AM, Laschewsky A, Müller-Buschbaum P, Papadakis CM (2008) Colloid Polym Sci 286:1079–1092
43. Bivigou-Koumba AM, Kristen J, Laschewsky A, Müller-Buschbaum P, Papadakis CM (2009) Macromol Chem Phys 210:565–578
44. Adelsberger J, Kulkarni A, Jain A, Wang W, Bivigou-Koumba AM, Busch P, Pipich V, Holderer O, Hellweg T, Laschewsky A, Müller-Buschbaum P, Papadakis CM (2010) Macromolecules 43:2490–2501
45. Jain A, Kulkarni A, Bivigou-Koumba AM, Wang W, Busch P, Laschewsky A, Müller-Buschbaum P, Papadakis CM (2010) Macromol Symp 291/292:221–229
46. Adelsberger J, Meier-Koll A, Bivigou-Koumba AM, Busch P, Holderer O, Hellweg T, Laschewsky A, Müller-Buschbaum P, Papadakis CM (2011) Colloid Polym Sci 289:711–720
47. Adelsberger J, Metwalli E, Diethert A, Grillo I, Bivigou-Koumba AM, Laschewsky A, Müller-Buschbaum P, Papadakis CM (2012) Macromol Rapid Comm 33:254–259
48. Adelsberger J, Grillo I, Kulkarni A, Sharp M, Bivigou-Koumba AM, Laschewsky A, Müller-Buschbaum P, Papadakis CM (2013) Soft Matter 9:1685–1699
49. Brown W, Mortensen K (eds) (2000) Scattering in polymeric and colloidal systems. Gordon and Breach, Amsterdam
50. Grillo I (2008) Chapter 13: Small-angle neutron scattering and application in soft condensed matter. In: Borsali R, Pecore R (eds) Soft matter characterization, vol 1. Springer, New York
51. Ewen B, Richter D (1997) Adv Polym Sci 134:1–129
52. Richter D, Monkenbusch M, Arbe A, Colmenero J (2005) Adv Polym Sci 174:1–221
53. Päch M, Zehm D, Lange M, Dambowsky I, Weiss J, Laschewsky A (2010) J Am Chem Soc 132:8757–8765
54. Skrabania K, Miasnikova A, Bivigou-Koumba AM, Zehm D, Laschewsky A (2011) Polym Chem 2:2074–2083
55. Wang W, Troll K, Kaune G, Metwalli E, Ruderer M, Skrabania K, Laschewsky A, Roth SV, Papadakis CM, Müller-Buschbaum P (2008) Macromolecules 41:3209–3218
56. Wang W, Kaune G, Perlich J, Papadakis CM, Koumba AMB, Laschewsky A, Schlage K, Röhlsberger R, Roth SV, Cubitt R, Müller-Buschbaum P (2010) Macromolecules 43:2444–2452
57. Zhong Q, Wang W, Adelsberger J, Golosova A, Koumba AMB, Laschewsky A, Funari SS, Perlich J, Roth SV, Papadakis CM, Müller-Buschbaum P (2011) Colloid Polym Sci 289:569–581
58. Zhong Q, Metwalli E, Kaune G, Rawolle M, Bivigou-Koumba AM, Laschewsky A, Papadakis CM, Cubitt R, Müller-Buschbaum P (2012) Soft Matter 8:5241–5249
59. Zhong Q, Adelsberger J, Niedermeier M, Golosova A, Bivigou-Koumba AM, Laschewsky A, Funari SS, Papadakis CM, Müller-Buschbaum P (2013). Colloid Polym Sci 291:1439–1451
60. Zhong Q (2012) Structure and transition behavior of novel thermoresponsive polymer films. Doctoral thesis. Technische Universität München, München
61. Miasnikova A (2012) New hydrogel forming thermo-responsive block copolymers of increasing structural complexity. Doctoral thesis. Universität Potsdam, Potsdam
62. Winnik FM, Davidson AR, Hamer GK, Kitano H (1992) Macromolecules 25:1876–1880
63. Kujawa P, Segui F, Shaban S, Diab C, Okada Y, Tanaka F, Winnik FM (2006) Macromolecules 39:341–348
64. Koga T, Tanaka F, Motokawa R, Koizumi S, Winnik FM (2008) Macromolecules 41:9413–9422
65. Cho EC, Lee J, Cho K (2003) Macromolecules 36:9929–9934
66. Aseyev V, Hietala S, Laukkanen A, Nuopponen M, Confortini O, Prez FED, Tenhu H (2005) Polymer 46:7118–7131
67. Junk MJN, Li W, Schlüter AD, Wegner G, Spiess HW, Zhang A, Hinderberger D (2011) J Am Chem Soc 133:10832–10838
68. Nuopponen M, Kalliomäki K, Laukkanen A, Hietala S, Tenhu H (2008) J Polym Sci, Part A: Polym Chem 46:38–46
69. Farago B, Monkenbusch M, Richter D, Huang JS, Fetters LJ, Gast AP (1993) Phys Rev Lett 71:1015–1018
70. Kanaya T, Monkenbusch M, Watanabe H, Nagao M, Richter D (2005) J Chem Phys 122:144905–144913
71. Harms S, Rätzke K, Faupel F, Egger W, Ravello L, Laschewsky A, Wang W, Müller-Buschbaum P (2010) Macromol Rapid Comm 31:1364–1367
72. Wang W, Metwalli E, Perlich J, Troll K, Papadakis CM, Cubitt R, Müller-Buschbaum P (2009) Macromol Rapid Comm 30:114–119
73. Wang W, Metwalli E, Perlich J, Papadakis CM, Cubitt R, Müller-Buschbaum P (2009) Macromolecules 42:9041–9051
74. Zhou S, Wu C (1996) Macromolecules 29:4998–5001
75. Müller-Buschbaum P, Stamm M (1998) Macromolecules 31:3686–3692
76. Müller-Buschbaum P, Gutmann JS, Lorenz C, Schmitt T, Stamm M (1998) Macromolecules 31:9265–9272
77. Tanaka T, Hocker LO, Benedek GB (1973) J Chem Phys 59:5151–5159
78. Landau LD, Lifshitz EM (1986) Theory of elasticity, 3rd edn. Pergamon Press, Oxford

79. Li Y, Tanaka T (1990) J Chem Phys 92:1365–1371
80. Yan Q, Hoffman AS (1995) Polymer 36:887–889
81. Kaneko Y, Nakamura S, Sakai K, Aoyagi T, Kikuchi A, Sakurai Y, Okano T (1998) Macromolecules 31:6099–6105
82. Kaholek M, Lee W-K, Ahn S-J, Ma H, Caster KC, LaMattina B, Zauscher S (2004) Chem Mater 16:3688–3696
83. Yim H, Kent MS, Mendez S, Lopez GP, Satija S, Seo Y (2006) Macromolecules 39:3420–3426
84. Tamai Y, Tanaka H, Nakanishi K (1996) Macromolecules 29:6750–6760
85. Grinberg VY, Dubovik AS, Kuznetsov DV, Grinberg NV, Grosberg AY, Tanaka T (2000) Macromolecules 33:8685–8692
86. Okada Y, Tanaka F (2005) Macromolecules 38:4465–4471
87. Kojima H, Tanaka F (2010) Macromolecules 43:5103–5113
88. Tanaka F, Koga T, Kojima H, Winnik FM (2011) Chin J Polym Sci 29:13–21

Visualization of Hydrogel Shrinkage Due to Ion Replacement by ^{27}Al and ^{23}Na Magnetic Resonance Imaging

M. Raue, J. Martins, M. Küppers, T. Mang, B. Blümich, and S. Stapf

Abstract

Ionic hydrogels are polymer networks that have the ability to store large amounts of water. Due to this behavior hydrogels based on sodium acrylate (SA) are mostly used in diapers and soil conditioners. In this work diffusion and exchange processes in polyelectrolyte hydrogels were followed by ^{23}Na and ^{27}Al magnetic resonance imaging (MRI) techniques. We present ^{23}Na and ^{27}Al MRI experiments of the replacement reaction in hydrogels (classical superabsorbers, hydrogels with an inherent charge gradient and with a higher charge) as a feasibility study of monitoring the combined diffusion/exchange process as well as the shrinking and local defect growth in macroscopic model hydrogels. It is shown that the contraction of the gel induced by increasing the number of physical crosslinks through aluminum ions is of non-linear nature and takes place in several steps in contrast to high charge-density hydrogels based on acrylic acid and maleic acid.

Keywords

Polyelectrolyte hydrogel • NMR ^{27}Al and ^{23}Na relaxation • MRI ^{27}Al and ^{23}Na • Dynamic exchange processes • Gradient hydrogel

Abbreviations

AA	Acrylic acid
AAS	Atomic absorption spectroscopy
χ_i	Molality
CPMG	Carr-Purcell-Meiboom-Gill
HD	Hexanediol
MA	Maleic acid
MRI	Magnetic resonance imaging (MRI)
NMR	Nuclear magnetic resonance
PEGDMA	Poly(ethylene glycol)-dimethacrylate
SA	Sodium acrylate
SCP	Sokolan CP
SR	Swelling ratio

Introduction

Artificial hydrogels have been known since the 1940s to be ion-exchange materials [1]. Since the 1960s, hydrogels have been used in diapers as well as to enhance the water

*Both the authors contributed equally to this work.

M. Raue
Institut für Technische und Makromolekulare Chemie, RWTH-Aachen University, Worringer Weg 1, 52074 Aachen, Germany

Institute for Applied Polymer Science, Aachen University for Applied Science Heinrich-Mußmann-Str. 1, 52428 Jülich, Germany

S. Stapf (✉)
Department of Technical Physics II, Technical University Ilmenau, PO Box 100 565, 98684, Ilmenau, Germany
e-mail: siegfried.stapf@tu-ilmenau.de

J. Martins
Institut für Technische und Makromolekulare Chemie, RWTH-Aachen University, Worringer Weg 1, 52074 Aachen, Germany

Department of Technical Physics II, Technical University Ilmenau, PO Box 100 565, 98684, Ilmenau, Germany

M. Küppers • B. Blümich
Institut für Technische und Makromolekulare Chemie, RWTH-Aachen University, Worringer Weg 1, 52074 Aachen, Germany

T. Mang
Institute for Applied Polymer Science, Aachen University for Applied Science Heinrich-Mußmann-Str. 1, 52428 Jülich, Germany

G. Sadowski and W. Richtering (eds.), *Intelligent Hydrogels*, Progress in Colloid and Polymer Science 140,
DOI 10.1007/978-3-319-01683-2_3,

retention capacity of dry soils in agriculture. However for the first time in the more recent development of gels, in 1978 Tanaka reported that the polymer network in polyacrylamide gels [2] collapses by decreasing the temperature or increasing the acetone concentration of an acetone-water solution. Since then, so-called "smart" hydrogels have evolved and attracted considerable scientific interest.

Hydrogels are called "smart" when they swell and shrink as a response to a range of external stimuli including magnetic and electric fields, thus enabling them to be applied as biodegradable drug delivery systems or simple drug release gels, biosensors, valves, filters, mechanical or optical on-off switches and artificial actuators [3–11]. However to apply smart ionic polymers as sensors and actuators mostly depends on the feasibility to generate fast responses to different stimuli such as, in particular, a change in temperature, pH, ion concentration in solutions and in the polarity of solvents. The response is generally expressed by a deformation of the chemically designed network resulting in a swelling or shrinking of the gel. Volume changes of two or three orders of magnitude have been observed in the past [12]. In most cases, the hydrogel network is a porous mesh of chemically cross-linked polymer chains with additional macroscopic properties of the bulk matrix.

The fast swelling response of ionic hydrogels, with poly(sodium acrylate) (SA) being one of the most well-known example, is the result of the osmotic pressure difference caused by the varying concentration of the sodium ions inside and outside the network matrix. Electrostatic interactions between the sodium ions and the anionic carboxylic groups in the polymer backbone lead to a repulsion in presence of water, adding a supplementary swelling force [13]. The swelling parameters can be controlled by the cross-linker type, its concentration and the charge density which affects the mechanical properties. A low crosslink density results in high mobility of the polymer backbone. This specifically designed network can store a considerable liquid volume in the matrix.

Foreign ions with polyvalent character such as calcium, magnesium and aluminum interact with the carboxyl groups present in the matrix. Due to a stronger affinity, they eventually replace the sodium ions that were initially present in the matrix. Thus, this ion exchange leads to restricted mobility of the polymer backbone. Since the osmotic equilibrium is also partially affected, the solvent is then expelled from the hydrogel until a new swelling equilibrium is reached. Consequently, from a macroscopic perspective, the sample exhibits a decreased storage capacity, the matrix collapses, and the hydrogel particles shrink. Meanwhile, the conformation of the polymer also changes thereby causing the chains to be less extended on average. Understanding the dynamics and the competing activities of the sodium and the foreign ions is fundamental for controlling the swelling parameters and the mechanical properties when designing high-performance hydrogel systems [1, 12].

A particular problem of the hydrogels is that diffusion processes, especially for large samples, limit their fast response rate. Another important factor is the fragility of ionic hydrogels: They are extremely soft and break easily when handled in the highly swollen state. They also tend to lose their structural integrity during swelling and shrinking if macroscopic heterogeneities of the gel structure are present. Repeated swelling-shrinking cycles will thus eventually destroy the original shape of the hydrogel. Therefore, the use of hydrogels is restricted for cycle tasks where lifespan and repetition factors are essential for applications such as actuators [14, 15]. This current study focusses on directly monitoring the sodium and aluminum ions by means of magnetic resonance for two reasons: On the one hand, their sensitivity to local dynamics has been found to be stronger than that of the ubiquitous protons where, in undeuterated and thus technically relevant samples, the signal is generated from backbone protons and water protons alike. On the other hand, applying NMR imaging techniques allows one to directly visualize local ion concentrations outside and, with some restrictions, inside the hydrogel, and to highlight heterogeneities as well as defects in the gel during swelling/shrinking. Sodium ions are bound to the carboxyl groups present in the acrylate backbone but retain a high degree of mobility due to exchange with the bulk. By monitoring the sodium signal, one therefore not only obtains a measure of localized concentration but is also able to assess the mobility of the ions and, indirectly, of the backbone they are attached to. Monitoring sodium and aluminum ions has the further advantage that they can, with little modifications, be observed with the same hardware. In addition, their resonance frequencies are similar to that of carbon-13, another important NMR-active nucleus for which standard hardware and acquisition procedures are available [16].

While the Larmor frequencies of ^{23}Na and ^{27}Al are similar and the natural abundance of both nuclei is 100 %, the main difference in measuring these nuclei lies in the much larger quadrupolar coupling constant of ^{27}Al which leads to extremely broad lines in an anisotropic environment which makes conventional magnetic resonance imaging (MRI) unfeasible. In solution, however, fast tumbling averages out this effect and leads to sufficiently narrow lines. Even so, only a handful of pioneering studies on ^{27}Al MRI have been reported in the literature either regarding the solution state [17, 18], or the solid state [19–24]. Nonetheless, ^{23}Na, having much higher biomedical and physiological significance than ^{27}Al, has been the subject of a wide range of fundamental and applied MRI studies, like Nielles-Vallespin

Table 1 ^{23}Na relaxation times T_1 and T_2 of the hydrogel sample based on poly(SA) and its derivatives

Sample	Ratio (AA:MA)	Cross-linker	χ_i [mol-%]	SR [%]	T_1 [ms]	T_2 [ms]
Poly(SA) 5	100:0	MBA	5	500	10.53 ± 0.05	4.47 ± 0.05
SCP 1	70:30	HD	1	400	3.15 ± 0.53	2.61 ± 0.04
SCP 5-1	70:30	HD	5	500	3.60 ± 0.02	0.89 ± 0.02
SCP 5-2	70:30	HD	5	200	2.21 ± 0.01	1.61 ± 0.02
SCP 10	70:30	HD	10	200	2.21 ± 0.02	0.58 ± 0.01
Poly(SA-co-MA) 7.5	50:50	HD	7.5	500	3.72 ± 0.04	1.16 ± 0.04

et al. whose goal was to implement ^{23}Na MRI in a clinical setting for neuro-oncological and muscular imaging [25–27].

Experimental Section

Materials

For the gel preparation, the following chemical compounds were used: poly(ethylene glycol)-dimethacrylate (PEGDMA, M_n~750 g/mol, Sigma-Aldrich), N,N- methylene-bis(acrylamide) (MBA, Merck), 1-[4-(2-Hydroxyethoxy)-phenyl]-2hydroxy-2-methyl-1-propane-1-one (Irgacure 2959, Ciba), hydroxyethyl methacrylate (HEMA, Acros Organics), Methoxy-poly(ethylene glycol)-meth-acrylate (MPEG 350MA, Bisomer), sodium peroxydisulfate (NPS, Merck), triethylamine, (TEA, Merck), poly(acrylic acid-co-maleic acid) (SA:MA = 50:50, M_n~5,000 g/mol, Sigma-Aldrich), Sokalan CP 45 (SA:MA = 30:70, M_n~70,000 g/mol, BASF), hexanediol (HD, Merck), p-toluenesulfonic acid (Merck), n-heptane (Sigma-Aldrich). These extra pure grade chemicals were used without any further purification. The acrylic acid (AA, Merck) reagent was equivalently neutralized with extra pure grade sodium hydroxide to achieve a 35 % SA solution.

Sample Preparation

Poly(SA) was prepared by free-radical cross-linking copolymerization of SA with PEGDMA in aqueous solution in a 200 ml flat vessel. The amount of substance was set to 100 mmol and the cross-linker to 5 mmol, respectively. The reaction time was 30 min at a temperature of 60 °C. The starter system contained a photo initiator Igarcure 2959 and a peroxide initiator NPS. After preparation, the samples were purified from the non-reacting monomers by equilibrium swelling in deionized water in a tea bag for 24 h while the water of the flask was changed several times and the hydrogel sample was dried at 80 °C overnight. This purification and drying operation was repeated twice, and small dried hydrogel blocks of 1 mm in diameter and around 2–3 mm in length were obtained.

Table 2 Composition of the triple-layer bulk hydrogel with an inherent charge-gradient

Substance	MPEG 350MA [mmol]	HEMA [mmol]	SA [mmol]	PEGDMA [mmol]	Irgacure [mmol]
1st layer	35	65	–	5	0.14
2nd layer	35	40	25	5	0.14
3rd layer	35	20	45	5	0.14

Hydrogels with a higher charge, based on prepolymers of poly(AA-co-MA), were manufactured via esterification with a HD as a cross-linker. In a 500 ml three-necked flask with a stirrer, thermometer and water separator, a composition of 100 mmol prepolymer according to the AA:MA ratio based on the average of the molecular weight (20 % solid content in aqueous solution), x mmol HD (Table 1), 0.1 mmol p-toluensulfonic acid and 200 ml heptane were mixed and stirred at 100 °C for 16 h until no further water was identified. After the reaction, the hydrogel was separated from the solvent and dried at room temperature overnight. Then, the hydrogel was neutralized with a double excess of sodium hydroxide according to the number of carboxyl groups.

After neutralization, the attained hydrogel was dried at 80 °C in an oven, milled with an IKA A11 basic to 300 μm particle size and washed with deionized water twice to remove the residual monomer, and subsequently dried again.

Hydrogels with an inherent charge gradient were produced by a layer method by photo-initiated radical polymerization. A mixture of SA, HEMA, MPEG 350 MA and PEGDMA, Irgacure 2959 was produced in which PEGDMA serves as a cross-linker and Irgacure 2959 as a photoinitiator. Table 2 presents the specific compositions of the three different layers, whereby each layer is 3 mm thick.

All layers were balanced and directly polymerized on top of each other under a UV-lamp. The first layer already had reached a jelly-like texture but had not been reacted to completion yet, when the next layer was placed on top.

By this common method, different kinds of layers could be constructed to generate a hydrogel with an inherent charge gradient. After the last step, the samples were dried first at room temperature and then at 50 °C in an oven.

Measurements

The dried cylindrical hydrogel blocks (poly(SA) and hydrogel with an inherent gradient) were weighted and then placed into a 5 mm NMR tube. A defined amount of deionized water was added to achieve a swelling ratio (*SR*) of 500 %. The SR was calculated by the following formula [12]:

$$SR\ (\%) = \left(\frac{W_{H_2O}}{W_{dried-Hydrogel}}\right) \times 100 \quad (1)$$

where W_{H2O} denotes the mass of water absorbed by the dried hydrogel sample, and $W_{dried\text{-}hydrogel}$ the mass of the dried hydrogel sample.

The theoretical amount of sodium ions was calculated for a sample with the amount of cross-linker of 5 mmol. The estimated mass of sodium for a 16 mg hydrogel sample was 2.78 mg. The control exchange-reaction with HCl, measured by AAS, showed that 75.68 % of the total amount of sodium ions was released. The prepared samples were left at room temperature for at least 1 h to achieve the equilibrium swelling ratio. Following relaxation time measurements of the ^{23}Na nuclei, a volume of 0.5 ml aqueous solution with 2.7 mg Al ions was added to the top of the swollen hydrogel. Then, the NMR imaging measurements were carried out, alternating between ^{27}Al and ^{23}Na resonances. Prior to each experiment, the resonance frequency was adjusted manually.

Particles (200–400 μm) of hydrogels with a higher charge, based on SA and poly (AA-co-MA), were weighed into a 5 mm NMR tube and were swollen to a SR of between 200 % and 400 % by injecting deionized water. The exchange reaction was started with ca. 0.1 g of a 0.5 M $Al_2(SO_4)_3$ aqueous solution.

The hydrogels' sodium mass concentration was determined by flame atomic absorption spectroscopy using an AAS-1100B instrument by Perkin Elmer. For this purpose, 200 mg of the dried sample was chemically treated with concentrated hydrochloric acid to exchange the bounded sodium ions in the sample. After a reaction time of 24 h, the sample was filtered and then washed with 500 ml of deionized water. In a next step, 2 g of KCl were dissolved in the eluate solution of the hydrogel's exchanged sodium ions before filling it up to the 1 l-mark in a volumetric flask. The calibration curve was obtained with a series of 0, 2.5, 5.0, 7.5, and 10.0 mg/l sodium ion solution, respectively. The emission of the sodium ion mass concentration was measured at a wavelength of 589.2 nm and a slit width of 0.7 nm.

The NMR measurements were recorded at room temperature (295 K) on a Bruker DSX 500 Spectrometer at 11.7 T 132.4 MHz NMR frequency for ^{23}Na and 130.3 MHz for ^{27}Al. The transverse relaxation times were determined with a Carr-Purcell-Meiboom-Gill (CPMG) sequence with an echo time of 55.8 μs and between 2.048 and 16.384 echoes to follow the fast decay. The longitudinal relaxation times were measured with a saturation recovery sequence with 24 time steps. The decays and magnetization build-up curves were fitted satisfactorily by single exponentials. The diffusion measurements were performed with a stimulated echo sequence with a fixed gradient-pulse duration $\delta = 1.1$ ms and a gradient pulse separation Δ of between 4 and 50 ms. The reported values represent averages of 3–6 measurements of each sample. The two-dimensional ^{23}Na and ^{27}Al imaging experiments were carried out using a spin echo imaging sequence. The field of view was 128×64 pixels with a spatial resolution of $39 \times 312.5\ \mu m^2$, in the respective dimensions (x : 5 mm; z : 20 mm). The data were Fourier-transformed following zero-filling to 256×256 points leading to an equivalent spatial resolution of $19.5 \times 78\ \mu m^2$. The recorded imaging echo-time was 14.8 ms to follow only the free ^{23}Na in solution; and the recycle delay between subsequent scans was set to 250 ms in order to recover the full sodium and aluminum equilibrium magnetization. The intensities of the ion spin density values were obtained by integration over the whole sample space and plotted against exchange time.

Results and Discussion

In the first experiment, MRI techniques were used to follow the exchange reaction between the poly(SA) hydrogel with PEGDMA as a cross-linker (5 mol-%) and an aluminum solution. Figure 1 shows 2D sodium and aluminum density images, which have been acquired with the imaging parameters mentioned above. The grey-filled area shows the part of the image used to integrate the signal to follow the evolution of the two species in the exchange experiment. This area corresponds to the volume within the rf-coil of detection.

Image (a) is the sodium density image of the ^{23}Na ions in the hydrogel matrix. Image (c) is the first aluminum density image recorded immediately, where the aluminum solution is visualized above the hydrogel. The situation after 24 h of exchange is represented by Images (b) and (d), acquired in short succession on first the ^{23}Na, then the ^{27}Al resonance. It can be seen that sodium has been expelled from the gel region and is distributed evenly in the solution part above the gel; sodium in the gel matrix has been almost completely replaced by aluminum, but a heterogeneous distribution of signal intensity remains to be seen in Image (b) which can be attributed to fluid filling cracks in the gel that had appeared due to shrinkage. Aluminum, on the other hand, has gained

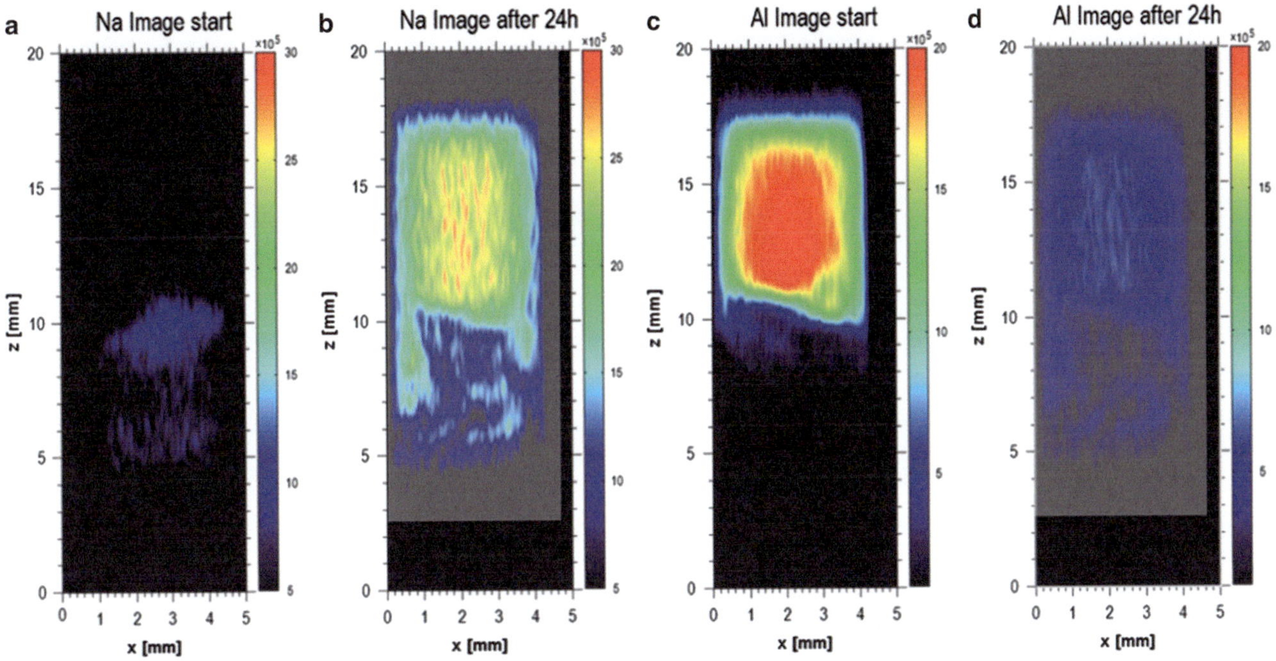

Fig. 1 Image (**a**) represents the sodium spin density and Image (**c**) the aluminum spin density before starting the aluminum exchange experiment of the cylindrical poly(SA) hydrogel block with PEGDMA as a cross-linker. Images (**b**) and (**d**) show the sodium and aluminum spin density after an exchange time of 24 h. The *grey* shaded regions in image (**b**) and (**d**) delineates the area from which the signal integral was calculated to monitor the intensities of aluminum and sodium during the experiment

ingress into the gel but is invisible due to its short relaxation time. However, it can be identified in the same crack regions that are filled with sodium. The volume above the gel remains homogeneously filled with aluminum but at a much reduced concentration. For a quantification of the actual amounts of ions, the weighting of the signal intensity with relaxation times need to be taken into account; these times were found to be weakly dependent on concentration in the bulk solution, but considerably reduced due to binding of the ions within the gel.

In the beginning of the experiment, the longitudinal relaxation time of the bound sodium inside the hydrogel ($SR = 500$ %) is $T_1 = 16.06 \pm 0.04$ ms and the transverse relaxation time is $T_2 = 8.82 \pm 0.04$ ms; these values are confirmed by a series of concentration- and swelling dependent hydrogel measurements. After 24 h of exchange, the relaxation time of the exchanged sodium of the hydrogel was measured and both a longitudinal relaxation time of $T_1 = 60.30 \pm 1.53$ ms and a transverse relaxation time of $T_2 = 23.17 \pm 0.07$ ms were observed which agree well with bulk solution results [28–30]. The bound sodium ions of the collapsed hydrogels are expected to possess much shorter transverse relaxation times due to the smaller pore sizes and the resulting decreased mobility. The signal does not contribute to the overall signal which is confirmed by the monoexponential nature of the decay functions. The sodium diffusion coefficient is $D_{\text{Na-gel}} = 0.34 \cdot 10^{-9}$ m^{2}s^{-1} in comparison to $D_{\text{Na-free}} = 1.29 \cdot 10^{-9}$ m^{2}s^{-1}in the bulk solution. This indicates that the ^{23}Na nuclei have a lesser mobility, but are not bound to corresponding acidic sites. Instead, they become exchanged with the surrounding counterions of the polymer backbone in the pores. Nevertheless, the longitudinal relaxation time of free ^{27}Al ions in solution, i.e. at the beginning of the experiment, is $T_1 = 78.50 \pm 2.00$ ms, and the transverse relaxation time is $T_2 = 18.24 \pm 0.05$ ms. After they have replaced sodium inside the gel, the aluminium relaxation times have become so short that they cannot be detected anymore. This infers that they are found in a strongly bound state and are restricted in mobility, thus showing solid-like behavior.

Figure 2 shows the time evolution of the relative integral intensities of sodium and aluminum recorded over a period of 24 h. In all experiments, one needs to consider that the signal intensity is weighted by a prefactor $\exp(-t_E/T_2)$ with the transverse relaxation time T_2 and the given echo time t_E of the imaging experiments. For bound aluminum, T_2 is so short that no signal above the noise level can be detected. Thus, it can be concluded that the observed ^{27}Al signal is truly proportional to the amount of aluminum ions in solution where the relaxation time is substantially longer. The remainder of aluminum nuclei has been incorporated into the hydrogel. Sodium, by contrast, has a much longer T_2 inside the gel so that, even at time zero before the release of sodium from the gel, the signal can still be detected.

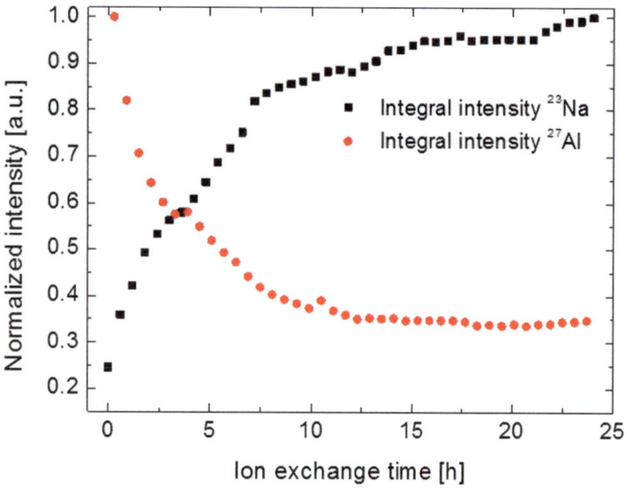

Fig. 2 Integral intensity calculated over the grey area shown in Fig. 1. Each point represents a recorded experiment in 20 min intervals. The *red dots* express the aluminium intensity, and the *black squares* illustrate the sodium intensity over 24 h of observed exchange time

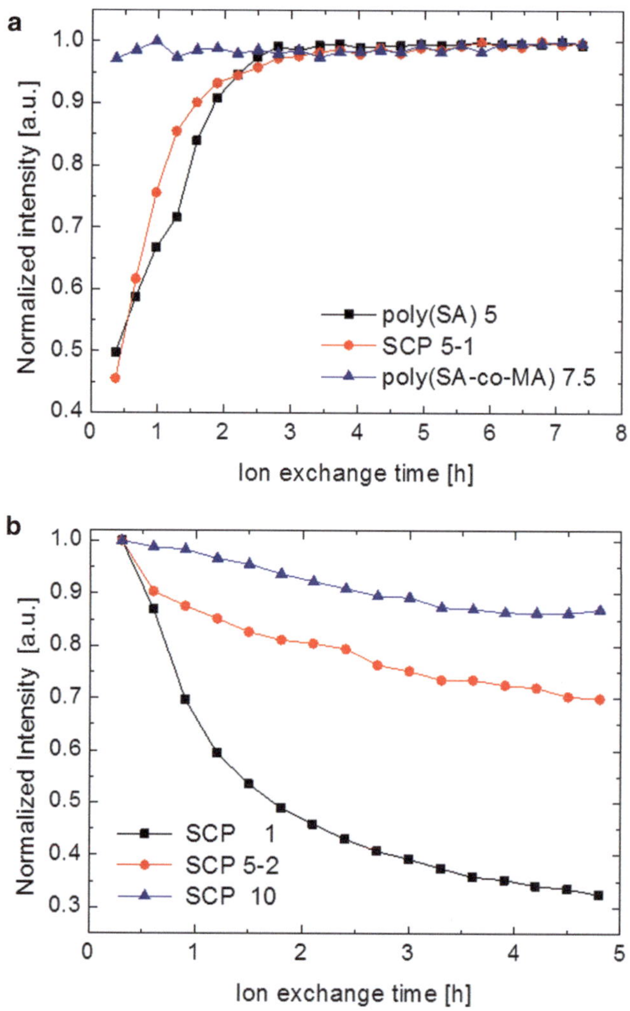

Fig. 3 (**a**) ^{23}Na-^{27}Al exchange reaction of hydrogels with a high charge followed by ^{23}Na intensity; (**b**) ^{23}Na-^{27}Al exchange reaction of full neutralized SCP 45 with different amounts of crosslink densities followed by the integrated ^{27}Al intensity

The difference of the signal at any time compared to its value at $t = 0$ corresponds to the ^{23}Na ions released from the hydrogel into the solution.

In general, three different exchange rates can be observed due to the ion replacement during the ^{23}Na-^{27}Al exchange process. In the first step of the exchange experiment, a relatively strong reduction of the amount of free aluminum ions can be observed which takes 3 h. Having replaced the sodium ions, the aluminum ions are now incorporated into the hydrogel network matrix near the carboxyl groups. Because of the positive trivalent character and a higher chemical affinity of the aluminum ions compared to sodium for the negatively charged carboxyl groups, the relaxation times of the now strongly bound and mobility-restricted aluminum ions species cannot be recorded any longer because their initial mobility has dramatically been reduced. As a consequence of both this reduced mobility and the large quadrupolar coupling constant, aluminum ions behave now "solid-like" and are surrounded by the neighboring carboxyl groups of the polymer backbone matrix. The electrostatic attraction induced by the positively trivalent charged aluminum ions and the negatively charged carboxyl groups not only reduces the aluminum mobility, but also leads to gel shrinkage which begins at the accessible surface of the gel. At the same time, sodium ions are being released, gain higher mobility which is accompanied by bulk-like relaxation times, and diffuse into the solution phase. The exchange in the deeper layer is therefore restricted by this density increase in the surface region, which delays the replacement of ^{23}Na by ^{27}Al inside the gel. The exchange process thus slows down which can be seen from the time-dependence of the Na and Al signals after about 3 h of exchange time. Due to the collapse of the first step, cracks arise in the hydrogel bulk, which yields another type of exchange (second step), where unreacted surfaces in the hydrogel start to exchange. This determines the actual exchange rate between the ^{23}Na and ^{27}Al ions. In the third step, the hydrogel network contracts and the free docking positions of the carboxyl network groups cannot be accessed as easily by the aluminum ions. The diffusion of the remaining free aluminum ions is now driven by the difference between the ion concentration inside and outside of the hydrogel. Figure 2 shows that the exchange process is practically completed after approximately 24 h when the signal intensities approach an asymptotic limit.

After almost 24 h of online monitoring, both the sodium and the aluminum ion concentrations remain approximately constant. It can be assumed that the majority of carboxyl

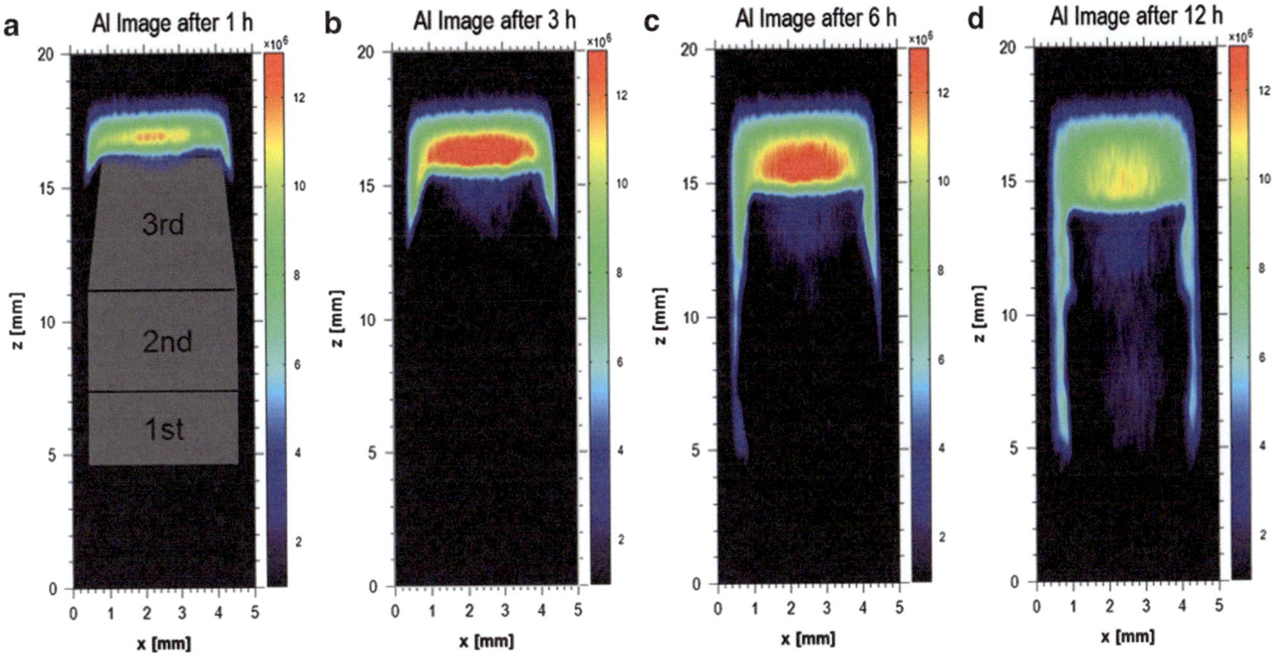

Fig. 4 Monitoring the ion density during the collapse of a triple-layer hydrogel with an inherent charge gradient followed by the ^{27}Al MRI technique. The first layer from the bottom contains no SA, in contrast to layer 2 (25 mmol) and layer 3 (45 mmol)

groups is now bound by aluminum counterions. Taking into account the initial intensity of each ionic species, one can conclude that about two carboxylic groups interact with one single aluminum ion at a time.

In another series of ^{23}Na-^{27}Al exchange experiments, the influence of a higher charge of poly(SA) and its derivatives with different kind of cross-linker was investigated by following the integrated ^{23}Na intensity; the influence of the cross-linker density was followed by means of the integrated ^{27}Al intensity. The corresponding relaxation times T_1 and T_2 are shown in Table 1. Hydrogels based on SA and MA show much shorter relaxation times than the poly(SA) due to the higher concentration of the counterions in the polymer matrix.

The higher-charged hydrogels (poly(AA-co-MA) 7.5) collapsed much faster than the poly(SA) 5 and SCP 5–1 (Fig. 3a) because of the higher concentration of ^{23}Na ions inside the hydrogel which yields a higher osmotic pressure. The sample poly(AA-co-MA) 7.5 collapsed so fast that it was not possible to follow the exchange process, while the ion release of sample SCP 5–1 was slightly faster than that of sample poly(SA) 5.

The influence of the cross-linker density was investigated by means of the exchange process. With decreasing the cross-linker density, the exchange capacity and thus the maximum quantity of bound cations increases (Fig. 3b). One reason for this behavior is the decrease in the number of charged carboxyl groups in the backbone; another may be the smaller pore sizes due to a higher physical crosslink density [26]. While the exchange speed is quite similar, the rate at the end of the ion exchange process increases with a decreasing amount of cross-linker. Due to higher charge and a faster ion-exchange in comparison to that of poly(SA), hydrogels based on poly(SA-co-MA) could be applied as a more effective ion-exchanger or as an intelligent valve in microfluidic systems to speed up the switching process.

In the following experiment, the collapse of a bulk hydrogel with an inherent charge gradient is shown using ^{27}Al imaging (Fig. 4) where only aluminum ions dissolved in solution contribute to the image. The hydrogel (Fig. 5a) is made up of three different hydrogel composites with different swelling behavior (Fig. 5b), whereby the layers are vertically arranged. All three hydrogel layers gradually start to collapse once the layer surface comes into contact with the aluminum sulfate solution. The images presented here show a negative image of the hydrogel volume, where three significant characteristic differences can be observed:

The decrease in the volume of the third layer is much more pronounced than that of the second layer due to the higher concentration of SA and the consequentially higher ability to create physical crosslinking points with the ^{27}Al ions. Compared to both upper layers the first layer (bottom) contains no SA and reaches a smaller swelling ratio due to the missing charge group. On the one hand, the first layer does not fill the NMR tube completely so that the ^{27}Al solution can surround the first layer, when the second layer has collapsed. On the other hand, no physical crosslinking takes place, so that the change in the volume between the swollen

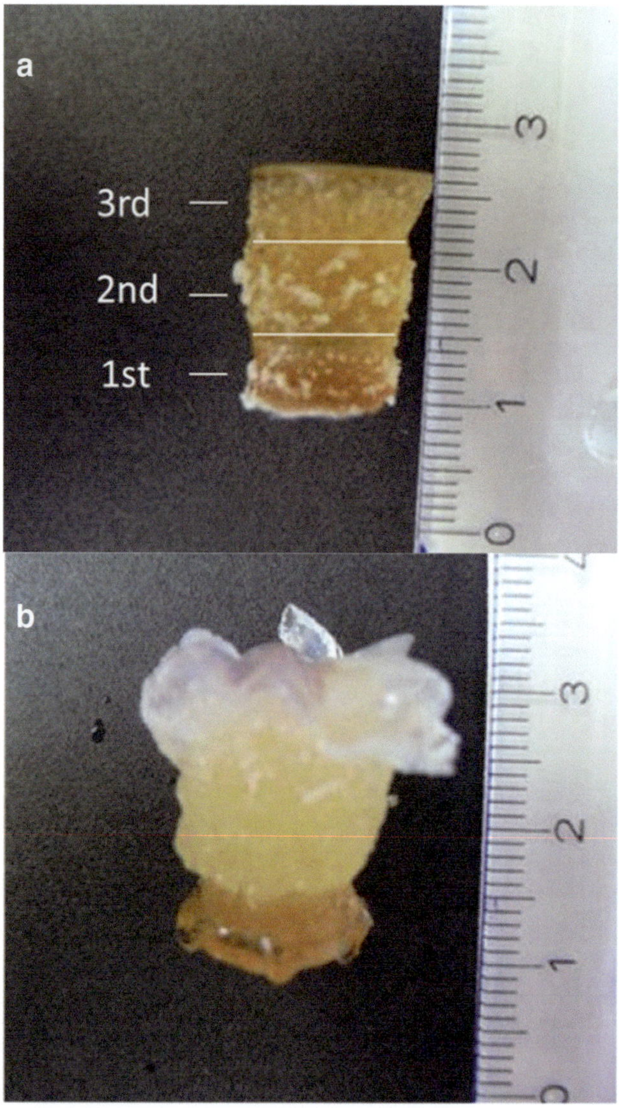

Fig. 5 Image (**a**) shows a dried bulk hydrogel with an inherent charge gradient with three different layers as shown in Table 2, while Image (**b**) presents the hydrogel bulk after 30 min swelling in deionized water

and the collapsed states remains relatively small. This kind of hydrogel with an inherent charge gradient layer-by-layer bears potential as an actuator in microfluidic systems, intelligent valves and sealing materials due to the chemical design of each layer and its resulting swelling/shrinking behavior.

Conclusion

NMR is a unique and suitable technique to record and monitor online exchange processes of foreign ions, in this case, of trivalent aluminum ions in a sodium acrylate hydrogel matrix. The exchange phenomenon of the collapsing hydrogel shows a non-linear behavior while the exchange process. Furthermore observed are three different ion exchange rates, which can be assigned to the different layers of the swollen cylindrical hydrogel samples in the NMR tube. The ion exchange of the inside layers is restricted by a shielding effect induced by the collapse of the hydrogel matrix and the embedded aluminum ions. It could also be shown that the cross-linker density and the charge-density strongly affect the ion exchange rate and the speed of shrinkage of hydrogels based on SA and MA with a high charge density. Due to the lower sensitivity of the bound aluminum ions in the hydrogel matrix, ^{27}Al MRI techniques are a useful tool to obtain two- or three-dimensional negative images of polyelectrolyte hydrogels during a shrinking or swelling process.

Acknowledgments The authors wish to thank the German Research Foundation (DFG) for its financial support (Sta 511/7-3 and Ma 3806/2-3).

References

1. Dorfner K (1991) Ion exchangers. Walter de Gruyter, Berlin/New York
2. Tanaka T (1978) Collapse of gels and the critical endpoint. Phys Rev Lett 40:820–823
3. Wichterkem O, Lim D (1960) Hydrophilic gels for biological use. Nature 185:117–118
4. Wood DA (1980) Biodegradable drug delivery systems. Int J Pharm 7:1–18
5. Ohya S, Nakayama Y, Matsuda T (2001) Thermoresponsive artificial extracellular matrix for tissue engineering: hyaluronic acid bioconjugated with poly(N-isopropylacrylamide) grafts. Biomacromolecules 2:856–863
6. Kondo A, Kamura H, Higashitani K (1994) Development and application of thermo-sensitive magnetic immunomicrospheres for antibody purification. Appl Microbiol Biotechnol 41:99–105
7. Rubio-Retama J, Zafeiropouls NE, Serafinelli C, Rojas-Reyna R, Voit B, Cabarcos EL, Stamm M (2007) Synthesis and characterization of thermosensitive PNIPAM microgels covered with superparamagnetic γ-Fe_2O_3 nanoparticles. Langmuir 23:10280–10285
8. Kwon SM, Kim HS, Jin HJ (2009) Multiwalled carbon nanotube cryogels with aligned and non-aligned porous structures. Polymer 50:2786–2792
9. Retama JR, López MS, Pérez JP, Cabanillas GF, López-Cabarcos E, López-Ruiz B (2005) Biosensors based on acrylic microgels a comparative study of immobilized glucose oxidase and tyrosinase. Biosens Bioelectron 20:2268–2275
10. Pérez JP, López MS, López-Cabarcos E, López-Ruiz B (2006) Amperometric tyrosinase biosensor based on polyacylamide microgels. Biosens Bioelectron 22:429–439
11. Otero TF (2008) Reactive conducting polymers as actuating sensors and tactile muscles. Bioinspir Biomim 3:35004
12. Buchholz FL, Graham AT (1998) Modern superabsobrent polymer technology. Wiley-VCH, New York
13. Fernández-Barbero A, Suárez IJ, Sierra-Martín B, Fernández-Nieves A, de las Nieves FJ, Marquez M, Rubio-Retama J, López-Cabarcos E (2009) Gels and microgels for nanotechnological application. Adv Colloid Interface Sci 147–148:88–108

14. Carpi F, Smela E (2009) Biomedical applications of electroactive polymer actuators. Wiley, Chichester
15. Gerlach G, Arndt KF (2009) Hydrogels sensors and actuators. Springer, Heidelberg
16. Blümich B (2009) NMR imaging of materials. Clarendon Press, Oxford
17. Schwarz R, Kaspar A, Seelig J, Künnecke B (2002) Gastrointestinal transit times in mice and humans measured with ^{27}Al and ^{19}F nuclear magnetic resonance. Magn Reson Med 48:255–261
18. Kaspar A, Bilecen D, Scheffler K, Seelig J (1996) Aluminum-27 nuclear magnetic resonance spectroscopy and imaging of the human gastric lumen. Magn Reson Med 36:177–182
19. Andre JP, Macke HR (2003) NMR spectroscopy of group 13 metal ions: biologically relevant aspects. J Inorg Biochem 97:315–323
20. Conradi M (1991) High-resolution imaging of a ceramic. J Magn Reson 93:419–422
21. Bodart P, Nunes T, Randall EW (1996) Stray-field imaging of quadrupolar nuclei of half integer spin in solids. Solid State Nucl Magn Reson 8:257–263
22. Koptyug IV, Sagdeev DR, Gerkema E, Van As H, Sagdeev RZ (2009) Solid-state ^{27}Al MRI and NMR thermometry for catalytic applications with conventional (liquids) MRI instrumentation and techniques. J Magn Reson 175:21–29
23. Koptyug IV, Lysova AA, Sagdeev RZ, Kirillov VA, Kulikov AV, Parmon V (2005) In situ MRI of the structure and function of multiphase catalytic reactors.Catal Today 105:464–468
24. Koptyug IV, Khomichev AV, Lysova AA, Sagdeev RZ (2007) Multinuclear MRI of solids: from structure to transport. Appl Magn Reson 32:321–331
25. Nielles-Vallespin S, Weber M, Bock M, Bongers A, Speier P, Combs SE, Wöhrle J, Lehmann-Horn F, Essig M, Schad LR (2007) 3D radial projection technique with ultrashort echo times for sodium MRI: clinical applications in human brain and skeletal muscle. Magn Reson Med 57:74–81
26. Aursand IG, Veliyulin E, Böcker U, Ofstad R, Rustad T, Erikson U (2009) Water and salt distribution in Atlantic salmon (salmo salar) studied by low-field ^{1}H NMR, ^{1}H and ^{23}Na MRI and light microscopy: effects of raw material quality and brine salting. J Agric Food Chem 57:46–54
27. Augath M, Heiler P, Kirsch S, Schad LR (2009) In vivo ^{39}K, ^{23}Na and ^{1}H MR imaging using a triple resonant RF coil setup. J Magn Reson 200:134–136
28. Kadota K, Takase K, Shimosaka A, Hidaka A (2006) Influence of habit modifiers on particle shape in a crystallization process. KONA 24:203–212
29. Rijniers LA, Magusin PCMM, Huinink HP, Pel L, Kopinga K (2004) Sodium NMR in porous materials. J Magn Reson 167: 25–30
30. Ohuchi M, Meadows P, Horiuchi H, Sakai Y, Furihata K (2000) Dynamics of sodium and lithium counter-ions and water molecules in cation-exchange resins as shown by NMR spectroscopy. Polym J 32:760–770

Sodium NMR Relaxation: A Versatile Non-invasive Tool for the Monitoring of Phase Transitions and the Estimation of Effective Pore Sizes of Supramolecular Hydrogels

M. Raue, A. Bernet, M. Küppers, S. Stapf, H.-W. Schmidt, B. Blümich, and T. Mang

Abstract

Nuclear magnetic resonance (NMR) relaxometry of liquids is a widely used tool to characterize porous media. In particular, ^{23}Na NMR is an especially suitable method when applied to gels and biological tissues. In this work we investigated the thermoreversible melting and gelation processes of supramolecular hydrogels formed by succinamic acid-based amphiphiles (**SAn**) in a saturated aqueous $NaHCO_3$ solution (sat. aq. $NaHCO_3$ sol.). We could show that it is not only possible to determine the melting points and to monitor the gelation process with ^{23}Na relaxometry, but also to estimate the effective pore size based on the expanded Brownstein-Tarr model. Our findings are in good agreement with data from differential scanning calorimetry (DSC) and transmission electron microscopy (TEM) experiments.

Keywords

^{23}Na relaxation • Succinamic acid • Amphiphile • Pore size • Porous material • Hydrogelator

Abbreviations

aq.	Aqueous
CPMG	Carr-Purcell-Meiboom-Gill
DSC	Differential scanning calorimetry
ff	Freeze-fracture replica
NMR	Nuclear magnetic resonance
RT	Room temperature
SAn	Succinamic acid-based amphiphiles (n: number of carbon atoms in linear alkyl chain)
sat.	Saturated
SEM	Scanning electron microscopy
sol.	Solution
TEM	Transmission electron microscopy

Introduction

Sodium ions exhibit high biological and ecological relevance. The ^{23}Na isotope is encountered with a natural abundance of 100 % and a nuclear spin of 3/2. Nuclear spins

M. Raue (✉)
Institut für Angewandte Polymerchemie, Aachen University for Applied Science, Heinrich-Mußmann-Str. 1, 52428 Jülich, Germany

Institut für Technische und Makromolekulare Chemie, RWTH-Aachen University, Worringer Weg 1, 52074 Aachen, Germany
e-mail: raue@fh-aachen.de

A. Bernet (✉) • H.-W. Schmidt
Macromolecular Chemistry I, Bayreuth Institute for Macromolecular Research (BIMF), Bayreuth Center for Colloids and Interfaces (BZKG), University of Bayreuth, Universitätsstr. 30, 95440 Bayreuth, Germany
e-mail: andreas.bernet@uni-bayreuth.de

M. Küppers • B. Blümich
Institut für Technische und Makromolekulare Chemie, RWTH-Aachen University, Worringer Weg 1, 52074 Aachen, Germany

S. Stapf
Department of Technical Physics II, Technical University Ilmenau, 100 565, Ilmenau, 98684 Germany

T. Mang
Institut für Angewandte Polymerchemie, Aachen University for Applied Science, Heinrich-Mußmann-Str. 1, 52428 Jülich, Germany

G. Sadowski and W. Richtering (eds.), *Intelligent Hydrogels*, Progress in Colloid and Polymer Science 140,
DOI 10.1007/978-3-319-01683-2_4,

greater than 1/2 possess a quadrupole moment. Thus ^{23}Na is subject to additional relaxation mechanisms beyond those applicable to ^{1}H nuclear magnetic resonance (NMR) and is sensitive to local electric field gradients. Electrical fields of molecular electrical dipoles or ion charges in the proximity of a ^{23}Na ion cause fluctuating electric field gradients leading to relaxation of the ^{23}Na nucleus. This nuclear quadrupole-interaction, which usually dominates the competing dipolar relaxation contribution, is an important source of information in electrolyte solutions when investigating ion solvation and association. Therefore, NMR relaxometry gives access to parameters such as concentration of sodium ions, local distributions of electric charges, their motion, and their associated changes affected by their quantities.

The relevant parameters obtained from the experiments are the relaxation times T_1 and T_2. The longitudinal relaxation time T_1 gives information about the molecular mobility in the sample, whereas the transverse relaxation time T_2 correlates with the binding strengths and therefore senses local properties such as exchange dynamics. A decrease in local molecular mobility therefore results in shorter T_1 and T_2 values, but has a stronger effect on the latter [1–4].

Rijniers et al. [5] have shown that it is possible to determine the pore size of silica particles filled with ^{23}Na salt solution via ^{23}Na relaxometry. They discussed the Brownstein-Tarr model:

$$\frac{1}{T_i} = \frac{1}{T_{i,b}} + \frac{S}{V}\rho_i \tag{1}$$

in which T_i is the relaxation time (i = 1 or 2), $T_{i,b}$ is the bulk relaxation, S/V is the surface to volume ratio of a pore and ρ_i is the surface relaxivity. To apply the Brownstein-Tarr Modell three criteria must be fulfilled. First: The experiment takes place in the fast-diffusion regime. This means that the time that the ions need to pass the pores is much shorter than the experimental timescale. Second: The ions are not affected by the electric field gradients while travelling through the pores. Third: The surface relaxation must influence the relaxation behavior. Rijniers et al. applied Eq. 1 to ^{23}Na and showed that the relaxation times T_1 and T_2 and the surface relaxivity depend on the ^{23}Na concentration inside the pores, leading to the following equation:

$$\frac{1}{T_i(C)} = \frac{1}{T_{i,b}(C)} + \frac{S}{V}\rho_i(C) \tag{2}$$

The results obtained for different concentrations of sodium chloride in aqueous solution correlated well with the known pore size of silica particles (Nucleosil) [5–7].

Supramolecular hydrogels that are based on organic low molecular weight compounds belong to the group of physical gels. The molecules with a molar weight typically below 3,000 g mol^{-1} self-assemble under certain conditions from solution by non-covalent intermolecular interactions thus yielding three-dimensional mesoscopic networks. Due to capillary forces, the aqueous solvent phase is entrapped and a macroscopic hydrogel is formed. Generally, only a rather low amount of hydrogelator (typically *ca.* 10 g L^{-1} and below) is necessary to allow hydrogel formation. Furthermore, the formed gels are in most cases thermoreversible due to the non-covalent intermolecular interactions [8, 9]. This area of research is rapidly expanding since the end of the twentieth century and bears potential for promising applications such as tissue engineering, controlled drug release, and pollutant capture and removal [10], to name just a few.

To determine the pore size of (supramolecular) gels is very difficult [11]. Although it is possible to examine dried gels (so-called xerogels) by SEM and TEM methods, one cannot exclude morphological changes due to drying effects. Furthermore, the supramolecular network is compressed laterally, thus suggesting higher network densities than those that are actually present in the native gel state. Cryogenic electron microscopy methods such as cryo-SEM or freeze-fracture replica TEM (ff-TEM) can overcome these limitations, but are highly sophisticated and demanding procedures. Therefore a non-invasive, reliable and straightforward method for the determination of effective pore sizes and distributions in the native gel state is desired.

Low molecular weight amphiphiles based on succinamic acid derivatives have been found to be excellent hydrogelators for aqueous solutions comprising high contents of alkaline sodium salts [12]. The resulting supramolecular hydrogels are thermoreversible, and detailed experimental studies indicated that the hydrogel formation includes several hierarchical steps spanning the length scale from nanometers to centimeters. Turco et. al. already showed that the determination of the mesh size of alginate hydrogels by ^{1}H NMR relaxometry via cross-correlation with mechanical measurements is possible [13]. An alternative method has been presented by Wallace *et al.* that estimated the mesh size of supramolecular hydrogels based on naphthalene diphenylalanine by pulsed field gradient NMR spectroscopy [14]. In our case hydrogel formation exclusively occurs in the presence of high amounts of sodium salts, and it seems that the sodium ions play a crucial role for the thermoreversible supramolecular self-aggregation and dissolution process. Therefore we aimed to utilize ^{23}Na NMR relaxometry as a non-invasive probe both for the examination of the thermally induced gel formation and dissolution, and for the determination of the effective pore sizes.

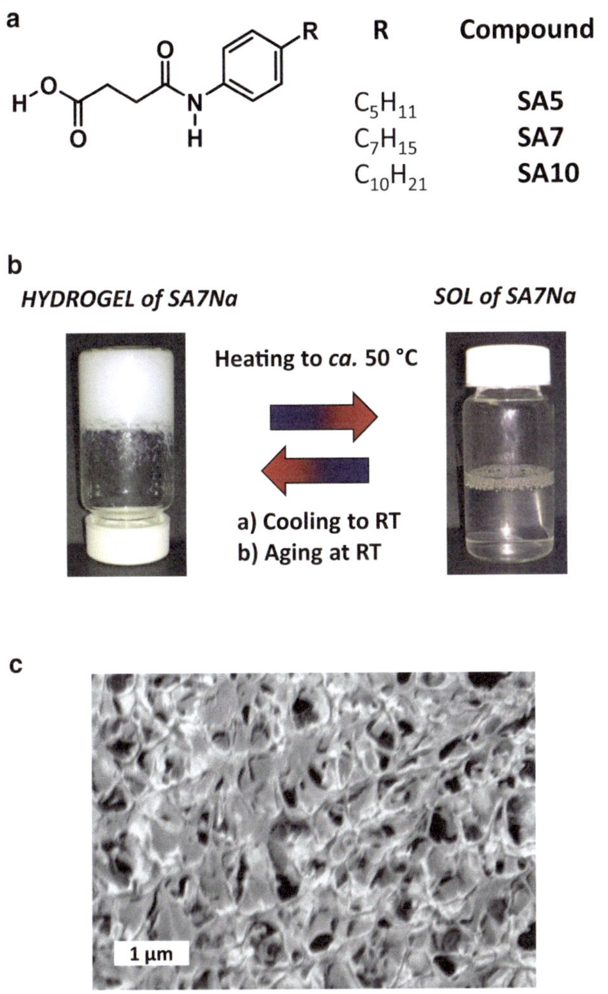

Fig. 1 (**a**) Chemical structure of the investigated succinamic acid-based low molecular amphiphiles **SA5**, **SA7**, and **SA10**. (**b**) Thermoreversible transformation between gel and sol states and *vice versa* of a sample prepared from 10 g L^{-1} of **SA7** in sat. aq. $NaHCO_3$ sol. Note that the **SAn** amphiphiles are in situ transformed into the corresponding sodium carboxylate salts **SAnNa** due to the high alkalinity of the solvent. (**c**) TEM micrograph of a freeze-fracture replica (ff-TEM) prepared from a shock-frozen sample of 10 g L^{-1} of **SA7** in sat. aq. $NaHCO_3$ sol. By applying the ff-TEM methodology the morphology of the sample at an initial stage of the hydrogel formation was visualized. A fibrillar network structure exhibiting voids with diameters of several hundreds of nanometers can be recognized

Experimental Section

Materials

Sodium hydrogen carbonate ($NaHCO_3$) was obtained from Sigma-Aldrich and was of high purity (p.a., ≥99.0 %). Succinamic acid-based amphiphiles **SA5**, **SA7**, and **SA10** (Fig. 1) were used and prepared as described elsewhere [12]. Silica particles (Nucleosil) with a diameter of 10 µm and defined pore sizes (30, 50, 100, and 400 nm) were received from Macherey-Nagel (Düren, Germany).

Sample Preparation

The amphiphiles **SAn** (n = 5, 7, 10) were dissolved in a sat. aq. $NaHCO_3$ sol. under stirring at concentrations of 5, 10, 20, 40, and 50 g L^{-1} at room temperature (RT, 24–25 °C). Due to the alkalinity of the aqueous solvent, the amphiphiles are *in situ* transformed into their corresponding sodium carboxylate salts which are the actual hydrogelators. After complete dissolution of amphiphiles the samples were stored at 8 °C for several hours to induce hydrogel formation before measurements were carried out.

A portion of 100 mg of silica particles (Nucleosil) with known pore sizes as mentioned above were stored for 24 h in a sat. aq. $NaHCO_3$ sol. in a snap cap vial. The mixtures were then centrifuged and the supernatant was decanted before measurements were carried out.

Measurements

NMR experiments were performed on a Bruker AV III 300 spectrometer at 7.05 T (^{23}Na resonance frequency 79.34 MHz) at certain temperatures (RT to 40 °C). The sample containing 300 mg of the thermally liquefied hydrogel was placed into an NMR tube (Ø: 7mm) and stored at 8 °C for several hours to induce hydrogel formation. The transverse relaxation times T_2 were determined with the Carr-Purcell-Meiboom-Gill (CPMG) sequence to follow the fast decay with an echo time between 70 and 130 µs. The longitudinal relaxation times T_1 were measured with a saturation recovery sequence. The decays and magnetization build-up curves were fitted by single or double exponentials.

The gelation and melting points of the **SAn** hydrogels were determined on a Netzsch DSC Maia. 10–20 mg of a gel sample were placed into an aluminum pan and sealed with a lid. The sample was heated from 4 °C to 60 °C under a nitrogen atmosphere and subsequently cooled to the starting temperature with a heating and cooling rate of 2 °C min^{-1}. This temperature program was repeated and the characteristic temperature values were determined from the peaks in the DSC curve.

The determination of the gelation time was performed with a Setaram MicroDSC III (Setaram France). A mixture of hydrogelator and aqueous sodium salt solution (*ca.* 0.7 mL) was filled into a Hastelloy C276 sample cell which was immediately sealed to avoid evaporation. The reference cell was filled with an equal amount of pure solvent. The sample was heated from 20 °C to 40 °C with a heating rate of 2 °C min^{-1} and tempered at 40 °C for 10 min. The sample was then cooled down to different holding temperatures (15, 20, 23, and 25 °C) and was subsequently stored at the holding temperature for at least 12 h to monitor the gelation process. The time at which the peak maximum of the corresponding gel formation occurred was considered for further evaluation (gelation time).

For freeze-fracture replica TEM, replicas were prepared by applying the thermally liquefied sample prepared from 10 g L^{-1} of **SA7** in sat. aq. $NaHCO_3$ sol. to a small gold holder. The sample was slowly cooled to RT during 5 min, shock-frozen with liquid nitrogen and fractured with a microtome plate in the cryo-fracture replication unit at −100 °C. The fresh fracture surface of the sample was shadowed with Pt/C at an angle of 45 ° and then coated with C at an angle of 90 ° for replica stabilization. Replicas were cleaned by immersing them several times in desalted water. The ff-TEM micrographs were taken at RT with a Zeiss CEM 902 electron microscope (Zeiss NTS Germany) operated at 80 kV.

Results and Discussion

In this work we have investigated amphiphilic succinamic acid-based low molecular weight hydrogelators with different alkyl chain lengths (Fig. 1a) which show a thermoreversible transformation between the gel and the sol state (Fig. 1b) in form of their corresponding sodium carboxylate salts. By ff-TEM methodology the mesoscopic morphology of a hydrogel sample prepared from 10 g L^{-1} of **SA7** in sat. aq. $NaHCO_3$ sol. at an initial stage of the hydrogel formation was visualized. The pores of such fibrillar hydrogel networks have diameters of several hundreds of nanometers (Fig. 1c).

Differential scanning calorimetry can be used to determine the phase transition temperature of a thermoresponsive gel [15]. During heating of a gel sample the melting temperature is detected, whereas the gel formation temperature is obtained during cooling of the corresponding sol. First, we investigated the influence of the length of the alkyl chain of the hydrogelators **SA5**, **SA7**, and **SA10** in a sat. aq. $NaHCO_3$ sol. (Fig. 2a). All samples show a relatively broad transition during melting. The melting process of a **SA7** hydrogel, for example, takes place from 25.5 °C to 30.5 °C with a DSC melting peak maximum at 28.6 °C. Furthermore, a pronounced hysteresis can be recognized when comparing melting and gelation temperature of each hydrogel sample. Additionally, the melting and gelation temperatures increase with increasing alkyl chain length due to pronounced hydrophobic interaction [16]. We also observed that in contrary to the gel melting, the gel formation is a process that is highly sensitive to the applied temperature program. The cooling rate and especially the final holding temperature strongly influence the time that is necessary for gel formation (gelation time). At low temperatures, gelation starts already after several minutes. If the holding temperature however is close to the melting temperature, up to 20 h can pass until gelation starts, as shown in Fig. 2b.

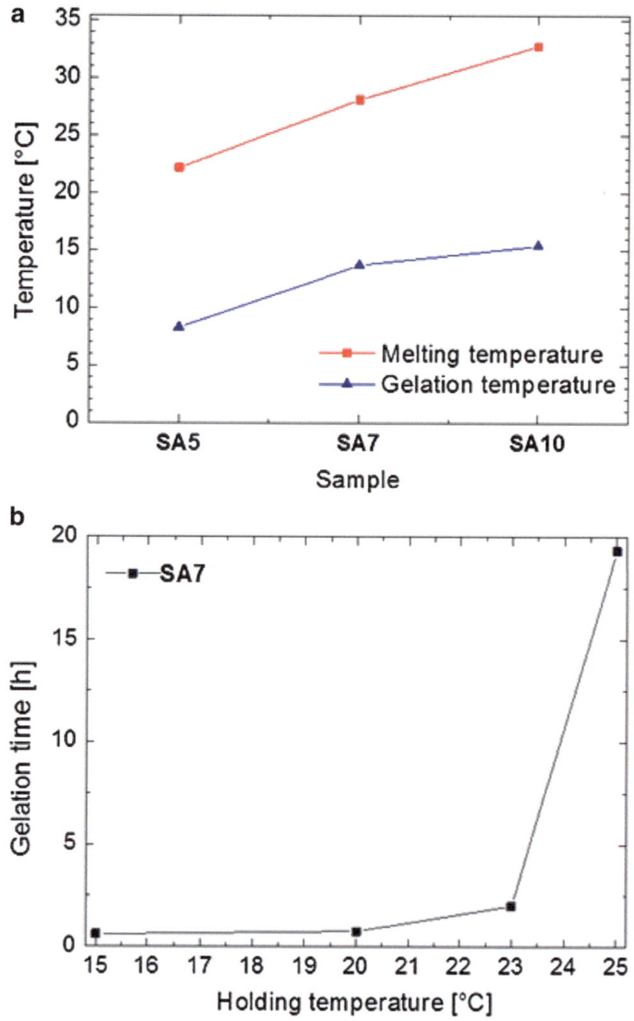

Fig. 2 (**a**) Melting and gelation temperatures of **SAn** samples measured by DSC as a function of the length of the alkyl chain. (**b**) Gelation time of a **SA7** sample *versus* holding temperature. For evaluation, the peak maxima of the corresponding μDSC curves were taken. All samples were prepared with **SAn** concentrations of 10 g L^{-1} in sat. aq. $NaHCO_3$ sol.

Due to the high concentration of ^{23}Na inside the hydrogels based on **SAn** hydrogelators, ^{23}Na NMR relaxometry allows for the investigation of such kind of systems in a contact-free and nondestructive manner. We first measured longitudinal relaxation times T_1 (Fig. 3a) and the transverse relaxation times T_2 (Fig. 3b) of different hydrogelator samples (**SA5**, **SA7**, and **SA10**) as a function of temperature to determine the correlation between the melting point and the length of the alkyl chain. With increasing temperature the T_1 values generally increase. Only **SA10** shows a behavior different from that of the other samples **SA5** and **SA7**, as T_1 increases until the melting temperature is reached. Upon further heating T_1 decreases again. One explanation for this behavior could be the high viscosity of the **SA10** sample above the melting temperature which leads to lower mobility of the

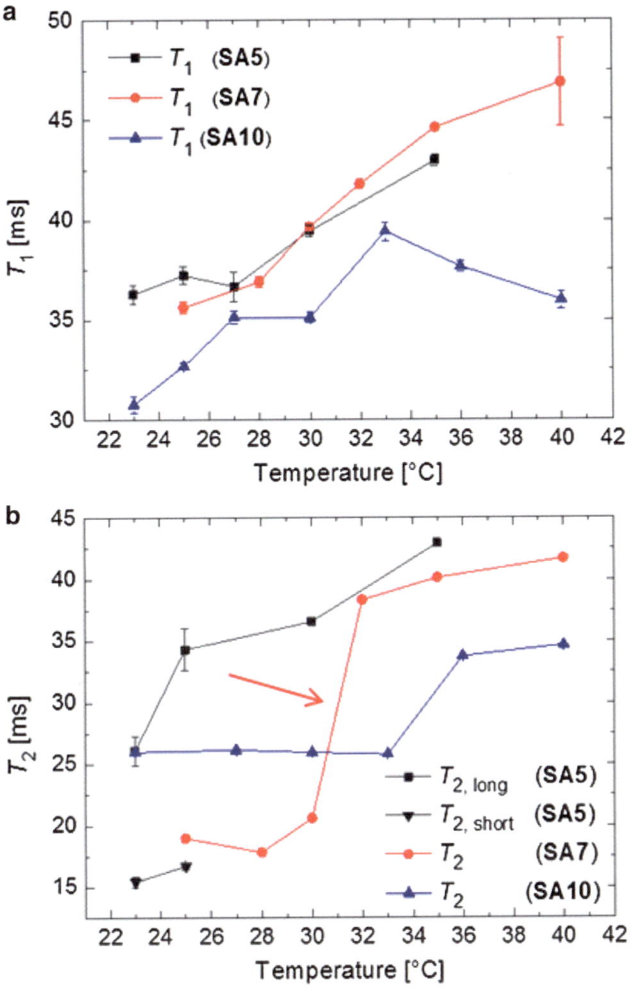

Fig. 3 The relaxation times T_1 (**a**) and T_2 (**b**) for ^{23}Na ions of different hydrogel samples (10 g L^{-1} of **SA5**, **SA7**, or **SA10** in sat. aq. $NaHCO_3$ sol.) as a function of the temperature. The *arrow* in Fig. 3b indicates the gel-sol transition of the **SA7** hydrogel sample

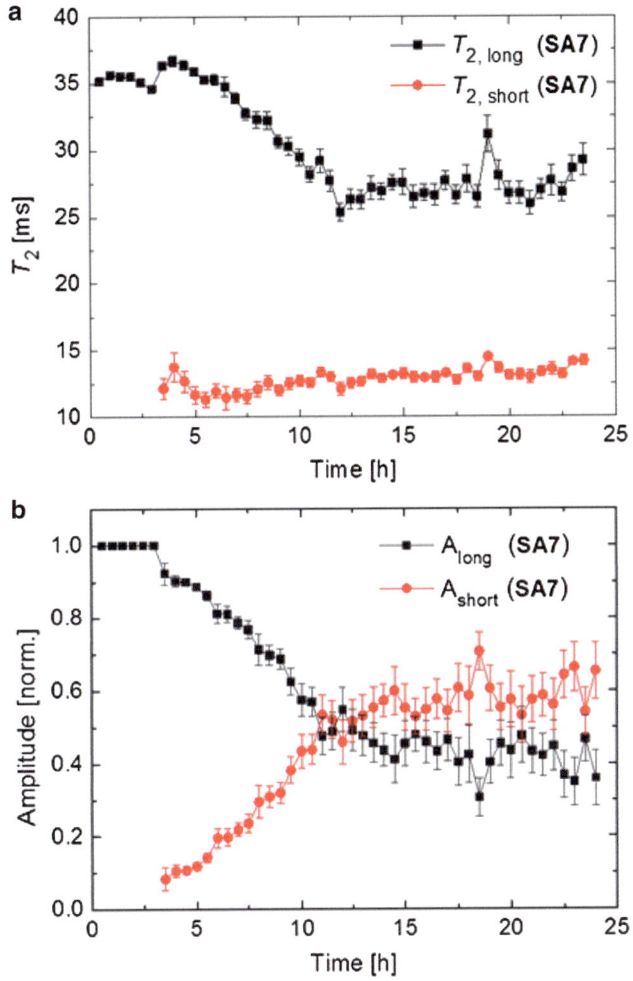

Fig. 4 (**a**) The transverse relaxation times $T_{2,\,long}$ and $T_{2,\,short}$ for ^{23}Na ions in a sample prepared from 10 g L^{-1} of **SA7** in sat. aq. $NaHCO_3$ sol. as a function of time during the gelation process. (**b**) The normalized amplitudes of the long (A_{long}) and short (A_{short}) relaxation time components as a function of time during the gelation process of the same sample

^{23}Na inside the solution. This phenomenon is not observed in case of the **SA5** and **SA7** samples, where the hydrogels transform into free-flowing aqueous solutions when the corresponding gel-sol-transition temperature T_{gel} is exceeded.

Changes in transverse relaxation times T_2 were even more pronounced. At the gel-sol transition of **SA7** the transverse relaxation time T_2 shows a jump from 20.6 ± 0.1 ms at 30 °C to 38.3 ± 0.2 ms at 32 °C (red arrow in Fig. 3b). The latter value is comparable to that of the free ^{23}Na ions in sat. aq. $NaHCO_3$ solution. In the case of sample **SA5**, we observed up to two different species $T_{2,\,long}$ and $T_{2,\,short}$. We attribute this to the presence of a hierarchically structured three-dimensional hydrogel network exhibiting two different types of pores. This assumption is supported by the fact that it has already been reported that sodium carboxylate amphiphiles are able to form fibrillar structures that further aggregate, thus yielding bundles of fibers exhibiting smallanisotropic pores in the interstices of the bundled fibers [17]. When performing the DSC experiments, a similar trend was observed, as the melting temperature also increases with increasing alkyl chain length. Furthermore, the phase transition temperatures measured by NMR (point of inflection of the T values) correlate very well with the values evaluated from the DSC experiments (melting peak maxima). Therefore, we can conclude that the reason for the change in the relaxation times T_1 and T_2 is the thermally induced change of the supramolecular mesoscopic structures.

The growth of a supramolecular network can also be monitored directly by NMR (Fig. 4). 10 g L^{-1} of the hydrogelator **SA7** were dissolved in sat. aq. $NaHCO_3$ sol. without any heating, and the T_2 components were monitored over time (Fig. 4a). Initially, only a $T_{2,\,long}$ component can be detected. After 3 h the gelation starts and a second component ($T_{2,\,short}$)

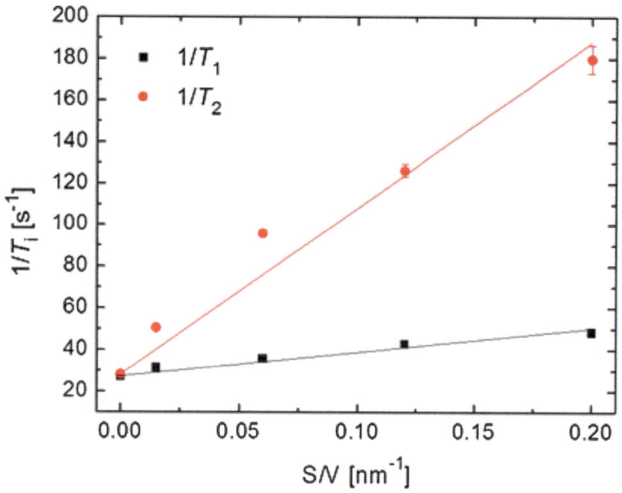

Fig. 5 Relaxation rates $1/T_1$ and $1/T_2$ for ^{23}Na ions (sat. aq. $NaHCO_3$ sol.) in Nucleosil particles as a function of the surface to volume ratio S/V

appears, while $T_{2,\text{long}}$ decreases with time. As the ratio of the virial coefficients of the biexponential fits correlates to the spin density of the ^{23}Na nuclei, quantitative information about the relative amount of the different species $T_{2,\text{long}}$ and $T_{2,\text{short}}$ can be obtained. Therefore, the formation of the supramolecular network can be followed quantitatively by comparing the relative amplitudes of the transverse relaxation times $T_{2,\text{long}}$ and $T_{2,\text{short}}$ (Fig. 4b). It can be recognized that under the applied conditions the gelation process is relatively slow and is not fully complete even after 24 h.

In several cases we also observed a weak third T_2 species in the sol state in the range of 100–400 μs that vanished upon hydrogel formation (not shown). The amplitude of this signal was very low and at the lower detection limit of the used setup. As it has already been postulated that the formation of hierarchical superstructures of amphiphilic sodium carboxylates involves a crystallization of the sodium carboxylate groups [17, 18], and due to the fact that the mentioned signal vanishes upon hydrogel formation, it would make sense to attribute this third T_2 species to the sodium ions that are relatively strongly bound in the sodium carboxylate headgroups. The low intensity of this signal can be explained by the relatively low amount of hydrogelator (usually 1 % w/v).

With the aid of the Brownstein-Tarr model (Eq. 2) the effective pore sizes of the hydrogels under investigation have been determined. First, a calibration curve based on silica particles with defined pore sizes in a sat. aq. $NaHCO_3$ sol. was prepared by plotting the relaxation rates $1/T_1$ and $1/T_2$ *versus* the corresponding pore size (Fig. 5). The slopes of the resulting linear fits provide the surface relaxivities ρ_i. The surface relaxivities from the data shown in Fig. 5 were found to be $\rho_1 = 799 \pm 45$ nm s^{-1} for T_1 and $\rho_2 = 121 \pm 13$ nm s^{-1} for T_2, respectively. With these

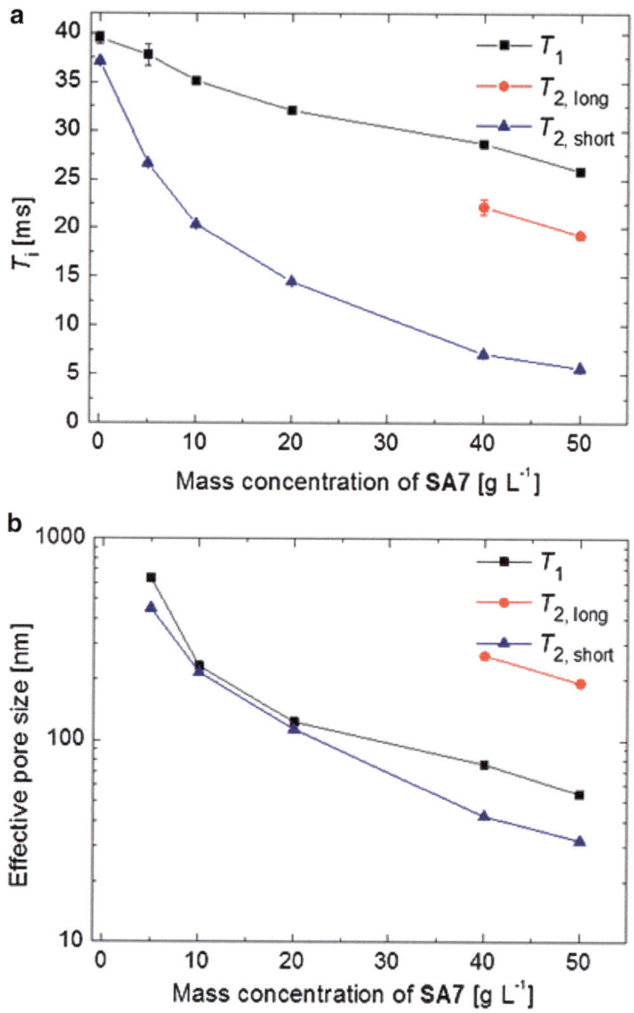

Fig. 6 Relaxation times and effective pore size *versus* mass concentration of **SA7** in sat. aq. $NaHCO_3$ sol. measured at RT. (**a**) Longitudinal and transverse relaxation times for ^{23}Na ions. (**b**) Effective pore size

surface relaxivities ρ_i and the known bulk values $1/T_{i,\text{ bulk}}$ ($T_{1,\text{ bulk}} = 39.5 \pm 0.6$ ms and $T_{2,\text{ bulk}} = 37.2 \pm 0.1$ ms) the effective pore size of the hydrogel can be estimated.

Afterwards, we investigated the influence of the amount of the hydrogelator on the relaxation times while maintaining a constant concentration of sodium ions in solution. For this purpose, we prepared samples containing different amounts of hydrogelator **SA7** in sat. aq. $NaHCO_3$ sol. and determined the corresponding T_1 and T_2 values (Fig. 6). With increasing hydrogelator concentration, both T_1 and T_2 decrease (Fig. 6a). By applying the Brownstein-Tarr model to our system, we obtain effective pore sizes (calculated from T_2) that decrease from 450 nm to 30 nm with increasing mass concentration (Fig. 6b). This is reasonable, as one would expect a higher network density with increasing amount of gelator, thus resulting in smaller effective pore sizes. Furthermore, the effective pore size estimated by the Brownstein-Tarr model for the hydrogel sample prepared from 10 g L^{-1}

of **SA7** correlates well with the pore sizes visualized from a similar sample by ff-TEM imaging (Fig. 1c). At low concentrations of **SA7** (5, 10, and 20 g L^{-1}) we generally observe a monomodal system, where T_1 and T_2 values give similar values for the effective pore size. At higher concentrations (40 and 50 g L^{-1}) we obtain a bimodal distribution for T_2 as shown in Fig. 6b. We relate this to the presence of a hierarchical superstructure of the hydrogel network exhibiting both bigger pores (the voids in between the fibrillar hydrogel network) and smaller pores (formed within the interior of the supramolecular network fibers).

Conclusion

We were able to demonstrate that ^{23}Na NMR relaxation methods can be a highly useful non-invasive tool to characterize supramolecular hydrogel systems. It was possible to determine the melting points, which are in good agreement with the DSC results, and to monitor both the melting and the formation of supramolecular hydrogels based on low molecular weight amphiphiles derived from succinamic acid. As the behavior of the relaxation times T_1 and T_2 accords to the Brownstein-Tarr model it was additionally possible to estimate the effective pore size of the hydrogels, which correlates well with results obtained by TEM experiments. The coexistence of two distinct T_2 relaxation times that was observed especially at higher gelator concentrations and at later stages of the hydrogel formation process (at longer gelation times) allowed for concluding a hierarchical superstructure of the supramolecular hydrogel assemblies exhibiting two different species of pores. This information is of essential importance for future applications, i. e. to incorporate active ingredients into supramolecular networks for drug delivery systems or for applying such hydrogel systems as filtration media.

Acknowledgments The authors wish to thank the German Research Foundation (DFG) for financial support in the frame of Programme SPP 1259 "Intelligente Hydrogele". We are indebted to Dr. M. Krekhova for the preparation of the ff-TEM images and M. Behr for fruitful discussions.

References

1. Tromp RH, van der Maarel JR, de Bleijser J, Leyte JC (1991) Counter-ion dynamics in crosslinked poly(styrene sulfonate) systems studied by NMR. Biophys Chem 41:81–100
2. Gustavsson H, Lindman B, Bull T (1978) Sodium-23 nuclear magnetic resonance in polyanion solution. Correlation times and quadrupole coupling constants of sodium(+) bound to poly(methacrylic acid). J Am Chem Soc 100:4655–4661
3. Levij M, de Bleijser J, Leyte JC (1981) Long-range electric field gradients in charged polymer solutions as probed by sodium-23 relaxation. Chem Phys Lett 83:183–191
4. Woessner DE (2001) NMR relaxation of spin-3/2 nuclei: effects of structure, order, and dynamics in aqueous heterogeous systems. Concepts Magn Reson 13:294–325
5. Rijniers LA, Magusin PCMM, Huinink HP, Pel L, Kopinga K (2004) Sodium NMR in porous materials. J Magn Reson 167:25–30
6. Brownstein KR, Tarr CE (1979) Importance of classical diffusion in NMR studies of water in biological cells. Phys Rev A 19:2446–2453
7. Valckenborg EMW, Pel L, Hazrati K, Kopinga K, Marchand J (2001) Pore water distribution in mortar during drying as determined by NMR. Mater Struct 34:599–604
8. de Loos M, Feringa BL, van Esch JH (2005) Design and application of self-assembled low molecular weight hydrogels. Eur J Org Chem 17:3615–3631
9. Dastidar P (2008) Supramolecular gelling agents: can they be designed? Chem Soc Rev 37:2699–2715
10. Sangeetha NM, Matira U (2005) Supramolecular gels: functions and uses. Chem Soc Rev 34:821–836
11. Estroff LA, Hamilton AD (2004) Water gelation by small organic molecules. Chem Rev 104:1201–1217
12. The self-assembly behavior and hydrogel formation potential of the succinamic acid-based class of amphiphiles will be reported in detail elsewhere.
13. Turco G, Donati I, Grassi M, Marchioli G, Lapasin R, Paoletti S (2011) Mechanical spectroscopy and relaxometry on alginate hydrogels: a comparative analysis for structural characterization and network mesh size determination. Biomacromolecules 12:1272–1282
14. Wallace M, Adams DJ, Iggo JA (2013) Analyses of the mesh size in a supramolecular hydrogel by PFG-NMR spectroscopy. Soft Matter 9:5483–5491
15. Höhne G, Hemminger W, Flammersheim HJ (1996) Differential scanning calorimetry. Springer, Berlin
16. Roy S, Dasgupta A, Das PK (2007) Alkyl chain length dependent hydrogelation of L-Tryptophan-based amphiphile. Langmuir 23:11769–11776
17. Wang D, Hao J (2011) Self-assembly fibrillar network gels of simple surfactants in organic solvents. Langmuir 27:1713–1717
18. Hao J, Liu W, Yuan Z (2008) Gel phase originating from molecular quasi-crystallization and nanofiber growth of sodium laurate-water system. Soft Matter 4:1639–1644

Tracer Mobility in Aqueous Poly(*N*-isopropylacrylamide) Grafted Networks: Effect of Interactions and Permanent Crosslinks

A. Vagias, P. Košovan, C. Holm, H.-J. Butt, K. Koynov, and G. Fytas

Abstract

Using fluorescence correlation spectroscopy, we measured the translational mobility of three molecular tracers in aqueous poly(*N*-isopropylacrylamide) (PNIPAAm) grafted networks, under good solvency conditions. The influence of permanent crosslinks and the interaction between polymer and tracers was revealed. In contrast to the athermal tracer Alexa 647, we observed strong deviation from single Fickian diffusion for two interacting tracers, Alexa 488 and Rhodamine 6G. The dynamics of the latter tracers were represented by a double Fickian diffusion in PNIPAAm solutions and hydrogels for volume fractions up to 30 %. A different slowdown was observed for the slow process in hydrogels and solutions. For the hydrogels, it becomes virtually concentration independent above about 10 %, whereas for solutions it increases below this threshold with PNIPAAm concentration. A pure crowding effect, free of interactions, was observed for both Alexa 647 and for the fast diffusion process of the interacting tracers, whereas their slow process was attributed to the interaction between tracer and polymer. The elucidation of effects due to crowding as well as due to the strength and nature of interactions on the molecular diffusion in hydrogels is needed for both, fundamental and practical perspectives, as for example in biosensors.

Keywords

Hydrogels • Crosslinks • Polymer solutions • Penetrant diffusion • FCS • Interacting tracers

A. Vagias • H.-J. Butt • K. Koynov (✉)
Max Planck Institute for Polymer Research, Ackermannweg 10, 55128 Mainz, Germany

C. Holm
Institut für Computerphysik, Universität Stuttgart, Allmandring 3, 70569 Stuttgart, Germany

P. Košovan
Institut für Computerphysik, Universität Stuttgart, Allmandring 3, 70569 Stuttgart, Germany

Department of Physical and Macromolecular Chemistry, Faculty of Science, Charles University in Prague, Hlavova 8, 120 00 Praha 2, Czech Republic

G. Fytas (✉)
Max Planck Institute for Polymer Research, Ackermannweg 10, 55128 Mainz, Germany

Department of Materials Science & Technology, University of Crete and F.O.R.T.H./IESL, P.O. Box 1527, 71110 Heraklion, Greece
e-mail: fytas@iesl.forth.gr

Introduction

Tracer diffusion in crowded environments has attracted strong interest, because it is relevant for many macromolecular systems [1–23] and its understanding is pivotal for several applications[13, 24, 25]. Such tight conditions are frequently encountered in biology, when examining for instance the diffusion of lipids [6] or proteins in cellular membranes or in the cytosol [7], while tracer dynamics may relate to viscoelastic properties in the polymer network [1, 18]. Today it is known that the mobility of a tracer in dense macromolecular environments can be significantly influenced by the tracer size and shape [8, 14, 19, 21], as well as the matrix concentration [8, 19], and in certain cases by the matrix molecular weight [10, 19]. Other important parameters include the presence of immobile moieties, such

G. Sadowski and W. Richtering (eds.), *Intelligent Hydrogels*, Progress in Colloid and Polymer Science 140,
DOI 10.1007/978-3-319-01683-2_5,

as crosslinks or the existence of specific tracer-polymer interactions.

Crosslinks, e.g. in hydrogels, represent a fundamental inherent feature of polymer networks that give rise to significant interest not only in view of the corresponding mechanical properties of the networks [26–28], but also with respect to the influence that crosslinks may exert on solute diffusivities [29]. One of the very pioneering works concerning bulk diffusivity measurements in hydrogels dates back to the 1980s [30], relating empirically solute diffusivity to solute size, matrix geometry and swelling features. The importance of crosslinks was recently addressed also by computer simulations of molecular tracer diffusion in polyethylene glycol hydrogels [31]. In another recent simulational work, deviations from Fickian diffusion in hydrogels were envisaged as modified tracer motions due to tracer coupling with specific conformational and dynamic modes of the network [32]. Most experimental works that deal with the mobility of molecular tracers [33, 34] or nanoparticles [15, 35, 36] in crosslinked matrices focus on non-interacting systems.

Only in few cases interactions between the diffusing tracer and the matrix were explicitly considered. The existence of domains with different polarity in fluorinated Polyethylene Glycol (PEG) gels can differentiate anticancer drug dynamics between either free diffusion in PEG-rich domains or hindered diffusion in hydrophobic pockets [13]. Additionally, crosslinking of PEG solutions with agents interacting with the probe tracers has also led to deviations from a single Fickian diffusion of fluorescence correlation spectroscopy (FCS) autocorrelation curves [16]. In a similar context, a recent electron paramagnetic resonance (EPR) study has revealed local heterogeneities in thermoresponsive PNIPAAm hydrogels, originating from hydrophobic nanoshelters near the collapse transition of the network [37]. In these studies, however, the involvement of changes in inherent polymer chain dynamics (presence of permanent crosslinks, or not) is missing. It is necessary to understand how these may influence tracer dynamics under similar polymer concentrations.

With this in mind, in this paper we study the mobility of molecular diffusants in hydrogels under the simultaneous effect of permanent crosslinks and varying interactions with the polymer matrix. Inspired by the large number of potential applications of thermoresponsive polymers, e.g. in drug delivery [13, 38] and biosensors [24, 25], we have chosen a grafted thermoresponsive PNIPAAm hydrogels as a model system for a crosslinked polymer network.

Even though NMR, dynamic light scattering (DLS) [39] or FCS may all satisfactorily resolve tracer dynamics in polymer networks, here we selected the technique of FCS that offers – compared to the other previously mentioned techniques – the additional advantages of single molecule sensitivity and concurrent probing of fluorescent tracers with different sizes and/or interactions with the matrix. The method is based on monitoring the fluorescent intensity fluctuations, caused by the diffusion of fluorescent tracers through a confocal volume-limited observation spot. A correlation analysis of these fluctuations yields the tracer's diffusion coefficient and local tracer concentration [40]. Even though FCS has been primarily applied in molecular and cellular biology [4, 7, 41–43], it has been lately extended to polymers and colloid science [44], especially addressing the tracer diffusion in macromolecular polymer solutions [5, 8, 45–47] and in polymer gels [33–35, 39, 48].

We have used FCS to monitor the mobility of different molecular tracers with varying attractive strength in aqueous networks – polymer solutions and grafted crosslinked hydrogels – of a terpolymer consisting of 94 % mol poly(*N*-isopropylacrylamide) (NIPAAm), 5 % mol poly-methacrylic acid and 1 % mol benzophenone methacrylate. The simultaneous presence of hydrophilic (charged methacrylic acid unit) and hydrophobic features (benzophenone groups) on it provides the advantage to possibly unmask tracer-polymer interactions of different nature. In addition to the given hydrophilic and hydrophobic comonomers present on the terpolymer, the lower critical solution temperature (LCST) of NIPAAm constituent ($\sim$32 °C) [34] is close to human body temperature. This renders the particular PNIPAAm terpolymer an appropriate hydrogel for thermoresponsive biosensors [25, 49]. To address the issue of attractive strength, we have employed a strongly (Rhodamine 6G), a weakly (Alexa 488) and a non-interacting (Alexa 647) tracer and compared their diffusion.

Our motivation is driven from the need to enrich current information available for thermoresponsive crosslinked networks [24, 25, 34, 48, 50] and is 3-fold: first, to quantify the effects on the corresponding tracer mobilities by the simultaneous interplay of increased matrix concentration and presence of crosslinks with the existing interactions; second, to rationalize the observed dependences and third, to discuss about possible scaling relations by comparing our findings with analogous results reported in the literature.

Experimental

A. **Samples:** The examined PNIPAAm terpolymer (Mw= 280 k) consisted of 94 % mol poly(*N*-isopropyacrylamide) as well as of hydrophilic (5 % mol of methacrylic acid) and hydrophobic (1 % mol of benzophenone methacrylate) groups. It was synthesized by free radical polymerization as described elsewhere [48]. Its polydispersity index (PI) was PI = 2.7, as obtained from gel permeation chromatography. The benzophenone groups served as the cross-linking agent

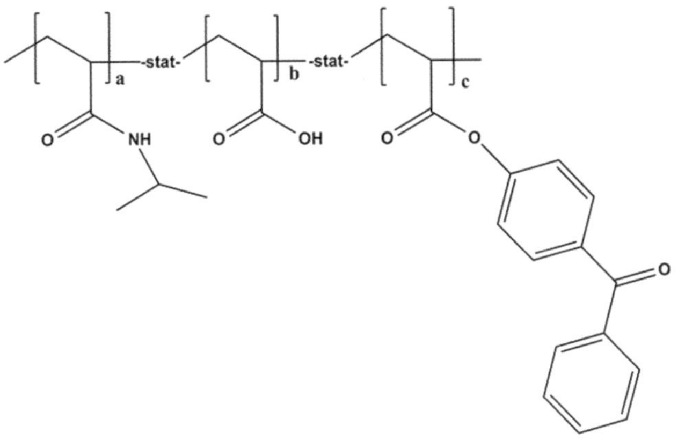

Scheme 1 Chemical structure of the poly(*N*-isopropylacrylamide) based terpolymer

between the polymer chains, upon illumination with UV light at wavelength $\lambda = 365$ nm. The chemical structure of the PNIPAAm terpolymer unit is shown in Scheme 1.

All fluorescent tracers, A488 (Alexa Fluor 488 5-TFP, A-30005), A647 (Alexa Fluor 647 cadaverine, disodium salt, A30679) and Rh6G (Rhodamine 6G Chloride, R634), were purchased from Invitrogen Inc. As A647 lacked any attractive interactions with the polymer system [34], it was present simultaneously in the studied polymer samples together with the other dyes, acting as a 'control' tracer. Regarding the tracer surface charges, Rh6G is a positively charged tracer [51], while zeta potential measurements in mili-Q water revealed that A488 and PNIPAAm are strongly negatively charged and slightly negatively charged, respectively.

B. **Sample preparation.** Grafted PNIPAAm layers were prepared as follows: A 10 wt% PNIPAAm solution in ethanol was spin coated at room temperature onto pre-functionalized round microscope glass slides (diameter 25 mm, thickness: 0.16 mm) at certain spinning speed (250 rpm) and spinning time (60 s). After spin coating, the slides were annealed for 1 h at 170 °C in vacuum (a temperature higher than the glass transition temperature of PNIPAAm), in order to relieve the polymer system from possible stresses. The slides were then dried at 50 °C overnight in vacuum, to remove any traces of the solvent and then subsequently crosslinked by UV irradiation (Stratalinker 2400, Stratagene) at $\lambda = 365$ nm (1 h of crosslinking corresponds to an irradiation energy dose of about 6.28 J cm^{-2}). Consecutively, the slides were rinsed 15 times in situ with absolute ethanol, to remove any uncrosslinked chains; between all steps before crosslinking, the slides were kept in argon atmosphere. The dry thickness was measured by a profilometer (KLA-Tencor Stylus P-16+) in 5–6 different locations of the dried sample, after the crosslinked polymer had been rinsed in ethanol. The measurements in the grafted hydrogels were performed 30′ after addition of the fluorescent tracer aqueous solution, to ensure that the gel has fully reached swelling equilibrium. In all samples of this study (solutions and grafted hydrogels), only ultrapure deionized water was used (filtered through a MilliQ purification system, resistivity 18.2 MΩ cm) without any buffers.

Several aqueous PNIPAAm solutions were prepared below and above the overlap concentration, c^* ($= M_w/(4\pi N_A R_H^3/3) = 0.003$ g ml^{-1}, N_A being the Avogadro number and R_H the hydrodynamic radius of the polymer, $R_H = 15$ nm). A certain volume of aqueous fluorescent dye solution (10 nM concentration for each dye: A647 and either Rh6G, or A488) was added in the sample. The samples were sealed with Teflon samples and were stirred overnight at 400 rpm at room temperature.

C. **Sample Holders.** The measurement of tracer mobility for all samples (polymer solutions and fully swollen grafted hydrogels) was accomplished at room temperature ($T = 25$ °C), ensuring good solvency conditions of PNIPAAm in water. Reusable Attofluor steel chambers were used as sample holders for the grafted PNIPAAm, while 8-well NUNC chambers (where the chamber consisted of polystyrene and the cover glass of borosilicate, provided by Thermo Scientific Inc.) were used for the PNIPAAm aqueous solutions. In order to prevent solvent evaporation during the experiments, the Attofluor sample chamber was covered with a round microscope glass slide.

D. **Fluorescence correlation spectroscopy (FCS).** The measurements were performed on a commercial FCS setup (Carl Zeiss, Jena, Germany) consisting of the module ConfoCor2, and an inverted microscope, Axiovert 200. A 40× Plan Neofluar objective was used, bearing the following features: high numerical aperture-NA (NA = 1.2), working distance 0.28 mm and water as immersion liquid. The A647 was excited by a He–Ne laser at $\lambda = 633$ nm, while an argon ion (Ar^+) laser at $\lambda = 488$ nm and a HeNe laser at $\lambda = 543$ nm were used to excite A488 and Rh6G, respectively. The fluorescent emission was collected by the same objective and after passing through an emission filter and a confocal pinhole, delivered to an avalanche photodiode detector capable of single-photon counting. The temporal fluctuations of the detected fluorescence intensity, $\delta F(t)$ were recorded and evaluated in terms of an autocorrelation function, $G(t) = \frac{<\delta F(t) \cdot \delta F(t+\tau)>}{<F(t)>^2}$. For each type of polymer solution or gel, at least three independent measurements (2–10 min each) at different positions in the sample were performed. The axial (z_o) and the lateral (r_o) dimensions of the Gaussian confocal observation volume for each excitation wavelength were calibrated by reference measurements in dilute (10 nM) aqueous solutions of A488, Rh6G and A647 using published values of their diffusion coefficients [52].

Table 1 Monomer volume fraction for the studied hydrogels at 25 °C

Code: HG-i	HG-1	HG-2	HG-3	HG-4	HG-5
ϕ (±10 %)	0.013	0.030	0.084	0.15	0.26

E. **Swelling ratio of grafted hydrogels.** The thickness of the fully swollen hydrogels was determined using the FCS setup described above. By shifting the microscope objective, the position of the FCS observation volume was scanned in z-direction (normal to the film plane) with a step of 1 μm and the average fluorescence intensity signal that is proportional to the local density of the tracers was recorded as a function of the focus position (z-scan). A typical z-scan (Fig. 2 below) depicts two transition regions, representing the hydrogel/water and glass/hydrogel interfaces. The distance between these regions represents the thickness of the fully swollen gel. The swelling ratio R_S, was determined as the ratio between the fully swollen thickness and the dry thickness. The monomer concentrations (volume fractions, $\phi = R_s^{-1}$) for the studied hydrogel (HG) are shown in Table 1.

The monomer volume fraction ϕ exhibits an inverse proportionality to the UV irradiation time: The longer the UV irradiation, the more free radicals from benzophenone moieties would be created, leading to more covalent benzophenone bonds and thus, to higher crosslink densities.

Results

Data Analysis

In order to obtain quantitative information for the tracer diffusion in the studied systems, the experimentally measured FCS autocorrelation curves have to be fitted with an appropriate model function. For an ensemble of n different types of freely diffusing (Fickian diffusion) fluorescent species, the autocorrelation function has the following analytical form [34]:

$$G(t) = 1 + \frac{1}{N}\cdot\sum_{i=1}^{n}\frac{F_i}{\left(1+\frac{t}{\tau_i}\right)\cdot\sqrt{\left(1+\frac{t}{S^2\cdot\tau_i}\right)}}\cdot\left(1+\frac{\mathrm{T}}{1-\mathrm{T}}\cdot e^{-\frac{t}{\tau_T}}\right) \quad (1)$$

Here, N represents the average number of species in the observation volume, T and τ_T are the fraction and the decay time of the triplet state, and $S = z_o/r_o$ is the so called structural parameter, with 2 z_o and 2 r_o the axial and lateral size of the observation volume, respectively. F_i and τ_i are the fraction and the lateral diffusion time of the ith type of species. Diffusion coefficient D_i for i^{th} species can be directly extracted from τ_i through: $D_i = r_0^2/(4\tau_i)$. While a single Fickian diffusion model (n = 1 in Eq. 1) represents well our experimental data for non-interacting tracers, it failed to describe the dynamics of the interacting tracers in both PNIPAAm solutions and hydrogels.

Quite frequently, deviations from single Fickian diffusion have been reported for a tracer that diffuses in crowded environments [20, 22]. To quantitatively describe such tracer dynamics, the approach of anomalous diffusion can be employed, where the mean square displacement $<\Delta r^2(t)>$ may be either non-linearly related to time $<\Delta r^2(t)> \sim t^\alpha, \alpha < 1$ [2, 3, 11, 20, 22], or it could be linearly related to time ($\alpha = 1$) but the tracer's displacements would follow a non-Gaussian distribution [11]. With respect to FCS studies, the subdiffusion model ($\alpha < 1$) has been often used to fit experimentally measured autocorrelation curves and account for deviations from simple Fickian diffusion of biomolecules or nanoparticles in crowded matrix environments [3, 12, 53, 54]. However, even though the description of the dynamics with such model is seemingly successful, the assignment of certain physical significance to the stretched exponential parameter 'α' is rather not straightforward and should be carefully justified [55]. For example, in the case of non-permanently crosslinked systems and non-interacting tracers with sizes larger than the matrix heterogeneity (e.g. mesh size), it has been shown that anomalous diffusion can reflect the viscoelastic properties of the matrix [23]. The typical mesh size of the swollen PNIPAAm gels that we studied is similar to semi-dilute solutions i.e. in the range 3–8 nm [34] that is almost one order of magnitude larger than the diameter of the used tracers. Thus, viscoelasticity or obstructed diffusion play no important role and non-interacting molecular tracers display Fickian diffusion controlled by crowding effects [8]. Accordingly, we did not attempt to interpret our data using subdiffusion models for the FCS autocorrelation curves.

Some alternative models developed for FCS autocorrelation curves that measured diffusion in presence of interactions, such as the diffusion and rare strong adsorption [56] or a diffusion and reaction model [4] failed to describe our experimental data. Therefore, all experimental autocorrelation curves reported in this work were fitted with Eq. 1 using either one or two components (n = 1 or 2), as appropriate. The physical meaning of the two component fits was recently justified for interacting tracers diffusing in PNIPAAm solutions [57] and is further discussed in section "Discussion" below. Furthermore, in all fits the triplet time and fraction were used as free fit parameters. We obtained triplet times in the range 1–3 μs as expected for the studied dyes.

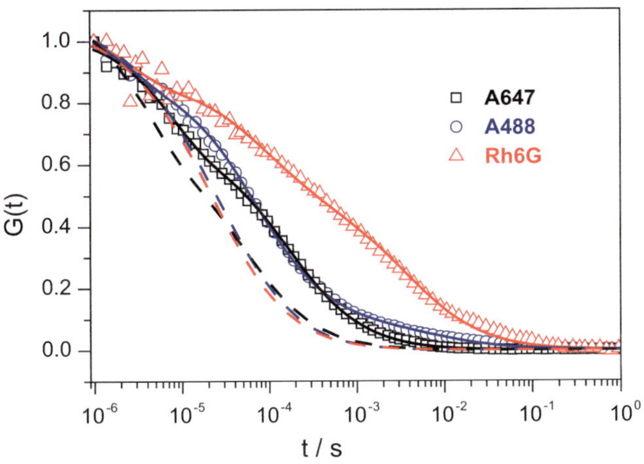

Fig. 1 Presence of interactions in PNIPAAm solutions: Normalized fluorescence intensity autocorrelation functions G(t) for the three molecular tracers in non-dilute aqueous solutions of PNIPAAm (280 k) at $c = 0.13$ g ml^{-1} and 25 °C: A647 (*squares*), A488 (*circles*) and Rh6G (*triangles*). *Solid lines* denote the representation by Eq. 1 using either n = 1 (for A647) or n = 2. *Dashed lines* represent single component fits (n = 1) to the experimentally measured G(t) (not shown) of the tracers in water

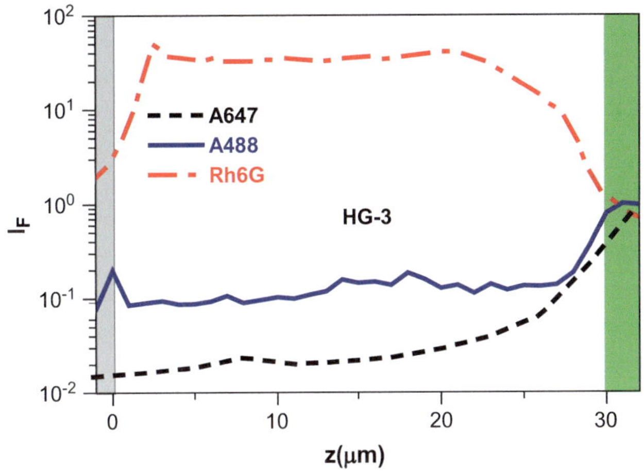

Fig. 2 Tracer density profile: Normalized fluorescence intensity (I_F) for the three tracers in HG-3 versus z-distance normal to the substrate. Supernatant solution, hydrogel and substrate are denoted with *green*, *white* and *grey* colors, respectively

Effects of Interactions in Temporary and Permanent PNIPAAm Networks

A recent work focused on the diffusion of A647 in grafted hydrogel layers of the same PNIPAAm terpolymer at different temperatures, up to the LCST of PNIPAAm [34]. In the current study, however, we have used A647 as an internal standard and have examined the mobility of two interacting tracers (A488, Rh6G) in PNIPAAm aqueous solutions and PNIPAAm grafted hydrogel layers. Each tracer exhibits different strength of attraction with the particular PNIPAAm terpolymer, as can be seen by the normalized FCS intensity autocorrelation curves G(t), and from the tracer density profiles (z-scans) shown in Fig. 2. In Fig. 1, G(t) for all three examined tracers in aqueous solutions of the PNIPAAm terpolymer at the same polymer concentration ($c = 0.13$ g ml^{-1}) are shown together with the fits (solid curves), along with the G(t) for each tracer in pure water (dashed curves), for comparison.

In PNIPAAm aqueous solutions, the presence of interactions leads to deviation from single Fickian diffusion for A488 and Rh6G. As the strength of interactions increases (Fig. 1), a stronger slowdown in the tracer's G(t) is observed relative to the respective tracer's G(t) in pure water. We therefore employed the two component Fickian model (Eq. 1, n = 2) that describes well the experimental G(t) for either A488 or Rh6G. Qualitative information about the strength of tracer-polymer interactions was acquired from the corresponding tracer density profiles in the grafted PNIPAAm hydrogels (HG) at ambient conditions, shown in Fig. 2.

On account of the chemical structures, it follows that Rh6G is positively charged [51], while both A647 and A488 are negatively charged. The PNIPAAm terpolymer is weakly negatively charged, due to the presence of 5 % mol acrylic monomer units. Therefore, Rh6G should be electrostatically attracted to the polymer, while A488 and A647 should be repelled from it. In addition to electrostatic interactions, the presence of hydrophobic substructures in all molecular tracers need to be considered. These should exhibit attractive interactions with the polymer backbone of the polymer and particularly with the hydrophobic benzophenone groups. The resultant effect is therefore an interplay of electrostatics and hydrophobicity, which work synergistically in the case of Rh6G, but compete in the cases of A488 and A647. Based on Fig. 2, we conclude that Rh6G is strongly attracted to the polymer, while both A488 and A647 are repelled. Furthermore, the less pronounced depletion of A488 as compared to A647 indicates that the hydrophobic attraction between A488 and the polymer can more efficiently counteract the electrostatic repulsion than in the case of A647. On account of the above discussion concerning Fig. 2, and of the shape and fits of the G(t) presented in Fig. 1, we classify our tracers as strongly interacting (Rh6G), weakly interacting (A488), and non-interacting (A647). However, one should keep in mind that the origin of the effective specific interactions may be rather complex.

The non-interacting A647 can be utilized to prove whether concentration and interaction effects can be separately addressed simply by examining the shape and the trend of G(t) in PNIPAAm solutions and grafted PNIPAAm hydrogels at the same monomer concentration. To make this point strong, we have deliberately chosen high PNIPAAm concentration and high crosslink density corresponding to

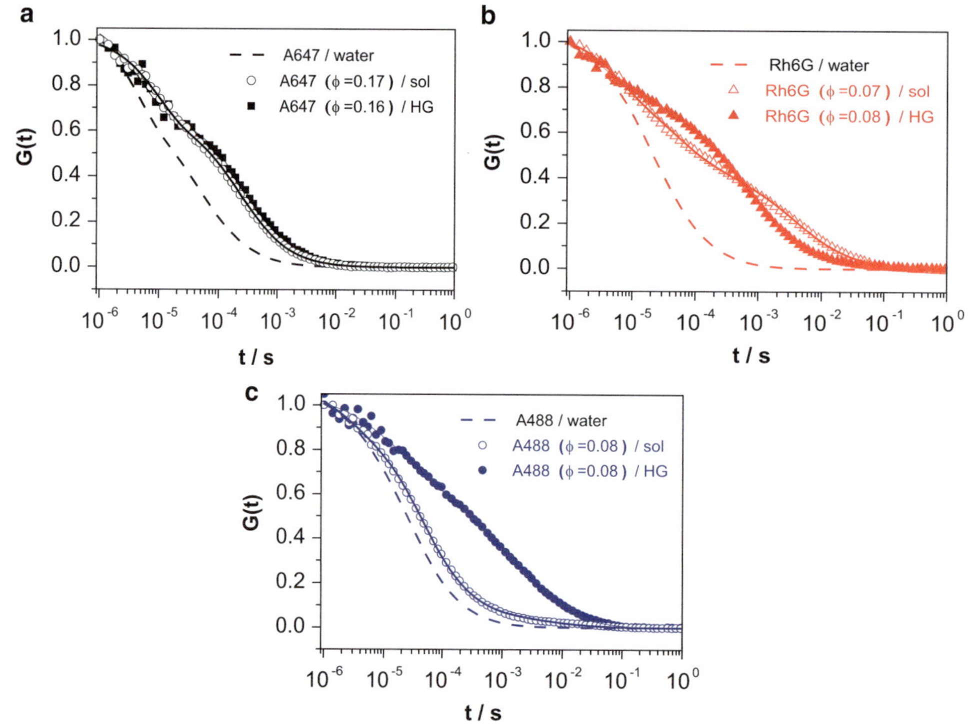

Fig. 3 Influence of permanent crosslinking. Normalized G(*t*) for the three molecular tracers in PNIPAAm aqueous solutions and grafted PNIPAAm hydrogels with very similar monomer concentration at 25 °C. The G(t) for each tracer was fitted by Eq. 1 (*solid lines*) using either n = 1 (A647, A488 in solution) or n = 2 (Rh6G, A488 in HG). *Dashed curves* represent single component fits (n = 1) to the experimentally measured G(*t*) (not shown) of the tracers in water

volume fractions $\phi \sim 0.17$, in order to contrast with the non-interacting A647 diffusion slowdown in the two network types.

It is clearly observed (Fig. 3) that in either network type the G(t) of A647 conforms to a single Fickian diffusion that is, however, slower relative to the free diffusion in pure water. Furthermore, the slowdown is slightly stronger in hydrogel than in polymer solution at the same overall monomer concentration, probably due to small additional drag in the former case, in agreement with an earlier study (see solid and dashed curves in the upper panel of Fig. 6 below) [34].

On the other hand, the strongly interacting Rh6G deviates from single Fickian diffusion in both hydrogel and solutions (Fig. 3b), even at low monomer concentration ($\phi = 0.08$). Apart from shape differences, the G(t) of Rh6G decays in the range of 10^{-2} s in both network types. Under the same conditions, G(t) of A488 (Fig. 3c) deviates from single Fickian diffusion only in the hydrogels but not in the polymer solutions. Therefore, it seems that the weakly repulsive A488 is a sensitive probe of the polymer network topology (i.e. presence of crosslinking). Based on Figs. 2 and 3, Rh6G should not only exhibit stronger attractive strength than A488, but the relative strength of hydrophobic and electrostatic interactions for the two tracers should be different. Further information about these differences will follow in the discussion (section “Discussion”).

Effect of Crosslink Density in Permanent Networks

Additional differences in the dynamics of the two tracers (A488, Rh6G) have been revealed, as the crosslink density of the hydrogel increases. The autocorrelation function G(t) of Rh6G is slower than that of A488 at relatively small monomer concentrations ($\phi = 0.01$–0.03) as seen in Fig. 4a. However, at higher crosslink densities (Fig. 4b; $\phi = 0.26$), the G(t) of A488 and Rh6G become comparable. Therefore, the evolution of tracer dynamics with crosslink density depends also on both the nature and relative strength of hydrophobic and electrostatic interactions, the latter being inevitable feature of the weak polyelectrolytes.

The autocorrelation curves measured for A488 and Rh6G in all studied hydrogels were fitted using Eq. 1 with n = 2. From the extracted diffusion coefficients (D) we then calculated the diffusion slowdown values (D/D_0), where D_0 is the diffusion coefficient of the respective tracer in pure water. Figure 5 shows the diffusion slowdown for the fast and the slow processes and the amplitude of the slow process (F_{slow}) as a function of the monomer concentration in the

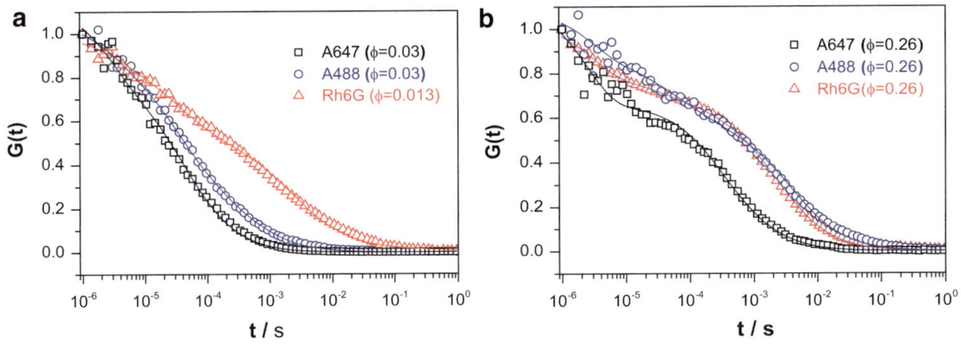

Fig. 4 Crosslink density: Normalized G(*t*) for the molecular tracers in PNIPAAm terpolymer hydrogels (HG) at 25 °C for extreme cases of volume fractions achieved through variation of the grafting densities. A: $\phi = 0.013$ (HG-1) and $\phi = 0.03$ (HG-2); B: $\phi = 0.26$ (HG-5)

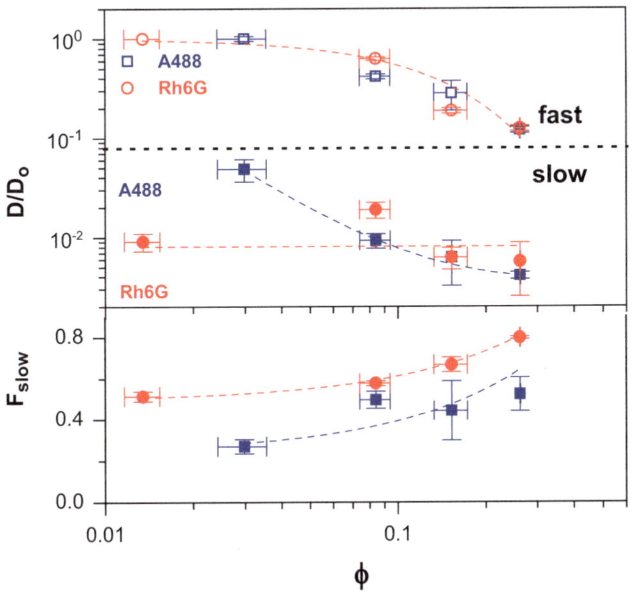

Fig. 5 Diffusion slowdown $D(\phi)/D_o$ – *upper panel* – and amplitude of the slow process (F_{slow}) – *lower panel* – as a function of monomer concentration (ϕ) for A488 and Rh6G in hydrogels. The *dashed lines* are drawn to guide the eye

four examined hydrogels. The slowdown of the fast diffusion process is almost identical for both tracers in all studied hydrogels. On the other hand, there is a 5-fold difference in the slow process slowdown between A488 and Rh6G at small ϕ (0.01–0.03), where Rh6G exhibits stronger slowdown. For the slow process, disparities between A488 and Rh6G slowdown at the same low ϕ (in the range 0.01–0.03) progressively diminish, with increasing ϕ. The observed differences in the slowdown between A488 and Rh6G may be attributed to differences in the underlying tracer-polymer interactions.

The different strength of tracer-polymer interactions can also be visualized by the larger F_{slow} for Rh6G (strongly interacting) compared to A488 at the same ϕ. We recall that the crosslink density in permanent PNIPAAm networks influences differently the tracer diffusion slowdown at low ϕ, based on the strength of the particular hydrophobic and electrostatic interactions: single Fickian for A647, deviations from single Fickian diffusion for the other tracers with stronger slowdown for Rh6G and intermediate slowdown for A488.

In summary:

(i) In PNIPAAm aqueous solutions, the tracer translational motion deviates from that of a single Fickian diffusion with increasing strength of tracer-polymer attractions, as judged by the tracer density profiles in grafted hydrogels (Fig. 2).

(ii) Solutions versus Hydrogels: The weakly repulsive A488 is a more sensitive probe of the polymer network type – presence of crosslinks-than the more strongly attractive tracer Rh6G (Figs. 2, 3, 4, and 5).

(iii) In PNIPAAm hydrogels: Crosslink density exerts distinct tracer diffusion slowdown (Fig. 5).

Discussion

Deviation from Single Fickian Diffusion: Rationalization of Tracer Mobility

In view of our results in (i), a number of different models can be employed to describe dynamics in cases where tracer mobility deviates from single Fickian diffusion [4], [43], [54], [56],[58]. We chose, however, the two-component diffusion model (Eq. 1, $n = 2$), namely on account of its good representation of the experimental G(t) in both PNIPAAm hydrogels and solutions and on the conformity of the adjustable parameters to a physically meaningful tracer mobility scenario.

The two-component model can be rationalized by dynamical equilibrium between states associated with different lifetimes: τ_f corresponds to the lifetime in the unbound state (free tracer), while τ_s corresponds to the lifetime the tracer is bound to the polymer. During a given lifetime, the tracer can diffuse a particular distance and the squared length scale l_i^2 of a travelled distance in a given lifetime is roughly proportional to the product of the diffusion coefficient D_i and the lifetime τ_i (i = f,s), of the particular state: $l_i^2 \sim \tau_i \cdot D_i$. If the exchange kinetics between the two states is not too fast, they should be resolved as independent diffusion processes. This kind of model has been demonstrated to give a coherent description of tracer dynamics in PNIPAAm solutions [57]. As recently shown [59], a process can be resolved by FCS only if the characteristic length scale associated with it is greater than the FCS focal spot size.

Slow Tracer Mobility in Hydrogels and Polymer Solutions

Figure 6 shows the slowdown $D_i(\phi)/D_o$ (i = f,s) for both the fast and slow process for A488 and Rh6G along with the amplitude F_{slow} of the slow process as a function of ϕ in PNIPAAm aqueous solutions and hydrogels at 25 °C. In the case of PNIPAAm solutions, the diffusion slowdown for the fast process conforms to a chain length independent 'master' curve observed for non-interacting tracers in polymer solutions (dashed curve) [8]. For the same process in hydrogels, the slowdown falls on the solid curve (Fig. 6) representing the single component diffusion of the non-interacting A647 in the same hydrogels. This slowdown is attributed solely to crowding effect expressed by the monomer concentration [34].

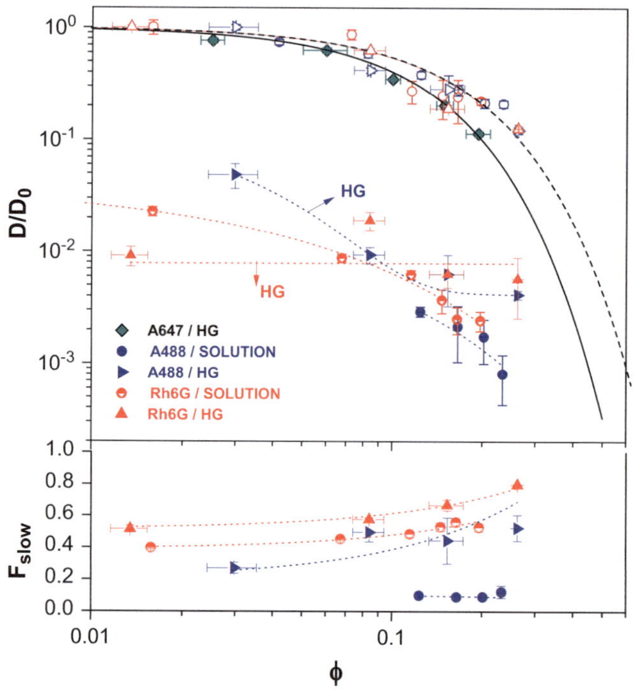

Fig. 6 Mobility slowdown $D(\phi)/D_o$ for the two diffusion processes in the case of weak and strong attractions exemplified by A488 and Rh6G (fast: *open symbols*; slow: *solid symbols*) in solutions (*circles*) and in hydrogels (*triangles*). *Arrows* denote the datasets for hydrogels. The *dashed* and *solid curves* denote stretched exponential concentration dependences recently reported for non-interacting molecular tracer diffusion slowdown, $D(\phi)/D_o$, in solutions [8] and hydrogels [34], respectively. Mobility slowdown data of non-attractive tracer (A647) in PNIPAAm hydrogels from an earlier study [34] are also shown (*green rhombi*). *Lower panel*: Amplitude of the slow process in G(t) of A488 and Rh6G in PNIPAAm solutions and hydrogels. The *dotted lines* are drawn to guide the eye

For the slow process, the diffusion slowdown of Rh6G and A488 in PNIPAAm solutions and hydrogels clearly deviates from a superposition to a single curve. Concerning the slow process in PNIPAAm aqueous solutions, D_s/D_o decreases monotonically with concentration ϕ and becomes similar for both dyes, while its concentration dependence resembles that of the polymer self-diffusion slopes [60] (see Fig. 7). In hydrogels, the slow process is distinctly different both in diffusion slowdown and in amplitude F_{slow}. It is worth noting that F_{slow} for A488 is measurable only above $\phi \sim 0.1$ in solutions, in contrast to the hydrogels with discernible slow process already above $\phi \sim 0.03$. Since the physical and chemical networks consist of the same polymer, the tracer-PNIPAAm interactions should be similar. The differences in diffusion in the two networks might reflect different durations of binding times τ_s for the interacting tracers in each network type.

Such behavior may stem from differences in the inherent polymer chain dynamics in the two types of networks. The arrest of tracer slow diffusion D_s/D_o ($\sim$0.01) in hydrogels appears to be tracer dependent as it occurs at $\phi \geq 0.1$ for A488, while it is present for Rh6G over all examined crosslinking densities. The tracer specificity of this effect excludes 'frozen-in' hydrogel dynamics as the lone reason, as indicated by the single diffusion of the non-interacting A647. We attribute the slowdown disparity between the two tracers at low ϕ to the different sign of Coulombic interactions between each tracer and the negatively charged PNIPAAm, being attractive for Rh6G and repulsive for A488.

With increasing crosslinking density and hence ϕ, the number of chargeable acrylic groups increases proportionally with the total monomer concentration. However, as the total monomer concentration approaches the dissociation constant of the acrylic acid ($10^{-4.2}$ *M*) [61], the degree of charging decreases. Beyond this concentration, the number of charged groups increases much slower than the total monomer concentration. In contrast, the total number of monomers available for hydrophobic binding increases proportionally with polymer concentration without constraints that would be due to dissociation constants.

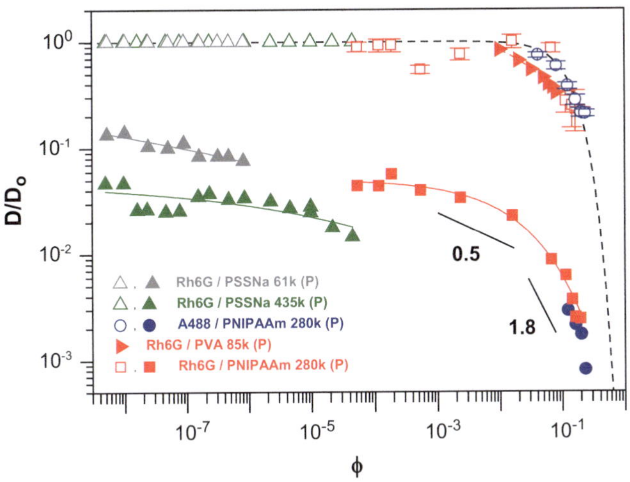

Fig. 7 Diffusion slowdown, $D(\phi)/D_o$, for the interacting tracers A488 and Rh6G in PNIPAAm aqueous solutions at 25 °C. Comparison with reported values for non-interacting [62] (*red triangles*) and interacting tracers (*green* and *grey triangles*) [47] in aqueous polymer solutions. *Open* and *solid symbols* denote fast and slow process, accordingly, whereas the *dashed (black) curve* denotes the concentration dependence of the single diffusion of non-interacting molecular tracer slowdown in polymer solutions [8]. The *solid curves* through the data are drawn to guide the eye and the slopes 0.5 and 1.8 are scaling predictions of polymer self-diffusion [60]

Therefore, we expect that the short range attractive hydrophobic interactions may dominate the electrostatic A488-PNIPAAm repulsions at high ϕ, rationalizing the observed behavior in Figs. 5 and 6. The anticipated increase of the slowdown in hydrogels with ϕ may be probably compensated by the concurrent increase of the PNIPAAm hydrogel's cooperative diffusion as ϕ increases [39]. Hence, the interplay among electrostatics, short-range attractions and the hydrogel cooperative diffusion might all control the complex slow tracer diffusivity.

Tracer Diffusion in Aqueous Polymer Solutions: Comparison with Recent Studies

Figure 7 shows the diffusion slowdown $D(\phi)/D_o$ for the molecular tracers in PNIPAAm aqueous solutions and the reported diffusion data for Rh6G in other aqueous polymer solutions at 25 °C in presence [47] or absence [62] of attractions. We consider the case of Rh6G in PSSNa aqueous solutions [47] as an example of strong attractions, where two diffusion coefficients have been recently reported. The fast component follows the master curve for non-interacting molecular tracers (dashed black curve) in polymer solutions [8], whereas the slow diffusion of Rh6G (D_s) depends on PSSNa molecular weight (435 k and 61 k in Fig. 7) [47] and decreases with concentration as in the case of Rh6G-PNIPAAm (280 k). However, the interaction mechanism observed in Ref.[47] was predominantly electrostatic and could be explained by counterion condensation due to high charge density of the highly sulfonated polystyrene used by the authors. This mechanism can only play a minor role in our samples that have only 5 % mol methacrylic acid > acrylic acid.

Additionally, Rh6G was found not to interact in aqueous PVA solutions, where the data have been accordingly well described by a single Fickian diffusion – red triangles in Fig. 7 [62]. The trend for the single Fickian diffusion data for Rh6G in aqueous PVA solutions (red triangles) approaches the master curve prediction for non-interacting tracers in solutions [8] (dashed black curve), in spite of some quantitative deviation.

Conclusions

By means of FCS, we assessed the effect of crosslink density, monomer volume fraction and interactions on the molecular tracer mobility in temporary and permanent PNIPAAm aqueous networks. Even though interactions and crosslink density effects may be separately addressed using a non-interacting tracer A647, a sensitive interplay between crosslinking density with strength and nature of the tracer-polymer interactions dictates the emerging tracer mobility. For the temporary networks, increase of the attractive strength led to more significant deviations from single Fickian behavior. A fast and a slow process have been resolved for the interacting tracers, suggesting that their length scales are larger than the fixed dimensions of the FCS illumination volume.

The unexpectedly resolved stronger diffusion slowdown in entangled temporary networks compared to permanent networks for the same tracer remains not fully understood, but it may relate with differences in inherent chain dynamics of the polymer network and with cooperative network dynamics. Moreover, the arrest in diffusion slowdown observed in permanent networks for Rh6G implies the simultaneous action of different competing mechanisms, among which osmotic, electrostatic and possibly short-range forces. Due to the repulsive Coulombic interactions with PNIPAAm, the weakly attractive A488 was found to be a more sensitive probe of the network type and the network crowding conditions (volume fraction ϕ).

The comparison with reported tracer diffusion studies in polymer solutions intimately relates the presented double diffusion process to tracer-polymer interactions, while the slow diffusion coefficient is polymer molecular weight dependent. To reveal the exact length scales of the associated processes of the existing study, it is considered necessary to support our results by either simulations or FCS experiments with varying observation volumes.

Acknowledgment We thank the Deutsche Forschungsgemeinschaft for financial support in the framework of SPP1259 "Intelligente Hydrogele". P.K. acknowledges support from grant "LK21302 Návrat" from the Czech Minitry of Education. We would like to thank Katja Nilles for synthesis of the polymer, Andreas Best and Gabi Hermann for technical support, and Uli Jonas for helpful discussions.

References

1. Mason TG, Weitz DA (1995) Phys Rev Lett 74:1250–1253
2. Schwille P, Korlach J, Webb WW (1999) Cytometry 36:176–182
3. Wachsmuth M, Waldeck W, Langowski J (2000) J Mol Biol 298:677–689
4. Pramanik A (2004) Curr Pharm Biotechnol 5:205–212
5. Grunwald D, Cardoso MC, Leonhardt H, Buschmann V (2005) Curr Pharm Biotechnol 6:381–386
6. Zhang L, Granick S (2005) Proc Natl Acad Sci U S A 102:9118–9121
7. Bacia K, Kim SA, Schwille P (2006) Nat Meth 3:83–89
8. Cherdhirankorn T, Best A, Koynov K, Peneva K, Muellen K, Fytas G (2009) J Phys Chem B 113:3355–3359
9. Cherdhirankorn T, Floudas G, Butt HJ, Koynov K (2009) Macromolecules 42:9183–9189
10. Cherdhirankorn T, Harmandaris V, Juhari A, Voudouris P, Fytas G, Kremer K, Koynov K (2009) Macromolecules 42:4858–4866
11. Wang B, Anthony SM, Bae SC, Granick S (2009) Proc Natl Acad Sci U S A 106:15160–15164
12. Szymanski J, Weiss M (2009) Phys Rev Lett 103:038102–038104
13. Mathias E, Aponte J, Kornfield J, Ba Y (2010) Colloid Polym Sci 288:1655–1663
14. Rusu L, Lumma D, Rädler JO (2010) Macromol Biosci 10:1465–1472
15. Susoff M, Oppermann W (2010) Macromolecules 43:9100–9107
16. Zustiak SP, Boukari H, Leach JB (2010) Soft Matter 6:3609–3618
17. Kalwarczyk T, Ziębacz N, Bielejewska A, Zaboklicka E, Koynov K, Szymański J d, Wilk A, Patkowski A, Gapiński J, Butt H-J r, Hołyst R (2011) Nano Lett 11:2157–2163
18. Felderhof BU (2011) J Chem Phys 134
19. Cai LH, Panyukov S, Rubinstein M (2011) Macromolecules 44:7853–7863
20. Hellmann M, Klafter J, Heermann DW, Weiss M (2011) J Phys Condens Matter 23:234113
21. Ochab-Marcinek A, Holyst R (2011) Soft Matter 7:7366–7374
22. Eliazar I, Klafter J (2011) J Phys A Math Theor 44:405006
23. Ernst D, Hellmann M, Kohler J, Weiss M (2012) Soft Matter 8:4886–4889
24. Chun Jen H, Dostalek J, Knoll W (2010) Biosens Bioelectron 26:1425–1431
25. Mateescu A, Yi W, Dostalek J, Jonas U (2012) Membranes 2:40–6969
26. Sakai T, Matsunaga T, Yamamoto Y, Ito C, Yoshida R, Suzuki S, Sasaki N, Shibayama M, Chung U-i (2008) Macromolecules 41:5379–5384
27. Hao J, Weiss RA (2011) Macromolecules 44:9390–9398
28. Sugimura A, Asai M, Matsunaga T, Akagi Y, Sakai T, Noguchi H, Shibayama M (2013) Polym J, 45(3), 300–306.
29. Reinhart CT, Peppas NA (1984) J Membr Sci 18:227–239
30. Ritger PL, Peppas NA (1987) J Control Release 5:37–42
31. Wu Y, Joseph S, Aluru NR (2009) J Phys Chem B 113:3512–3520
32. Tabatabaei F, Lenz O, Holm C (2011) Colloid Polym Sci 289:523–534
33. Modesti G, Zimmermann B, Borsch M, Herrmann A, Saalwachter K (2009) Macromolecules 42:4681–4689
34. Raccis R, Roskamp R, Hopp I, Menges B, Koynov K, Jonas U, Knoll W, Butt HJ, Fytas G (2011) Soft Matter 7:7042–7053
35. Michelman-Ribeiro A, Boukari H, Nossal R, Horkay F (2005) Macromol Symp 227:221–230
36. Seiffert S, Oppermann W (2008) Polymers 49:4115–4126
37. Junk MJN, Jonas U, Hinderberger D (2008) Small 4:1485–1493
38. Ward MA, Georgiou TK (2011) Polymers 3:1215–1242
39. Gianneli M, Roskamp RF, Jonas U, Loppinet B, Fytas G, Knoll W (2008) Soft Matter 4:1443–1447
40. Rigler R, Elson ES (2001) Fluorescence correlation spectroscopy: theory and applications. Springer, New York
41. Schwille P, Meyer-Almes FJ, Rigler R (1997) Biophys J 72:1878–1886
42. Haustein E, Schwille P (2003) Methods 29:153–166
43. Michelman-Ribeiro A, Mazza D, Rosales T, Stasevich TJ, Boukari H, Rishi V, Vinson C, Knutson JR, McNally JG (2009) Biophys J 97:337–346
44. Koynov K, Butt H-J (2012) Curr Opin Colloid Interface Sci 17:377–387
45. Lumma D, Keller S, Vilgis T, Rädler JO (2003) Phys Rev Lett 90:218301
46. Zustiak SP, Nossal R, Sackett DL (2011) Biophys J 101:255–264
47. Jia P, Yang Q, Gong Y, Zhao J (2012) J Chem Phys 136:084904
48. Gianneli M, Beines PW, Roskamp RF, Koynov K, Fytas G, Knoll W (2007) J Phys Chem C 111:13205–13211
49. Toma K, Dostalek J, Knoll W (2011) Opt Express 19:11090–11099
50. Berger J, Reist M, Mayer JM, Felt O, Peppas NA, Gurny R (2004) Eur J Pharm Biopharm 57:19–34
51. Gear ARL (1974) J Biol Chem 249:3628–3637
52. http://www.picoquant.com/technotes/appnote_diffusion_coefficients.pdf
53. Guigas G, Weiss M (2008) Biophys J 94:90–9494
54. Fritsch CC, Langowski J (2010) J Chem Phys 133:025101–025111
55. Sokolov IM (2012) Soft Matter 8:9043–9052
56. Wirth MJ, Ludes MD, Swinton DJ (2001) Appl Spectrosc 55:663–669
57. Vagias A, Raccis R, Koynov K, Jonas U, Butt H-J, Fytas G, Košovan P, Lenz O, Holm C (2013) Complex tracer diffusion dynamics in polymer solutions. Phys Rev Lett 111 (8):088301. doi:10.1103/PhysRevLett.111.088301
58. Mahurin SM, Dai S, Barnes MD (2003) J Phys Chem B 107:13336–13340
59. Enderlein J (2012) Phys Rev Lett 108:108101
60. Gennes PGd (1979) Scaling concepts in polymer physics. Cornell University Press, Ithaca
61. Echeverria C, Peppas NA, Mijangos C (2012) Soft Matter 8:337–346
62. Michelman-Ribeiro FHA, Nossal R, Boukari H (2007) BioMacromolecules 8:1595–1600

Poly-NIPAM Microgels with Different Cross-Linker Densities

Scaling Behavior of the Network Fluctuations in the Vicinity of the Volume Phase Transition

Matthias Karg, Sylvain Prévost, Astrid Brandt, Dirk Wallacher, Regine von Klitzing, and Thomas Hellweg

Abstract

Thermoresponsive microgel particles made of the monomer *N*-isopropylacrylamide (NIPAM) and the cross-linker molecule *N,N'*-methylenebisacrylamide (BIS) were synthesized using three different cross-linker molar ratios. The volume phase transition behavior of these colloids was investigated by means of dynamic light scattering (DLS) and small angle neutron scattering (SANS) covering the different length scales of interest. Both methods provide the temperature of the volume phase transition in good agreement. The volume change as followed by DLS is described using the Flory-Rehner theory, leading to the determination of the spinodal temperature. Furthermore, the network correlation length ξ, which is available from appropriate fits of the measured SANS profiles, was used to study the critical behavior in terms of scaling laws.

The results from DLS and SANS show a strong cross-linker density dependence.

Keywords

Microgel • Volume phase transition • Thermoresponsive • SANS • DLS • Cross-linker • Scaling behavior

Introduction

Responsive microgels and especially responsive microgels made of the monomer *N*-isopropylacrylamide (NIPAM) were studied intensely for more than two decades, which can be mainly attributed to their interesting volume phase transition (VPT) behavior [1–9]. The change of the overall particle dimensions as a function of external parameters like temperature, pH and ionic strength was often studied by dynamic light scattering (DLS) [10–18]. This method allows to investigate the center of mass diffusion of the colloidal gel particles in highly diluted dispersion. The diffusion coefficient depends directly on the size of the scattering objects. Therefore, DLS can be used to determine the VPT behavior in terms of the hydrodynamic radius evolution.

In addition to DLS, imaging techniques such as atomic force microscopy (AFM), transmission electron microscopy (TEM) and scanning electron microscopy (SEM) are

The online version of this chapter (doi: 10.1007/978-3-319-01683-2_6) contains supplementary material, which is available to authorized users.

M. Karg (✉)
Physikalische Chemie – Kolloidale Systeme, Universität Bayreuth, Universitätsstraße 30, 95440 Bayreuth, Germany
e-mail: matthias.karg@uni-bayreuth.de

S. Prévost • A. Brandt • D. Wallacher
Helmholtz-Zentrum Berlin für Materialien und Energie GmbH, Hahn-Meitner-Platz 1, 14109 Berlin, Germany

R. von Klitzing
Stranski-Laboratorium für Physikalische und Theoretische Chemie, Institut für Chemie, TU Berlin, Straße des 17. Juni 124, 10623 Berlin, Germany

T. Hellweg
Physikalische und Biophysikalische Chemie, Universität Bielefeld, Universitätsstraße 25, 33615 Bielefeld, Germany

G. Sadowski and W. Richtering (eds.), *Intelligent Hydrogels*, Progress in Colloid and Polymer Science 140,
DOI 10.1007/978-3-319-01683-2_6,

frequently used to investigate the shape, size, and size distribution of microgel particles [16–22].

Recently, the responsive character of microgel particles was studied by cryo-TEM [23] and AFM using a liquid cell [24–29]. While the before mentioned techniques only allow to study the overall particle size and shape, information on the internal structure is rather difficult to gain from these methods.

Therefore, small angle neutron scattering (SANS) is generally used to study the local network morphology of macrogels [14, 16, 30, 31], microgels [15, 32–37] and more recently for chemically uncross-linked PNIPAM [38]. It is well known that SANS measurements can resolve the liquid-like fluctuations of such soft polymer networks [14, 16, 30, 31, 33]. However, the internal network structure of cross-linked microgels is still a matter of debate.

The aim of the present work is to study the volume phase transition of poly-NIPAM microgels with different connectivities on global and local length-scales by means of DLS and SANS, respectively. Here, the focus of the SANS experiments lies on the temperature evolution of the network correlation length ξ in the vicinity of the phase transition temperature. Only a rather small number of publications tries to address this problem in the context of microgels. Furthermore, in these works often large deviations from the theoretical predictions were found. However, these discrepancies were probably related to imprecisions in temperature control during these experiments and the too low number of data points available [14, 16, 39].

The correlation length ξ can be interpreted as the average mesh size between the entanglements in the internal network of a microgel under good solvent conditions (swollen state). In this state, attractive hydrogen-bonds between the solvent H_2O (D_2O, respectively) and the polymer backbone lead to a high degree of swelling and solvent contents higher than $\approx$85 % can be found. Surpassing the VPT temperature, ξ is a measure for the network fluctuations [14, 30, 40, 41] and **not directly related to the actual distance between the chemical cross-links**. It is predicted by theory that the correlation length of the network fluctuations diverges close to the critical point. This behavior follows a scaling law with a scaling exponent ν. The 3D Ising model predicts a value of $\nu = 0.63$ and in some cases this corresponds to the experimental findings [42–45]. However, in the present work we show that the scaling behavior of ξ strongly depends on the network connectivity and deviations from theory can be found.

Within this work SANS measurements were done using a custom built high precision temperature cell. This allowed to determine the temperature of the VPT with high accuracy and most important with better precision compared to previous works. On this basis, the scaling behavior of ξ could be reliably investigated. In addition analysis of the radially averaged scattering intensity at different values of momentum transfer could be used to determine the VPT temperatures and to visualize the critical behavior in the vicinity of the VPT.

Theoretical Background

Flory-Rehner Theory of the Volume Phase Transition

The swelling capacity of a gel can be expressed by calculating the ratio between the gel volume in the collapsed state and its volume in the fully swollen state. For microgel particles with their spherical shape the so-called deswelling ratio β can be calculated using the respective hydrodynamic radii determined from DLS:

$$\beta = \frac{V_T}{V_{swollen}} = \left(\frac{R_h(T)}{R_{h,swollen}}\right)^3 . \qquad (1)$$

Here, $R_h(T)$ is the hydrodynamic radius at a certain temperature and $R_{swollen}$ represents the particle radius in the fully swollen state.

A detailed theoretical description of the swelling behavior of gels provides the Flory-Rehner theory [46, 47]. In this framework it is assumed, that thermodynamic equilibrium for a gel is attained, if no net transfer of solvent takes place across the gel-solvent interface. Therefore, the macroscopic state of a homogeneous neutral gel can be described by means of the osmotic pressure in the gel. According to the classic Flory-Rehner theory [46] the osmotic pressure Π consists of two terms. The first contribution represents the mixing free energy of the polymer and the surrounding solvent; the second term describes a change in the elastic free energy due to the deformation of the polymer network upon swelling.

$$\Pi = \underbrace{-\frac{N_A k_B T}{\upsilon}\left[\phi + \ln(1-\phi) + \chi\phi^2\right]}_{mixing\ term} + \underbrace{\frac{N_c k_B T}{V_0}\left[\left(\frac{\phi}{2\phi_0}\right) - \left(\frac{\phi}{\phi_0}\right)^{1/3}\right]}_{elastic\ free\ energy\ term} \qquad (2)$$

Here, N_A is Avogadro's number, T the absolute temperature, υ the molar volume of solvent, N_c the number of chains contained in the gel network, V_0 the volume of the relaxed Gaussian gel network, ϕ the volume fraction of the polymer, ϕ_0 the volume fraction at reference conditions (fully collapsed state) and χ the polymer-solvent interaction parameter [41], also called Flory-Huggins-Stavermann interaction parameter. χ describes the change in free energy ΔF per solvent molecule (divided by $k_B T$) caused by the change from a solvent-solvent contact to a solvent-polymer contact. Using two further parameters $A = \frac{2\Delta S + k_B}{2k_B}$ and $\theta = \frac{2\Delta H}{2\Delta S + k_B}$, the polymer-solvent interaction parameter χ can be defined as

$$\chi = \frac{\Delta F}{k_B T} = \frac{\Delta H - T\Delta S}{k_B T} = \frac{1}{2} - A\left(1 - \frac{\theta}{T}\right) \quad (3)$$

ΔS and ΔH are the changes in entropy and enthalpy per monomeric unit of the network related to the volume phase transition.

For some polymer-solvent systems it was experimentally found, that χ depends on the concentration of the polymer in a nonlinear manner [48]

$$\chi(T, \phi) = \chi_1(T) + \chi_2\phi + \chi_3\phi^2 + \ldots \quad (4)$$

In addition in the case of electrostatically charged systems, the osmotic pressure Π has to be extended by an electrostatic term.

Dividing Eq. 2 by $k_B T$ and using Eqs. 3 and 4 up to the second term leads to

$$\frac{\Pi}{k_B T} = \frac{N_A}{\upsilon}\left[-\phi - \ln(1-\phi) - \left\{\frac{1}{2} - A\left(1 - \frac{\theta}{T}\right) + \chi_2\phi\right\}\phi^2 \right] + \frac{N_c}{V_0}\left[\left(\frac{\phi}{2\phi_0}\right) - \left(\frac{\phi}{\phi_0}\right)^{1/3}\right] \quad (5)$$

Assuming that the temperature dependent swelling of the microgels occurs under isobar conditions, the osmotic pressure Π within the gel becomes zero and Eq. 5 can be rewritten [49]

$$T_{\Pi=0} = \frac{A\,\phi^2\,\theta}{\frac{\phi_0}{N_{gel}}\left[\frac{1}{2}\left(\frac{\phi}{\phi_0}\right) - \left(\frac{\phi}{\phi_0}\right)^{\frac{1}{3}}\right] - \phi - \ln(1-\phi) + \left(A - \frac{1}{2}\right)\phi^2 - \chi_2\phi^3} \quad (6)$$

with N_{gel} is the average degree of polymerization of the polymer chain between cross-links, θ is the temperature at which the hydrodynamic radius approximates to its minimum and $A = \frac{2\Delta S + k_B}{2k_B}$, which roughly comprises the entropy change of the system. The polymer volume fraction is accessible from DLS as follows:

$$\frac{\phi}{\phi_0} = \frac{V_{collapsed}}{V(T)} = \left(\frac{R_{h,collapsed}}{R_h(T)}\right)^3 \quad (7)$$

Equation 5 quantitatively describes the volume phase transition of macroscopic gels [46, 50, 51]. In contrast to macroscopic gels, microgels have a different network structure due to their different synthesis procedure. The most important difference is the gradient of the cross-linker density from the center to the surface. Hence, the average number of monomers between the chemical cross-links, N_{gel}, also exhibits a gradient and grows from the inside to the microgel surface. Despite of the fact that this is not incorporated in the Flory-Rehner approach, the theory was already successfully used to at least qualitatively describe the swelling curves of microgels [39, 52].

Small Angle Scattering from Macrogels

Two contributions can be expected to influence the scattering profile of macroscopic gels (macrogels), a static one and a fluctuating one:

$$I(q) = I_{stat}(q) + I_{dyn}(q) \quad (8)$$

While, $I_{stat}(q)$ arises from the cross-linking, $I_{dyn}(q)$ accounts for dynamic liquid-like fluctuations of the gel network. Both, the cross-links and the network fluctuations cause local inhomogeneities.

q is the magnitude of the scattering vector given by:

$$|\boldsymbol{q}| = q = \frac{4\pi}{\lambda}\sin\left(\frac{\theta}{2}\right)$$

The fluctuations of the gel network can be characterized by the correlation length ξ, which is often referred to as the blob or mesh size of a swollen gel [53]. It can be computed by fitting the high q part of the SANS curve using an Ornstein-Zernicke (OZ) function, which has the form of a Lorentzian [30]:

$$I_{dyn}(q) = \frac{I_L(0)}{1 + q^2\xi^2} \quad (9)$$

According to theoretical predictions for gels, ξ diverges in the vicinity of the VPT temperature and the following scaling law is found:

$$\xi = \left|\frac{T - T_c}{T_c}\right|^{-\nu} \quad (10)$$

Here, T is the absolute temperature, T_c the temperature of the VPT and ν the critical exponent. For gels, ν is expected to have a value of 0.63 (3D Ising model [42–44]) or 0.5 (mean field calculation [40, 41]).

For the prefactor $I_L(0)$, which denotes the scattering intensity in the $\boldsymbol{q} = 0$ limit, a similar scaling behavior with a scaling exponent γ is expected:

$$I_L(0) = \left| \frac{T - T_c}{T_c} \right|^{-\gamma} \tag{11}$$

Here, the 3D Ising model would give $\gamma = 1.24$ while mean field theory predicts $\gamma = 1.0$.

A macrogel is expected to have a homogeneous distribution of cross-linker molecules over its whole volume. This is due to the conditions during the synthesis which is done under good solvent conditions. Hence, in addition to the already discussed liquid-like inhomogeneities caused by network fluctuations, static inhomogeneities due to the presence of cross-links in the network contribute to the scattering profile. This static contribution $I_{stat}(q)$ can be described by a simple Gaussian:

$$I_{stat}(q) = I_G(0) \exp\left(-\frac{q^2 \Lambda^2}{3} \right) \tag{12}$$

Λ is a measure for the length of the linear polymer segments between two cross-linker points.

Therefore, SANS curves for macrogels can be well described by a sum of a Lorentzian and a Gaussian (sum of Eqs. 9 and 12) [30].

Small Angle Scattering from Microgels

Compared to their macroscopic homologues, microgels are different since they are colloidal particles having in addition to the internal network structure also a particle form factor. Moreover, their solutions might also exhibit an interparticle structure factor.

However, in the present case only the internal structure is studied and only the Porod part of the form factor is found to contribute to the SANS data in the used q range.

Moreover, microgels are made under poor solvent conditions and due to the differences in reactivity might have a non-statistical distribution of the cross-linker inside the particles.

Microgels usually have diameters in the range from 100 nm to several μm [1]. Hence, at low q scattering from the particle interface and the surrounding medium contributes significantly to the scattering behavior. In this case the static contribution $I_{stat}(q)$ from the internal network inhomogeneities cannot be resolved. If a q range down to 10^{-4} Å^{-1} can be covered rather sophisticated models for the form factor and the internal structure can be used [37]. However, in the present study only local fluctuations inside the network were of interest.

The low q scattering can then be described by the Porod law, as suggested by Mears et al. [32]:

$$I_{stat}(q) \propto \frac{A}{V} \frac{1}{q^4} \tag{13}$$

In the past, slight deviations from the ideal exponent of –4.0 were found for microgels. This effect can be described using an approach introduced by Wong [54]. In this work deviations from ideal Porod behavior are interpreted in terms of surface roughness and the scattering data are described introducing a modified version of Porod's law:

$$I_{stat}(q) \propto \frac{a}{q^{(3+x)}} + \frac{b}{q^4} \tag{14}$$

Here, x accounts for the surface roughness of the spherical objects. For microgels typical values of x are around 1.3 [16, 55].

In sum the scattering profiles for microgels can therefore be described by:

$$I(q) \propto \frac{a}{q^{(3+x)}} + \frac{b}{q^4} + \frac{I_L(0)}{1 + \xi^2 q^2} + I_0 \tag{15}$$

Here, I_0 is the baseline parameter. This equation was used to fit all scattering curves treated within this work except of curves for the fully-collapsed state (at 45 °C). In the latter case, no reasonable contribution for the liquid-like network fluctuations could be found and the prefactor of the Lorentzian $I_L(0)$ was set a 0.

Experimental

N-isopropylacrylamide (NIPAM; Aldrich, 97 %), *N,N'*-methylenebisacrylamide (BIS; Fluka, ≥99.5 %) and potassium peroxodisulfate (KPS; Fluka, ≥99.0 %) were used as received.

Water was purified using a MilliQ system (Millipore). The final resistance was 18 MΩ.

Microgel Preparation

The preparation of the microgels is mainly based on the procedure reported first by Pelton and Chibante in 1986 [19]. Briefly, 566 mg of NIPAM and the desired amount of the cross-linker BIS were dissolved in 50 mL water in a 100 mL three-neck flask. Under continuous stirring with a magnetic

Table 1 Recipes used for the synthesis of the employed poly-NIPAM microgels

Sample	Batch size (mL)	m(NIPAM) (mg)	m(BIS) (mg)	mol-% BIS	Yield (%)
2 % BIS	50	566	15	2	72
5 % BIS	50	566	39	5	82
15 % BIS	50	566	116	15	65

stirrer, the reaction solution was heated to 70 °C and purged with nitrogen in order to remove oxygen. At 70 °C the solution was allowed to equilibrate for ≈30 min. Afterwards, the polymerization was started by the addition of 0.3 mg KPS dissolved in 0.3 mL of water. The clear and colorless solution became turbid within the first 5–10 min of reaction time. The polymerization was allowed to proceed for 4 h. Afterwards, the strongly white dispersion was cooled down to room temperature over night under continuous stirring.

In order to remove unreacted monomer as well as low molecular weight oligomers, the raw microgel dispersions were dialyzed over 14 days with daily solvent exchange (MilliQ water). The used membranes for dialysis had an average pore size of 1.0–1.5 nm (Visking, type 36/32). Finally, the cleaned microgel dispersions were freeze-dried.

The gravimetrically determined yields of the polymerizations were ranging from 65 to 82 %. A summary of the experimental details is given in Table 1.

DLS

DLS measurements were performed with a standard goniometer setup (ALV GmbH, Langen, Germany) and a toluene bath as temperature and matching bath. The temperature was controlled by a PT100 thermocouple, which was placed in the toluene bath close to the sample position. By this, a stability in temperature of ±100 mK was achieved. A HeNe laser with $\lambda = 632.8$ nm was used as light source. The constant output power was 35 mW.

Intensity time autocorrelation functions were recorded at different angles and subsequently analyzed by inverse Laplace transformation (ILT). For this the FORTRAN based program CONTIN [56, 57] was employed.

Each swelling curve was recorded beginning at the lowest temperature and the temperature has been risen stepwise with at least 20 min of equilibration time. To guarantee a complete swelling of the microgels, the samples were stored for at least 24 h at 7 °C prior to each run. For each temperature at least three intensity time autocorrelation functions were measured at one angle. This allowed to determine hydrodynamic radii with errors less than 5 %.

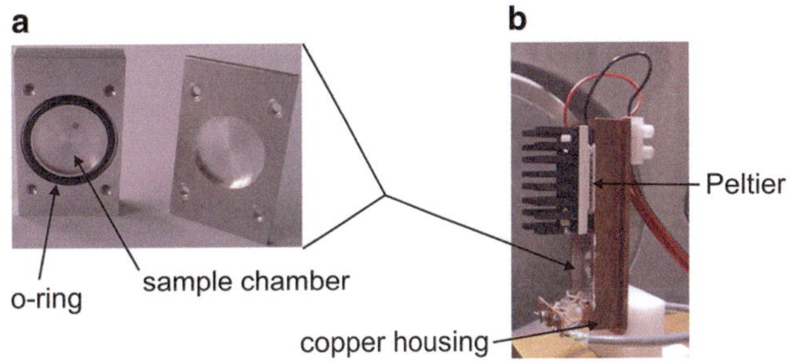

Fig. 1 Photographs of the setup used for SANS measurements. (**a**) Sample cell made of aluminum. The thickness of the sample chamber (neutron pathway) is 1 mm, while the diameter of the circular chamber is 15 mm. (**b**) Copper housing with Peltier element and heat sink. The housing is build in a way that the aluminum cells shown in (**a**) fit perfectly in the housing, in order to guarantee a good thermal contact. The neutron beam comes from the *right*, the detector is placed on the *left*

SANS

SANS experiments were performed on the V4 [58, 59] instrument of the Helmholtz Zentrum Berlin für Materialien und Energie GmbH (Berlin, Germany). The measurements were aiming at the determination of the correlation length ξ from the Ornstein-Zernicke contribution (see Eq. 9). Hence, only a rather small q range has been of interest. Therefore, all measurements were only done at sample-to-detector distances of 2.5 and 8.0 m.

In order to study the temperature evolution of ξ measurements with high precision in temperature are required. Hence, we used a custom-build sample environment consisting of aluminum cells of 1 mm pathway and a massive copper housing with a Peltier temperature element. The temperature was measured by PT 100 thermocouples placed in two positions, one close to the sample chamber, directly in the aluminum cell and the other one in the copper housing. The difference of both temperatures was ≈100 mK. The fluctuations of the temperature measured by each thermocouple were less than 30 mK. Photographs of the setup are shown in Fig. 1.

Due to the isotropic character of the scattering patterns, the collected data were circularly averaged. Correction was done accounting for electronic noise, detector efficiency and solvent/empty cell scattering. All of the initial data treatment was done using BerSANS [60, 61].

The data analysis for the determination of the correlation length ξ of the microgel networks was done using the SASfit program by J. Kohlbrecher [62]. Adjustable prefactors for the particle contrast were used for all of the fits presented within this work.

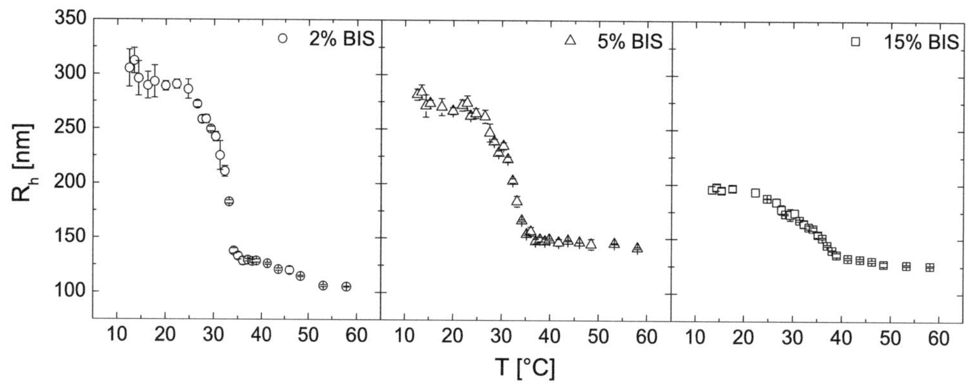

Fig. 2 Swelling curves obtained by temperature-dependent DLS measurements in D_2O. Shown are the results from a heating cycle starting at low temperatures, where the particles are fully swollen. The shown hydrodynamic radii were determined by applying the Stokes-Einstein equation using the diffusion coefficients D_T

Results and Discussion

This work aims at the investigation of the volume phase transition behavior of PNIPAM microgels with different connectivities. While DLS and AFM (images shown in the supplementary material) were used to study the overall size, polydispersity and shape of the particles, the internal structure was studied by SANS. Related to the latter experiments, special care was taken to exactly control and stabilize the temperature during the measurements.

Overall Particle Dimensions

In order to investigate the particle size and the thermoresponsive behavior of the microgels, DLS measurements were performed. DLS allows to study the translational diffusion of the colloidal microgels as a function of temperature. Applying the Stokes-Einstein equation, the hydrodynamic radii of the colloids are obtained. The temperature evolution of the R_h values of the three different microgels dispersed in D_2O is shown in Fig. 2. These curves were obtained during a heating cycle starting at the lowest temperature, where the microgels are fully swollen. These so-called swelling curves clearly reveal the volume phase transition behavior of the particles. The temperatures of the VPTs and representative hydrodynamic radii for the swollen and collapsed state are listed in Table 2.

The VPT temperatures are in the range of 31.8–33.7 °C, which is in agreement with values from the literature [63]. No systematic change in VPTT could be observed. Several cooling/heating cycles lead to a pronounced hysteresis especially during the first cycle for 2 % crosslinker. With increasing amount of cross-linker the hysteresis decreases [64].

In addition to the swelling curves, angular-dependent measurements at 20 °C were performed to proof wether the

Table 2 Results of DLS measurements from dilute dispersions in D_2O. Given are the hydrodynamic radii R_h and deswelling ratios β of microgels with different cross-linker densities (2, 5 and 15 mol-%) dispersed in D_2O. Hydrodynamic radii are given for the swollen state, at 15 °C, and the collapsed state, at 45 °C. The temperature of the volume phase transition (VPPT) for each microgel was determined from the respective swelling curves of the microgels

Sample	R_h at 15 °C (nm)	R_h at 45 °C (nm)	β	VPTT (°C)
2 % BIS	294	120	0.07	32.3
5 % BIS	272	147	0.16	31.8
15 % BIS	196	130	0.29	33.7

DLS resolves purely translational motions of the microgels. The left diagram in Fig. 3 shows the obtained average relaxation rates $\langle \Gamma \rangle$ as a function of q^2 according to the relationship $\langle \Gamma \rangle = D_T q^2$. The solids lines are linear fits providing the translational diffusion coefficient D_T. The results show clearly that only purely diffusive motion is resolved.

In order to describe the possible uptake of solvent, we determined the deswelling ratios β according to Eq. 1 using the radius at T = 15 °C as $R_{swollen}$. The temperature evolution of the deswelling ratios for the three investigated microgels are compared in Fig. 3. The difference in the deswelling behavior becomes clearly visible at higher temperatures (above 32 °C). The discontinuous character of the volume phase transition increases with decreasing connectivity while at the same time the possible uptake of solvent increases (lower values of β). The dependence of β on the connectivity is shown in Fig. 4. Here, also data for a PNIPAM microgel without BIS as cross-linker are shown. The latter particles are so-called self-cross-linked [65, 66]. As result from these experiments a linear dependence of the deswelling ratio on the cross-linker density is found. However, at higher cross-linker densities than the ones used within this study we expect the deswelling ratios to reach a plateau.

In Fig. 5 the swelling ratio α is converted into volume fraction and plotted with the corresponding temperature.

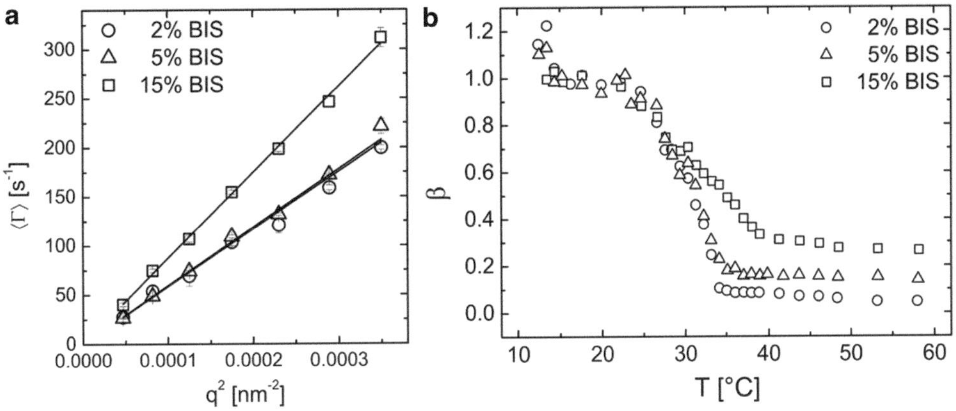

Fig. 3 (**a**) Results from angular-dependent DLS measurements at 20 °C. Plotted is the average relaxation rate $\langle \Gamma \rangle$ as a function of the *square* of the scattering vector q. The *solid lines* are linear fits. (**b**) Deswelling ratios β as a function of temperature for the poly-NIPAM microgels with different cross-linker densities dispersed in D_2O using the radii at T = 15 °C as R_0

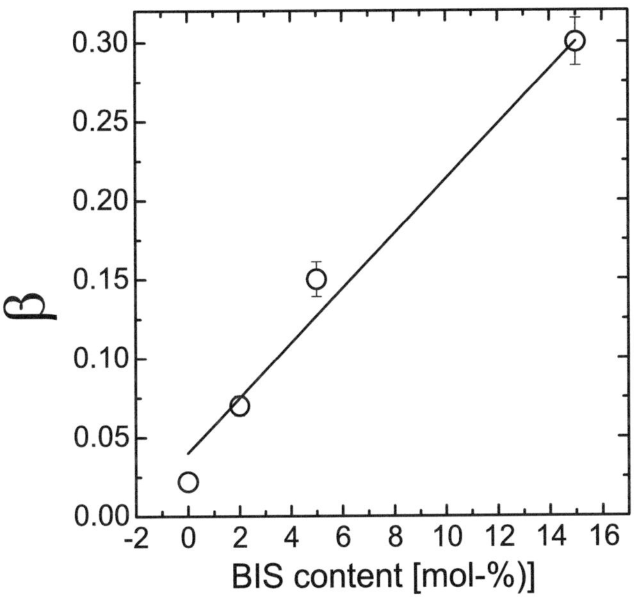

Fig. 4 Deswelling ratios β as a function of cross-linker density (BIS content in mol-%) using the radii in the swollen state R_0 at T = 15 °C and the collapsed state at T = 45 °C

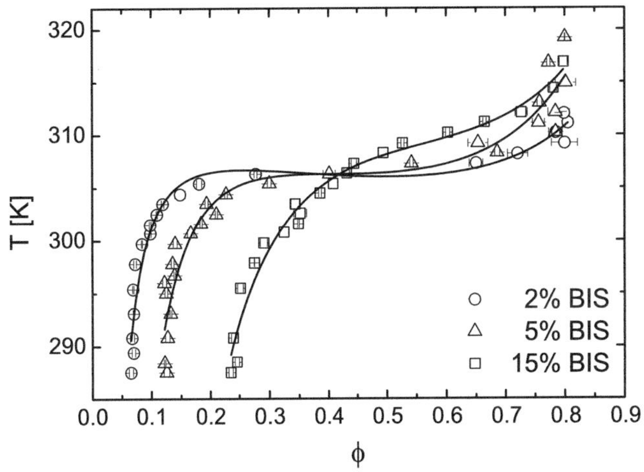

Fig. 5 Fits according to the Flory-Rehner model for the swelling of a gel network (Eq. 6)

The solid lines in the figure are fits according to Eq. 6. Using a ϕ_0 of 0.8 all experimental data can be fitted and the otained values for parameters like N_{gel} at least reveal the expected change if the amount of BIS in the particles is increased.

The average number of monomers between 2 cross-links, N_{gel}, goes down from 189 for microgels with 2 % BIS to 13 for microgels with 15 % BIS. However, since the Flory-Rehner approach does not account for the cross-linker gradient in the microgels, the absolute values are only a kind of average. The obtained results are in good agreement with previous works by Crassous et al. [52] and Fernández-Barbero et al. [39]. The results from the fits are given in Table 3.

The investigations by AFM and DLS show that the overall dimensions, the particle morphology and the volume phase transition behavior depend rather strongly on the connectivity.

Internal Particle Morphology

An important parameter for the characterization of a gel network is the so-called network correlation length ξ. This parameter can be obtained from SANS data of microgel particles. As already discussed in section "Small Angle Scattering from Macrogels", ξ is the lengthscale of the network fluctuations [14, 30, 38, 40, 41].

Already the temperature evolution of the scattering intensity at one fixed value of q can provide the phase transition behavior (see Fig. 6).

Table 3 Parameters as obtained by fitting the data in Fig. 5 using Eq. 6. ϕ_0 was kept at 0.8 for samples. N_{gel} is found to decrease with increasing cross-linker density, whereas the temperature θ increases

	2 % BIS	5 % BIS	15 % BIS
A	−13.5	−7.7	−9.6
χ_2	0.976	0.94	1.19
N_{gel}	196	90	13
Θ (K)	311	315	324

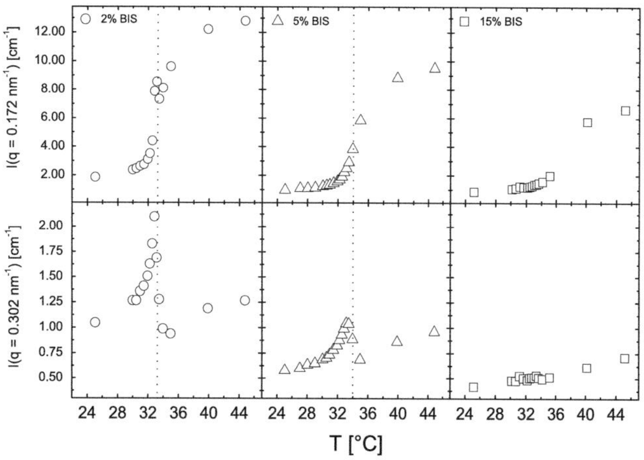

Fig. 6 Absolute scattering intensities at two different values of momentum transfer. Shown are the temperature evolutions for the three investigated microgel systems. The *dotted lines* indicate the position of the volume phase transition temperature (VPTT)

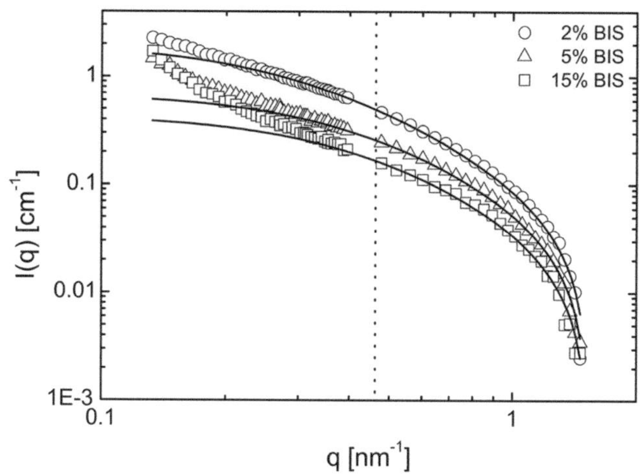

Fig. 7 SANS profiles for microgels with different cross-linker densities at 25 °C (swollen state). The profiles were corrected for the incoherent background. The *solid lines* are fits using just the OZ-part of Eq. 15. The *dotted line* indicates the minimum value of q where the Ornstein-Zernicke function satisfyingly (see Eq. 9) describes the scattering curve

At $q = 0.172\,\text{nm}^{-1}$, the Porod contribution dominates the scattering behavior and hence just a continuous increase of $I(q)$ is observed for all cross-linker contents. For the lowest amount of cross-linker this increase is slightly more pronounced than in the case of 5 and 15 % cross-linker density. Surpassing the VPT, the microgels decrease in size and hence the scattering from the particle/solvent interface changes. In addition, the scattering length density of the overall microgel particles changes, since water is expelled from the interior of the microgels and hence the polymer volume fraction of a particle increases. These two effects are attributed to cause the increase of the scattering intensity at low q when the microgels undergo the VPT.

The values of $I(q)$ at $q = 0.302\,\text{nm}^{-1}$, which is in a q-range where the OZ part contributes significantly to the scattering signal, show a rather pronounced divergence of $I(q)$ close to the volume phase transition temperature. This divergence of the scattering intensity can be directly related to the critical behavior of $I_L(0)$, which will be discussed in the next section. In comparison only the highly cross-linked microgel shows a different behavior. In the latter case, the intensity increase is less pronounced with increasing temperature and no divergence is found. The results of this rather simple comparison suggest, that the OZ contribution and the VPT behavior become less pronounced for increasing connectivity of the microgels. This is in agreement with the shown results from DLS, which reveal that the possible degree of solvent uptake decreases for increasing cross-linker density. In other words, a higher connectivity leads to a more solid-like network, with smaller mesh-sizes and hence a more static structure. The temperatures of the VPT T_c used in the analysis of the critical behavior for the microgel with 2 and 5 % BIS discussed in the next section were determined from the $I(q)$ evolution at $q = 0.302\,\text{nm}^{-1}$. Since the values of $I(q)$ diverge strongly in a very narrow temperature range, T_c could be determined with a much higher precision than from DLS data. Furthermore, due to the unique setup used for the SANS measurements, the precision in T is in the range of 30 mK.

The difference of the network morphologies is presented in Fig. 7, where SANS profiles in the swollen state (at 25 °C) are compared. The solid lines in the two diagrams are fits accounting just for the OZ contribution, and hence for the dynamic fluctuations of the network, according to Eq. 9.

As the diagram shows, the Lorentzian fits nicely describe the high q-part of the spectra. In contrast, at lower values of q the used model does not lead to a satisfying description of the data. For $q \leq 0.25\,\text{nm}^{-1}$ scattering from the interface between the particles and the surrounding medium contributes significantly to the scattering profiles. Therefore, at low q just the OZ contribution cannot describe the scattering profiles and additionally a Porod law has to be taken into account (see Eq. 15). The results of the obtained fits to the Ornstein-Zernicke function plotted in Fig. 7 show a significant difference with respect to $I_L(0)$, which decreases with increasing network connectivity. This trend can also clearly be seen from the fit parameters given in the table in the supplementary information.

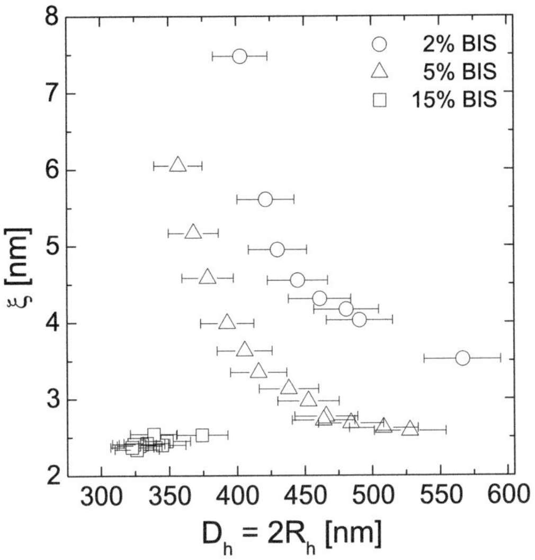

Fig. 8 Correlation length ξ as a function of the hydrodynamic diameter D_h for the different microgels. The errors for ξ are less than the symbol size

Figure 8 compares the obtained values for ξ with the hydrodynamic diameter D_h from DLS measurements. The respective diameters, at the temperatures were SANS experiments have been performed and hence values for ξ could be determined, were calculated by linear interpolation within the swelling curves shown in Fig. 2. For this only the two diameters in the vicinity of the temperatures for the desired ξ value have been used. This approach leads to a higher uncertainity in D_h as for the results shown in Fig. 2 and hence errors of 5 % for D_h were assumed, shown by the error bars in Fig. 8. The results from this analysis clearly show a decrease of ξ with increasing particle size for the microgels with 2 and 5 % BIS. For 15 % again no significant change of ξ was found.

Saunders investigated the structure of PNIPAM microgels with 0.7 and 7.3 mol-% off cross-linker (BIS) by SANS [67]. He suggested a core-shell structure of the microgel particles with a low density shell with a thickness of ≈20 nm. Furthermore, he concluded that the dynamic mesh size (correlation length ξ) of the microgels is independent of the cross-linker content. This is in contrast to our findings. An explanation for this might be the different temperature ranges used in the two works. While Saunders performed SANS measurements at one temperature below the VPT temperature and at a temperature in the very close vicinity of the VPT, in the present work only ξ values for temperatures below the VPT were determined.

Scaling Behavior of the Correlation Length

The temperature evolution of the network correlation length ξ and the prefactor of the OZ contribution $I_L(0)$ are shown in Fig. 9.

The presented values were determined from fits of the SANS profiles of the three microgels according to Eq. 15. Additionally, the fit parameters of the OZ contribution, accounting just for the liquid-like fluctuations of the network, are listed in the supplementary material.

For the two lowly cross-linked microgels (2 and 5 mol-% BIS), an increase of $I_L(0)$ and ξ with increasing temperature reaching the VPT temperature is observed. Interestingly, for the most cross-linked microgel (15 mol-% BIS nominal cross-linker feed), only slight changes and no significant increase of $I_L(0)$ and ξ are found. Hence, it can be assumed, that the volume phase transition of this particular microgel cannot be treated as a critical one.

The scaling behavior of $I_L(0)$ and ξ is shown in Fig. 10 in terms of a scaling plot according to Eqs. 11 and 10.
The temperatures of the VPT T_c were determined from the $I(q)$ evolutions at $q = 0.302\,\mathrm{nm}^{-1}$ shown in Fig. 6. In case of the highly cross-linked microgel, scaling plots do not at all follow the theoretical predictions and hence the data were not included in the figure. At present the only explanation is that the "distance" to the critical point increases with the BIS content and therefore, the respective boundary conditions for a mean field or 3D Ising description are no longer fulfilled. However, it is well-known that the liquid-like character of gels decreases with increasing connectivity. Hence, it is expected that the amplitude of the internal network fluctuations goes down.
In contrast, for the lowly cross-linked microgels the expected scaling-behavior is found. The scaling exponents in case of the system with 5 % connectivity ($\gamma = 0.82 \pm 0.03$ and $\nu = 0.51 \pm 0.02$) are in satisfying agreement with theoretical predictions from mean field theory, which would give $\nu = 0.5$ and $\gamma = 1.0$ [40, 41]. Calculations on basis of the 3D Ising model would give $\gamma = 1.24$ and $\nu = 0.63$ for responsive gels [42–45]. Shibayama et al. determined values close to these predicted ones for PNIPAM macrogels [30]. Recently, Meier-Koll et al. investigated the scaling behavior of chemically uncross-linked PNIPAM from semidilute dispersions using SANS [38]. The authors determined critical exponents of $\nu = 0.44$ and $\gamma = 0.81$, which are also close to values predicted by mean field theory.
For 2 % connectivity $\gamma = 0.50 \pm 0.03$ and $\nu = 0.29 \pm 0.01$ can be determined, which is significantly smaller than the theoretical values. This difference might be explained by a much higher degree of linear polymer segments and a less homogeneous distribution of the cross-linker points.

Structure in the Collapsed State

In order to investigate the network structure in the fully collapsed state, SANS measurements were also performed at 45 °C. This temperature is in a regime, where no more significant changes in particle dimensions and minimum

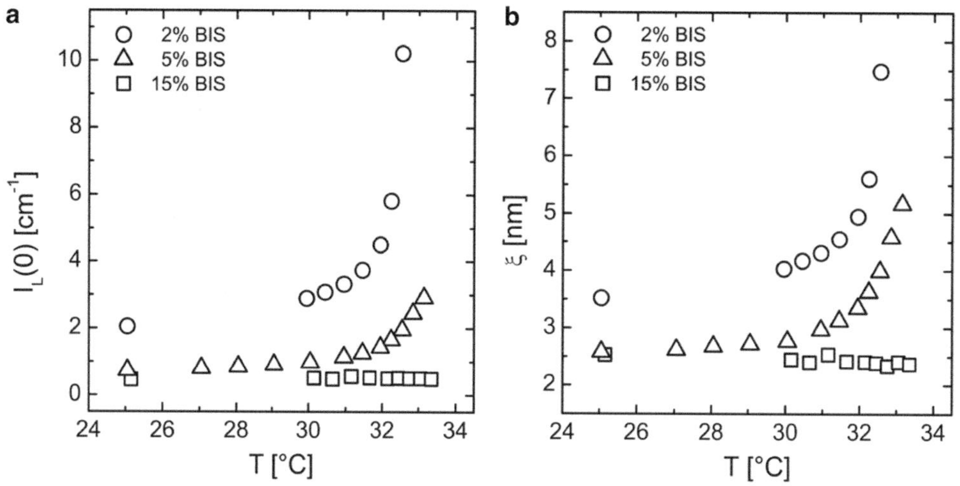

Fig. 9 Ornstein-Zernicke parameters determined by respective fits of the SANS profiles for the different microgels and different temperatures according to Eq. 15. (**a**) $I_L(0)$ as a function of temperature. (**b**) Correlation length ξ as a function of temperature

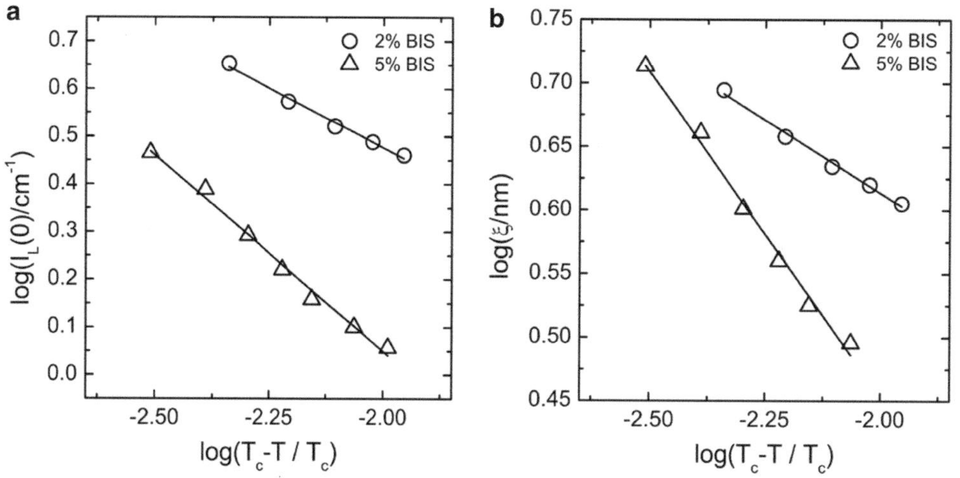

Fig. 10 Scaling behavior of microgels with 2 and 5 mol% BIS as cross-linker. (**a**) Scaling plot of $I_L(0)$. The *solid lines* are linear fits providing the scaling exponents γ. (**b**) Scaling plot of ξ. The *solid lines* are linear fits providing the scaling exponents ν

hydrodynamic radii are observed (see e.g. swelling curves in Fig. 2). The SANS profiles for the three microgels measured under these bad solvent conditions are shown in Fig. 11. In case of the collapsed particles, the model used for the fit includes only the Porod law modified according to Wong [54] and the incoherent background scattering. The values for x, accounting for the surface roughness of the particles are $x = 1.25$ (2 % BIS), $x = 1.23$ (5 % BIS) and $x = 1.16$ (15 % BIS). Values, which can be found in the literature for microgels are in the order of $x = 1.3$ [16,55].

Conclusions

In this work, we studied pure PNIPAM microgels with three different cross-linker densities. Results from AFM and DLS, focusing on the overall particle dimensions as well as results from SANS, focusing on the internal structure show a strong dependence on the cross-linker density. The swelling curves as obtained from DLS experiments were converted into microgel volume and the obtained curves were described using the "classical" Flory-Rehner theory for network swelling. The theory was found to qualitatively describe also the swelling of microgels despite of the different inhomogeneities inside these particles compared to macrogels. The degree of polymerization between the cross-links is systematically going down with increasing cross-linker content and hence, the outcome of the Flory-Rehner analysis seems to yield meaningful trends also in the case of microgels.

The reported SANS measurements were performed with high precision in temperature. Hence, the temperature of the VPT of the microgel system dispersed in D_2O could be determined with high accuracy and in good agreement with results from DLS. A large number of SANS curves was measured in the vicinity of the VPT temperature. The obtained profiles were fitted by Eq. 15 and the correlation length ξ was determined from the OZ part. The expected

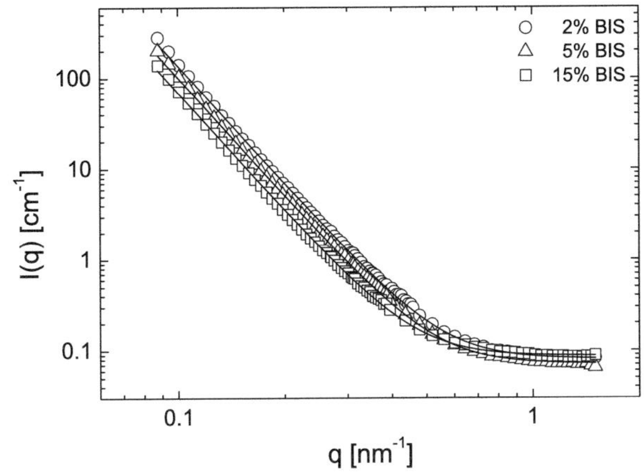

Fig. 11 SANS profiles for microgels with different cross-linker densities at 45 °C (collapsed state). The *solid lines* are fits using Eq. 14 accounting only for the. Porod scattering (modified version as disccused in the article) and an incoherent background. Note: The profiles were not corrected for the incoherent background

scaling law behavior of $I_L(0)$ and ξ is found for the two weakly cross-linked microgels (2 and 5 mol% BIS). However, only the scaling exponents for the PNIPAM microgel with 5 % BIS are in agreement with predictions by mean field theory. The values found for the 2 % BIS system are significantly lower than exponents predicted by both, the mean field and 3D Ising approach.

For the microgel with the highest cross-linker density (15 mol%) no significant change of $I_L(0)$ and ξ with temperature was found. This might be related to the distance to the critical point in the phase diagram and is in line with previous publications [14, 16]. It is often observed that the VPT of microgels broadens strongly with increasing amount of cross-linker.
Furthermore, SANS measurements at 45 °C, where the particles are expected to be in the fully collapsed state, revealed a rather high surface roughness of the microgels.

The present work clearly shows the limits of the present theoretical descriptions available. Due to the completely different synthesis and the thus caused gradient of the cross-linker inside the microgels only a qualitative description can be achieved at present.

Acknowledgements This work has been supported by the Deutsche Forschungsgemeinschaft within the framework of the priority program SPP 1259.

References

1. Pelton R (2000) Temperature-sensitive aqueous microgels. Adv Colloid Interface Sci 85:1–33
2. Wu C, Zhou S (1997) Volume phase transition of swollen gels: discontinuous or continuous. Macromolecules 30:574–576
3. Saunders BR, Vincent B (1999) Microgel particles as model colloids: theory, properties and applications. Adv Colloid Interface Sci 80:1–25
4. Shibayama M, Tanaka T (1993) Volume phase transition and related phenomena of polymer gels. In: Dusek (ed) Responsive gels: volume transitions I. Volume 109 of advances in polymer science, 1 edn. Springer, Berlin, pp 1–62
5. Andersson M, Maunu SL (2006) Volume phase transition and structure of poly(N-isopropylacrylamide) microgels studied with ^{1}H-NMR spectroscopy in d$_2$o. Colloid Polym Sci 285:293–303
6. Dingenouts N, Nordhausen Ch, Ballauff M (1998) The volume transition in thermosensitive micronetworks as observed by small-angle x-ray scattering. Ber Bunsenges Phys Chem 102(11):1594–1596
7. Nayak S, Lyon LA (2005) Soft nanotechnology with soft nanoparticles. Angew Chem Int Ed 44:7686–7708. Review
8. Hoare T, Pelton R (2007) Functionalized microgel swelling: comparing theory and experiment. J Phys Chem B 111:11895–11906
9. Das M, Zhang H, Kumacheva E (2006) Microgels: old materials with new applications. Annu Rev Mater Res 36:117–142
10. Wu C, Zhou S (1997) Light scattering study of spherical poly(N-isopropylacrylamide) microgels. J Macromol Sci B36(3):345–355
11. Kim J-H, Ballauff M (1999) The volume transition in thermosensitive core-shell latex particles containing charged groups. Colloid Polym Sci 277:1210–1214
12. Varga I, Gilanyi T, Meszaros R, Filipcsei G, Zrinyi M (2001) Effect of cross-link density on the internal structure of poly(N-isopropylacrylamide) microgels. J Phys Chem B 105:9071–9076
13. Hoare T, Pelton R (2004) Highly pH and temperature responsive microgels functionalized with vinylacetic acid. Macromolecules 37:2544–2550
14. Kratz K, Lapp A, Eimer W, Hellweg T (2002) Volume phase transition and structure of TREGDMA, EGDMA, and BIS cross-linked PNIPA microgels: a small angle neutron and dynamic light scattering study. Colloids Surf A 197(1–3):55–67
15. Hellweg Th (2003) Properties of NIPAM-based intelligent microgel particles: investigated using scattering methods. In: Liz-Marzán LM, Kamat PV (eds) Nanoscale materials, 1st edn. Kluwer, Dordrecht, pp 209–225
16. Kratz K, Hellweg Th, Eimer W (2001) Structural changes in pnipam microgel particles as seen by SANS, DLS, and EM techniques. Polymer 42(15):6531–6539
17. Snowden MJ, Chowdhry BZ, Vincent B, Morris GE (1996) Colloidal copolymer microgels of N-isopropylacrylamide and acrylic acid: pH, ionic strength and temperature effects. J Chem Soc Faraday Trans 92(24):5013–5016
18. Karg M, Pastoriza-Santos I, Rodriguez-González B, von Klitzing R, Wellert S, Hellweg T (2008) Temperature, pH, and ionic strength induced changes of the swelling behavior of PNIPAM-poly(allylacetic acid) copolymer microgels. Langmuir 24:6300–6306
19. Pelton RH, Chibante P (1986) Preparation of aqueous latices with N-isopropylacrylamide. Colloids Surf 20:247–256
20. Schmidt S, Motschmann H, Hellweg T, von Klitzing R (2008) Thermoresponsive surfaces by spin-coating of PNIPAM-co-PAA microgels. A combined AFM and ellipsometry study. Polymer 49:749–756
21. Schmidt S, Hellweg T, von Klitzing R (2008) Packing density control in P(NIPAM-co-AAc) microgel monolayers: effect of surface charge, pH, and preparation technique. Langmuir 24:12595–12602
22. Nerapusri V, Keddie JL, Vincent B, Bushnak IA (2006) Swelling and deswelling of adsorbed microgel monolayers triggered by changes in temperature, pH, and electrolyte concentartion. Langmuir 22:5036–5041
23. Crassous JJ, Ballauff M, Drechsler M, Schmidt J, Talmon Y (2006) Imaging the volume transition in thermosensitive core-shell particles by cryo-transmission electron microscopy. Langmuir 22(6):2403–2406

24. Wiedemair J, Serpe MJ, Kim J, Masson JF, Lyon LA, Mizaikoff B, Kranz C (2007) In-situ AFM studies of the phase-transition behavior of single thermoresponsive hydrogel particles. Langmuir 23:130–137
25. Höfl S, Zitzler L, Hellweg T, Herminghaus S, Mugele F (2007) Volume phase transition of smart microgels in bulk solution and adsrobed at an interface: a combined AFM, dynamic light, and small angle neutron scattering study. Polymer 48:245–254
26. FitzGerald PA, Dupin D, Armes SP, Wanless EJ (2007) In situ observations of adsorbed microgel particles. Soft Matter 3:580–586. No copy
27. Tagit O, Tomczak N, Vancso GJ (2008) Probing the morphology and nanoscale mechanics of single poly(N-isopropylacrylamide) microgels across the lower-critical-solution temperature by atomic force microscopy. Small 4:119–126
28. Burmistrova A, von Klitzing R (2010) Control of number density and swelling/shrinking behavior of P(NIPAM–AAc) particles at solid surfaces. Mater Chem 20:3502–3507
29. Burmistrova A, Richter M, Eisele M, Üzüm C, von Klitzing R (2011) The effect of co-monomer content on the swelling/shrinking and mechanical behaviour of individually adsorbed PNIPAM microgel particles. Polymers 3:1575–1590
30. Shibayama M, Tanaka T, Han CC (1992) Small angle neutron scattering study on poly(N-isopropylacrylamide) gels near their volume-phase transition. J Chem Phys 97(9):6829–6841
31. Shibayama M, Tanaka T, Han CC (1992) Small-angle neutron scattering study on weakly charged temperature sensitive polymer gels. J Chem Phys 97(9):6842–6854
32. Mears SJ, Deng Y, Cosgrove T, Pelton R (1997) Structure of sodium dodecyl sulfate bound to a poly(NIPAM) microgel particle. Langmuir 13:1901
33. Crowther HM, Saunders BR, Mears SJ, Cosgrove T, Vincent B, King SM, Yu G-E (1999) Poly(NIPAM) microgel particle de-swelling: a light scattering and small-angle neutron scattering study. Colloids Surfaces A Physicochem Eng Asp 152: 327–333
34. Berndt I, Richtering W (2003) Doubly temperature sensitive core-shell microgels. Macromolecules 36:8780–8785
35. Berndt I, Pedersen JS, Lindner P, Richtering W (2006) Inlfuence of the shell thickness and cross-link density on the structure of temperature-semsitive poly-N-isopropylacrylamide-poly-N-isopropylmethacrylamide core-shell microgels investigated by small-angle neutron scattering. Langmuir 22:459–468
36. Dingenouts N, Seelenmeyer S, Deike I, Rosenfeldt S, Ballauff M, Lindner P, Narayanan T (2001) Analysis of thermosensitive core-shell colloids by small-angle neutron scattering including contrast variation. Phys Chem Chem Phys 3:1169–1174
37. Stieger M, Richtering W, Pedersen JS, Lindner P (2004) Small-angle neutron scattering study of structural changes in temperature sensitive microgel colloid. J Chem Phys 120(13):6197–6206
38. Meier-Koll A, Pipich V, Busch P, Papadakis CM, Müller-Buschbaum P (2012) Phase separation in semidilute aqueous poly(N-isopropylacrylamide) solutions. Langmuir 8: 8791–8798
39. Fernandez-Barbero A, Fernandez-Nieves A, Grillo I, Lopez-Cabarcos E (2002) Structural modifications in the swelling of inhomogeneous microgels by light and neutron scattering. Phys Rev E 66(5):051803/1–10
40. Dusek K (1993) Responsive gels: volume transitions I. Volume 109 of advances in polymer science, 1 edn. Springer, Berlin
41. Dusek K (1993) Responsive gels: volume transitions II. Volume 110 of advances in polymer science, 1 edn. Springer, Berlin
42. Bagnuls C, Bervillier C (1985) Nonasymptotic critical-behavior from field theory at d = 3 the disordered-phase case. Phys Rev B 32(11):7209–7231
43. Bagnuls C, Bervillier C (1985) A new theoretical constraint on the static critical-behavior of the specific heat. Phys Lett A 107(7):299–304
44. Bagnuls C, Bervillier C (1987) Nonuniversal power laws and crossover from critical to classical behavior. Phys Rev Lett 58(5):435–438
45. Li Y, Tanaka T (1989) Study of the universality class of the gel network system. J Chem Phys 90(9):5161–5166
46. Flory PJ, Rehner J (1943) Statistical mechanics of cross-linked polymer networks. i. Rubberlike elasticity. J Chem Phys 11(11):512–520
47. Flory PJ (1953) Principles or polymer chemistry. Cornell University Press, Ithaca/London
48. Eichinger BE, Flory PJ (1968) Thermodynamics of polymer solutions. J Chem Soc Faraday Trans 64:2035–2052
49. Hirotsu S, Hirokawa Y, Tanaka T (1987) Volume-phase transitions of ionized N-iospropylacrylamide gels. J Chem Phys 87(2):1392–1395
50. Flory PJ, Rehner J (1943) Statistical mechanics of cross-linked polymer networks. ii. Swelling. J Chem Phys 11(11):521–526
51. Flory PJ, Erman B (1982) Theory of elasticity of polymer networks. 3. Macromolecules 15(3):800–806
52. Crassous JJ, Wittemann A, Siebenbürger M, Schrinner M, Drechsler M, Ballauff M (2008) Direct imaging of temperature-sensitive core-shell latexes by cryogenic transmission electron microscopy. Colloid Polym Sci 286:805–812
53. de Gennes P-G (1979) Scaling concepts in polymer physics. Cornell University Press, Ithaca/London
54. Wong P (1985) Scattering by inhomogeneous systems with rough internal surfaces: porous solids and random-field Ising systems. Phys Rev B 32(11):7417–7424
55. Karg M, Wellert S, Pastoriza-Santos I, Lapp A, Liz-Marzán LM, Hellweg T (2008) Poly(N-isopropylacrylamide) microgels with silica nanoparticle core: the volume phase transition/collapse of the polymer shell as seen by small angle neutron scattering and dynamic light scattering. Phys Chem Chem Phys 10: 6708–6716
56. Provencher SW (1982) A constrained regularization method for inverting data represented by linear algebraic or integral equations. Comput Phys Commun 27:213–217
57. Provencher SW (1982) CONTIN: a general purpose constrained regularization program for inverting noisy linear algebraic and integral equations. Comput Phys Commun 27:229–242
58. Keiderling U, Wiedenmann A (1995) New SANS instrument at the BER II reactor in Berlin, Germany. Physica B 213 and 214:895–897
59. Keller T, Krist T, Danzig A, Keiderling U, Mezei F, Wiedenmann A (2000) The polarized neutron small-angle scattering instrument at BENSC Berlin. Nucl Instrum Methods Phys Res Sect A 451:474–479
60. Keiderling U (1997) A new software package for SANS data processing at the Hahn-Meitner-Institut in Berlin, Germany. Physica B 234:1111–1113
61. Keiderling U (2002) The new BerSANS-PC software for reduction and treatment of small angle neutron scattering data. Appl Phys A-Mater Sci Process 74:1455–1457
62. Kohlbrecher J (2008) SASfit: a program for fitting simple structural models to small angle scattering data. Paul Scherrer Institut, Laboratory for Neutron Scattering, CH-5232, Villigen
63. Heskins M, Guillet JE (1968) Solution properties of poly(N-isopropylacrylamide). J Macromol Sci Chem A2 2:1441–1455
64. Burmistrova A, Richter M, Uzum C, von Klitzing R (2011) Effect of cross-linker density of P(NIPAM-co-AAc) microgels at solid surfaces on the swelling/shrinking behaviour and the youngs modulus. Colloid Polym Sci 289:613–624

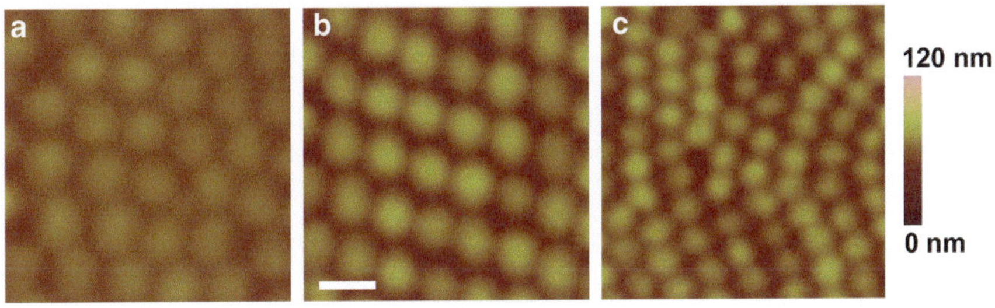

Fig. 12 AFM height profiles of poly-NIPAM microgels with 2 mol-% BIS (**a**), 5 mol-% BIS (**b**) and 15 mol-% BIS (**c**). These images were recorded against air at 25 °C in tapping mode. Sample preparation was done by spin-coating. The scale bar is 400 nm

65. Kratz K, Hellweg Th, Eimer W (1998) Effect of connectivity and charge density on the swelling and local structural properties of colloidal PNIPA microgels. Ber Bunsenges Phys Chem 102:1603–1608
66. Gao J, Frisken BJ (2003) Cross-linker-free N-isopropylacrylamide gel nanospheres. Langmuir 19:5212–5216
67. Saunders BR (2004) On the structure of poly(N-isopropylacrylamide) microgel particles. Langmuir 20:3925–3932

Supporting Information

AFM images were taken on a multimode AFM equipped with a Nanoscope III controller. For all samples, height profiles were recorded in tapping mode (against air) using tips with spring constants of 42 N/m from Olympus (OMCL-AC160TS).

Samples were prepared by spin-coating (300 s at 2,000 RPM) on $1 \times 1\ cm^2$ pieces silicon wafers (bare). The wafers were first cleaned from organic impurities by etching with H_2O_2:$H_2SO_4 = 1$:1 and then treated with a plasma cleaner. The used microgel dispersions had a concentration of 0.5 % by weight. The recorded profiles were smoothed (first order).

Fitting Parameters from SANS Analysis

Table 4 Results of SANS investigations of microgels with different cross-linker densities (2, 5 and 15 mol-%). Given are the values of ξ and $I_L(0)$ as a function of temperature as obtained from the OZ contribution of the SANS spectra

	2 % BIS		5 % BIS		15 % BIS	
T (°C)	ξ (nm)	$I_L(0)$ (cm^{-1})	ξ (nm)	$I_L(0)$ (cm^{-1})	ξ (nm)	$I_L(0)$ (cm^{-1})
25.05	3.52	2.05	2.58	0.73	–	–
25.15	–	–	–	–	2.53	0.46
27.05	–	–	2.62	0.80	–	–
28.05	–	–	2.68	0.85	–	–
29.05	–	–	2.72	0.91	–	–
29.95	4.03	2.89	–	–	–	–
30.05	–	–	2.77	0.98	–	–
30.15	–	–	–	–	2.45	0.51
30.45	4.17	3.08	–	–	–	–
30.65	–	–	–	–	2.40	0.48
30.95	4.31	3.32	2.97	1.14	–	–
31.15	–	–	–	–	2.54	0.56
31.45	4.55	3.74	3.13	1.26	–	–
31.65	–	–	–	–	2.42	0.52
31.95	4.95	4.50	3.35	1.44	–	–
32.15	–	–	–	–	2.41	0.50
32.25	5.61	5.82	3.63	1.66	–	–
32.45	–	–	–	–	2.39	0.51
32.55	7.49	10.23	3.99	1.96	–	–
32.75	–	–	–	–	2.34	0.51
32.85	–	–	4.58	2.45	–	–
33.05	–	–	–	–	2.41	0.50
33.15	–	–	5.17	2.92	–	–
33.35	–	–	–	–	2.37	0.49
33.45	–	–	5.35	2.82	–	–

Tailored Macromolecules Versus Nanoparticles as Additives for Mechanical Reinforcement of NCO-sP(EO-*stat*-PO) Hydrogels

Konstantina Harrass, Haika Hildebrandt, Martin Moeller, and Juergen Groll

Abstract

Hydrogels are three dimensional networks of hydrophilic polymers that exhibit high equilibrium water contents (up to 99 w/w-%) in aqueous environment. Beside the manifold advantages such as intrinsic biocompatibility, low friction and enormous swelling behaviour, this high water content also results in an intrinsically low mechanical stability. Different reinforcement strategies have been developed that rely on the strengthening by a second component. In this study we present the improvement of six arm star-shaped poly(ethylene oxide-stat-propylene oxide) (sPEOPO) primary networks by two different strategies: incorporation of amino-functionalized silica nanoparticles (nanocomposites, NC) and generation of interpenetrating networks (IPNs) by addition of tailored linear triblock copolymers. The compression moduli and mechanical stress-at-break of both hydrogel systems are compared and discussed. In comparison to sPEOPO, the NC hydrogels demonstrate a strengthening effect with a 2.5 times higher compression stress-at-break whereas the IPN increases the modulus up to a factor of 1.5.

Keywords

Hydrogel • Interpenetrating Polymer Network (IPN) • Nanocomposite • Mechanical stress-at-break • Compression strength

Introduction

Hydrogels are fascinating materials that combine high water content with low surface friction and that may be tailored in a wide range through alteration of the chemical nature of the constituting network. Their high water content usually makes them intrinsically cytocompatible and thus interesting for biomedical applications [1–9]. One general disadvantage of hydrogels that significantly affects their practical relevance is their usually low mechanical stability. Hence, a variety of approaches have been explored for improving the mechanical stress-at-break of hydrogels. This is usually achieved by heterogeneous systems, for example Interpenetrating Polymer Networks (IPNs) [10–21]. IPNs consist of two or more polymer networks which are partially entangled on a molecular scale but not covalently bonded to each other. This usually leads to additive combination of the characteristics of each single network to the overall properties of the IPN. This way, for example, poly(ether urethane)/poly(methyl methacrylate) (PEU/PMMA) IPN hydrogels exhibiting significant compression stress-at-break of 3.5 MPa have been developed and are presently evaluated as artificial cartilage substitutes [22]. Further, biocompatible counterparts of 3,6-anhydro-α-L-galacto-β-D-galactan (agarose, type VII) and poly(ethylene glycol) diacrylate (PEG-DA) were recently integrated to design a IPN hydrogel material as implant in cartilage regeneration that comprises a compression strength of approximately 0.8 MPa [23].

K. Harrass • H. Hildebrandt • M. Moeller
DWI e.V. and Institute of Technical and Macromolecular Chemistry, RWTH Aachen University, Forckenbeckstr. 50, 52074 Aachen, Germany

J. Groll (✉)
Department of Functional Materials in Medicine and Dentistry, University of Wuerzburg, Pleicherwall 2, 97070, Wuerzburg, Germany
e-mail: juergen.groll@fmz.uni-wuerzburg.de

G. Sadowski and W. Richtering (eds.), *Intelligent Hydrogels*, Progress in Colloid and Polymer Science 140,
DOI 10.1007/978-3-319-01683-2_7,

Another way to achieve hydrogels with improved mechanical characteristics is reinforcement with usually nanometer sized fillers. The properties of such nanocomposites (NC's) depend on the characteristics of constituent phases and are strongly influenced by the filler/matrix interphase as well as the interaction between homogeneous and disperse phase [24, 25]. Hence, polymer composites may exhibit significantly enhanced polymer properties at very low volume fractions of nano-fillers. Motivated through this, a variety of NC hydrogels that possess improved mechanical properties have been developed. For example, NC hydrogels consisting poly(N-isopropylacrylamide) (PNIPAAm) as a matrix and Laponite XLG (Clay-G)/Laponite XLS (Clay-S) as a dispersed phase may reach a compressive strength of up to 20 MPa [26]. Also various types of clay particles have been used as a cross linker agent during a free-radical polymerization of N-isopropyl-acrylamide (NIPAM) [27–30]. Depending on the concentration of clays, the compressive strength of the resulting hydrogels changed from 0.7 to 5 MPa. Recently, also hydrophilic reactive microgels (HRM) have been presented in the literature as soft fillers for hydrogels [31]. These HRM hydrogels did not break even at 90 % strain and sustain high stresses between 1.88 and 4.6 MPa.

We have reported before that six armed star shaped isocyanate (NCO) functional hydrophilic poly(ethylene oxide-stat-propylene oxide) prepolymers ((NCO-sP(EO-*stat*-PO); for clarity abbreviated as **sPEOPO**) may serve as basis for the formation of hydrogels in water [32]. Network formation occurs by addition reaction of amines, which are formed in situ hydrolysis of the NCO groups, with isocyanates. These gels are rubbery and flexible as xerogels, but brittle in the swollen state. Accordingly, the purpose of the present study was to develop reinforcing agents capable to improve the mechanical performance of **sPEOPO** hydrogels via tailored functionality and reactivity. A macromolecular approach based on linear triblock copolymers of poly(*tert*-butyl glycidyl ether)-block-poly(ethylene oxide)-block-poly(*tert*-butyl glycidyl ether) (P(*t*-BuGE$_X$-b-EO$_Y$-b-*t*-BuGE$_X$); P1) and poly(glycidol)-block-poly(ethylene oxide)-block-poly(glycidol) (P(G$_X$-b-EO$_Y$-b-G$_X$); P2), relied on additional cross-linking due to hydrophobic interaction (P1) or covalent bond formation through reaction with the NCO groups (P2). In both cases, the middle PEO-block was intended to serve as molecular spring for additional energy dissipation during mechanical load. The polymers were synthesised with anionic polymerization, characterized by Nuclear Magnetic Resonance (NMR) spectroscopy and Size Exclusion Chromatography (SEC), and they used as additive for hydrogel formation. In addition, a nanocomposite strategy comprising amino-functionalized silica nanoparticles as reinforcement agent for **sPEOPO** hydrogels followed. With the amino-groups on the nanoparticle surface, the particles were intended to act as additional multifunctional covalent cross-linking point for the hydrogels. Nanoparticles were analysed using transmission electron microscopy (TEM), dynamic light scattering (DLS) and zeta potential. Hydrogels resulting from both strategies were prepared, characterized by Fourier transform infrared spectroscopy (FTIR spectroscopy), Cryo-field emission scanning electron microscopy (Cryo FE-SEM) and regarding equilibrium water content (EWC) and lastly compared regarding the improvement of the mechanical strength by quasi-static compression measurements.

Experimental Section

Material

Six armed, NCO-terminated star-shaped polyether's with a backbone of statistically copolymerized ethylene oxide (EO, 80 %) and propylene oxide (PO, 20 %) in the ratio 4:1 having a molecular weight of 12 kDa – (**sPEOPO**) were prepared as described elsewhere [33]. *tert*-Butyl glycidyl ether (*t*-BuGE, 99 %) and toluene (99.8 %; both from Sigma-Aldrich, Germany) were dried over calcium hydride (CaH_2) and sodium (Na), respectively, distilled and stored subsequently under a nitrogen atmosphere. Tetraethyl orthosilicate (TEOS, 98 %; VWR, Germany) was distilled prior usage. Poly(ethylene oxide) (PEO, $M_n = 35{,}000$, $PDI = M_w/M_n = 1.08$; Polymer Source, USA), cesium hydroxide monohydrate (CsOH, 99.5 %) and trifluoroacetic acid (TFA, 99.5 %; both from Acros Organics, USA), ammonium hydroxide (NH_4OH, 25 % of NH_3 in aqueous solution; VWR, Germany), (3-aminopropyl)triethoxysilane (APTES, 99 %), ethanol (C_2H_5OH, 99.5 %) and buffer solution HPCE (20 mM sodium phosphate, $pH = 8.5$; all procured by Sigma-Aldrich, Germany) were used as received.

Synthesis

Poly(tert-butyl glycidyl ether)-block-poly(ethylene oxide)-block-poly(tert-butyl glycidyl ether)

PEO (3 g) was dissolved in distilled water (12 ml) and CsOH monohydrate (2-fold molar excess with regard to the hydroxyl groups of PEO) was added. After the reaction mixture was stirred at ambient temperature for 15 h, dry toluene (60 ml) was inserted to the system. Water was eliminated by 4 days continuous azeotropic distillation under nitrogen atmosphere. Fresh dry toluene (40 ml) was delivered each day to the mixture in order to replace the amount removed the previous day. In consequence, the obtained solution was transported under nitrogen atmosphere into a glove box and the excess of CsOH monohydrate was separated by filtration. The activated caesium dialcoholate was isolated through vacuum distillation of toluene, and subsequently the residual was dried for 3 h at 40 °C. Subsequently, the calculated amount of protected glycidol monomer, *t*-BuGE (0.33 g) was added to the macroinitiator and the polymerization was completed for 5 days at 80 °C. The obtained poly(*tert*-butyl glycidyl ether)-block-poly(ethylene oxide)-block-poly(*tert*-butyl glycidyl ether) (P(*t*-BuGE$_X$-b-EO$_Y$-b-*t*-BuGE$_X$)) was dissolved in a mixture of ethanol/distilled water = 1/7 and the solution was placed into a cellulose dialysis membrane (Spectra/Por® – MWCO 500, Carl ROTH). After 2 days of purification against distilled water and lyophilisation in continuance of 1 day, the structure of the polymer was determined.

Poly(glycidol)-block-poly(ethylene oxide)-block-poly(glycidol)

The tertiary butyl ether groups in protected triblock copolymers were cleaved by TFA according to the procedure described priory [34]. P(*t*-BuGE$_X$-b-EO$_Y$-b-*t*-BuGE$_X$) (2 g) was dissolved in TFA (0.5 ml) and consequently stirred at ambient temperature for 3 h. The excess of TFA was isolated by vacuum distillation and subsequently, the obtained poly(glycidyl trifluoroacetate)-block-poly(ethylene oxide)-block-poly(glycidyl trifluoroacetate) was mixed with 20 ml distilled water at ambient temperature for 24 h. In the further step, the generated TFA was subtracted under vacuum and consequently the deprotected copolymer of poly(glycidol)-block-poly(ethylene oxide)-block-poly(glycidol) (P(G$_X$-b-EO$_Y$-b-G$_X$)) was dried for 3 h at 40 °C.

Synthesis of Amino-Functionalized Silica Nanoparticles

Monodisperse silica nanoparticles with diameters of 130 nm were prepared by hydrolysis and condensation reactions of TEOS in presence of ethanol as a solvent and ammonia as a basic catalyst [35]. Firstly, ethanol (220 ml) was mixed with NH_4OH (12 ml) and stirred for 5 min. Throughout a sequence step, TEOS (7.7 ml) was added to the above composition, and the reaction took place during 24 h at ambient temperature. The obtained colloidal solution was separated by the aid of a high speed centrifuge. Silica nanoparticles were cleaned using ethanol for 3 times to eliminate non-desirable constituents and thereafter, they were dried continuously for 2 h at 100 °C. The average particle diameter and PDI were evaluated by dynamic light scattering (DLS) measurements. The results got are d = 119 nm and PDI = 0.024.

For functionalization of silica nanoparticles, APTES (20 μl) was mixed with a previously prepared suspension (10 ml) and stirred for 24 h. Subsequently, the obtained nanoparticles were centrifuged for 10 min at 1,400 rpm whereas the supernatant was removed, and solution of hydrochloric acid with pH 4 (10 ml) was added. Lastly, the nanoparticles possessing amino groups on the surface were re-dispersed by sonication procedure in a sonication bath for 10 min.

Hydrogel Preparation

Synthesis of Interpenetrating Polymer Network (IPN) Hydrogels

For preparation of Interpenetrating Polymer Network (IPN) hydrogels, aqueous triblock copolymers of P(*t*-BuGE$_X$-b-EO$_Y$-b-*t*-BuGE$_X$) and P(G$_X$-b-EO$_Y$-b-G$_X$) in concentration of 0.5 and 1.0 w/w-% were prepared in degased water. The **sPEOPO** (12 kDa) samples were dissolved in the desired amount of each copolymer solution. Hence, the absolute concentration of the obtained gels was 10, 15 and 20 w/w-%, respectively. Then, the solutions were injected between a parallel pair of glass plates separated by a rubber gasket (thickness – 3 mm), and subsequently placed in an autoclave vessel for 24 h.

Synthesis of Nanocomposite (NC) Hydrogels

Dispersions of amino-functionalized silica nanoparticles possessing concentrations of 4.12, 5.0, 7.5 and 10.0 mg/ml, respectively, were prepared in buffer solutions of hydrochloric acid (HCl) with pH 4. Proper quantities of the colloidal dispersion and **sPEOPO** prepolymer were mixed together. After stirring for a short time, sodium phosphate buffer solutions were added, so that the absolute pH value of 8.5 was reached for each of the compositions. This procedure had to be performed in attempt to prevent a protonation of amino groups and to ensure a cross-linking by polymer precursor. The concentrations of nanoparticles inside nanocomposite (NC) hydrogels were 1.7, 2.0, 3.0 and 4.0 w/w-%. The use of relatively high concentrations of nanoparticles is a consequence of better dispersions and the purpose will be explicated later in the chapter for mechanical properties.

Methods for Characterization

Nuclear Magnetic Resonance (NMR) Spectroscopy

The composition of P(*t*-BuGE$_X$-b-EO$_Y$-b-*t*-BuGE$_X$) and P(G$_X$-b-EO$_Y$-b-G$_X$) copolymers was determined by proton NMR (^{1}H NMR) spectroscopy. The spectra were recorded on a Bruker DPX-300 spectrometer operating at 300 MHz. Deuterated chloroform ($CDCl_3$) and deuterated dimethyl sulfoxide (DMSO, CD_3SOCD_3) served as a solvent in case of protected and deprotected copolymers, respectively, whereas tetramethylsilane (TMS) worked as internal standard.

Size Exclusion Chromatography (SEC)

SEC analysis of PEO, P(*t*-BuGE$_X$-b-EO$_Y$-b-*t*-BuGE$_X$) and P(G$_X$-b-EO$_Y$-b-G$_X$) was performed at 30 °C by applying an isocratic pump (Agilent 1100) and a refractive index detector (Wyatt Optilab DSP). The eluate was water (HPLC gradient grade, Roth) with 0.1 M sodium nitrate ($NaNO_3$, p.a. AppliChem) as salt and methanol (p.a. Gheyer) as internal standard. The flow rate was 1 ml/min. Four PSS suprema columns filled with modified acrylate-copolymer-network particles were applied. The length of the first column was 50 mm whereas of the other three separation columns were 300 mm. The diameter of each column was 8 mm, the diameter of the copolymer particles 10 μm, and the nominal pore width of the separation columns was 30, 1,000, and 3,000 Å, correspondingly. The calibration was implemented by narrow distributed poly(ethylene glycol)/poly(ethylene oxide) or polysaccharide polymer pullulan as standards.

Fourier Transform Infrared Spectroscopy (FTIR Spectroscopy)

FTIR spectra of dry *t*-BuGE, P(*t*-BuGE$_X$-b-EO$_Y$-b-*t*-BuGE$_X$) and P(G$_X$-b-EO$_Y$-b-G$_X$) were recorded on a Nicolet Nexus 470 FTIR spectrometer (Thermo Nicolet, Madison, USA). *t*-BuGE, P(*t*-BuGE$_X$-b-EO$_Y$-b-*t*-BuGE$_X$) and P(G$_X$-b-EO$_Y$-b-G$_X$), 0.2–1 mg of each substances was mixed with 100 times excess of potassium bromide (KBr). The obtained compositions were homogenised, pressed into tablets at a pressure of 50–75 MPa and subsequently exposed to IR measurement, applying the transmission technique. The spectra represent an average of 1,000 scans.

Moreover, the FTIR spectra of **sPEOPO** before gel formation as well as dry samples of IPN and NC hydrogels were recorded on the same spectrometer, however using the photo-acoustics spectroscopy method (PAS). The PAS-FTIR measurements were performed in atmosphere of helium and similarly, the spectra represent the average of 1,000 scans.

Equilibrium Water Content (EWC) Measurements

Hydrogel samples of IPN and NC with defined dimensions were weighed after drying at ambient temperature under vacuum for 24 h. Then, they were immersed into distilled water (200 ml) for 7 days in order to reach the equilibrium swelling state. After 7 days swelling, the weight of the samples practically did not change further. The EWC was determined by applying the following relationship:

$$\mathrm{EWC}\,[\%] = \left(\frac{w_{sw} - w_d}{w_{sw}}\right) \cdot 100 \tag{1}$$

where w_{ws} and w_d are the weight of the sample in the swollen and in the dry state, respectively.

Cryo-Field Emission Scanning Electron Microscopy (Cryo-FESEM)

The morphology of amino-functionalized silica nanoparticles as well as IPN and NC hydrogels was studied by Cryo-FESEM (Hitachi, S-4800). At first, a small droplet of dispersion of amino-functionalized silica nanoparticles, respectively, a thin equilibrium swollen hydrogel film was shock-frozen to −196 °C either in liquid nitrogen (IPN hydrogels) or methane cooling down in presence of liquid nitrogen (nanoparticle dispersions and NC hydrogels), and subsequently placed inside of the sample holder. In every case, the top section of the frozen droplet, respectively hydrogel was cut, and the SEM images were taken with a secondary electron (SE) detector.

Mechanical Properties

Static compression test measurements were performed on equilibrium swollen **sPEOPO** SN, IPN and NC hydrogels using dynamic mechanical analysis equipment (DMA Q800, TA Instruments). Cylindrical-shaped specimens were used with diameter to thickness ratio according ISO 604 (Plastics – Determination of compressive properties) [36]:

$$\left(\frac{d}{t}\right)^2 \cdot 0.4 > \epsilon^* \tag{2}$$

where d, t and ε^* are diameter, thickness of the specimen and maximum strain, respectively. With "strain" is meant the actual compressive strain which is defined by

$$\epsilon = \frac{t_0 - t}{t_0} \tag{3}$$

where t_0 and t describes the initial and current thickness of the sample. Therefore, the value of ε can range theoretically between 0 and 1.

Compression test measurements were conducted using a submersion compression clamp which allows retaining the swollen condition during the measurement. The crosshead speed was 0.5 N/min. Three samples of each kind of hydrogel were tested at 37 °C in water environment. The average value for compression modulus (E_c), stress-at-break (σ_B) and strain-at-break (ε_B) were calculated by the aid of DMA software. For stress calculation initial diameters of the samples are used irrespective of the strong flattening of the samples during loading. E_c is determined using following equation:

$$\frac{(\sigma_2 - \sigma_1)}{(\epsilon_2 - \epsilon_1)} \tag{4}$$

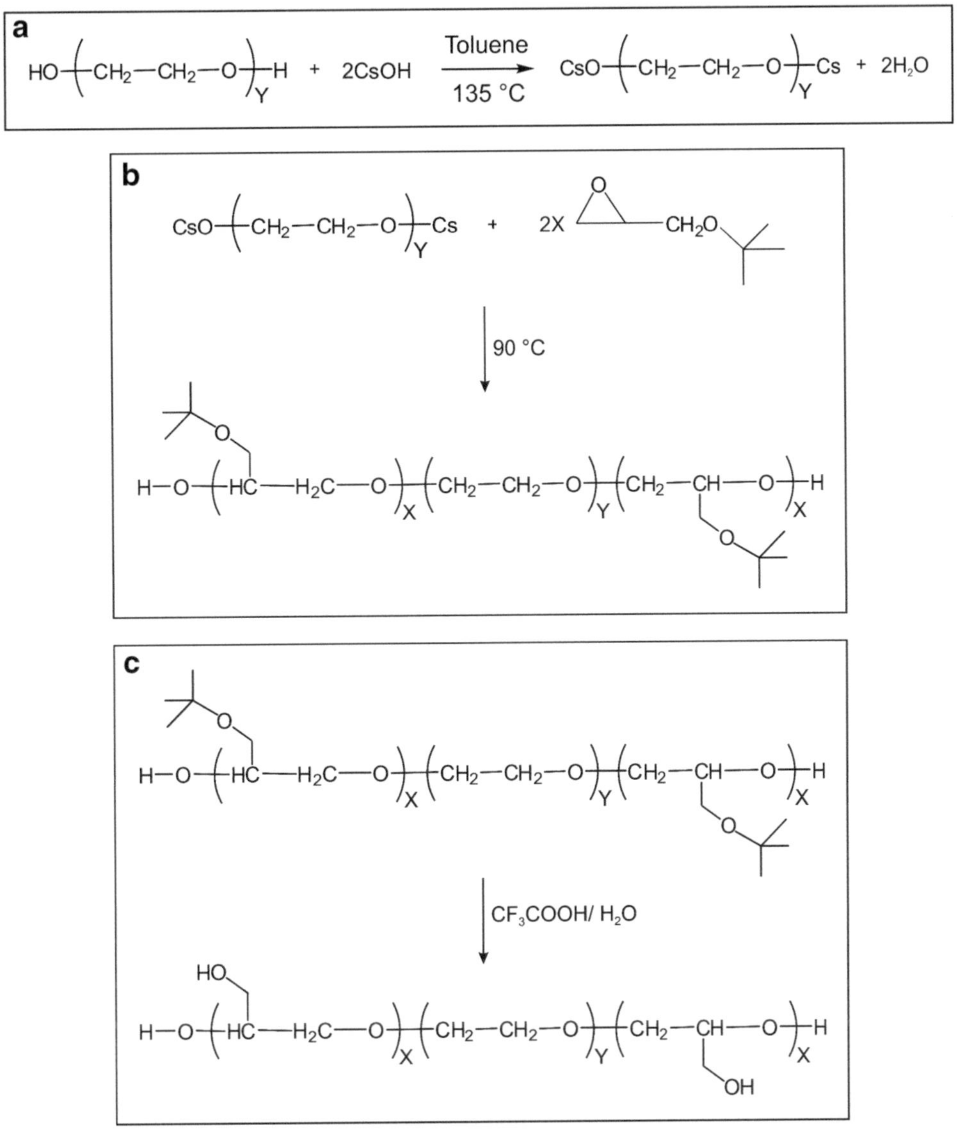

Fig. 1 Synthesis of linear triblock copolymers – protected and deprotected: (**a**) Activation reaction of PEO macroinitiator by CsOH. (**b**) Anionic ring-opening polymerization of PEO-dialcoholate by tert-butyl protected glycidol. (**c**) Cleavage of the tert-butyl ether groups

where $\varepsilon_2 = 0.0025$ and $\varepsilon_1 = 0.0005$ are the limits of strain, as well as σ_1, σ_2 are their corresponding stress values (according to ISO 604 to ensure the elastic behaviour).

Results and Discussion

Linear High Molecular Weight Triblock Copolymers

Our synthetic strategy to create non-branched poly(ethylene oxide-block-glycidol) copolymers was based on the use of PEO macromonomers and tert-butyl ether protected glycidol monomer. The preparation route of non-branch copolymers comprises several single steps: (a) activation the hydroxyl group of difunctional macroinitiator PEO by azeotropic distillation under presence of CsOH (Fig. 1a), (b) subsequent anionic ring-opening polymerization of PEO-dialcoholate by protected glycidol in absence of solvent (Fig. 1b) and (c) lastly, deprotection of the tert-butyl ether groups (Fig. 1c). ^{1}H NMR analysis of P(*t*-BuGE$_X$-b-EO$_Y$-b-*t*-BuGE$_X$) and P(G$_X$-b-EO$_Y$-b-G$_X$) reveals the structures expected. In case of P(*t*-BuGE$_X$-b-EO$_Y$-b-*t*-BuGE$_X$), the signal at $\delta = 1.17$ ppm (18H, C(CH$_3$)$_3$ – 2; denoted as d on Fig. 2a is instigated by the tert-butyl ether groups. However, the peak intensity at $\delta = 3.2$–4 ppm (14H, CH$_2$CH$_2$O and CH$_2$CHCH$_2$ refer to a, b and c on Fig. 2a designates the protons of the PEO and *t*-BuGE back bonds. Moreover, the absence of a signal at $\delta = 1.17$ ppm and presence of a characteristic peak at $\delta = 5.6$–6.4 ppm originated by the hydroxyl groups is a proof for complete deprotection. The polymer structure of protected and deprotected triblock copolymers are

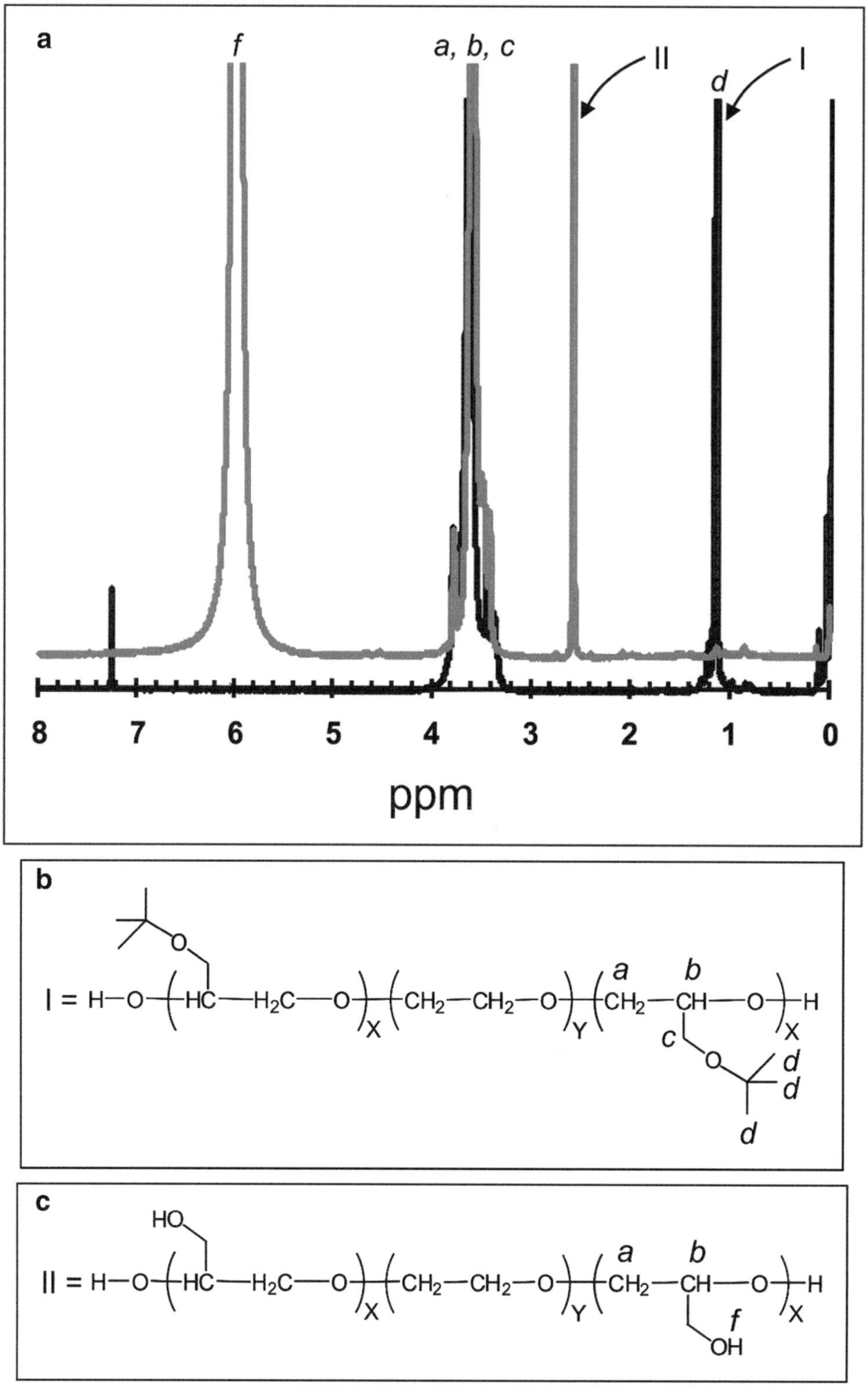

Fig. 2 Determination the structure of linear triblock copolymers: (**a**) ^{1}H NMR of protected triblock copolymer in chloroform-d and deprotected triblock copolymer in dimethyl sulfoxide-d6. (**b**) and (**c**) structural formulas of protected and deprotected triblock copolymers with respectively signals

Table 1 Molecular weights and polydispersity index of P(*t*-$BuGE_X$-b-EO_Y-b-*t*-$BuGE_X$) and P(G_X-b-EO_Y-b-G_X)

	NMR			SEC
Polymer	DP[a]	M_n[b]	M_n[c]	PDI = M_w/M_n
PEO	–	–	–	1.19
P(*t*-$BuGE_X$-b-EO_Y-b-*t*-$BuGE_X$)	24	38,127	38,906	1.2
P(G_X-b-EO_Y-b-G_X)	24	36,396	36,745	1.22

[a]Degree of polymerization (DP) of *t*-BuGE repeat group established by ^{1}H NMR spectroscopy
[b]M_n of copolymer determined by ^{1}HNMR spectroscopy
[c]Theoretical predictable M_n of copolymer

presented in Fig. 2b, c. The amount of *t*-BuGE repeating groups and average number molecular weight (M_n) of the copolymers were additionally evaluated by the peak areas at $\delta = 3.2$–4 ppm with the integral at $\delta = 1.17$ ppm set to 1 (Table 1). The M_n quantity of copolymers determined this way by ^{1}H NMR spectroscopy is in agreement with the theoretically predictable values (Table 1). In addition, FTIR spectroscopy of *t*-BuGE, P(*t*-$BuGE_{12}$-b-EO_{796}-b-*t*-$BuGE_{12}$) and P(G_{12}-b-EO_{796}-b-G_{12}) were performed to prove the structure of the copolymers (Fig. 3a). The presence of characteristic peaks between 1,375–1,355 cm^{-1} and 1,395–1,380 cm^{-1} in the protected triblock copolymer shows evidence that the polymerization occurred. Further, the absence of bands in the area in the P(G_{12}-b-EO_{796}-b-G_{12}) demonstrates successful deprotection. The SEC analysis of PEO, P(*t*-$BuGE_{12}$-b-EO_{796}-b-*t*-$BuGE_{12}$) and P(G_{12}-b-EO_{796}-b-G_{12}) shows unimodal distributions (Fig. 3b) and PDI of 1.19, 1.2 and 1.22 (Table 1).

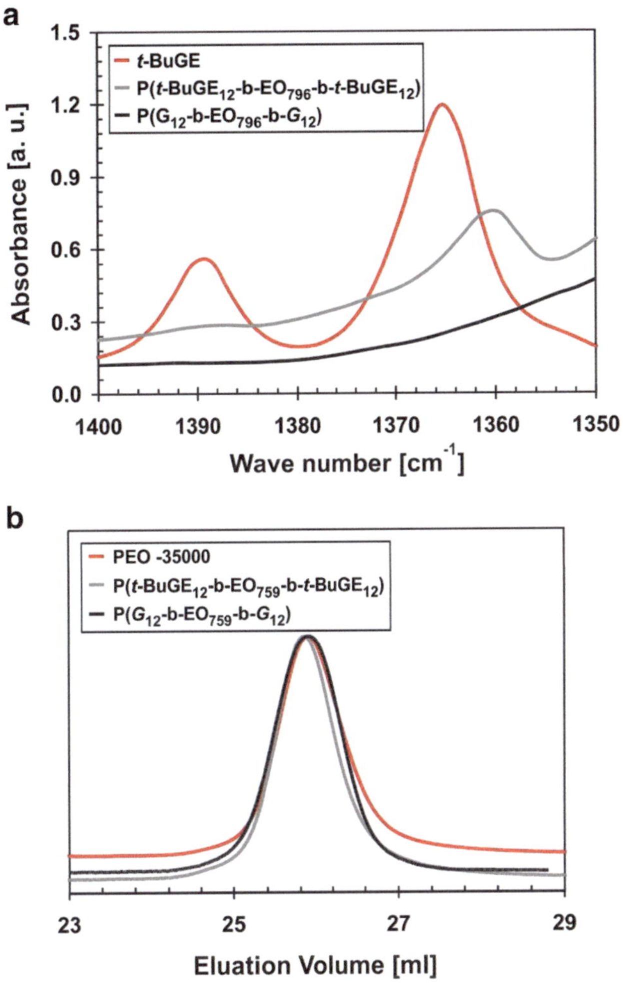

Fig. 3 Analysis of monomer, macroinitiator and linear triblock copolymers – protected and deprotected: (**a**) FTIR spectroscopy and (**b**) SEC use water as eluate solvent

Hydrogel Preparation and Chemical Composition

After characterization, the synthesized high molecular weight linear triblock copolymers were used as additive to improve the mechanical properties of **sPEOPO** SN hydrogels, either by pure hydrophobic interaction (block-copolymer P1) or through covalent integration into the **sPEOPO** network by reaction between the alcohol groups of the deprotected glycidols in block-copolymer P2 with aliphatic NCO groups of the reactive **sPEOPO** prepolymers. Gels were prepared by adding aqueous solutions of copolymers P(*t*-$BuGE_{12}$-b-EO_{796}-b-*t*-$BuGE_{12}$)(Y) and P(G_{12}-b-EO_{796}-b-G_{12})(Y), possessing various concentrations of **sPEOPO** prepolymer. Homogeneous viscous mixtures were originally created and subsequently injected between two glass plates, kept at equal distance by a rubber gasket. Complete gelation occurred in continuation of 24 h under high pressure in autoclave and thus, smooth and transparent IPN hydrogels were obtained.

In case of NC hydrogels with aminofunctional silica nanoparticles, the gelation process occurs via interactions of NCO groups of prepolymer and amino groups of silica nanoparticles. As a consequence, urea bridges result due to connecting of star-shaped polymer precursor and reinforcement agent. Following both strategies, the cross-linking reaction between sPEOPO through hydrolysis of NCO to amines and subsequent reaction of these amines with remaining isocyanates occurs in parallel to the mechanisms described above.

Because of an easier understanding when going more in detail to explain the characteristics of IPN and NC hydrogels, hereafter, they are designated as follows: **$sPEOPO_X$/P(*t*-$BuGE_{12}$-b-EO_{796}-b-*t*-$BuGE_{12}$)(Y)** as **$sPEOPO_X$/TC-P(Y)**, **$sPEOPO_X$/P(G_{12}-b-EO_{796}-b-G_{12})(Y)** as **$sPEOPO_X$/TC-D(Y)** and **$sPEOPO_X$/Nanoparticles(Y)** as **$sPEOPO_X$/NP (Y)**, respectively. The symbols X and Y represent the

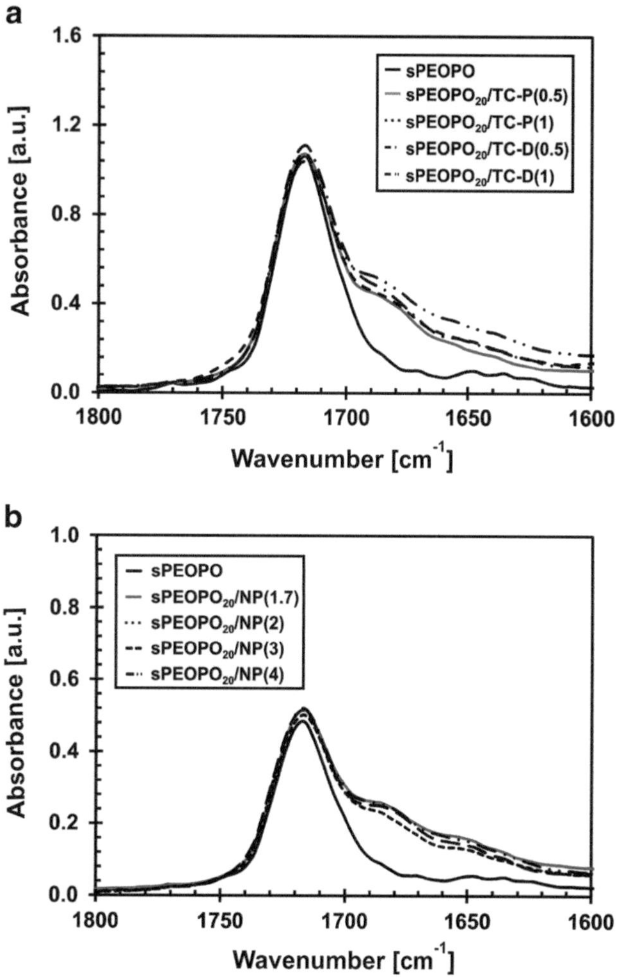

Fig. 4 FTIR spectra of non-hydrolysed **sPEOPO** compared with each IPN (**a**) and NC (**b**) xerogels (the subscripts show various concentrations of triblock copolymers – protected and deprotected and amino functionalized silica nanoparticles, respectively)

w/w-% of **sPEOPO** prepolymer as well as copolymers and nanoparticles, respectively.

FTIR spectroscopy of unreacted **sPEOPO** prepolymer, IPN and NC hydrogels after extraction of the water was performed to assess the chemical composition of the gels (Fig. 4). The spectra of the IPN gels with various end concentrations of **sPEOPO** (10, 15 and 20 w/w-%) are nearly equal, therefore, only the data for **sPEOPO$_{20}$/TC-P(Y)** and **sPEOPO$_{20}$/TC-D(Y)** IPN are presented. The peak at 2,264 cm^{-1} originated from the NCO groups in the non-hydrolysed **sPEOPO** prepolymer. However, the absence of a signal in the spectra of IPN and NC gels (data not shown) is a proof of a chemical cross-linking and hydrolysis of the isocyanate.

Figure 4a, b illustrate the FTIR spectra of **sPEOPO** prepolymer with dry IPN and NC gels in the region of wavenumbers 1,800–1,600 cm^{-1}. The peak at 1,716 cm^{-1} is the consequence of the reactions between IPDI and OH-terminated, star-shaped prepolymer at the time of **sPEOPO** synthesis as well as also between IPDI- and OH-groups of the triblock copolymers. In comparison to **sPEOPO**, the normalized spectra of IPN gels display a slight rise in the relative intensity of the urethane band by an increase the concentration of the copolymers that can be assigned to a realisation of a chemical reaction between star prepolymer and OH groups of the synthesised triblock copolymers.

The FTIR spectra of IPN and NC xerogels demonstrate a presence of a shoulder at 1,680 cm^{-1} as a consequence of biocompatible urea bridges. In comparison to IPN gels where the shoulder can be designated to the inter-prepolymer cross-linking of **sPEOPO** in case of NC, the formation of urea bonds is due to reaction between the **sPEOPO** and amino groups of silica nanoparticles. Moreover, except for **sPEOPO$_{20}$/NP(2)** NC gel, an enlarge the concentrations of reinforcement causes a gently intensify of the bands due to a greater possibility of cross-linking.

The EWC of all kinds of IPN and NC hydrogels prepared in the study was determined in a series of three independent measurements per gel. Each IPN and NC hydrogels display an analogous high EWC of more than 90 %. Changes in the hydrogel constitutions led to non-significant effects on EWC that are thus not discussed in more detail.

Morphology of IPN and NC Hydrogels

The internal morphology of IPN and NC hydrogels was investigated by Cryo-FESEM technique. Figure 5 displays the images of IPN hydrogels originates from **sPEOPO$_{20}$** and various concentrations of triblock copolymers – protected and deprotected. Each four types of IPN hydrogels exhibit macroporous structure and a great diversity of pore sizes (Table 2). In case of **sPEOPO$_{20}$/TC-P(Y)** IPN hydrogels (Fig. 5 a, b), an increase the concentration of protected triblock copolymer causes a creation of pores that possess slight diameters (300–700 nm) whereas the amount of the largest one (>1,100 nm) stay persistent. Contrary, IPN hydrogels of **sPEOPO$_{20}$/TC-D(Y)** (Fig. 5c, d) demonstrate a decrease the quantity of little pores (300–700 nm) by a rise the concentration of deprotected triblock copolymer. However, the percentage of the big ones (>1,100 nm) is kept a constant. Moreover, the presence of P(G_{12}-b-EO_{796}-b-G_{12}) leads to a double increase the amount of pores that possess diameter >1,100 nm, compared to the IPN hydrogels originates from P(*t*-BuGE_{12}-b-EO_{796}-b-*t*-BuGE_{12}).

Depending on the type of polymer, the average pore size of the system decreases or increases by the rise the concentration of copolymer, for example from 738 to 690 nm for (P(*t*-BuGE_{12}-b-EO_{796}-b-*t*-BuGE_{12})) and 770–816 nm for (P(G_{12}-b-EO_{796}-b-G_{12})), respectively. Thus, in case of **sPEOPO$_{20}$/TC-P(Y)** IPN hydrogels a higher

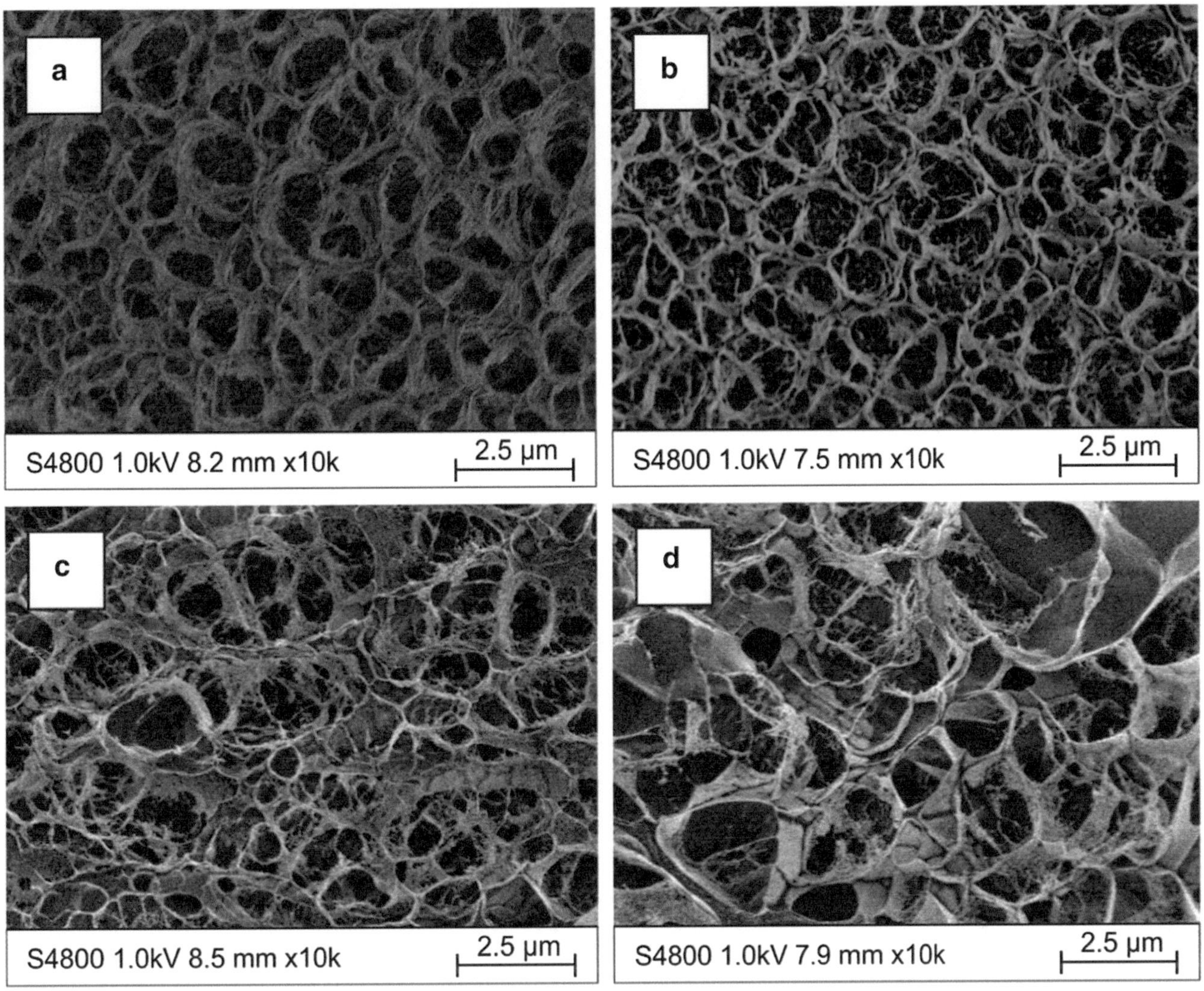

Fig. 5 Cryo-FESEM images of IPN hydrogels with various concentrations of triblock copolymers – protected and deprotected. (**a**, **b**) **sPEOPO$_{20}$/TC-P(Y)** and (**c**, **d**) **sPEOPO$_{20}$/TC-D(Y)** with Y = 0.5 w/w-% (**a**, **c**) and Y = 1 w/w-% (**b**, **d**)

Table 2 Determination of the pore size distributions in IPN hydrogels

	Relative amount of pores [%] in IPN hydrogels			
Pore diameter	**sPEOPO$_{20}$/TC-P(Y)**		**sPEOPO$_{20}$/TC-D(Y)**	
[nm]	Y = 0.5	1.0	0.5	1.0
300–700	45.7	51.4	48.6	34.3
700–1,100	45.7	40	34.3	48.6
>1,100	8.6	8.6	17.1	17.1

copolymer concentration causes a greater possibility of NCO star prepolymer to react through the hydroxyl groups, and correspondingly enhances the entanglements among the components of the network. In consequence, the cross-linking density and the pore size of the hydrogels are reduced. On opposite, the presence of larger pores in **sPEOPO$_{20}$/TC-D(Y)** IPN hydrogels due to rising copolymer concentration may be attributed to the higher hydrophilicity and the effect of steric constrains in the system.

Figure 6 presents the Cryo-FESEM images of **sPEOPO$_{20}$/NP(Y)** NC hydrogels. An increase the concentration of nanoparticles causes a development and enlargement of agglomerates inside the hydrogels, i.e. the average amount of silica particles in ratio to **sPEOPO** prepolymer decreases significantly. Except for **sPEOPO$_{20}$/NP(3)** NC, the photos of all other hydrogels demonstrate a presence of aggregates. The diameters of reinforcement vary depending on the concentrations and range up to 1 μm or even more. Due to the preparation strategy, the cutting of frozen hydrogel samples for NC gels causes a detaching of nanoparticles and remaining of residual voids (Fig. 6a, b, and d).

Mechanical Properties

Static compression test measurements were accomplished on equilibrium swollen IPN and NC hydrogels using dynamic

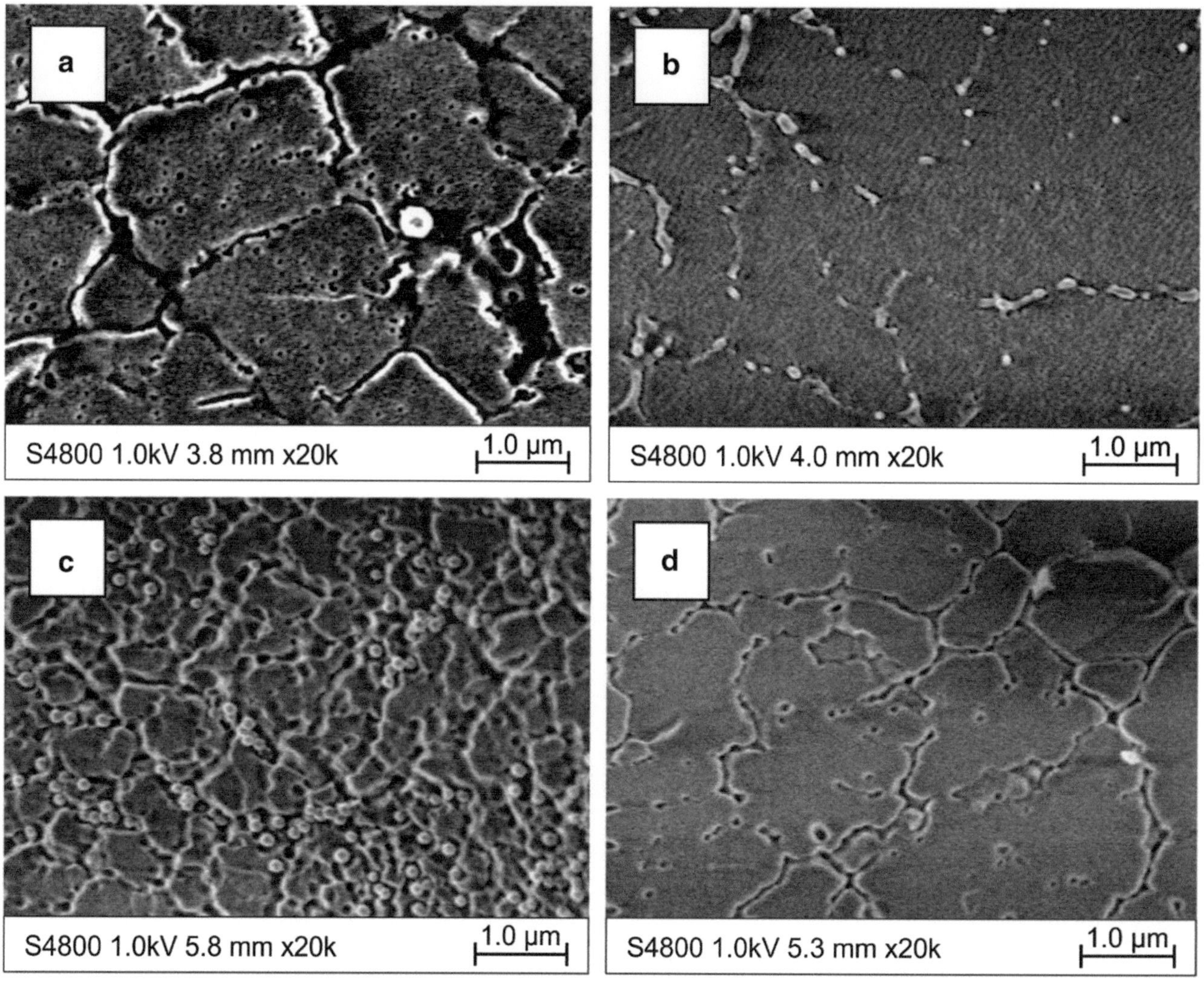

Fig. 6 Cryo-FESEM images of **sPEOPO$_{20}$ /NP(Y)** NC hydrogels. (**a**) Z = 1.7 w/w-%; (**b**) Z = 2.0 w/w-%; (**c**) Z = 3.0 w/w-% and (**d**) Z = 4.0 w/w-%

mechanical analysis equipment (DMA Q800, TA Instruments). In every experiment, samples with diameter 10 mm and thickness according to Eq. 2 were used.

The principal purpose of reinforcement agents is the improvement of the mechanical properties of polymers. The feature of fortifying is mainly carried on the high resistance of the reinforcing factor against strain as a consequence of high modulus. Belonging to stress conditions, the reinforcement agent plays a significant role in the materials load, if the interfacial interactions between reinforcement and polymer matrix are sufficient. The effect of fortifying is stronger whenever the interface between reinforcement agent and polymer matrix are higher pronounced. Therefore, modulus and stiffness of materials are improved significantly by homogeneous dispersal/entanglement of reinforcement.

Table 3 presents the comparison of the compression moduli (E_c) of SN as well as IPN and NC hydrogels fortified by various reinforcement agents. Clearly, an increase the concentrations of **sPEOPO** polymer precursor causes a gently enhancement of the average E_c values. However, a raise the concentrations of reinforcement instigates a slight decrease (P(*t*-BuGE$_{12}$-b-EO$_{796}$-b-*t*-BuGE$_{12}$)) or minor enlarge (P(G$_{12}$-b-EO$_{796}$-b-G$_{12}$) and nanoparticles) of the compression moduli. Despite of the effect generally observed for reinforced polymers, the compression moduli of IPN and NC hydrogels do not increase proportionally with additive concentration, probably due to the high water content that easily allows the motion of the polymer chains.

The compression modulus of IPN hydrogels fortified by 0.5 w/w-% P(*t*-BuGE$_{12}$-b-EO$_{796}$-b-*t*-BuGE$_{12}$) is higher in comparison to the 1 w/w-% concentrations of the reinforcement agent, that is opposite to the IR data priory designated. These consequences can be attributed not only to the concentrations of the secondary components but also to the interactions between the reactants and steric hindrance

Table 3 Compression moduli of IPN and NC hydrogels

sPEOPO c [w/w-%]	Kind of reinforcement	c (reinforcement) [w/w-%]	Compression modulus E_c [kPa]
10	–	–	10.1 ± 1.5
15			16.1 ± 1.0
20			28.3 ± 0.8
10	TC-P	0.5	16.5 ± 4.5
15			28.1 ± 0.9
20			33.0 ± 2.9
10		1	15.6 ± 1.2
15			20.1 ± 0.6
20			32.7 ± 5.8
10	TC-D	0.5	13.3 ± 3.3
15			17.9 ± 1.9
20			29.6 ± 3.2
10		1	15.1 ± 3.6
15			24. 7 ± 2.4
20			40.1 ± 6.5
20	NP	1.7	27.2 ± 1.9
		2	31.4 ± 5.2
		3	31.7 ± 3.2
		4	38.7 ± 3.5

Table 4 Compression stress-at-break of IPN and NC hydrogels

sPEOPO c [w/w-%]	Kind of reinforcement	c (reinforcement) [w/w-%]	Stress-at-break σ_B [kPa]
10	–	–	6.1 ± 0.9
15			18.2 ± 2.5
20			45.9 ± 2.3
10	TC-P	0.5	22.8 ± 3.6
15			40.5 ± 0.7
20			71.6 ± 8.3
10		1	23.1 ± 3.3
15			46.9 ± 5.1
20			65.3 ± 2.3
10	TC-D	0.5	39.1 ± 1.4
15			66.1 ± 5.3
20			79.4 ± 5.5
10		1	31.6 ± 3.2
15			48.5 ± 2.0
20			74.0 ± 0.5
20	NP	1.7	78.9 ± 6.1
		2	92.9 ± 11.3
		3	109.2 ± 12.0
		4	75.1 ± 3.9

effects. A higher concentrations does not yield further improvement of the modulus.

The highest E_c values of ~40 kPa and ~39 kPa are obtained for the cases of **sPEOPO$_{20}$/TC-D(1)** and **sPEOPO$_{20}$/NP(4)**. The values for compression stress-at-break (σ_B) of SN as well as IPN and NC hydrogels reinforced by different fortifying agents are presented in Table 4. Except for **sPEOPO$_{20}$/NP(4)** NC, an increase the content of reinforcement at constant concentrations of **sPEOPO** prepolymer prompts a decrease of hydrogel's stress-at-break. IPN originated by **sPEOPO** prepolymer and diverse concentrations of P(*t*-BuGE$_{12}$-b-EO$_{796}$-b-*t*-BuGE$_{12}$) displays an approximately equivalent stress-at-break value, whereas hydrogel fortified by 0.5 w/w-% of P(G$_{12}$-b-EO$_{796}$-b-G$_{12}$) demonstrates slightly higher values of σ_B. However, in all cases, the sPEOPO prepolymer reinforced by different concentrations of P(G$_{12}$-b-EO$_{796}$-b-G$_{12}$) displays higher stress-at-break in comparison to the SN hydrogels fortified by P(*t*-BuGE$_{12}$-b-EO$_{796}$-b-*t*-BuGE$_{12}$). These results can be assigned to the higher number of hydroxyl groups in case of deprotected triblock copolymer that causes a greater possibility to generate chemical and physical bonding between the constituents of the system.

Commonly, the high surface area of nanoparticles instigates a reinforcement of interfacial interactions between polymer matrix and particles. As a consequence, the mechanical properties of NC hydrogels tend to be significantly improved by enlarging the concentrations of fortifying agent. Conversely, the nanoparticles can easily create agglomerates at specific contents due to their high surface energy. Therefore, the mechanical characteristics of composites can decrease in case agglomeration occurs. In our study, a raise the concentrations of amino-functionalized silica nanoparticles up to 3.0 w/w-% causes enhancement of compressive strengths for **sPEOPO$_{20}$/NP(Y)** NC hydrogels. However, a further increase the content of reinforcing phase leads to a decline of σ_B. The non-appearance of agglomerates and uniformly dispersal of particles (Fig. 6c) shows the highest value of stress-at-break (109 kPa), that was obtained for **sPEOPO$_{20}$/NP(3)** NC hydrogels. In this system, apparently the chemical cross-linking between amino-groups on the nanoparticle surface and the isocyanate groups of **sPEOPO** was still able to prevent nanoparticle aggregation and let to a hydrogel in which the nanoparticles act as real multivalent cross-linking points that significantly enhance the mechanical strength of the hydrogel.

Conclusion

The study presents a strategy to improve the mechanical properties of **sPEOPO$_X$** hydrogels by incorporating various reinforcement agents – linear high molecular weight triblock copolymers and amino-functionalized silica nanoparticles. The concentrations of IPN and NC hydrogel constituents – **sPEOPO** polymer precursor and reinforcement agents – change as 10, 15, 20 w/w-% and 0.5, 1.0, 1.7, 2.0, 3.0, 4.0 w/w-% respectively. The cross-linking of **sPEOPO** pre-polymers and various reinforcement agents occurs through the reactions of isocyanates with hydroxyl/amino groups of copolymers/particles. Moreover, possible lateral interactions ensued by the hydrolysis of the NCO groups, and intermolecular forces between the prepolymers can promote the gelation process, too.

The IPN and NC strategies investigate a development of hydrogels possessing high equilibrium water content values of more than 90 w/w-%. Moreover, all IPN hydrogels demonstrate greatly porous micro structures – average pore size of ~0.6 to ~1.1 μm, whereas the morphology of NC presents a development of agglomerates at specific concentrations of reinforcement. The highest values of compression modulus and stress-at-break – ~40 kPa, respectively ~109 kPa are obtained for the case of **sPEOPO$_{20}$/TC-D(1)** IPN and **sPEOPO$_{20}$/NP(3)** NC which is approximately 1.5 and 2.5 times greater compared to the **sPEOPO$_{20}$** SN hydrogels – ~28 kPa, respectively ~46 kPa. Hence, nanoparticle reinforcement with covalent binding of the hydrogel to the particles showed stronger effects in this study than the tailored macromolecular approach. It should be mentioned, however, that a further increase of the energy-dissipating middle block as well as introduction of amino-groups instead of alcohols to increase the chemical reactivity towards **sPEOPO** remain as further possibilities to improve the effects of the block-copolymer strategy.

References

1. Hoare TR, Kohane DS (2008) Hydrogels in drug delivery: progress and challenges. Polymer 49:1993–2007
2. Peppas NA (1987) Hydrogels in medicine and pharmacy: properties and applications, vol 3. CRC Press, Boca Raton
3. Corkhill PH, Hamilton CJ, Tighe BJ (1989) Synthetic hydrogels VI. Hydrogel composites as wound dressings and implant materials. Biomaterials 10:3–10
4. Peppas NA, Huang Y, Torres-Lugo M, Ward JH, Zhang J (2000) Physicochemical foundations and structural design of hydrogels in medicine and biology. Annu Rev Biomed Eng 2:9–19
5. Lee KY, Mooney DJ (2001) Hydrogels for tissue engineering. Chem Rev 101:1869–1880
6. Hoffman AS (2002) Hydrogels for biomedical applications. Adv Drug Deliv Rev 54:3–12
7. Peppas NA, Hilt JZ, Khademhosseini A, Langer R (2006) Hydrogels in biology and medicine: from molecular principles to bionanotechnology. Adv Mater 18:1345–1360
8. Darwis D (2009) Role of radiation processing in production of hydrogels for medical applications. Atom Indonesia 2:85–104
9. Gong J, Katsuyama Y, Kurokawa T, Osada Y (2003) Double-network hydrogels with extremely high mechanical strength. Adv Mater 15:1155–1158
10. Millar JR (1960) Interpenetrating polymer networks. Styrene–divinylbenzene copolymers with two and three interpenetrating networks, and their sulphonates. J Chem Soc 1311–1317. http://pubs.rsc.org/en/content/articlelanding/1960/JR/jr9600001311#!divAbstract
11. Shibayama K, Suzuki Y (1967) Viscoelastic properties of multiple network polymers. IV. Copolymers of styrene and divinylbenzene. Rubber Chem Technol 40:476–483
12. Sperling LH, Friedman DW (1969) Synthesis and mechanical behavior of interpenetrating polymer networks: Poly(ethyl acrylate) and polystyrene. J Polym Sci Part A-2 7:425–427
13. Sperling LH, George HF, Huelck V, Thomas DA (1970) Viscoelastic behavior of interpenetrating polymer networks: poly(ethyl acrylate)–poly(methyl methacrylate). J Appl Polym Sci 14:2815–2824
14. Sperling LH, Taylor DW, Kirpatrich ML, George HF, Bardman RD (1970) Glass-rubber transition behaviour and compatibility of polymer pairs: poly(ethyl acrylate) and poly(methyl methacrylate). J Appl Polym Sci 14:73–78
15. Klempner D, Frisch HL, Frisch KC (1971) Topologically interpenetrating polymeric networks. J Elastoplastics 3:2–18
16. Klempner D, Frisch HL, Frisch KC (1970) Topologically interpenetrating elastomeric networks. J Polym Sci Part A-2 8: 921–935
17. Frisch HL, Klempner D, Frisch KC, Kwei TK (1970) Topologically interpenetrating elastomeric networks. Polym Prepr 11:483
18. Matsuo T, Kwei TK, Klempner D, Frisch HL (1970) Structure–property relationships in polyacrylate-poly(urethane-urea) interpenetrating polymer networks. Polym Eng Sci 10:327–331
19. Curtius AJ, Covitch MJ, Thomas DA, Sperling LH (1972) Polybutadiene/polystyrene interpenetrating polymer networks. Polym Eng Sci 12:101–108
20. Huelck V, Thomas DA, Sperling LH (1972) Interpenetrating polymer networks of poly(ethyl acrylate) and poly(styrene-co-methyl methacrylate). I. Morphology via electron microscopy. Macromolecules 5:340–347
21. Huelck V, Thomas DA, Sperling LH (1972) Interpenetrating polymer networks of poly(ethyl acrylate) and poly(styrene-co-methyl methacrylate). II. Physical and mechanical behaviour. Macromolecules 5:348–353
22. Rakovsky A, Marbach D, Lotan N, Lanir Y (2009) Poly(ethylene glycol)-based hydrogels as cartilage substitutes: synthesis and mechanical characteristics. J Appl Polym Sci 112:390–401
23. Dekosky BJ, Dormer NH, Ingavle GC, Roatch CH, Lomakin J, Detamore MS, Gehrke SH (2010) Hierarchically designed agarose and poly(Ethylene Glycol) interpenetrating network hydrogels for cartilage tissue engineering. Tissue Eng Part C 16: 1533–1542
24. Gupta RK, Kennel E, Kim K-J (2010) Polymer nanocomposites handbook. CRS Press, Boca Raton
25. Hussain F, Hojjati M, Okamoto M, Gorga RE (2006) Polymer-matrix nanocomposites, processing, manufacturing, and application: an overview. J Compos Mater 40:1511–1575
26. Liu Y, Zhu M, Liu X, Zhang W, Sun B, Chen Y, Adler H-JP (2006) High clay content nanocomposite hydrogels with surprising mechanical strength and interesting deswelling kinetics. Polymer 47:1–5
27. Haraguchi K, Li H-J (2006) Mechanical properties and structure of polymer-clay nanocomposite gels with high clay content. Macromolecules 39:1898–1903
28. Haraguchi K (2007) Nanocomposite hydrogels. Curr Option Solid State Mater Sci 11:47–54
29. Schexnailder P, Schmidt G (2009) Nanocomposite polymer hydrogels. Colloid Polym Sci 287:1–11
30. Haraguchi K (2011) Stimuli-responsive nanocomposite gels. Colloid Polym Sci 289:455–473
31. Qin X, Zhao F, Liu Y, Wang H, Feng S (2009) High mechanical strength hydrogels preparation using hydrophilic reactive microgels as crosslinking agents. Colloid Polym Sci 287:621–625
32. Dalton PD, Hostert C, Albrecht K, Moeller M, Groll J (2008) Structure and properties of urea-crosslinked star poly[(ethylene oxide)-*ran*-(propylene oxide)] hydrogels. Macromol Biosci 8:923–931

33. Goetz H, Uwe B, Bartelink CF, Gruenbauer HJ, Moeller M (2002) Preparation of isophorone diisocyanate terminated star polyethers. Macromol Mater Eng 287:223–230
34. Green TW, Wuts PG (1999) Protective groups in organic synthesis. Wiley, New York, pp 65–67 and 404–408
35. Stober W, Fink A, Bohn E (1968) Controlled growth of monodisperse silica spheres in the micron size range. J Colloid Interface Sci 26:62–69
36. ISO 604: 2002(E): Plastics – determination of compressive properties

Temperature-Sensitive Composite Hydrogels: Coupling Between Gel Matrix and Embedded Nano- and Microgels

Judith Meid, Swen Lehmann, and Walter Richtering

Abstract

Temperature sensitive composite hydrogels have been prepared by incorporating poly (N-isopropylacrylamide) (PNiPAM) nano- and microgels into a polyacrylamide hydrogel matrix. Composite gels with different compositions, i.e. contents of microgel, polyacrylamide and cross-linker have been prepared with different initiation reactions and at different preparation temperatures (above and below the volume phase transition temperature of PNiPAM). The size of the embedded nano- and microgels was determined by means of small angle neutron scattering and confocal laser scanning microscopy. The results show that the temperature sensitivity of the embedded PNiPAM particles can be retained depending on the size of the embedded microgels and the composition of the matrix. A corset-like functionality of the hydrogel matrix can be achieved when the matrix is highly cross-linked or an interpenetrating network (IPN) is formed. The internal structure of the composite gel depends on the preparation conditions as well as on the size of the PNiPAM particles. IPNs can be formed with microgel beads, but hardly with nanogels. Furthermore the mobility of dissolved fluorescently labeled dextran tracers was probed spatially resolved inside the composite gels by means of 2-focus fluorescence correlation spectroscopy. Tracer mobility depends on the local environment and depends on the coupling between embedded microgel and hydrogel matrix e.g. by IPN formation, and can be tailored by the composition of the system. The composite gel does not reveal a temperature dependent volume change because the matrix takes up the water released by the embedded particles. Nevertheless, the temperature dependent swelling of the PNiAM nano- and microgels affects the properties of the composite hydrogel.

Keywords

Microgel • PNiPAM • Neutronscattering • Gel • Fluorescence correlation spectroscopy • Rheology • Temperature sensitive

Introduction

Hydrogels are three-dimensional hydrophilic polymer networks, which are highly swollen in water. On the one hand they are viscoelastic solids on the other hand soluble molecules can diffuse through the hydrogel matrix. Often applications, as e.g. in the bio-medical field require specific mechanical properties as well as sufficient mobility of soluble species [1, 2]. Thus the mesh size of the network and the interactions between different components need to be controlled in order to enable the design of gels with tailor-made properties [3–5].

A very interesting group of materials are stimuli sensitive polymers [6]. Such materials show major changes in conformation in response to the change of external stimuli such

J. Meid • S. Lehmann • W. Richtering (✉)
Institute of Physical Chemistry, RWTH Aachen University, Landoltweg 2, 52056 Aachen, Germany
e-mail: richtering@rwth-aachen.de

G. Sadowski and W. Richtering (eds.), *Intelligent Hydrogels*, Progress in Colloid and Polymer Science 140, DOI 10.1007/978-3-319-01683-2_8,

as temperature, salt concentration, pH or co-solvent. Poly (N-isopropylacrylamide) (PNiPAM) is one famous example for a temperature and solvent sensitive polymer [7, 8]. Water being a good solvent at temperatures below the volume phase transition temperature (VPTT) becomes a poor solvent above the VPTT.

Composite hydrogels constitute an important approach that combines properties of hydrogels with those of functional particles [9–11]. The embedded species can be rigid particles as well as micro- and nanogels [12, 13].

The preparation of composite hydrogels with tailor-made properties requires knowledge on the micro- and nanostructural complexity of the gel environment close to the encapsulated particles. In addition, the dynamics of functional molecules and substrates inside such systems is relevant especially for applications of gels e.g. for controlled uptake and release or for cell stimulation.

In this context, it is necessary to perform investigations on model systems that are less complex, e.g. in terms of chemical composition, and provide more flexibility for the variation of experimental parameters. For this purpose, composite gels that contain temperature-sensitive nano- and micrometer-sized hydrogel particles are of particular value [14–18]. The embedded responsive microgel particles react to changes in their surrounding by pronounced swelling and deswelling. Due to their size and sensitivity to external stimuli, they provide a perfect model for cells that contract or relax upon stimulation [19].

In this contribution we focus on composite hydrogels consisting of a polyacrylamide matrix with embedded PNiPAM nanogels and microgel beads, respectively. The term "microgel" is often – but not exclusively – used for particles with a size <1 μm, whereas large microgel particles are sometimes called "microgel beads". In this contribution, both types of microgels are employed and in the following we will employ the term "nanogel" for small PNiPAM particles and "microgel bead" for large particles. The former were prepared via precipitation polymerization at T > VPTT, whereas the latter were prepared by inverse suspension polymerization at T < VPTT. One needs to keep in mind that the different polymerization techniques do not only lead to different particle sizes but also to a different internal structure [20–22].

The different size of nanogels and microgel beads requires different experimental techniques to be employed in order to determine the temperature dependent swelling: The mutual influence of hydrogel matrix and incorporated nanogels is probed by means of small angle neutron scattering (SANS). The size of large microgel beads is obtained from confocal laser scanning microscopy. In addition we probe the diffusion of tracer polymers inside the composite gels by means of 2-focus fluorescence correlation spectroscopy (2fFCS).

We will discuss the influence of polymerization conditions on the coupling between embedded particles and gel matrix. Different parameters as e.g. microgel size, cross-link density of the gel matrix and polymerization temperature are varied. The latter determines the size of the embedded microgels during gel preparation, as illustrated in Scheme 1. It is obvious that the preparation conditions and the coupling of the two networks will have profound influence on the structure and properties of the responsive composite gels.

Experimental Part

Materials: *N,N′*-methylenebisacrylamide (BIS), *N,N,N′,N′*-tetramethylethylenediamine (TEMED) and potassium peroxodisulfate (KPS) were obtained from Merck. *N*-isopropylacrylamide (NiPAM) was purchased from Acros. Acrylamide (AAm) was commissioned from Fluka/Serva. Rhodamine B was obtained from Polyscience Inc..

All chemicals were used without further purification. The prepared solutions and the reactions were all preformed in double distilled water or in D_2O (SANS samples), which was obtained from Deutero. For the SANS samples, also deuterated acrylamide purchased from Cambridge Isotopes Laboratories Inc. was used. The PNiPAM nanogel particles used for all measurements were synthesized and characterised by Berndt et al. [23]. They have a molar ratio of cross-linker (BIS) to monomer (NiPAM) of 1:71.

Hydrogel Synthesis: The filled and unfilled hydrogels were prepared under argon atmosphere in a glove box. Nanogels were redispersed in the pre-gel solution together with monomer and cross-linker for the preparation of composite hydrogels. A drop of saturated KPS solution was added and the polymerization was started by the addition of TEMED.

Fluorescently labeled PNIPAM microgel beads were prepared from NIPAM, BIS, and methacryloxyethyl thiocarbonyl rhodamine B (Polysciences Inc.) in two different ways: type-A beads were fabricated by droplet-based microfluidic templating, whereas type-B beads were synthesized by inverse suspension polymerization [24].

Dextrans labeled with Alexa Fluor 647 (10 kDa and 3 kDa, Invitrogen) or Alexa Fluor 488 (70 kDa, Invitrogen) were dissolved in water (LiChroSolv, Merk). 3-kDa dextran has a hydrodynamic radius of $R_h = 1$ nm, 10-kDa dextran has $R_h = 2.3$ nm, and 70-kDa dextran has $R_h = 6.5$ nm. All hydrodynamic radii were calculated via the Stokes–Einstein equation from the diffusion coefficient measured at infinite dilution.

Composite hydrogel samples for fluorescence studies were prepared with a final bead volume fraction of 0.05 at 25 °C. Less than 1 mg of the UV-initiator VA-086 (2,2′-azobis[2-methyl-*N*-2-hydroxyethyl) propionamide], Wako)

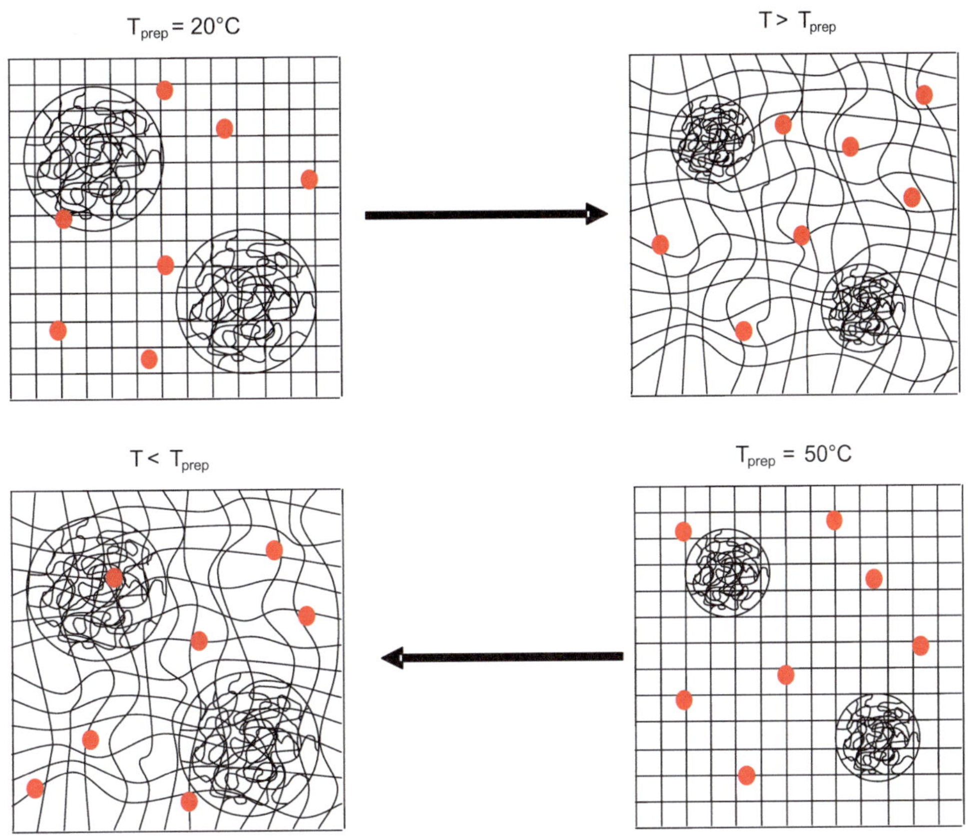

Scheme 1 This scheme illustrates the concept of the study: A polyacrylamide hydrogel is prepared in the presence of temperature sensitive PNiPAM microgels and tracer molecules labelled with a fluorescent dye. The composite gels are prepared at different temperatures and thus at different degree of swelling of the PNiPAM microgels

was added, and the solutions were filled into a temperature-controlled sample cell.

Spatially Resolved 2fFCS: The apparatus for the 2-focus fluorescence correlation spectroscopy and especially the temperature-controlled cell have been described before [25, 26].

Small Angle Neutron Scattering: SANS measurements were performed at the D11 beam line of the Institute Laue-Langevin (ILL) in Grenoble (France). Hellma quartz glass cells with a sample thickness of 1 mm were used. The data were corrected for background scattering (scattering of the microgel-free hydrogel matrix) and were calibrated on absolute scale using the incoherent scattering of H_2O according to the standard procedures at the ILL. All experiments were carried out using fully deuterated acrylamide for the preparation of the hydrogel matrix and D_2O as a solvent. However, the cross-linker was not deuterated.

Results and Discussion

Properties of temperature sensitive composite hydrogels will strongly depend on whether or not the incorporated PNiPAM gel particles will be able to respond to temperature changes. First we will show how composite hydrogels can be prepared that retain the T-dependent swelling and deswelling of the embedded nanogels.

The size of PNiPAM nanogels is probed by SANS using deuterated acrylamide as monomer for hydrogel formation and D_2O as solvent. The scattering intensity is dominated by the embedded – non-deuterated – nanogels and thus their size can be determined both prior and after formation of the hydrogel matrix [27].

Figure 1 shows scattering curves of nanogels in aqueous solution and embedded into a hydrogel matrix. The cross-linker to monomer ratio of the matrix was 1:60 and the preparation temperature was 20 °C. SANS data were recorded at two different temperatures: 25 °C, below the VPTT and 50 °C, above the VPTT. All samples show similar scattering curves independent of the surrounding of the nanogels. Obviously, the hydrogel matrix has only little effect on the temperature dependent swelling of the nanogels. The nanogels swell and deswell to the same size independent of their surroundings that is whether they are embedded inside the hydrogel or in free solution. Turbidity measurements showed that the gel matrix does

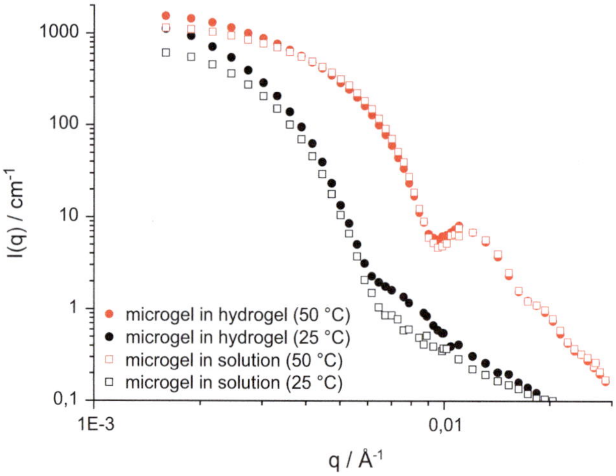

Fig. 1 SANS curves of samples prepared with fully swollen nanogels at 20 °C and measured at 25 and 50 °C. The SANS intensity is plotted versus momentum transfer q for PNiPAM nanogels in a cross-linked deuterated polyacrylamide matrix with a cross-linker to monomer ratio of 1:60 and in solution of deuterated acrylamide in D_2O

not affect the VPTT of the PNiPAM nanogels. These experiments clearly demonstrate that composite hydrogels can be prepared such that the temperature sensitivity of the embedded nanogels is fully retained [26]. Furthermore, the temperature induced swelling/deswelling transitions are fully reversible.

Different properties of hydrogel matrix and embedded nanogels can affect the mutual influence of both components. Figure 2 illustrates the influence of increased polymer concentration and of cross-link density in the hydrogel matrix on the swelling of the nanogels.

Scattering data obtained from a sample with an increased acrylamide concentration of 15 wt% are shown in Figure 2a, b. The nanogel size is still almost unaffected by the surrounding polyacrylamide matrix. Nanogels embedded into the gel at 20 °C can collapse upon heating and nanogels embedded in the collapsed state at 50 °C are able to reswell almost completely upon cooling [28].

Plots 2c and 2d show the influence of cross-link density. Again two samples are shown, which were prepared at the two different temperatures, below (20 °C) and above (50 °C) the VPTT of PNiPAM. However, the hydrogel matrix was prepared with a cross-linker to monomer ratio of 1:1, while the total polymer concentration is the same as in Fig. 2a, b. The scattering curves clearly show that nanogels embedded at 20 °C are able to collapse upon heating to 50 °C, indicating that no interpenetrating network was formed that could prevent a temperature-induced nanogel collapse. However, the hydrogel matrix is able to restrict the swelling of nanogels when they are incorporated in their collapsed state, see Fig. 2d. The scattering curves obtained at 20 °C from gels that were prepared at 50 °C, reveal a well pronounced form factor minimum, which is only slightly shifted to smaller q-values ($q_{min} = 0.0085$ Å^{-1}) as compared to the measurements at 50 °C; thus the nanogels are still almost completely collapsed. Obviously, the gel matrix prevents the reswelling of nanogels, which were incorporated in their collapsed state; in other words the polyacrylamide matrix acts like a corset. On the other hand, nanogels incorporated at low temperatures are able to collapse freely; indicating that there are only weak interactions between the nanogels and the surrounding matrix. Thus the corset-effect is caused by the mechanical properties of the highly cross-linked matrix and does not occur for less densely cross-linked matrixes with a high polymer concentration.

These experiments reveal a characteristic difference between these composite hydrogels and core-shell nanogels. There are no indications for a significant covalent or topological connection of the nanogel and hydrogel networks in the case of the composite gels. However, core-shell nanogels prepared by two-step precipitation polymerization are known to show shell restricted or enhanced swelling of the core, which indicates strong coupling of core and shell network via chemical bonds and/or IPN formation at the interface [29–34].

In order to enhance the formation of chemical bonds between the nanogels and the polyacrylamide matrix, we prepared samples by means of polymerization initiation via γ-irradiation. The synthesis of hydrogels consisting of acrylamide based monomers via γ-ray induced polymerisation and the structure of such hydrogels is well studied [35–37]. An initiation of polymerisation by γ-irradiation leads to covalent bonds between nanogels and hydrogel via transfer reactions. However, the exact chain initiating and cross-linking processes are only partially known [38]. Monomer radicals are generated by the direct effect of irradiation as well as by an indirect effect based on the reaction of the products of water radiolysis with the monomers [39]. However, not only the monomers are activated, the loose ends of the nanogels are activated as well enabling cross-linking between nanogels and polyacrylamide. This leads to chemical bonds between the nanogels and the surrounding hydrogel matrix and the nanogels themselves act as cross-linker. Therefore it is possible to prepare composite gels even without an additional chemical cross-linker.

In Fig. 3 we compare scattering results of composite hydrogels initiated via γ-irradiation with and without additional chemical cross-linker (BIS). The macroscopic gels were inherently stable in all cases. The scattering curves show that the nanogels are still able to swell and shrink with temperature. For both samples, the q-values of the form factor minima are at about the same positions for both temperatures. A structure factor maximum and a stronger increase in the forward scattering in the SANS curve is found at 20 °C for the composite gel without additional

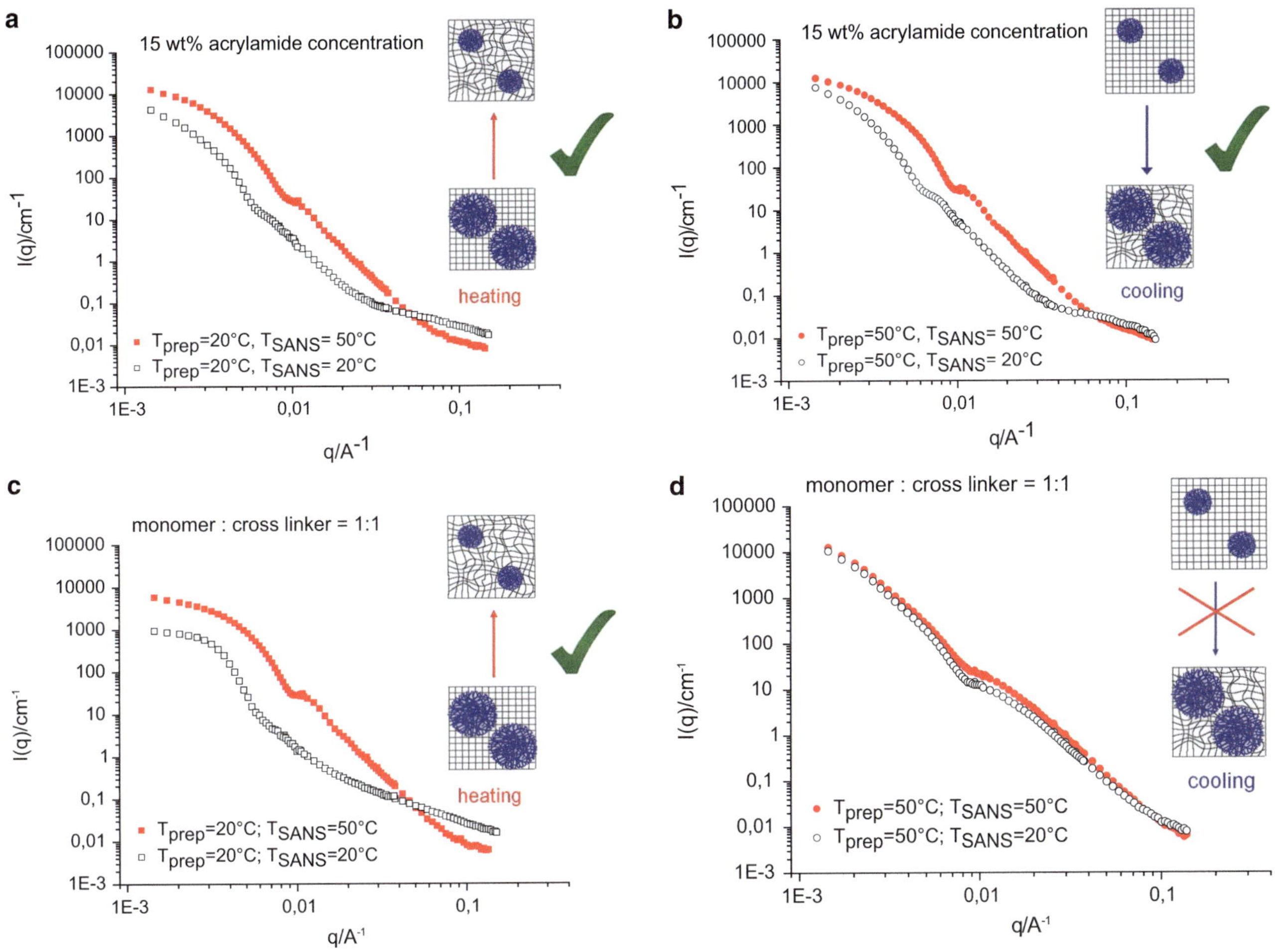

Fig. 2 SANS intensity versus q for PNiPAM nanogels in the cross-linked deuterated polyacrylamide matrix, swollen in D_2O. (**a**) and (**b**): SANS curves of composite gels with an increased polyacrylamide concentration (15 wt%); (**c**) and (**d**): cross-linker to monomer ratio of 1:1. The gels were prepared with swollen nanogels (at 20 °C, *squares*: (**a**) and (**c**)) and collapsed nanogels (at 50 °C, *circles*: (**b**) and (**d**)) and both measured at 20 °C (*black, open symbols*) and 50 °C (*red, full symbols*)

cross-linker. The distance between the nanogels is smaller and there might also be some aggregation. One reason for an aggregation may be the different polymerisation time. While polymerisations initiated with a radical starter only last a few minutes, the polymerisation started with γ-rays lasts several days. As the end groups of the nanogels are activated by the γ-irradiation it is possible that covalently connected nanogel clusters are formed during this very slow polymerisation since the nanogels are able to diffuse inside the sample. Thus the nanogels are no longer homogeneously distributed in the sample. If, however, an additional cross-linker is present during the polymerisation, the formation of the gel matrix is much faster, and consequently the diffusion of the nanogel particles through the sample is reduced and the formation of aggregates is less probable.

SANS data obtained at 50 °C shown in Fig. 3 as well as from other samples [27] reveal a form factor minimum shifted to higher *q* as compared to 20 °C. Thus the nanogels are able to shrink upon heating, however the stronger coupling to the matrix leads to higher size polydispersity in the collapsed state. This indicates that the hydrogel matrix in these samples affects the collapse of the nanogels.

Our second approach for studying composite hydrogels employs large PNiPAM microgel beads that are labelled with a fluorescent dye. This allows investigating the temperature dependent swelling of embedded microgel beads via imaging by confocal laser scanning microscopy. Scheme 2 shows the concept of these studies, and indicates the y-position of the z-x-plane that is shown in the following figures.

We prepared microgel beads with different polymer content and cross-link density and incorporated them in hydrogel matrices with different polyacrylamide concentration as given in Table 1 [23].

Fluorescence micrographs were recorded at 25 °C and 36 °C and are shown in Fig. 4. In water, the loosely cross-linked "type-A beads" have a swelling ratio of V_c/

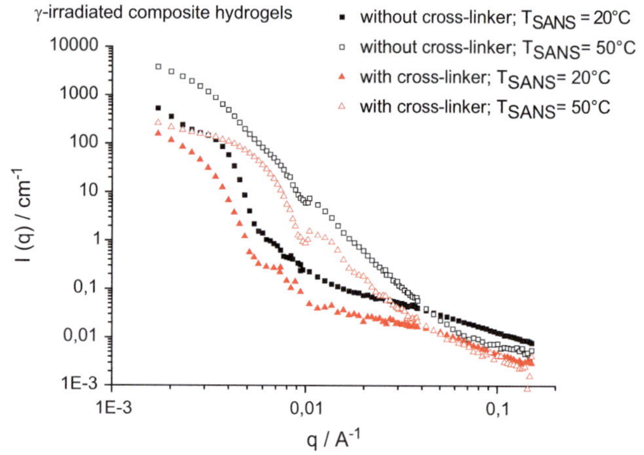

Fig. 3 SANS curves of nanogels in composite hydrogels, which were polymerized upon γ-initiation with (*red triangles*) and without (*black squares*) additional cross-linker (BIS). All samples were irradiated and polymerized at room temperature (20 °C). The samples were measured at 20 °C (*full symbols*) and 50 °C (*open symbols*). The cross-linker to monomer ratio of the composite gel sample with cross-linker was 1:60, the nanogel concentration for both samples was 0.2 wt%

$V_s = 0.09 \pm 0.01$, whereas stiff, densely cross-linked "type-B beads", have a swelling ratio of $V_c/\ V_s = 0.42 \pm 0.15$. V_c and V_s denote the microgel volume in collapsed and swollen state, respectively.

The swelling ratio of the less cross-linked, soft type-A beads (HG-50-A) becomes $V_c/\ V_s = 0.61$ when they are embedded a 50 gL^{-1} PAAM hydrogel matrix, thus they can deswell much less upon heating as compared to water. This observation strongly indicates the formation of an interpenetrating PAAM gel inside the PNiPAM beads. The dense type-B beads (sample HG-50-B) on the other hand, are able to collapse as in water, see Fig. 4. Apparently, the formation of an interpenetrating gel network is prevented by the dense structure of the beads. Figure 4 also shows that the center of mass of the type-B beads moves downwards when they collapse. Again, this indicates that there is hardly any coupling between the type-B beads and the surrounding hydrogel matrix.

A different result is obtained when the hydrogel matrix is made softer by reducing the PAAM concentration to 25 gL^{-1} (sample HG-25-A): the soft type-A beads are now able to collapse inside the PAAM matrix as in water. However, the center of the bead does not move when the bead collapses. This marked difference as compared to sample HG-50-B is an indication for the formation of an interpenetrating network that is not strong enough to prevent bead collapse, but sufficiently strong to hold the bead at its position upon heating. These experiments demonstrate again that composite gels with different thermosensitivity of the embedded microgel particles can be prepared.

It is interesting to compare the behavior of the soft microgel beads with the nanogels discussed above. The total monomer and cross-linker concentration inside the swollen gel particles is rather similar. However, indications for the formation of an interpenetrating network after polymerization of the polyacrylamide matrix were only obtained in the case of microgel beads. The different behavior of nanogels and microgel beads, respectively, can be caused by their different structure. The beads are synthesized by inverse suspension polymerization at low temperature whereas the nanogels are prepared by precipitation polymerization above the VPTT. The latter is known to lead to an inhomogeneous cross-linker distribution with less cross-linker content near the particle surface [40]. The fuzzy surface may prevent the formation of an interpenetrating network of nanogel particle and hydrogel matrix.

The mobility of molecules inside gels is also sensitive to the local network structure and allows probing the formation of interpenetrating networks [41, 42]. Furthermore the mobility of molecules is relevant in applications of composite hydrogels for uptake and release or cell culturing. Therefore we investigated how the heterogeneous local environment of composite hydrogels affects the mobility of tracer molecules. We use fluorescently labeled dextrans as tracer particles because they are known to hardly interact with polymers and are frequently used in model studies [43–45].

Fluorescence correlation spectroscopy (FCS) is well suited to measure tracer diffusion in gels, whereas photon correlation spectroscopy probes collective diffusion modes [46–48]. In this context, potential changes in the confocal volume, which can occur due to changes of the refractive index of the sample, need to be considered [49]. This is particularly relevant for the present study, as the local polymer density varies with the position in composite gels, leading to different refractive indexes and thus different sizes and shapes of the confocal volume.

To account for these complications, we use 2fFCS [50]. In 2fFCS, two laterally shifted but overlapping laser foci are used to determine correlation functions of each focus and of the cross-section of the two foci. The high accuracy of 2fFCS in crowded environments has been demonstrated previously [51]. The correlation function in 2fFCS is given by [48, 49]:

$$g\left(t,\delta,\nu\right)=\frac{c}{4}\sqrt{\frac{\pi}{Dt}}\int dz_1\int dz_2\left(\frac{\kappa(z_1)\kappa(z_2)}{8Dt+w^2(z_1)+w^2(z_2)}\times\exp\left[-\frac{(z_2-z_1)^2}{4Dt}-\frac{2\delta^2}{8Dt+w^2(z_1)+w^2(z_2)}\right]\right) \quad (1)$$

With D the translational diffusion coefficient, t the lag time of the correlation, c the concentration of the fluorescent particles, δ is the shift distance between the two foci

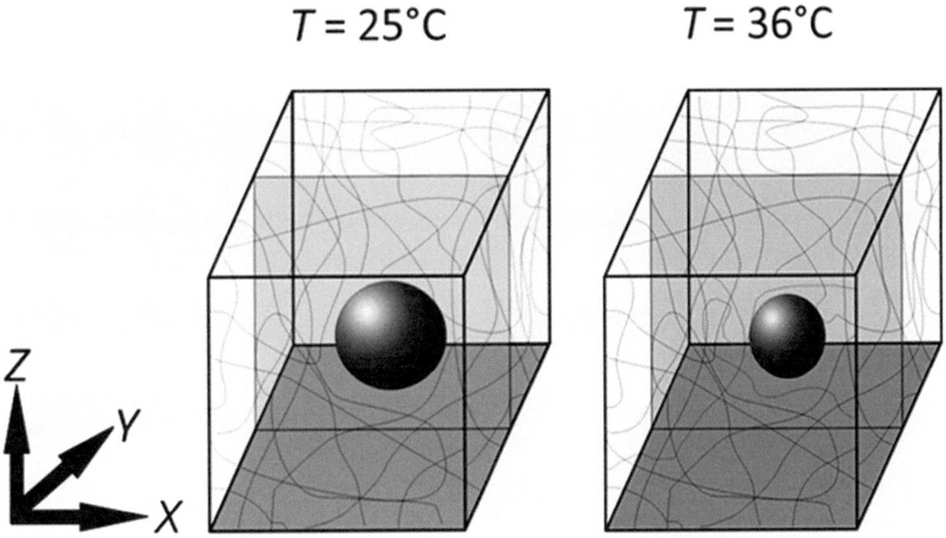

Scheme 2 3-D scheme of the composite hydrogels. The *sphere* represents a microgel bead, the *lower grey* plane indicates the *lower glass* cover slide of the inverted microscope; the *light grey* plane indicates the vertical confocal scanning plane; *left*: the swollen bead at 25 °C; *right*: the collapsed bead at 36 °C, in the same hydrogel matrix

(determined independently) and x, y, and z are Cartesian coordinates with z along the optical axis. The functions $w(z)$ and $\kappa(z)$ describe the molecule detection function and are given by

$$w(z) = w_0\left[1 + \left(\frac{\lambda_{ex}z}{\pi w_0^2 n}\right)^2\right]^{1/2} \quad (2)$$

$$\kappa(z) = 2\int_0^a \frac{d\rho\rho}{R^2(z)}\exp\left(-\frac{2\rho^2}{R^2(z)}\right) = 1 - \exp\left(-\frac{2a^2}{R^2(z)}\right) \quad (3)$$

$$R(z) = R_0\left[1 + \left(\frac{\lambda_{em}z}{\pi R_0^2 n}\right)^2\right]^{1/2} \quad (4)$$

λ_{ex} is the excitation wavelength, λ_{em} is the emission wavelength, n the refractive index, and a is the confocal pinhole radius. The correlation function can be calculated numerically, and w_0 and R_0 are fit parameters [48, 49]. The accuracy of single focus FCS measurements (in contrast to 2fFCS) is strongly affected by several factors as, e.g. refractive index mismatch between sample and immersion medium, absolute position of the focus laser intensity which hamper the determination of shape and size of the molecule detection function and thus the correct determination of diffusion coefficients [48, 49, 52].

Even pure hydrogels are known to exhibit heterogeneities on length scales of 10–100 nm [53, 54]. This is smaller than the optical resolution of our 2fFCS measurements, which probes the mobility of labeled molecules on length scales that are limited by the optical resolution [55]. The shift distance between the two foci in our 2fFCS setup is of the order of 300 nm. Thus, the mobility of tracer molecules probed by 2fFCS is averaged over both mesh size fluctuations and heterogeneities. It is important to notice that all two-foci cross-correlation data acquired in this study could be fitted with Eq. 1, i.e. the data are well described by normal diffusion processes. In other words, we did not find any indications for anomalous diffusion of the tracers inside the gels.

Table 1 *Compositions of composite hydrogel samples that consist of PNIPAM microgel beads embedded within PAAM hydrogel matrixes:* HG in the sample name indicates the existence of a PAAM hydrogel; 25 or 50 indicates the PAAM concentration in the hydrogel; A and B denote the bead type; c_{PAAM} is the matrix concentration; c_{NIPAM} the PNIPAM concentration during bead preparation; $n_{BIS} : n_{PAAM}$ is the molar ratio of cross-linker to monomer in the matrix; $n_{BIS} : n_{PNIPAM}$ is the molar ratio of cross-linker to monomer in the beads. Note that the type-A beads are less cross-linked and contain less polymer than the type-B beads. Hence, the type-A beads are referred to as soft beads, whereas the type-B beads are referred to as dense beads

Sample name	c_{PAAM} (gL^{-1})	c_{PNIPAM} (gL^{-1})	n_{BIS}/n_{PAAM}	n_{BIS}/n_{PNIPAM}
HG-25-A	25	100	1:60	1:70
HG-50-A	50	100	1:60	1:70
HG-25-B	25	143	1:60	1:13
HG-50-B	50	143	1:60	1:13

In Fig. 5 we summarize the diffusion coefficients of two dextran tracers of different molar mass (3 and 10 kg/mol) in different environments, namely water, hydrogels with PAAM concentration of 25 and 50 g/L, respectively, and at three

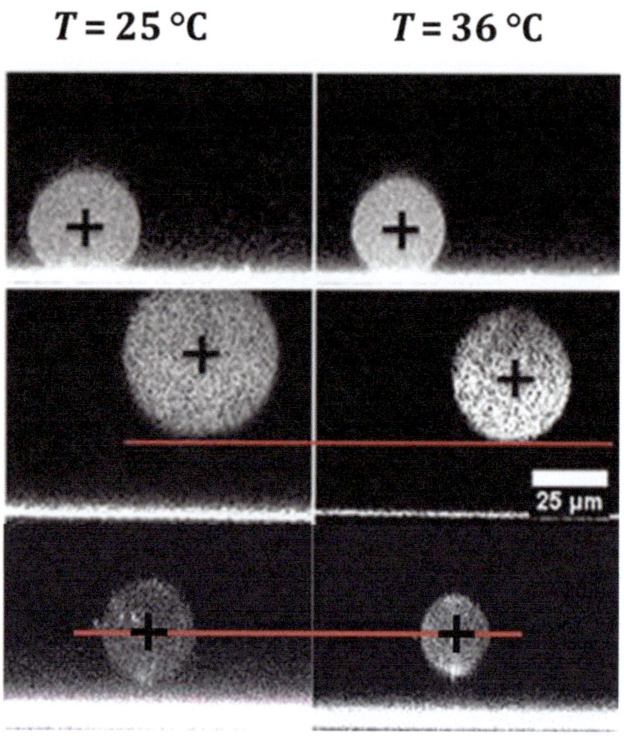

Fig. 4 Side view of PNIPAM beads embedded in a PAAM hydrogel. The *white line* at the bottom of the pictures is the glass cover slide. The beads are scanned in the center plane of the particle along the x–z-directions. *Black crosses* indicate the center of the beads. *Top*: sample HG-50-A, PNIPAM bead of type A embedded in a PAAM hydrogel with an AAM concentration of 50 gL^{-1} at 25 °C (*left*) and at 36 °C (*right*). The bead collapse is much less pronounced than in water. *Center*: sample HG-50-B, PNIPAM bead of type B embedded in a PAAM hydrogel with an AAM concentration of 50 gL^{-1} at 25 °C (*left*) and at 36 °C (*right*). The bead collapses as in water, and the center of the bead moves down until the bottom of the collapsed bead reaches the level of the bottom of the swollen bead. *Bottom*: sample HG-25-A, type-A PNIPAM bead embedded in a PAAM hydrogel with an AAM concentration of 25 gL^{-1}. The bead collapses almost to the same extend as in water, and the bead center does not move. The scale bar denotes 25 μm and applies to all panels

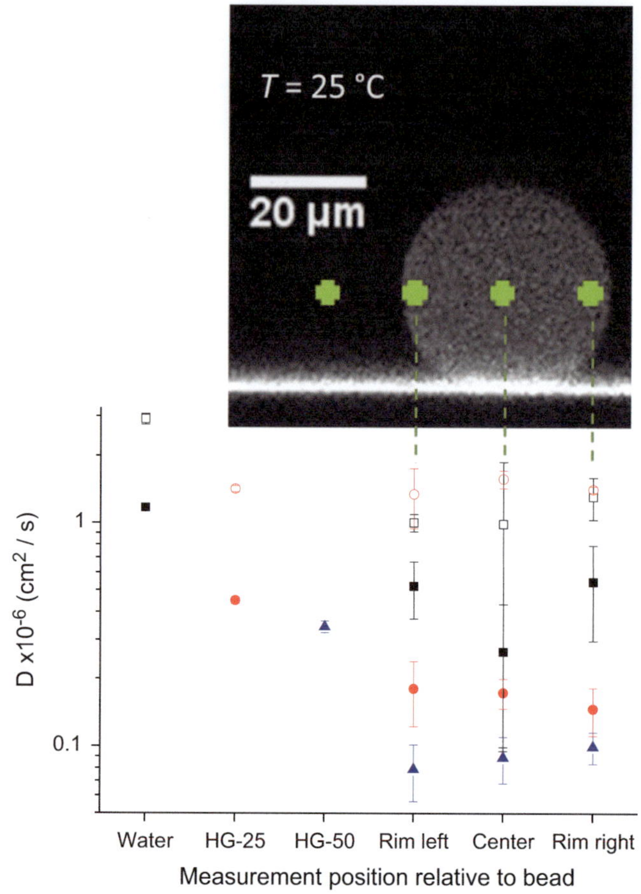

Fig. 5 Diffusion coefficient of (i) 10-kDa dextran out- and inside of type-A PNIPAM microgel beads, either embedded into a 25-gL^{-1} PAAM hydrogel matrix (*red circles*), in a 50 gL^{-1} PAAM hydrogel matrix (*blue triangles*) or suspended in water (*black squares*) at 25 °C and (ii) of 3 kDa dextran out- and inside of type-A PNIPAM microgel beads, either embedded into a 25-gL^{-1} PAAM hydrogel matrix (*open red circles*) or suspended in water (*open black squares*). The diffusion coefficients were obtained by averaging over three different beads. The *upper* picture is a confocal micrograph showing the different positions of 2fFCS measurements. The diffusion coefficient of the 3 kDa dextran is almost not affected by embedding the bead into a 25 gL^{-1} hydrogel. Whereas the diffusion coefficient of the 10 kDa dextran shows a strong effect of the hydrogel matrix

positions inside type-A PNIPAM beads embedded in these environments. The mobility of the tracers inside the beads agrees well with those in macroscopic PNIPAM hydrogels of same composition [23]. This indicates that the preparation of microgel beads inside emulsion droplets leads to the same network structure as polymerization of bulk gels.

Embedding the beads into a 25 g/L hydrogel, does not lead to a significant change in the diffusion coefficient of the 3 kg/mol dextran as compared to beads suspended in water. The diffusion coefficient of the 10 kg/mol dextran, however, is strongly influenced. This indicates first, that the 3 kg/mol dextran is too small to be influenced. Secondly, the decrease of the diffusion coefficient of the 10 kg/mol dextran indicates the formation of an interpenetrating network. The diffusion coefficient of the 10 kDa dextran inside beads decreases further when the beads are embedded into a 50 g/L PAAM hydrogel. This indicates, that a tighter IPN is formed, with a lower mesh size as compared to beads embedded into a 25 g/L hydrogel. A detailed discussion of the influence of hydrogel and bead composition on tracer diffusion has been provided recently [23].

Finally, we discuss how the coupling between beads and matrix affects tracer diffusion within the surrounding hydrogel matrix outside the beads. For this purpose, we probe the diffusion coefficients of 3-kDa and 70-kDa dextrans in the hydrogel matrixes at 25 °C and 36 °C with sample HG-25-A and HG-25-B. At 25 °C, the preparation temperature of pure PAAM and composite gels, we find identical tracer diffusion coefficients in the matrix. This is expected and demonstrates

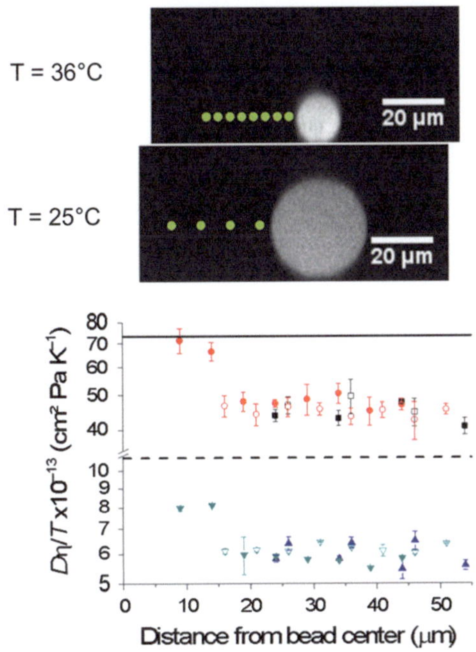

Fig. 6 Normalized diffusion coefficients of 70-kDa dextran (*triangles*) and 3 kDa dextran (*squares* and *circles*) in the hydrogel matrix close to the bead surface in sample HG-25-A (*filled symbols*) and HG-25-B (*open symbols*). *Squares* and *triangles* denote measurements at T = 25 °C, *circles* and *inverted triangles* at 36 °C. The diffusion coefficients were averaged over three different beads. The *horizontal lines* indicate the normalized diffusion coefficients of dextran in water; 70 kDa *horizontal dashed line*; 3 kDa dextran *solid line*

that the incorporation of the microgel beads does not affect the formation of the PAAM gel.

Figure 6 displays the diffusion coefficient of 70-kDa dextran in the hydrogel matrix of sample HG-25-A. In this plot, the diffusion coefficient is normalized by solvent viscosity, η, and T, to account for the trivial temperature dependence of diffusion. The normalized diffusion coefficients do not change when the sample is heated from 25 °C to 36 °C, and there is no dependence on the distance to the bead surface. For this sample, we have shown that an interpenetrating network is formed. However, the collapse of the bead does not affect the tracer diffusion in the surrounding matrix. The same result is found for 3-kDa dextran.

In contrast to the presence of an interpenetrating network in sample HG-25-A, there are no indications for an interpenetrating network in sample HG-25-B. The normalized tracer diffusion coefficients in the gel matrix of this sample are enhanced near the particle surface upon heating, as shown in Fig. 6.

We interpret this finding as follows: The data at 25 °C are taken at positions outside the beads, and when these positions are studied at 36 °C, the diffusion coefficient, normalized by the change of η and T, is not affected. Due to the strong collapse of the beads, however, it is possible to measure tracer diffusivities at positions that were occupied by the bead at 25 °C but are no longer occupied at 36 °C, see Fig. 6. These tests show that the tracer diffusivity increases in the volume formerly occupied by the bead when the bead is deswollen at 36 °C. The value of the increased normalized diffusion coefficient of 3-kDa dextran is the same as in water at 36 °C (Fig. 6, center). This suggests that the PAAM hydrogel is locally swollen so much by the water released from the collapsed bead that the small 3-kDa dextran is no longer restricted in its diffusion. The 70-kDa dextran also shows accelerated diffusion inside the volume previously occupied by the beads (Fig. 6, bottom). The normalized diffusion is, however, always smaller than in water.

These results show that the PAAM hydrogel matrix has swollen into the volume that was formerly occupied by the bead at 25 °C. The local segmental density of the PAAM hydrogel in these regions is lower than that in the bulk gel away from the bead, indicating that the gel matrix swells heterogeneously. The segmental density is still sufficiently high to hinder the diffusion of 70-kDa dextran compared to its diffusion in pure water, but the gel does no longer obstruct the diffusion of 3-kDa dextran. As the swelling of the matrix is localized to the volume close to the beads, it might be possible to form channels across the composite hydrogel when the bead concentration is so high that the beads are close together in the swollen state but apart from each other at higher temperatures when they are deswollen.

Summary

Incorporating thermosensitive PNiPAM nano- and microgels into a polyacrylamide matrix leads to temperature sensitive composite hydrogels. Different compositions have been prepared with different initialization reactions and at different preparation temperatures (above and below the VPTT). The results show that the temperature sensitivity of the embedded PNiPAM particles can be retained. A corset-like functionality of the hydrogel matrix can be achieved when the matrix is highly cross-linked or an interpenetrating network is formed.

The internal structure of the composite gel depends on the preparation conditions as well as on the size of the PNiPAM particles. IPNs can be formed with microgel beads, but hardly with nanogels. The composite gel does not reveal a temperature dependent volume change because the matrix takes up the water released by the embedded particles. Nevertheless, the temperature dependent swelling of the PNiPAM nano- and microgels affects mechanical [56] and optical properties of the composite hydrogel.

Furthermore the mobility of dissolved molecules depends on the local environment and is thus temperature dependent and can be tailored by the composition of the system. Microgels can be functionalized such that they provide specific binding sites or act as nanocontainers for enzymes [57–63]. In such cases, the local mobility of binding partners and substrates is of major relevance for applications.

Such composite gels provide new properties for sophisticated applications, e.g., in controlled uptake and release or sensing. It is possible to incorporate microgels of different size and functionality such that their sensitivity to different stimuli is retained as in solution, however, flocculation due to attractive particle interaction can be avoided.

Acknowledgements We thank the Deutsche Forschungsgemeinschaft for financial support within SPP 1259 and the Institute Laue-Langevin for SANS beam time at D11.

References

1. Langer R, Vacanti JP (1993) Science 260:920–926
2. McMahon TT, Zadnik K (2000) Cornea 19(5):730–740
3. Kuckling D (2009) Colloid Polym Sci 287(8):881–891
4. Haraguchi K (2011) Colloid Polym Sci 289:455–473
5. Yoshida R (2011) Colloid Polym Sci 289:475–487
6. Cohen Stuart MA, Huck WTS, Genzer J, Müller M, Ober C, Stamm M, Sukhorukov GB, Szleifer I, Tsurkruk VV, Urban M, Winnik F, Zauscher S, Luzinov I, Minko S (2010) Nat Mater 9:101–113
7. Heskins M, Guillet JE (1968) J Macromol Sci A2:1441–1455
8. Schild HG, Tirrel DA (1990) J Phys Chem 94:4352–4356
9. Peak CW, Wilker JJ, Schmidt G (2013) Colloid Polym Sci 291:2031
10. Haraguchi K, Xu Y (2012) Colloid Polym Sci 289:1627
11. Okay O, Oppermann W (2007) Macromolecules 40:3378–3387
12. Liu R, Milani AH, Freemont TJ, Saunders BR (2011) Soft Matter 7:4696–4704
13. Liu R, Milani AH, Saunders JM, Freemont TJ, Saunders BR (2011) Soft Matter 7:9297–9306
14. Lynch I, Dawson KA (2003) J Phys Chem B 107:9629–9637
15. Lynch I, de Gregorio P, Dawson KA (2005) J Phys Chem B 109:6257–6261
16. Salvat A, Söderman O, Lynch I (2007) J Phys Chem B 111:7367–7376
17. McGillicuddy FC, Lynch I, Rochev YA, Burke M, Dawson KA, Gallagher WM, Keenan AK (2006) J Biomed Mater Rest A 79:923–933
18. Soddemann M, Richtering W (2004) Prog Polym Sci 129:88–94
19. Raz N, Li JK, Fiddes LK, Tumarkin E, Walker GC, Kumacheva E (2010) Macromolecules 43:7277–7281
20. Pich A, Richtering W (2010) Adv Polym Sci 234:1–37
21. Landfester K, Musyanovych A (2010) Adv Polym Sci 234:39–63
22. Meyer S, Richtering W (2005) Macromolecules 38:1517–1519
23. Berndt I, Richtering W (2003) Macromolecules 36:8780–8785
24. Lehmann S, Seiffert S, Richtering W (2012) J Am Chem Soc 134:15963–15969
25. Müller CB, Richtering W (2008) Colloid Polym Sci 286:1215–1222
26. Müller CB, Loman A, Pacheco V, Koberling F, Willbold D, Richtering W, Enderlein J (2008) Eur Phys Lett 83(46001):1–5
27. Musch J, Schneider S, Lindner P, Richtering W (2008) J Phys Chem 112:6309–6314
28. Meid J, Friedrich T, Tieke B, Lindner P, Richtering W (2011) Phys Chem Chem Phys 13:3039–3047
29. Jones CD, Lyon LA (2003) Macromolecules 36:1988–1993
30. Jones CD, McGrath JG, Lyon LAJ (2004) Phys Chem B 108:12652–12657
31. Berndt I, Pedersen JS, Lindner P, Richtering W (2006) Langmuir 22:459–468
32. Gernandt J, Frenning G, Richtering W, Hansson P (2011) Soft Matter 7:10327–10338
33. Keerl M, Pedersen JS, Richtering W (2009) J Am Chem Soc 131:3093–3097
34. Richtering W, Pich A (2012) Soft Matter 8:11423–11430
35. Kishi R, Matsuda A, Miura T, Matsumura K, Iio K (2009) Colloid Polym Sci 187:505
36. Takekawa M, Takekawa M (2009) Colloid Polym Sci 287:323–334
37. Friedrich T, Tieke B, Meyer M, Pyckhout-Hintzen W, Pipich V (2010) J Phys Chem B 114:5666–5667
38. Strauss P, Knolle W, Naumov S (1998) Macromol Chem Phys 199:2229–2235
39. Chapiro A (1962) Radiation chemistry of polymeric systems. Interscience, New York
40. Stieger M, Richtering W, Pedersen JS, Lindner P (2004) J Chem Phys 120:6197–6206
41. Tabatabaei F, Lenz O, Holm C (2011) Colloid Polym Sci 289:523–534
42. Poggendorf S, Adama Mba G, Engel D, Sadowski G (2011) Colloid Polym Sci 289:545–559
43. Ponder E, Ponder RVJ (1960) Gen Physiol 43:753–758
44. Seiffert S, Oppermann W (2008) Polymer 49:4115–4126
45. Antoniou E, Themistou E, Sarkar B, Tsianou M, Alexandridis P (2010) Colloid Polym Sci 288:1301–1312
46. Modesti G, Zimmermann B, Borsch M, Herrmann A, Saalwachter K (2009) Macromolecules 42:4681–4689
47. Raccis R, Roskamp R, Hopp I, Menges B, Koynov K, Jonas U, Knoll W, Butt HJ, Fytas G (2011) Soft Matter 7:7042–7053
48. Burchard W, Richtering W (1989) Progr Colloid Polym Sci 80:151–163
49. Enderlein J, Gregor I, Patra D, Dertinger T, Kaupp UB (2005) Chem Phys Chem 6:2324–2336
50. Dertinger T, Pacheco V, von der Hocht I, Hartmann R, Gregor I, Enderlein J (2007) Chem Phys Chem 8:433–443
51. Müller CB, Eckert T, Loman A, Enderlein J, Richtering W (2009) Soft Matter 5:1358–1366
52. Enderlien J, Gregor I, Patra D, Fitter J (2004) Curr Pharm Biotechnol 5:155–161
53. Shibayama M (1998) Macromol Chem Phys 199:1–30
54. Nie J, Du B, Oppermann W (2004) Macromolecules 37:6558–6564
55. Enderlein J (2012) Phys Rev Lett 108:108101
56. Meid J, Dierkes F, Cui J, Messing R, Crosby AJ, Schmidt A, Richtering W (2012) Soft Matter 8:4254–4263
57. Karg M (2012) Colloid Polym Sci 290:673–688
58. Xu Y, Pharand L, Wen Q, Gonzaga F, Li Y, Ali MM, Filipe CDM, Pelton R (2010) Colloid Polym Sci 289:659–666
59. Schachschal S, Adler H-J, Pich A, Wetzel S, Matura A, Pee K-H (2011) Colloid Polym Sci 289:693–698
60. Sawada S, Sasaki Y, Nomura Y, Akiyoshi K (2011) Colloid Polym Sci 289:685–691
61. Welsch N, Becker AL, Dzubiella J, Ballauff M (2012) Soft Matter 8:1428–1436
62. Gawlitza K, Wu C, Georgieva R, Wang D, Ansorge-Schumacher MB, Klitzing RV (2012) Phys Chem Chem Phys 14:9594–9600
63. Wiese S, Spiess AC, Richtering W (2013) Angew Chem 125:604–607. Angew Chem Int Ed 52:576–579

In-situ Generation of Gold, Platinum and Palladium Nanoparticles in N-isopropylacrylamide-surfmer Copolymer Hydrogels and Catalytic Activity of the Gels

Katharina Kraus, Tamara Mielke, Tatjana Friedrich, and Bernd Tieke

Abstract

Preparation and catalytic activity of N-isopropylacrylamide (NIPAM)/surfmer copolymer hydrogels containing gold, palladium and platinum nanoparticles are described. The functional hydrogels were obtained in three steps. First a thermoresponsive copolymer gel was prepared upon ^{60}Co-gamma irradiation of a micellar aqueous solution of NIPAM and (2-methacryloyloxyethyl)dodecyldimethylammonium bromide (MEDDAB) or (11-methacryloyloxyundecyl)trimethylammonium bromide (AUTMAB). Secondly tetrachlorometalate anions were introduced in the gel upon ion exchange of the bromide counterions of the positively charged surfmer head groups. For this purpose, the gel was de-swollen at 50 °C and re-swollen in aqueous solution of tetrachloroauric acid, sodium tetrachloropalladate, or sodium tetrachloroplatinate at 20 °C. Thirdly, the metal nanoparticles were in-situ generated by exposing the de-swollen gel to an aqueous solution of sodium borohydride, which led to re-swelling and reduction of the complex anions. The formation of nanoparticles caused an immediate colour change of the gel from yellow or brown to golden black or black. TEM studies indicate aggregates of nanoparticles of approximately 3.5–5.3 nm in diameter. The metal-containing gels catalyze the reduction of 4-nitrophenolate (4-NP) into 4-aminophenolate below the lower critical solution temperature (LCST) and at high osmotic pressure, where the gels are highly swollen. NIPAM/MEDDAB gels are active at 20 °C and 30 °C, whereas at 40 °C (readily above the LCST of 29 °C) the activity is low and comparable with the metal-free gel. NIPAM/AUTMAB gels with large counterion dissociation are still active at 40 °C although the LCST is only 33.8 °C. Further studies indicate that the gold-containing gels exhibit the highest catalytic activity, and that the gels can be repeatedly used for catalysis because the metal nanoparticles are only slowly washed out.

Keywords

Hydrogel • Gold • Palladium • Platinum • Nanoparticle • Catalytic activity • N-isopropylacrylamide • Surfactant monomer

Introduction

In recent years intelligent hydrogels have been subject to numerous research studies [1] because of their reversible volume phase transitions [2] in response to external stimuli such as temperature [3], pH [4], magnetic field [5], light [6], solvent compositions [7] and ionic strength [8]. Most of the research activities were concerned with hydrogels based on cross-linked poly-N-isopropylacrylamide (pNIPAM), which exhibit a reversible volume transition from a clear, swollen, hydrophilic state to a turbid, shrunken, hydrophobic state, if heated to temperatures higher than 32 °C [9]. As recently demonstrated, the poor mechanical stability of

K. Kraus • T. Mielke • T. Friedrich • B. Tieke (✉)
Department of Chemistry, University of Cologne, D-50939 Cologne, Germany
e-mail: tieke@uni-koeln.de

G. Sadowski and W. Richtering (eds.), *Intelligent Hydrogels*, Progress in Colloid and Polymer Science 140,
DOI 10.1007/978-3-319-01683-2_9,

pure pNIPAM hydrogels can be considerably improved if copolymer hydrogels of NIPAM and surfactant monomers (surfmers) are prepared [10–13]. The copolymer gels can be obtained in a single-step reaction by exposing a micellar aqueous comonomer solution to gamma irradiation.

The use of cationic surfmers (surfactant monomers, polymerizable surfactants) as comonomer results in the formation of positively charged gels. As recently shown, the halide counterions can be easily exchanged for other anions [14–17]. By means of the ion exchange, functional hydrogels with interesting photophysical [14] and photochemical properties [15] could be prepared. Inorganic nanoparticles could be generated in-situ such as Prussian Blue particles [16] or catalytically active palladium nanoparticles [16]. Especially the in-situ generation of metallic nanoparticles is of interest, because a new type of catalytic gel with versatile and repeated use may become accessible. In a previous publication [16] we briefly communicated the catalytic activity of Pd-containing NIPAM-MUTMAB copolymer gels prepared upon radical copolymerization of NIPAM and MUTMAB (11-methacryloyloxyundecyl)trimethylammonium bromide in micellar solution. In our present article, a comprehensive study on the in-situ preparation of gold, platinum and palladium nanoparticles in NIPAM-surfmer copolymer gels and their use as catalysts in hydrogenation reactions is presented. Different from the previous communication, the gels were prepared upon radiation-induced copolymerisation of NIPAM and (2-methacryloyloxyethyl) dodecyldimethylammonium bromide [18, 19] (MEDDAB) or (11-acryloyloxyundecyl)trimethylammonium bromide (AUTMAB) [20] in an aqueous micellar solution. The two surfmers were chosen because of their different structure with the reactive unit either at the tail or head of the molecule (T- or H-type surfmer), which is accompanied by either a strong or weak counterion dissociation [20] leading to hydrogels with either strong or weak osmotic pressure and swelling behaviour. Use of gamma irradiation is advantageous because it homogeneously penetrates the sample and no additives such as initiator, accelerator, or cross-linker are needed. After in-situ preparation of the particles, the gels were studied on their catalytic activity in the reduction of 4-nitrophenol (4-NP) to yield 4-aminophenol. Spectroscopic and microscopic measurements were used in order to characterize the gel structure and morphology, and their catalytic activity.

Experimental Part

Materials

(2-Methacryloyloxyethyl)dodecyldimethylammonium bromide (MEDDAB) and (11-acryloyloxyundecyl)trimethylammonium bromide (AUTMAB) were synthesized according to the literature [18–20]. N-Isopropylacrylamide (NIPAM) was recrystallized from toluene and n-hexane (volume ratio: 1:2). MilliQ water was used for all experiments. Sodium tetrachloropalladate, sodium tetrachloroplatinate, and tetrachloroauric acid were purchased from Acros and used without further purification. MilliQ water was used for all experiments.

Preparation of hydrogels. Hydrogel samples were prepared in screw-capped glass tubes at 20 °C. Monomer solutions containing 10 wt% NIPAM and 1 wt% MEDDAB in degassed MilliQ water were homogenized using a Vortex mixer, aged for 24 h and copolymerized in a one-step reaction. The gels were copolymerized using ^{60}Co-gamma-irradiation, the dose rate being 300 Gy/h, and the total radiation dose being 80 kGy. The irradiation was carried out at Beta Gamma Service in Wiehl, Germany.

Exchange of counterions. The exchange of bromide counterions for inorganic complex ions was accomplished by using the thermosensitive swelling and shrinking of the gels as described in Scheme 1 [14–16]. First the gels were heated to 50 °C for 8 h to cause shrinking. Then they were re-swollen in an aqueous solution containing the inorganic complex anions at 20 °C for time periods between 1 and 21 days. The concentration of the aqueous solutions containing Na_2PdCl_4, $HAuCl_4$ or Na_2PtCl_4 was always 1 mM. The shrinking/re-swelling cycle was followed by two further shrinking/re-swelling cycles in pure MilliQ water in order to wash out all anions which were not fixed to the gel matrix.

In situ generation of Au^0, Pd^0, and Pt^0 nanoparticles. Hydrogels containing $AuCl_4^-$, $PdCl_4^{2-}$, and $PtCl_4^{2-}$ were shrunken at 50 °C for 8 h and re-swollen in an aqueous $NaBH_4$ solution at 20 °C for 24 h, the sodium borohydride concentration being 13 mM. The shrinking/re-swelling cycle was followed by two further shrinking/re-swelling cycles in pure MilliQ water in order to remove excess sodium borohydride or reaction products from the reduction of the inorganic complex anions.

Reduction of 4-nitrophenol with sodium borohydride. The reduction of 4-nitrophenol was carried out in an aqueous solution of 4-nitrophenol, the concentration being 0.05 mmol/L. The catalytic activity was studied at 20 °C and 30 °C, i.e., below the volume phase transition temperature of the gel, and at 40 °C, i.e., above the transition temperature. Prior to the experiments, the copolymer gels containing the nanoparticles were re-swollen to their original size in MilliQ water. Then the gel was put in the aqueous 4-nitrophenol solution and 5 mmol/L sodium borohydride were added. In order to study the catalytic reduction quantitatively, every 10 min a sample was taken from the 4-nitrophenol solution and a UV-spectrum was recorded. For comparison, the reaction was also studied without addition of the catalytic gel.

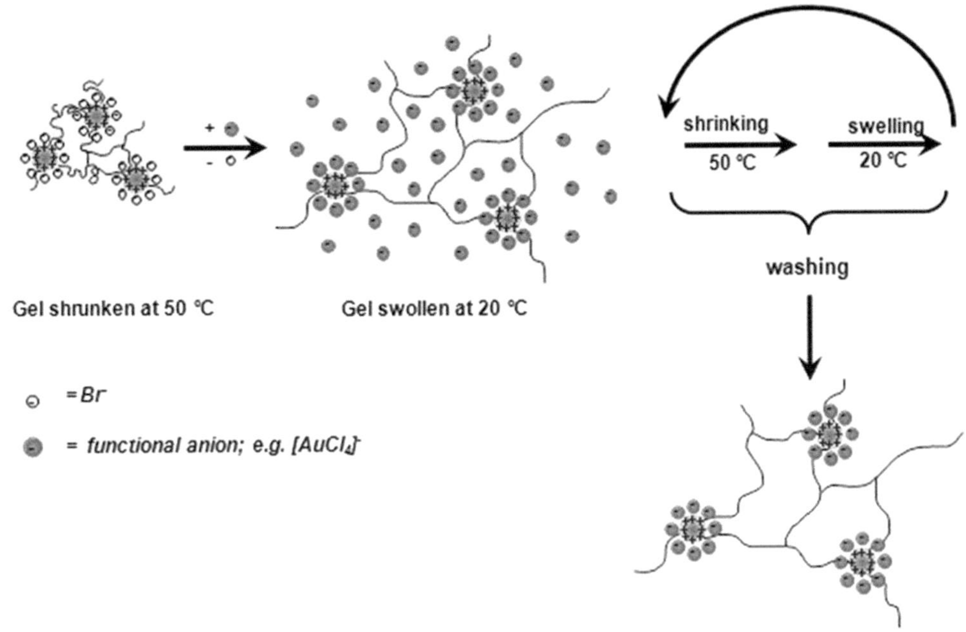

Scheme 1 Model of electrostatic self-assembly of functional anions

Methods

UV/Vis absorption spectra were recorded using a Perkin Elmer Lambda 14 spectrometer. The spectra were corrected by subtracting the background signal. Energy dispersive X-ray (EDX) spectroscopy was performed using a Zeiss Neon 40 scanning electron microscope (SEM) equipped with an Inca Dry Cool EDX detector for microanalysis. The gels were dried in an oven at 50 °C, fixed to the microscope holder using conductive silver paste, and coated with carbon (EMITECH K950X). The software Inca was used for the EDX measurements, and the experimental error was about 5 % of the measured value. TEM measurements were carried out using a LEO EM 912 Omega electron microscope.

For the study of the swelling/de-swelling behaviour the hydrogels were immersed in 150 ml Milli-Q-water (at 20 °C or 50 °C) for different time periods. The water uptake (or release) was determined gravimetrically by weighing the hydrogel prior and subsequent to the water treatment. The degree of swelling S was calculated as

$$S = \frac{W_t - W_0}{W_0}, \quad (1)$$

where W_t and W_0 are the weights of the sample subsequent and prior to swelling in water. S was determined as a function of time at constant temperature.

Results and Discussion

Preparation of the Hydrogels

Let us first describe the preparation and functionalization of the gels. The copolymer hydrogels were prepared from a micellar aqueous solution of NIPAM and one of the two surfactant comonomers (2-methacryloxyethyl)dodecyldimethyl ammonium bromide (MEDDAB) or (11-acryloyloxy-undecyl)trimethylammonium bromide (AUTMAB) using γ-irradiation. The radiation dose was always 80 kGy, the dose rate was ca. 300 Gy/h. The low dose rate was chosen after it was found that dose rates higher than 1 kGy/h lead to brittle gels with high cross-linking density, recognizable by the formation of voids and crazes in the samples. At dose rates lower than 500 Gy/h homogeneous and defect-free gels were obtained. Due to the copolymerization with the cationic surfmer, the resulting hydrogels were positively charged. The hydrogels were always prepared from comonomer solutions containing 10 wt% NIPAM and 1 wt% surfmer.

In-situ Generation of Metal Nanoparticles

After preparation the NIPAM/MEDDAB gels were used for in-situ generation of gold, palladium, and platinum nanoparticles, whereas the NIPAM/AUTMAB gels were only used for generation of palladium particles. Let us first describe

Fig. 1 Photographic images of NIPAM/MEDDAB copolymer hydrogel before (**a–c**, *left*) and after electrostatic adsorption of tetrachlorometalate anions (*middle*), the anions being tetrachloroaurate (**a**), tetrachloropalladate (**b**), and tetrachloro-platinate (**c**), and after reduction with $NaBH_4$ to yield gold (**a**), palladium (**b**), and platinum nanoparticles (**c**) in the gel (*right*)

the generation of gold particles in NIPAM/MEDDAB gels. A method recently applied for preparation of gold nanoparticles in polyelectrolyte hydrogels was adapted [21]. First the freshly prepared gels were de-swollen at 50 °C for 8 h. Then the gels were immersed and re-swollen in an aqueous solution of tetrachloroauric acid. The hydrogel samples changed their color from originally colorless to red orange indicating the ion exchange (Fig. 1). It took several days until the exchange process was complete. After approximately a week the gels were repeatedly shrunken and re-swollen in pure water in order to remove excess tetrachloroaurate ions, which were not electrostatically bound to the gel matrix (see also Scheme 1). For preparation of the nanoparticles, the gels were first de-swollen at 50 °C, and then swollen in an aqueous solution containing sodium borohydride. The borohydride anions reduced the $AuCl_4^-$ anions to form Au^0 nanoparticles. This process was accompanied by a color change of the gel from red orange to black-golden (Fig. 1a). The distribution of the nanoparticles in the gels was strongly a function of the exposure time to the tetrachloroaurate solution. Cross-sections of gels immersed in the solution for only 24 h indicate a very inhomogeneous distribution of the gold particles after the reduction step, the center of the gel being nearly colorless without any particles present (Fig. 2). However, if the gel was immersed for 3 weeks, the distribution of the particles was homogeneous.

For generation of Pd^0 and Pt^0 nanoparticles the NIPAM/MEDDAB gels were swollen in aqueous sodium tetrachloropalladate or sodium tetrachloroplatinate solutions, and subsequently reduced with sodium borohydride. The method was analogous to the preparation of palladium particles in core-shell microgels [22], or multilayered polyelectrolyte films reported recently [23]. With tetrachloropalladate a brown gel was obtained, which upon treatment with borohydride anions turned black (Fig. 1b), while with tetrachloroplatinate a yellow gel was obtained, which after reduction also turned black (Fig. 1c). The method for generation of palladium nanoparticles in the NIPAM/AUTMAB gels was the same as described above for the MEDDAB-based gels.

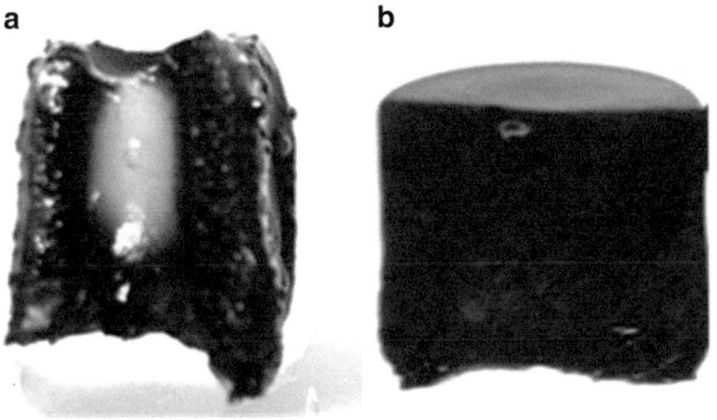

Fig. 2 Photographic images of NIPAM/MEDDAB copolymer hydrogels functionalized with tetrachloroaurate for 24 h (**a**) and 3 weeks (**b**) and subsequent treatment with sodium borohydride

In order to determine the concentration of the metal nanoparticles at or near to the surface of the gels, EDX measurements were carried out. The N:metal ratio (atom-%) was determined after reduction of the tetrachlorometalate ions with sodium borohydride. For the gold-containing NIPAM-MEDDAB gel, the N:Au ratio was 77.3: 22.7, while for the other MEDDAB-containing gels, the N:Pd and N:Pt ratios were 97.2:2.8, i.e., the exchange of bromide for tetrachlorometalate ions was essentially complete. Also, EDX-measurements of Pd-containing NIPAM/AUTMAB hydrogels indicated a high N:Pd ratio of 97:3. Corresponding EDX spectra are shown in Fig. 3.

Swelling Behavior of Gels Containing Metal Nanoparticles

The question now was, if the presence of the metal nanoparticles changes the reversible swelling behavior of the hydrogels. In order to answer the question, the swelling behavior of the metal-containing gels was investigated. For this purpose, the gels were immersed in 150 mL of water at 20 °C, and the weight was determined every 30 min for a time period of 8 h. Prior to weight determination the gel was put in a sieve and dried at the surface. In order to study de-swelling, the gel was kept in 150 mL of water at 50 °C and the weight decrease was determined every 30 min as described above for the swelling. For comparison, a metal-free gel was treated analogously and the swelling and de-swelling behavior was also studied. The swelling degree S was determined as indicated in the Experimental part. In Fig. 4, the swelling degrees of the metal-free and metal-containing gels are plotted versus time t. The metal-free NIPAM-MEDDAB gel exhibits a strong swelling at 20 °C, the change in S, ΔS, being 3.1 g/g after 8 h. The swelling is fully reversible, if the gel is kept at 50 °C. The palladium-containing NIPAM-MEDDAB gel exhibits a lower uptake of water at 20 °C, ΔS being 1.7 g/g after 8 h. It is reversible, if the gel is subsequently kept at 50 °C. Consequently the weight difference between the swollen and the shrunken state is smaller for the palladium-containing gel. For the gold and platinum-containing NIPAM-MEDDAB gels, the water uptake was even lower (Au: $\Delta S = 0.25$ g/g after 6 h; Pt: $\Delta S = 0.35$ g/g after 6 h). The reason might be that the metal nanoparticles impart hydrophobic interactions with the polymer chains in the gel so that the swelling becomes less pronounced. The metal-free NIPAM/AUTMAB hydrogels exhibit a strong swelling at 20 °C with ΔS being 5 g/g after 8 h. Again the S-value of the metal-containing hydrogel is much lower (ΔS being = 1.6 g/g after 8 h). If the gels are subsequently kept at 50 °C, the swelling is fully reversible.

Morphology of Gels Containing Metal Nanoparticles

The morphology of the metal-containing gels was studied for the example of the palladium containing NIPAM/MEDDAB gel. In order to determine the size, the nanoparticles were leached out from the network. For this purpose, the gel was mechanically crushed and ultrasonicated. Subsequently the nanoparticle suspension was dropped on a copper grid with carbon net and dried. Then TEM pictures were taken at room temperature. As shown in Fig. 5a, small partially aggregated particles with average diameter of 3.6 nm can be detected. NIPAM/AUTMAB hydrogels containing Pd^0 nanoparticles were treated in the same manner as the metal-containing MEDDAB-based gels. In Fig. 5a, b TEM picture of the nanoparticles suspension on a copper grid with carbon net is shown. The picture was taken at room temperature. Small aggregated particles with average diameter of 5.3 nm can be detected. They are slightly larger than the particles leached out from the NIPAM/MEDDAB hydrogel.

Catalytic Activity of Copolymer Hydrogels Containing Metal Nanoparticles

The catalytic activity of the metal-containing copolymer hydrogels was studied by investigating the reduction of 4-nitrophenol (4-NP) with sodium borohydride to yield 4-aminophenol. For mechanistic details of the process see a recent review [24]. The conversion was followed UV spectroscopically in aqueous solution by monitoring the maximum absorbance of the 4-nitrophenolate anion at $\lambda = 400$ nm.

NIPAM/MEDDAB copolymer hydrogels. The catalytic activity of metal-containing NIPAM/MEDDAB copolymer hydrogels was studied at 20 °C, 30 °C and 40 °C, i.e., at

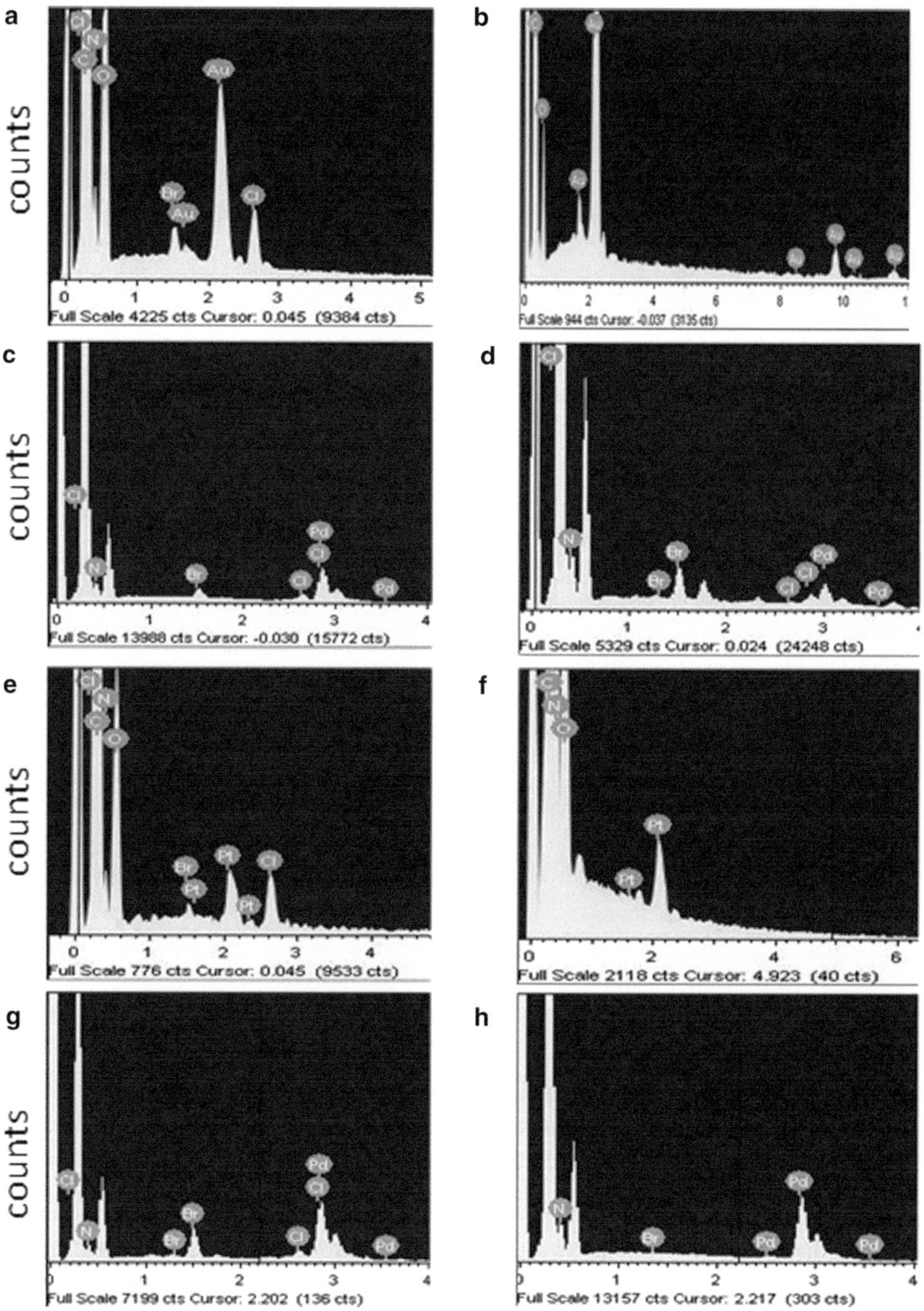

Fig. 3 EDX measurements of a NIPAM/surfmer copolymer hydrogels after exchange of bromide for metal complex anions, and formation of metal nanoparticles. Samples: NIPAM/MEDDAB hydrogels with tetrachloroaurate (**a**), tetrachloro-palladate (**c**), and tetrachloroplatinate anions (**e**), and after formation of Au^0 (**b**), Pd^0 (**d**), and Pt^0 (**f**) nanoparticles, and NIPAM/AUTMAB hydrogel with tetrachloropalladate anions (**g**), and after formation of Pd^0 nanoparticles (**h**)

temperatures below and above the volume phase transition temperature (lower critical solution temperature, LCST) of the metal-free hydrogel at about 29 °C. Aqueous solutions of 4-NP containing hydrogels with gold, palladium, platinum, or no particles were studied. In Fig. 6, absorption spectra of 4-NP in aqueous solution of pH 7 are shown. The spectra indicate the decrease of the absorbance due to the proceeding reduction at 30 °C. For the gel without metal nanoparticles the decrease was less than 30 % even 3 h after addition of the reducing agent (Fig. 6a). However, if metal nanoparticles are present, the reduction rate was accelerated. In presence of Pd reduction was almost complete within 80 min (Fig. 6b),while

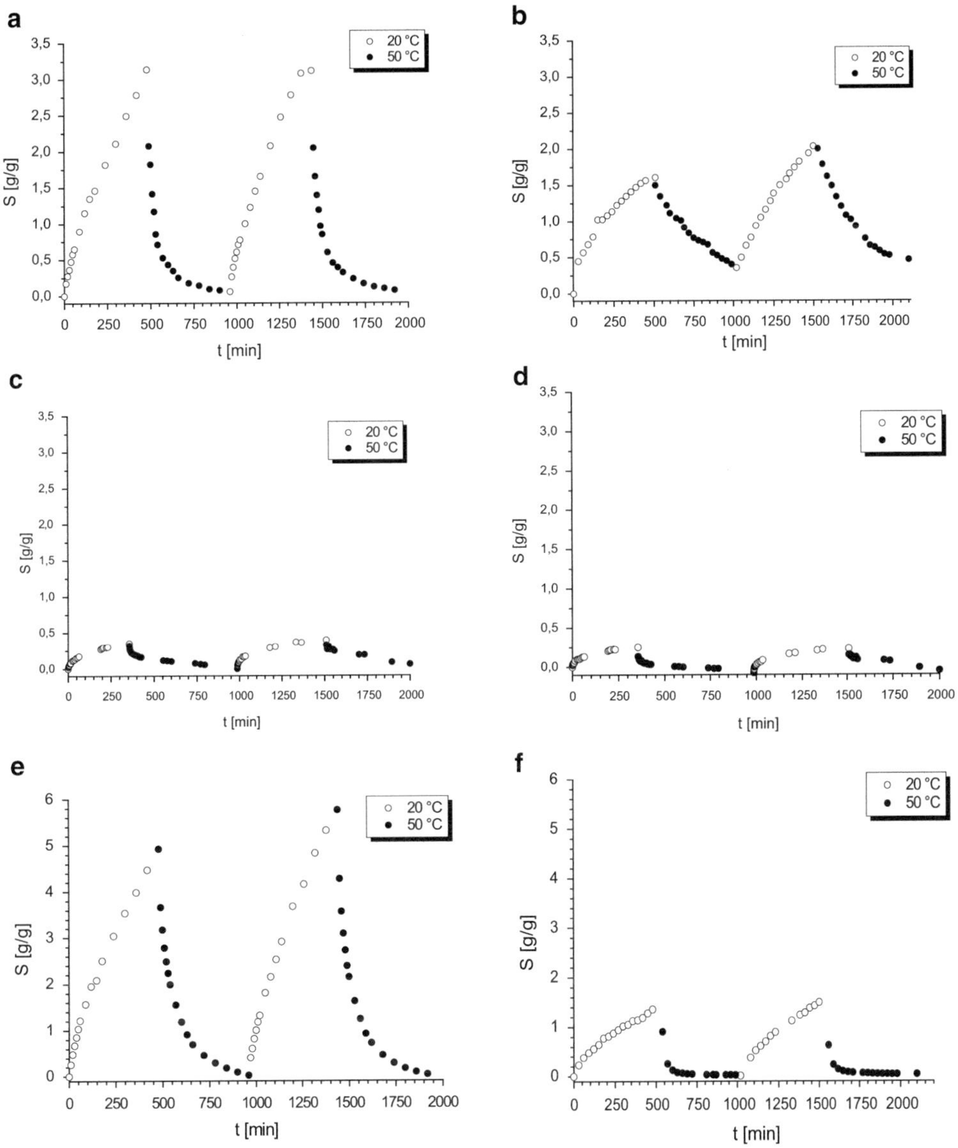

Fig. 4 Alternate swelling at 20 °C and shrinking at 50 °C of NIPAM/surfmer hydrogels without and with metal nanoparticles. Samples: NIPAM/MEDDAB gel without (**a**), and with Pd^0 (**b**), Pt^0 (**c**), and Au^0 nanoparticles (**d**), and NIPAM/AUTMAB gel without (**e**), and with Pd^0 nanoparticles (**f**)

in presence of Au or Pt it was complete after 30 or 12 min, respectively (Fig. 6c, d). At 20 °C and 40 °C, the decrease was considerably slower (spectra are not shown).

Since sodium borohydride was added in excess, a pseudo-first order kinetics can be assumed for the reduction. In this case the conversion can be calculated using Eq. 2:

$$\ln\left(c_t/c_o\right) = \ln\left(A_t/A_o\right) = -kt \tag{2}$$

with c_o and c_t being the concentration of 4-NP before and after reduction time t, k being the rate constant, and A_o and A_t being the absorbance of 4-NP before and after reduction time t. From the linear slope in the plot of $\ln(A_t/A_o)$ versus t (Fig. 7) the rate constant k could be easily determined. For the catalyst-free system at 30 °C, k was 0.0045 s^{-1}, while for the gold-, palladium- and platinum-containing gels it was 0.228, 0.051, and 0.132 s^{-1}, respectively.

Similar experiments were carried out at 20 °C and 40 °C. At 20 °C the rate constants were significantly lower than at 30 °C (Table 1) because counterion dissociation and osmotic pressure π originating from the charged surfmer head groups are lower [20, 25]. With increasing π the swelling increases and more 4-NP molecules are introduced in the gel. At 20 °C, S of the metal-free gel is 2.1 g/g after 6 h swelling, while at 30 °C it is 2.5 g/g. If the reduction is carried out at 40 °C, i.e. above the LCST of the gel (LCST of the metal-free

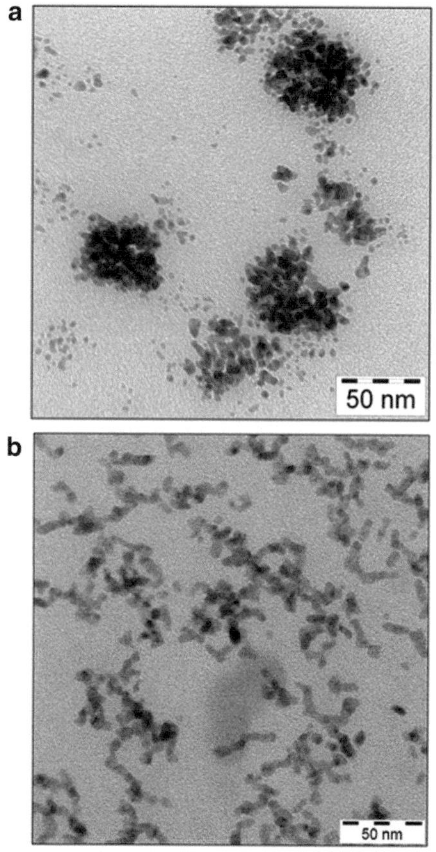

Fig. 5 TEM image of Pd^0 nanoparticles of a NIPAM/MEDDAB (**a**) and a NIPAM/AUTMAB (**b**) copolymer hydrogel network

Table 1 Rate constants k for reduction of 4-NP at different temperatures in presence of NIPAM/surfmer hydrogels with and without metal nanoparticles

		k [s^{-1}] at			
Hydrogel	Metal	20 °C	30 °C	40 °C	50 °C
NIPAM/ MEDDAB	Pd^0	0.025	0.051	0.0048	–
	Au^0	0.033	0.228	0.0071	–
	Pt^0	0.032	0.132	0.0034	–
	No metal	0.0008	0.0045	0.0080	–
NIPAM/ AUTMAB	Pd^0	0.016	0.048	0.064	0.020
	No metal	0.0009	0.0049	0.0078	0.022

gel is 29 °C), the gels are hydrophobic and shrunken, the nanoparticles are shielded by the polymer, and 4-NP cannot enter the gel. Consequently the rate constants of the metal-containing gels are very low, comparable with the metal-free gel (Table 1).

In Fig. 8a, an Arrhenius plot of the rate constants k versus $1/T$ is shown. Three main results are obtained. First, the metal-containing gels are catalytically active below the LCST, the activity increasing with temperature because of the increasing diffusion of the aqueous phase through the gel and improved contacts between the 4-NP molecules with the metal nanoparticles. Second, above the LCST the catalytic activity drops to the low value found for the metal-free hydrogel because the aqueous solution is expelled from the gel. The contact between 4-NP molecules and metallic nanoparticles is interrupted. Third, the catalytic activity decreases in the series Au->Pt->Pd-containing gel. The high activity of the gold-containing gel can be ascribed to the higher concentration of the gold nanoparticles as indicated by the EDX measurements (see above). The different particle concentration may originate from different interactions of the metals with the polymer matrix, metals with weak interactions being washed out more easily.

NIPAM/AUTMAB copolymer hydrogels. The catalytic activity of palladium-containing NIPAM/AUTMAB copolymer hydrogels was studied at 20 °C, 30 °C, 40 °C and 50 °C, i.e., below and above the LCST of the gel at 34 °C. The reaction conditions were the same as described above for the NIPAM/MEDDAB hydrogels. Again the decrease of the absorption maximum at $\lambda = 400$ nm of 4-NP at different temperatures (20 °C, 30 °C, 40 °C and 50 °C) was evaluated and $\ln(A_t/A_0)$ was plotted versus the reaction time (spectra at 30 °C are shown in Fig. 6e). Linear plots were obtained (Fig. 7e) from which the k values were determined (see Table 1). As for the metal-containing NIPAM/MEDDAB gels, a high rate constant k at 30 °C ($k = 0.048\ s^{-1}$) is found for the Pd-containing AUTMAB-based gels, but surprisingly the highest rate constant was found at 40 °C ($k = 0.064\ s^{-1}$). Small angle neutron scattering (SANS) measurements carried out previously [17] indicated a strong head group dissociation for AUTMAB in micellar aggregates, much stronger than for MEDDAB micelles. Thus the AUTMAB-containing gels exhibit a stronger osmotic pressure, which compensates the de-swelling at 40 °C, i.e., above the LCST of the metal-free gel at 33.8 °C [25]. Only at 50 °C the NIPAM/AUTMAB gels exhibit the expected decrease of k, because the de-swelling now dominates and 4-NP molecules are no longer in contact with the catalyst particles within the gel. In Fig. 8b an Arrhenius plot of the rate constants k as a function of 1/T is shown for the Pd-containing and Pd-free NIPAM/AUTMAB hydrogels. The experiment indicates that the catalytic activity is strongly related to the swelling of the gels. For charged hydrogels the swelling is both a function of their LCST and their osmotic pressure.

Conclusions

In our study we report on the successful functionalization of NIPAM/MEDDAB and NIPAM/AUTMAB copolymer hydrogels with catalytically active metal nanoparticles.

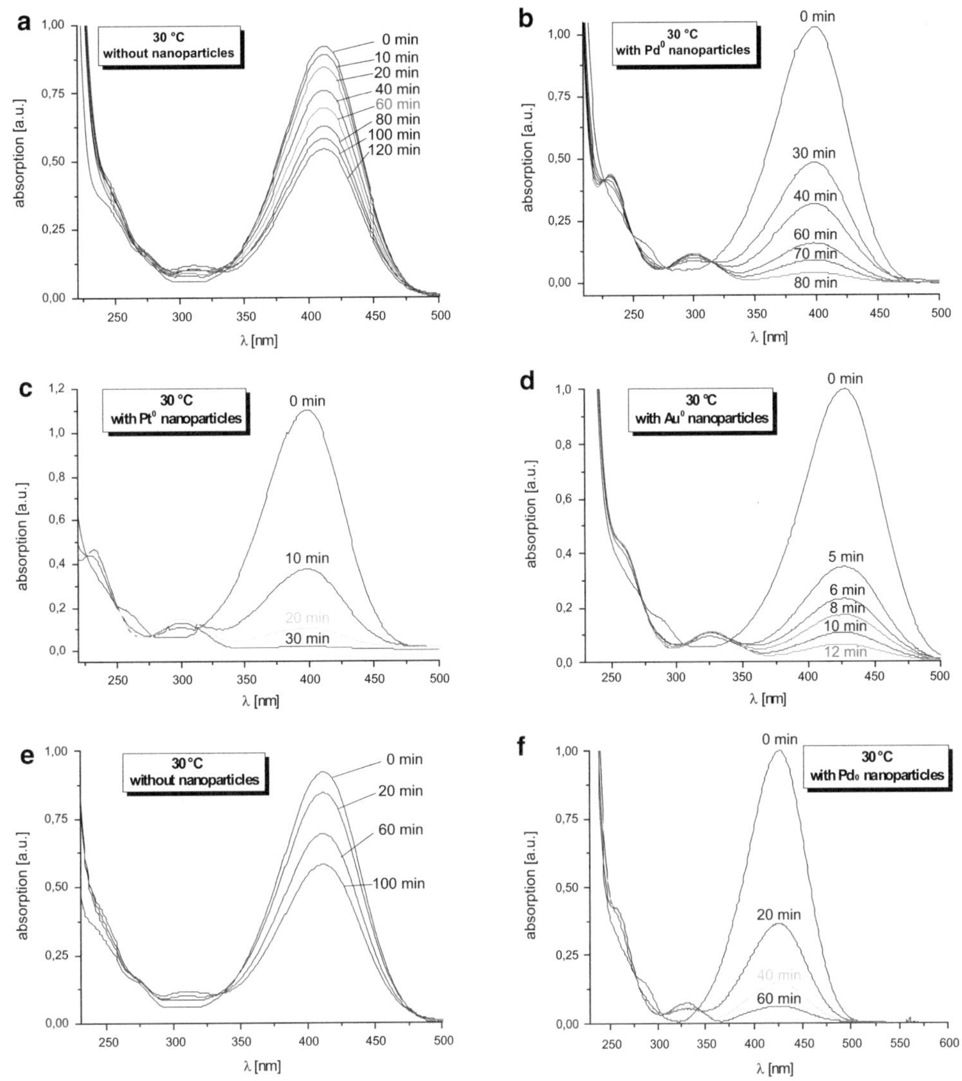

Fig. 6 Reduction of 4-nitrophenol with excess $NaBH_4$ in presence of NIPAM/surfmer hydrogels with and without metal nanoparticles. Samples: NIPAM/MEDDAB hydrogels without (**a**), and with Pd^0 (**b**), Pt^0 (**c**) or Au^0 nanoparticles (**d**), and NIPAM/AUTMAB hydrogel without (**e**) and with Pd^0 nanoparticles (**f**). Reaction conditions: [4-NP] = 0.05 mmol/L, [$NaBH_4$] = 5 mmol/L, T = 30 °C

Gold, platinum and palladium particles of approximately 3–5 nm in size could be prepared in-situ upon reduction of inorganic complex anions with sodium borohydride. The formation was proven by UV spectroscopy, EDX- and TEM-measurements.

The catalytic activity of the metal nanoparticles was demonstrated by studying the reduction of 4-nitrophenol with an excess of sodium borohydride. It was found that the catalytic activity is strongly controlled by the hydrophilicity of the NIPAM/surfmer-based copolymer gels. The hydrophilicity depends on the volume phase transition temperature, and the osmotic pressure of the gels. Below the phase transition temperature the gels are highly swollen and 4-NP molecules can easily diffuse through the gel and become reduced. Above the phase transition a penetration into the gel is rendered difficult, the gels are hydrophobic and the aqueous 4-NP solution is expelled from the gel, the catalyst is shielded by the polymer. The only exception is a high osmotic pressure originating from a strong head group dissociation, which may compensate the loss in hydrophilicity above the LCST in a small temperature range. This behavior was observed for the NIPAM/AUTMAB gels.

The catalytic activity of the gold containing NIPAM/MED DAB gels was highest because more gold than platinum or palladium nanoparticles could be prepared in the gel. The reason might be that these particles are better fixed to the gel matrix via coordinative interactions. Due to interactions with the gel matrix the metal particles were generally leached out from the gels rather slowly so that the gels could be repeatedly used for reduction experiments. Because of their excellent properties the metal nanoparticle-containing hydrogels are well suited for catalytic applications.

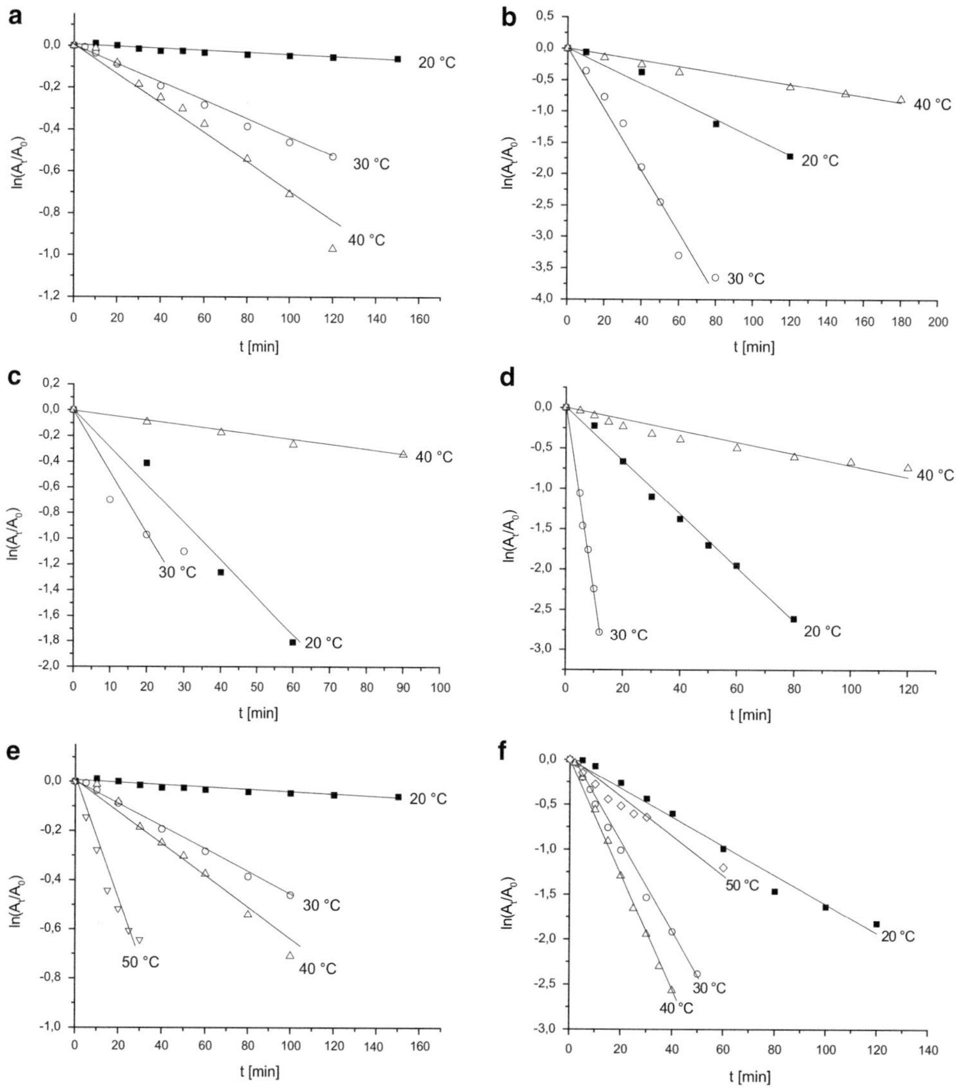

Fig. 7 Plot of $\ln(A_t/A_o)$ versus time for reduction of 4-nitrophenol with excess $NaBH_4$ in presence of NIPAM/surfmer hydrogel at different temperatures. Samples: NIPAM/MEDDAB hydrogel without (**a**), and with Pd^0 (**b**), Pt^0 (**c**) or Au^0 nanoparticles (**d**), and NIPAM/AUTMAB hydrogel without (**e**) and with Pd^0 nanoparticles (**f**). Reaction conditions: [4-NP] = 0.05 mmol/L, [$NaBH_4$] = 5 mmol/L. *Solid lines* are linear fits through the experimental points

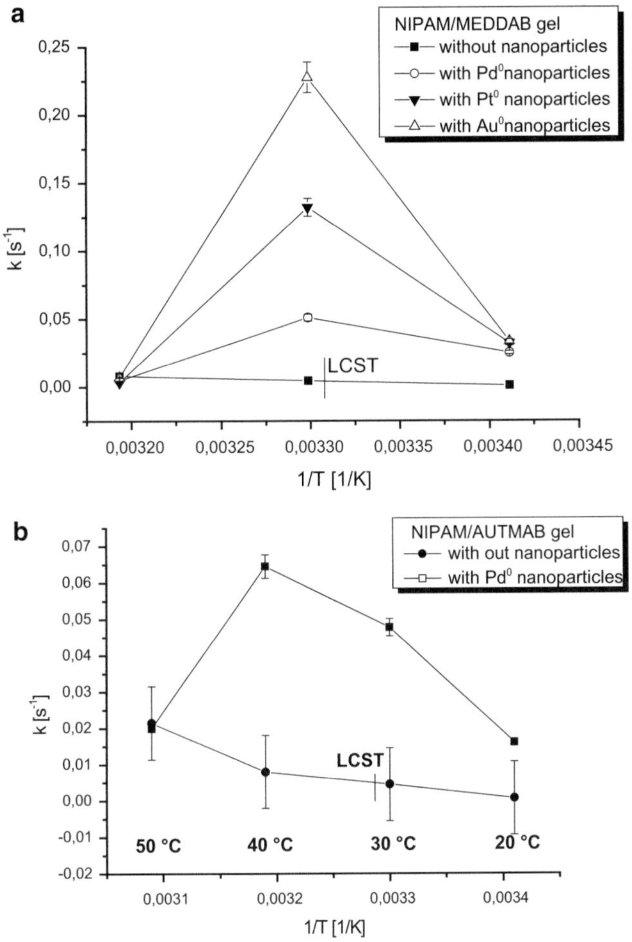

Fig. 8 Arrhenius plot of the rate constant k versus $1/T$ for reduction of 4-nitrophenol with $NaBH_4$ in presence of NIPAM/surfmer hydrogels with and without metal nanoparticles. Samples: NIPAM/MEDDAB gels (**a**), and NIPAM/AUTMAB gel (**b**). Reaction conditions: [4-NP] = 0.05 mmol/L, [$NaBH_4$] = 5 mmol/L. The LCSTs of the metal-free gels are indicated. *Solid lines* are connections of the data points

Acknowledgements The DFG is thanked for financial support (project TI 219/10-3, priority program 1256 'Intelligent hydrogels'). Prof. Dr. K. Meerholz is thanked for providing the SEM and Ruth Bruker is thanked for carrying out the EDX measurements. Dr. L. Belkoura is thanked for the TEMmeasurements.

References

1. Schild HG (1992) Prog Polym Sci 17:163
2. Tanaka T (1978) Phys Rev Lett 40:820
3. Hirokawa Y, Tanaka T (1986) J Chem Phys 81:6379
4. Li Y, Tanaka T (1992) Annu Rev Mater Sci 22:243
5. Xulu PM, Filipcsei G, Zrinyi M (2000) Macromolecules 33:1716
6. Suzuki A, Tanaka T (1990) Nature 346:345–347
7. Katayama S, Hirokawa Y, Tanaka T (1984) Macromolecules 17:2641
8. Ohmine I, Tanaka T (1982) J Chem Phys 77:5725
9. Heskins M, Guillet JE (1968) J Macromol Sci Chem A2:1441–1455
10. Noguchi Y, Okeyoshi K, Yoshida R (2005) Macromol Rapid Commun 26:1913–1917
11. Li W, Au H, Tan Y, Lu C, Liu C, Li P, Xu K, Wang P (2012) Soft Matter 8:5078–5086
12. Friedrich T, Tieke B, Stadler FJ, Bailly C (2011) Soft Matter 7:6590–6597
13. Friedrich T, Tieke B, Stadler FJ, Bailly C, Eckert T, Richtering W (2010) Macromolecules 43:9964–9971
14. Friedrich T, Tieke B (2010) Colloid Polym Sci 288:1479–1484
15. Friedrich T, Mielke T, Hentschel M, Domogalla M, Kraus K, Tieke B (2013) Macromol Rapid Commun 34:393–398
16. Friedrich T, Trinker A, Tieke B (2010) Macromol Symp 298:57
17. Li H, Cui X, Shen S, Hu D (2011) J Appl Polym Sci 122:509–516
18. Hamid S, Sherrington D (1987) Polymer 28:332
19. Nagai K, Obishi Y, Inaba H, Kudo S (1985) J Polym Sci Polym Chem Ed 23:1221
20. Friedrich T, Tieke B, Meyer M, Pyckhout-Hintzen W, Pipich V (2010) J Phys Chem B 114:5666
21. Yoon I, Zimmerman AM, Tester CC, DiCiccio AM, Jiang Y, Chen W (2009) Chem Mater 21:3924
22. Mei Y, Lu Y, Pelzer F, Ballauf M (2007) Chem Mater 19:1062
23. Kidambi S, Bruening ML (2005) Chem Mater 17:301
24. Hervés P, Pérez-Lorenzo M, Liz-Marzán LM, Dzubiella J, Lu YS, Ballauff M (2012) Chem Soc Rev 41:5577–5587
25. Friedrich T (2011) Intelligente Hydrogele durch strahleninduzierte Copolymerisation von mizellaren Lösungen aus Tensidmonomeren und N-Isopropylacrylamid oder Acrylsäure. Dissertation Universität zu Köln, Verlag Dr Hut, München. ISBN 978-3-8439-0007-2

Core-Shell Microgels as Nanoreactors

Yan Lu, Nicole Welsch, Joachim Dzubiella, and Matthias Ballauff

Abstract

Core-shell microgel particles consisting of a solid core and a shell of crosslinked PNIPA network present model systems of high interest. Here we review our recent work on the core-shell microgel particles and their application as nanoreactors for the immobilization of catalytic active metal nanoparticles or enzymes. The catalytic activity of nanoparticles can be modulated both by the volume transition and the change of polarity of the thermosensitive shell of the carrier system. Special emphasis is put on recent work on the kinetic analysis of protein adsorption onto microgel particles, which was complemented with the thermodynamic study done by ITC. All results herein demonstrate that core-shell microgel particles may serve as "active" nanoreactor for catalytically active nanostructures, namely for metal nanoparticles and enzymes.

Keywords

Core-shell microgel • Nanoreactor • Metal nanoparticles • Protein adsorption • Catalysis

Introduction

Environmentally responsive microgels, which can react to external stimuli such as temperature, pH, light, etc., have become the subjects of great interest in many research fields in the last two decades due to their versatile applications in drug delivery [1, 2], biosensing [3], chemical separation, and catalysis [4, 5]. Networks composed of thermoresponsive polymers as e.g. poly(N-isopropylacrylamide) (PNIPA) have been one of the most studied microgel systems, which have been widely used as model system for the understanding of the basic of microgels as well as potential applications of such systems [6–8]. These networks can undergo a volume transition: Immersed in cold water these networks swell by uptake of the solvent. At elevated temperature a phase transition takes place in which most of the water is expelled. Thus these systems have become the basis of responsive systems [9].

Recently, efforts have been made towards the development of functional hybrid systems with controlled structures and tunable properties. Microgel with stimuli-responsive behaviour is thus one of the most promising candidates for the deposition of metal nanoparticles or other molecules, which may have several advantages over others, namely, ease of synthesis, colloidal stability and easy functionalization with a stimulus-sensitive behaviour [10]. The use of microgel particles as nanoreactors for the deposition of functional materials has opened new and exciting possibilities for the

Y. Lu (✉) • N. Welsch
Soft Matter and Functional Materials, Helmholtz-Zentrum Berlin für Materialien und Energie, 14109 Berlin, Germany

Helmholtz Virtual Institute — Multifunctional Materials in Medicine, 14109 Berlin, Teltow, Germany
e-mail: yan.lu@helmholtz-berlin.de

J. Dzubiella • M. Ballauff (✉)
Soft Matter and Functional Materials, Helmholtz-Zentrum Berlin für Materialien und Energie, 14109 Berlin, Germany

Helmholtz Virtual Institute — Multifunctional Materials in Medicine, 14109 Berlin, Teltow, Germany

Institut für Physik, Humboldt-Universität zu Berlin, 12489 Berlin, Germany
e-mail: matthias.ballauff@helmholtz-berlin.de

G. Sadowski and W. Richtering (eds.), *Intelligent Hydrogels*, Progress in Colloid and Polymer Science 140,
DOI 10.1007/978-3-319-01683-2_10, © Springer International Publishing Switzerland 2013

synthesis and fabrication of various materials with advanced properties [11–14].

Here we review our recent work on the core-shell microgel particles and their application as nanoreactors for the immobilization of catalytic active metal nanoparticles or enzymes. At first, the synthesis of thermosensitive core-shell microgel particles will be briefly reviewed. An extension for the synthesis of classic spherical particles to anisotropic dumbbell-shaped microgel particles will be discussed. Then the immobilization of metal nanoparticles into microgel carrier systems will be reviewed. Special emphasis is put on recent work using core-shell particles as nanoreactor for the in situ formation of (bi)metallic nanoparticles with defined structure. The consequences of the thermosensitive behavior for possible catalytic applications will be reviewed. The catalytic activity of nanoparticles can be modulated both by the volume transition and the change of polarity of the thermosensitive shell of the carrier system. In the end, microgel particles can work as efficient carrier systems for the immobilization of proteins as well. A kinetic analysis of protein adsorption onto microgel particles was complemented with the thermodynamic study done by ITC.

Synthesis of Core-Shell Microgels

Core-shell microgels consisting of a solid core as e.g. polystyrene or silica, onto which a shell of crosslinked poly(N-isopropylacrylamide) (PNIPA) network is affixed, present systems of particular interest with properties that differ markedly from the ones of conventional microgels [15]. It can swell only along the radial direction; hardly any expansion is possible along the perpendicular directions. Hence, single core-shell microgels present interesting objects for thermodynamic studies of finite size effects [16].

Thermosensitive core-shell particles were first synthesized by Makino et al. [17], Okubo et al. [18], and by us [19, 20] with a two-stage approach. These particles consist of a PS core onto which a shell of poly(N-isopropylacrylamide) (PNIPA) crosslinked by N, N′-methylenebisacrylamide (BIS) was grafted. The synthesis of core-shell type microgel particles is performed in two steps via either seeded emulsion polymerization [19] or photo-emulsion polymerization [21]. By synthesis, microgel particles with different surface charge can be made by using different initiator during polymerization. It is worth to note that most of the charges are affixed on the PS core since the PS core particles have a residual surface charge originating from the first emulsion polymerization [20]. These charges lead to active repulsion at all temperatures. Thus, suspensions of the core-shell particles remain stable even at temperatures above the volume transition.

Ionic strength and temperature responsive microgels can be also fabricated by copolymerization of NIPA with functional monomers, which can be applied as carriers for proteins due to the charged surface and similar chemical functionalities to the protein [22]. Thus, acidic or basic comonomers, such as acrylic acid (AA) [23], vinylimidazole [24], or 4-vinylpyridine [25] have been often chosen for the formation of microgels with pH sensitivity. By incorporating charges into the microgel particles, electrostatic interactions will influence the volume phase transition and the immobilization of proteins and enzymes. This will be discussed in detail in section "Binding of Proteins to Core-Shell Microgels".

Recently, we have extended our previous work on spherical particles to dumbbell-shaped core–shell microgel particles. Comparison to the spherical microgel particles, much less work, especially experimental work, has been done on the phase behavior and the dynamics of concentrated colloidal suspensions of the non-spherical microgel systems due to the difficulties in fabrication of such particles with large scale and high monodispersity. In our study, we have synthesized core-shell microgel particles with dumbbell morphology [26]. The dumbbell-shaped core particles were first synthesized by seeded growth emulsion polymerization using PMMA as seeds and styrene monomer added under starved conditions [27]. A relatively high surface tension between the PMMA surface and styrene monomer as well as the low reaction temperature (60 °C) and the low concentration of styrene in the water phase (kinetic control) favor the dumbbell morphology. In the second step, a thermosensitive shell of PNIPA was grafted to the dumbbell-shaped core. The polydispersity of the core particles prepared by this method, however, was still appreciable and no fluid-to-crystal transition could be observed in these systems.

More recently, we have developed a new strategy for the fabrication of thermosensitive core-shell dumbbells characterized by a narrow size distribution [28]. These dumbbell-shaped microgel particles consist of a polystyrene (PS) core and a shell of thermosensitive PNIPA networks. They are synthesized in a two-step process: First, dumbbell-shaped core particles consisting of PS are made using the method of Park et al. [29]. This route leads to nearly monodisperse core particles which presents an important improvement of the synthetic route used previously [26]. In a second step, a shell of PNIPA is grafted onto these cores as outlined in ref. [26]. As shown in Fig. 1, the dumbbell-shaped PS core particles are well-defined and have a narrow size distribution with the major axis length of 261.0 ± 7.9 nm and R_C of 90.5 ± 6.0 nm. The images from cryo-TEM (Fig. 1b) show that the PNIPA shell was successfully attached to the PS core. The shell thickness L_H is 93.1 ± 8.4 nm at room temperature (T ca. 25 °C).

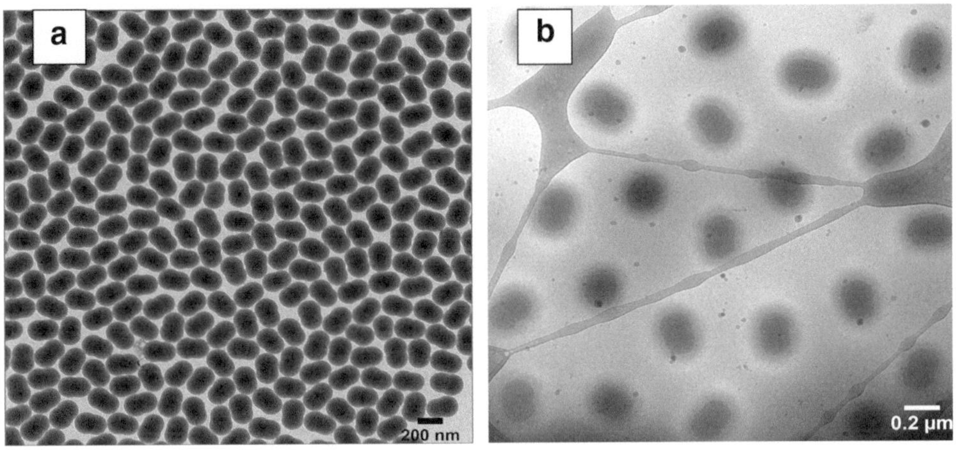

Fig. 1 TEM image of (**a**) dumbbell-shaped PS- core; cryo-TEM image (**b**) of the dumbbell-shaped core-shell microgels (Ref. [28])

It is worth to note that using cryo-TEM the aqueous suspension is shock-frozen turning the water phase into the glassy state. Thus, this method allows us to investigate the morphology of microgel particles in situ, no staining is necessary [30, 31]. Some time ago we have observed the transition behaviour of thermosensitive core-shell particles clearly by cryo-TEM [32]. In addition, quantitative evaluation of the gray scale of the cryo-TEM pictures yields the electron density of the sample. This in turn leads directly to the SAXS intensities that can be compared to experimental measurements. In this way, the cryo-TEM results can be compared to studies of core-shell particles done by scattering methods as small-angle neutron scattering (SANS) [33] and small-angle X-ray scattering (SAXS) [34].

Synthesis of Microgel-Metal Nanocomposite Particles

Immobilization of Metal Nanoparticles in Microgels

Microgel particles can be applied as carrier system for the immobilization of different inorganic nanoparticles, which could introduce optical [35], magnetic [36] and catalytic [37] features into the hybrid materials. More interestingly, these features are tunable due to the stimuli-responsive behaviour of the microgel particles [38]. Thus, these hybrid particles exhibit a combination of "inorganic" and "organic" characteristics [39, 40].

In our studies, thermosensitive core-shell microgel particles have been used as the template and different metal nanoparticles, such as Au, Ag, Rh and Pt, have been immobilized into these microgels [41, 42]. Different from the highly charged microgel particles, which were prepared by adding a positive or negative comonomer together with NIPA to the polymerization, microgels with a soft hydrated surface layer was used in our study that its charge stems from the initiator (V50 as cationic initiator and SDS as anionic initiator). The reduction to metallic nanoparticles in the presence of microgel particles was done at room temperature via the addition of $NaBH_4$. The immobilization of metal nanoparticles may be due to the strong localization of the metalate ions within the network, which is most probably caused by a complexation of the metalate ions by the nitrogen atoms of the PNIPA.

Figure 2 show the cryo-TEM images for the microgel-metal composite particles, from which it can be observed that well-defined metal nanoparticles have been homogeneously immobilized into the microgel particles [43]. No secondary particles or aggregates of the metal nanoparticles can be found. This indicates that microgel particles can work as "nanoreactors" for the immobilization of metal nanoparticles. More interestingly, the cross-linked PNIPA shell undergoes a volume transition at around 30 °C. Measurements of the composite particles by dynamic light scattering at different temperatures indicate that the original thermosensitivity of the PNIPA network is not hindered by the incorporation of metal nanoparticles into the network.

In-situ Formation of Au-Pt Bimetallic Nanorods

Morphological control of metal nanoparticles has become increasingly important, as many of their physical and chemical properties are highly shape dependent [44]. Among them, considerable attentions have been paid to the one-dimensional (1-D) inorganic nanomaterials such as nanorods [45], nanotubes [46], and nanowires [47], stemming from their unique size- and shape-dependent optoelectronic and surface properties [48]. For example, Au nanorods (Au-NRs) have been intensively studied, which show interesting

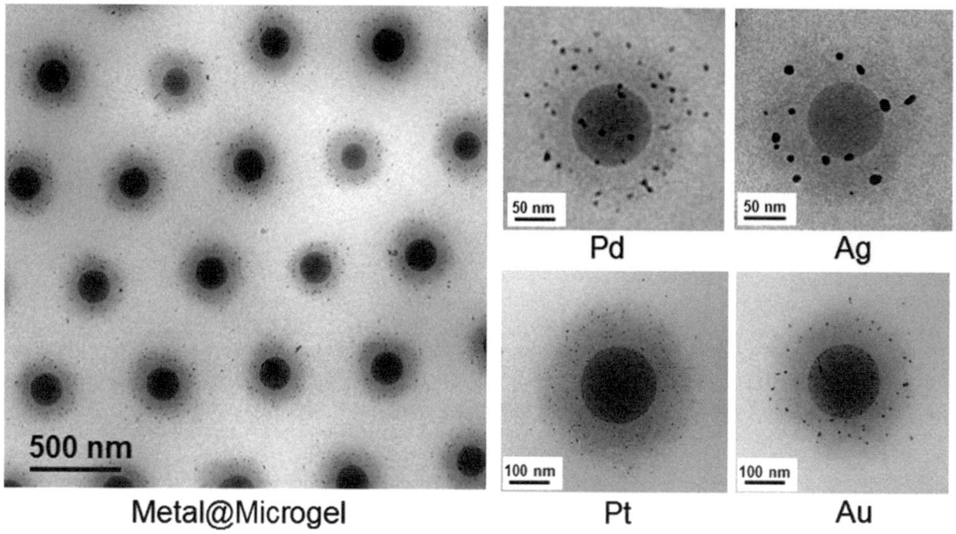

Fig. 2 Cryo-TEM images of microgel-metal nanocomposite particles (Ref. [43])

optical response in the visible and near-infrared (NIR) region because of the strong anisotropy in the shape [49].

Recently, we have demonstrated that bimetallic Au-Pt nanorods (NRs) can be in-situ grown into thermosensitive core-shell microgel particles by a novel two-step approach [50]. Using pre-formed spherical Au nanoparticles immobilized into thermosensitive microgel particles as seeds, Au NRs are grown directly inside the microgel carrier system via seeded-mediated growth procedure. As shown in Fig. 3, Au NRs with an average width of 6.6 ± 0.3 nm and length of 34.5 ± 5.2 nm (aspect ratio 5.2 ± 0.6) are homogeneously embedded into the shell of PNIPA networks. As a second step, platinum has been preferentially deposited onto the tips of Au NRs to form dumbbell-shaped bimetallic nanoparticles. The cryo-TEM image shown in Fig, 3 indicates that the obtained metallic Au-Pt NRs have an average width (middle) of 7.4 ± 0.8 nm and length of 39.5 ± 6.5 nm (aspect ratio 5.3 ± 0.6). No secondary Pt nanoparticles were observed either in the microgel template or in the solution. This demonstrates that core-shell microgels can be used as "smart nanoreactors" for the in situ growth of bimetallic nanoparticles with controlled morphology and high colloidal stability.

One of the most attractive properties for metal nanoparticles is their strong surface plasmon resonance absorption due to the high electronic polarizability of nanoparticles. The volume transition of microgel network leads to a strong red shift of the longitudinal plasmon band of the Au NRs, which are immobilized in microgel networks. In addition, the optical property of Au NRs is much more sensitive than that of spherical metal nanoparticles: A shift of plasmon band of ca. 30 nm is detected by increasing the temperature from 20 °C to 40 °C for Au NRs, while the respective shift for the microgel-Ag nanocomposites is only 6 nm [41].

Synthesis of Au-PNIPA Yolk-Shell Particles

Yolk-shell type nanostructures that metal nanoparticles are encapsulated in hollow porous shells have show excellent stability of the metallic nanoparticle during catalytic reactions [51]. In addition, comparison to inorganic mesoporous shell [52, 53], using crosslinked PNIPA networks as shell may have several important advantages, namely, colloidal stability, ease of synthesis and controllable pore size of PNIPA shell. Due to the swelling-deswelling behaviour of PNIPA networks, it is possible to effectively control the molecular diffusion through PNIPA shell via temperature. Meanwhile, the polarity of the shell changes with the volume phase transition of the PNIPA networks that the hydrophilicity of the microgels will be reduced dramatically with the increase of temperature [54]. More recently, we have fabricated new functional hybrid particles with yolk-shell structure that contains a movable metallic gold core completely isolated by a stimuli-sensitive PNIPA shell [55].

For the fabrication of yolk-shell particles, Au-SiO_2 - PNIPA trilayer composites with thermosensitive PNIPA shell have been first prepared through free radical polymerization using Au-SiO_2 nanoparticles modified by 3-(trimethoxysilyl) propylmethacrylate (MPS) as the seeds [56]. The SiO_2 layer can be further removed by NaOH treatment. Thus, Au-PNIPA particles with yolk-shell structures are formed as shown in Fig. 4a. Furthermore, the Au-PNIPA yolk-shell particles can work as "nanoreactors" for the in situ growth of Au core via ascorbic acid reduction of $HAuCl_4$ in the presence of CTAB [57]. Using CTAB, it can efficiently avoid secondary nucleation in the solution and control the final size of Au nanoparticles [58]. TEM images shown in Fig. 4

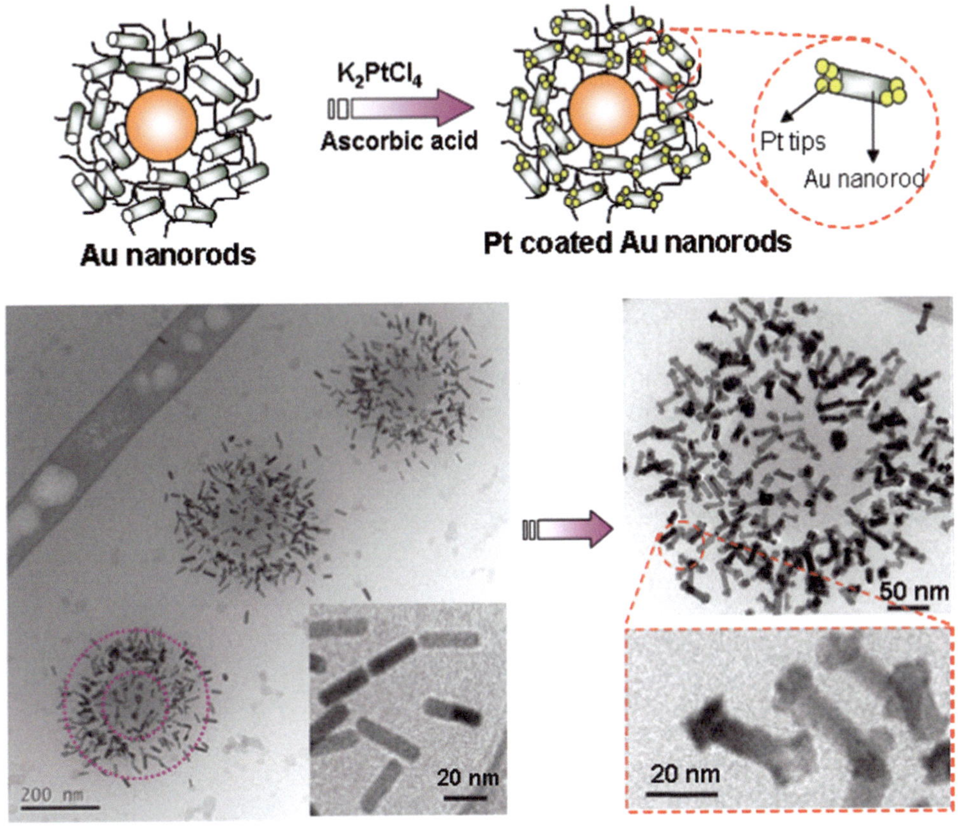

Fig. 3 Cryo-TEM images of Au-Pt bimetallic nanorods immobilized in PS-PNIPA microgels (Ref. [50])

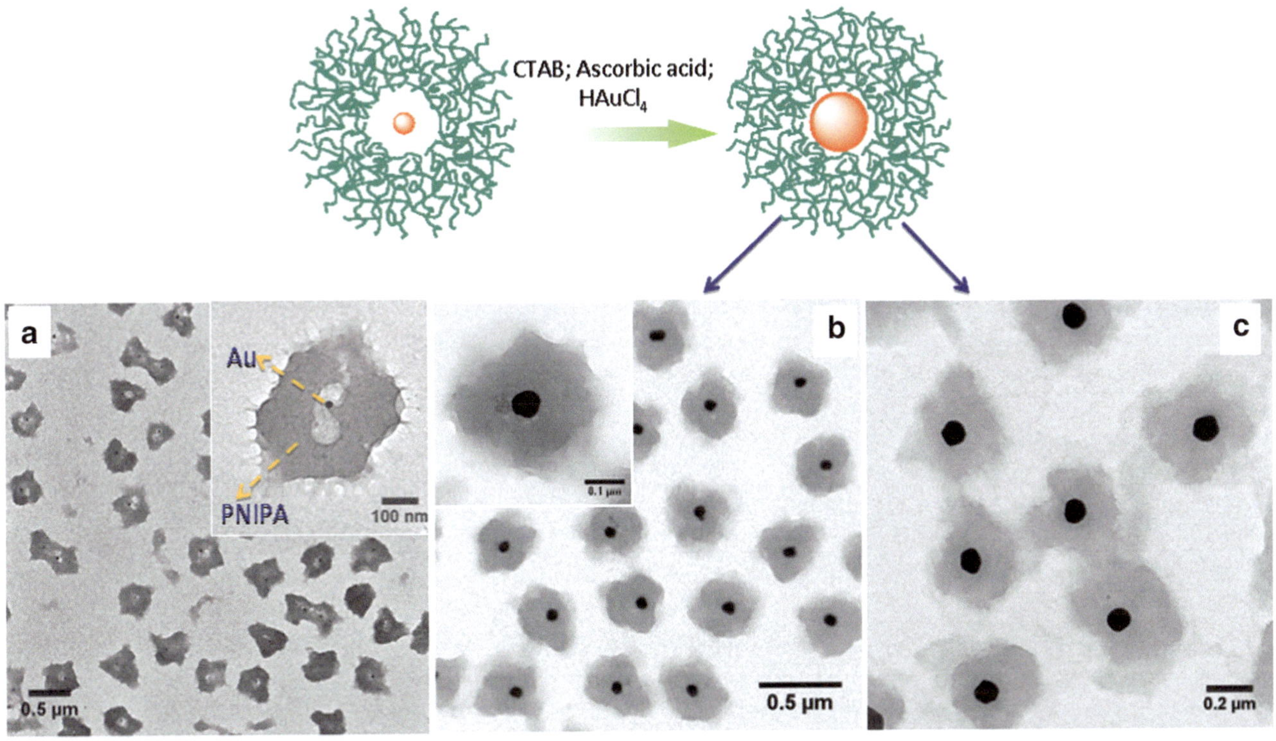

Fig. 4 TEM images of Au-PNIPA yolk-shell particles (**a**), Au (67 nm)-PNIPA (**b**) and Au (96 nm)-PNIPA (**c**) after further growth of 15 nm Au cores from the yolk-shell particles (Ref. [57])

reveal a noticeable increase in Au size, which varied from 67 ± 5 nm (Fig. 4b) up to 95 ± 6 nm (Fig. 4c) dependent on the reactant concentration. No secondary Au nanoparticles have been observed.

The UV-vis spectra for Au(95 nm)-PNIPA particles at different temperatures indicates clearly a red-shift as large as 10 nm of surface plasmon absorption band between 15 °C and 50 °C. In addition, an increase in absorbance has been also shown. Both effects are likely caused by an increase in refractive index of the microgel during its collapse as well as an increase of the local refractive index around the Au core nanoparticles [58]. This demonstrates that Au-PNIPA yolk-shell nanoparticles can act as an efficient, stable, thermoresponsive active "nanoreactors" for the in situ growth of Au cores that the optical properties of Au core can be modulated by the volume transition of PNIPA shell.

Microgel as Nanoreactor for Catalysis

Catalytic Activity for Reduction of 4-Nitrophenol

Metal nanoparticles have been widely used as catalysts for different reactions [59]. In our study, the reduction reaction of 4-nitrophenol (4-NP) by sodium borohydride has been applied as a benchmark reaction for monitoring the catalytic activity of microgel-metal hybrid particles [60, 61]. It is found that the rate constants k_{app} obtained at different temperatures for metal nanocomposites immobilized in the thermosensitive PS-PNIPA mcirogel particles do not follow the typical Arrhenius-type dependence on temperature, which is due to the volume transition of PNIPA-network as shown in Fig. 5. At low temperatures, the composite particles are suspended in water, which swells the thermosensitive network attached to the surface of the core particles. There is no limitation of diffusion for the reactant molecules within the network in this case. At higher temperatures (T > 32 °C), the PNIPA-network undergoes a volume transition, in which most of the water is expelled [38, 42]. This leads to a concomitant slowing down of the diffusion of reactants within the network. Thus, the catalytic activity of immobilized metal nanoparticles can be switched "on" and "off" by the swelling and collapse of the thermosensitive microgels. This demonstrated that the catalytic activity of the metal nanoparticles immobilized within the thermosensitive carrier particles can be modulated by the volume transition of the carrier system. Similar phenomenon has been also proved for Au nanoparticles encapsulated in PNIPA microgels as catalyst for the electron-transfer reaction between hexacyanoferrate (III) and borohydride [62].

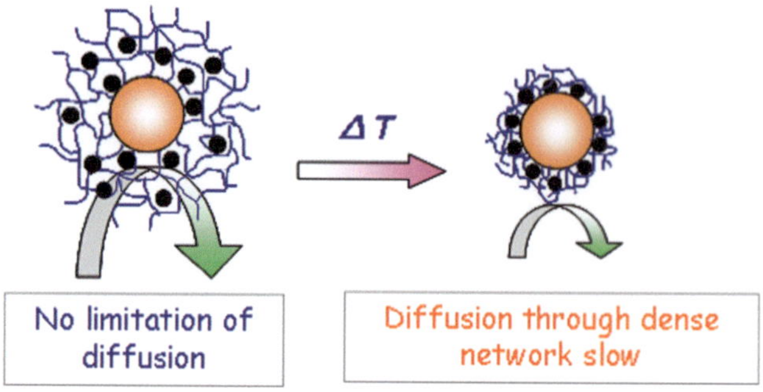

Fig. 5 Schematic representation of thermosensitive microgel-metal nanocomposite particles. The composite particles are suspended in water, which swells the thermosensitive network attached to the surface of the core particles. In this state the reagents can diffuse freely to the nanoparticles which act as catalysts. At higher temperatures (T > 32 °C) the network shrinks and the catalytic activity of the nanoparticles is strongly diminished in this state

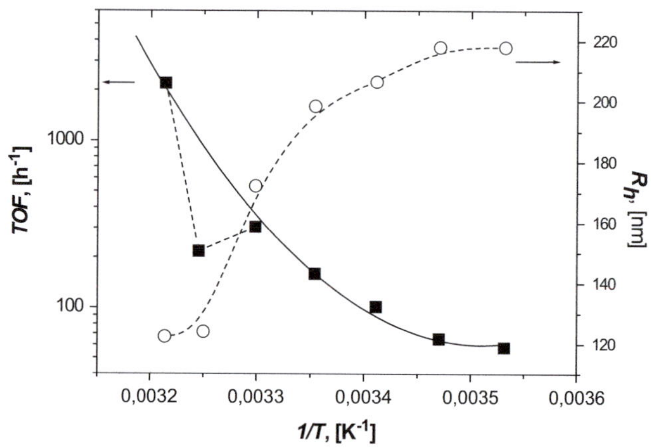

Fig. 6 TOF versus $1/T$ for the oxidation of benzyl alcohol in the presence of microgel-Au nanocomposite particles. The broken lines are guide lines for the beholder (Ref. [43])

Catalytic Activity for Oxidation of Alcohols

Metal nanoparticles embedded in thermosensitive core-shell microgel particles can also work efficiently as catalyst for the oxidation reaction of benzyl alcohol to benzaldehyde in aqueous media [43]. All reactions are carried out at room temperature using aerobic conditions and the conversions of the reaction can be checked by gas chromatography (GC) [63]. It has been found that microgel-metal nanocomposites efficiently catalyze the aerobic oxidation of benzyl alcohol at room temperature. No by-products have been detected by GC after the reaction and water is the only product formed besides the aldehyde.

Figure 6 shows the influence of temperature on the catalytic activity in an Arrhenius-plot of the Turnover Frequencies (TOFs), which can be compared to the data obtained for the reduction of 4-nitrophenol with an excess of $NaBH_4$ in presence of the metal nanocomposite particles. In principle,

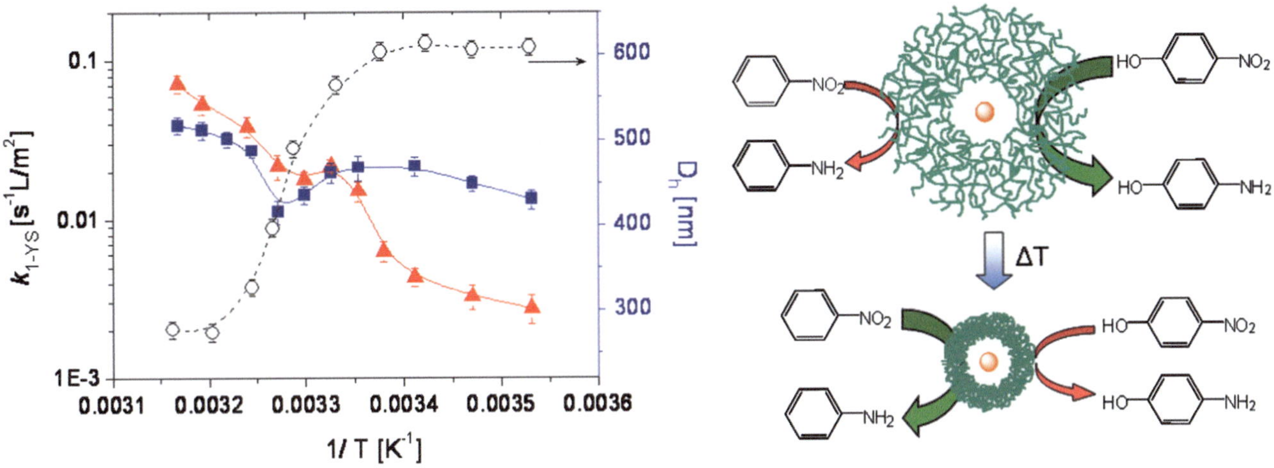

Fig. 7 *Left*: Arrhenius plots of the reaction rate constant k_1 (the apparent rate constant k_{app} normalized to surface area of Au nanocatalyst) of the yolk-shell particles. *Quadrangles*: reduction of 4-NP; *triangles*: reduction of NB; *open circles*: hydrodynamic diameter (D_h) of Au-PNIPA yolk-shell particles. *Right*: Schematic representation of the catalytic selectivity of Au-PNIPA yolk-shell nanoparticles that can be tuned between hydrophilic and more hydrophobic molecules through the volume transition of PNIPA shell for the catalytic reduction of 4-NP and NB (Ref. [55])

there are two factors that will influence the catalytic activity: Firstly, the collapsed PNIPA layer likely presented a higher steric barrier for benzyl alcohol molecules to diffuse from bulk aqueous dispersion to the surface of Au nanoparticles, causing the reaction to slow down. Secondly, with the increase of temperature, water-soluble hydrophilic PNIPA networks will become more hydrophobic which will be favorable for the diffusion of hydrophobic benzyl alcohol onto Au nanoparticle surface [54]. As shown in Fig. 6 the second effect prevails in most of the temperature range, the TOF value we got at 40 °C is much higher than the value following from the conventional Arrhenius kinetics. This is different from the behavior that we found for the reduction reaction of 4-nitrophenol. The results demonstrate that the catalytic activity of oxidation reaction of benzyl alcohol is in general more sensitive to the change of polarity than the diffusional barrier brought about by the volume transition of cross-linked PNIPA network.

Catalytic Activity for Reduction of Nitrobenzene

The catalytic reduction of nitrobenzene (NB) by $NaBH_4$ has been also applied as model reaction to test the catalytic activity of metal nanocomposite particles [64]. Here Au-PNIPA yolk–shell particles, that is, a single Au nanoparticle is encapsulated in a hollow thermosensitive PNIPA shell, have been used as catalyst as shown in Fig. 7 [55]. Yolk-shell structures have the clear advantages that individual metal nanoparticles are enclosed in a compartment that prevents aggregation with other nanoparticles. Moreover, the permeability of the shell may be tuned to a certain extend and yolk-shell systems may be regarded as true nanoreactors that allow us to study the catalytic activity of single nanoparticles in a defined environment.

As shown in Fig. 7, the catalytic activity of Au-PNIPA yolk-shell particles has been measured as a function of temperature T for the reductions of 4-NP and NB, respectively. The rate constants do not follow a simple Arrhenius law with constant activation energy, which shows a minimum, similar to the results reported for metal nanoparticles immobilized in PS-PNIPA core-shell particles [38, 42]. However, the overall rate for the catalytic reduction of NB dramatically increases when temperature is above the LCST of PNIPA microgel, such that a remarkable inversion takes places from a 4-NP-favoured reduction at lower $T <$ LCST to a NB-favoured reduction at higher $T >$ LCST.

This behavior has been further quantitatively modeled, which is schematically shown in Fig. 8. A spherical gold nanoparticle with radius R_0 is enclosed in a spherical nanoreactor of radius R_2. Diffusional control is achieved by a thermosensitive network extending between R_1 and R_2. If $R_1 > R_0$, the model describes yolk-shell-type nanoreactors where a single nanoparticle is enclosed in a cavity made up by a polymeric or an organic shell [65].

With full generality the rate constant k_{obs} in the system can be split into two parts: a term k_{surf} related to the rate constant of the surface reaction and a term k_D related to diffusion:

$$k_{obs}^{-1} = k_{surf}^{-1} + k_D^{-1} \tag{1}$$

The diffusion-controlled rate constant k_D for spherical sinks is given by the Smoluchowski-Debye expression for steady-state reactions, given by [66]

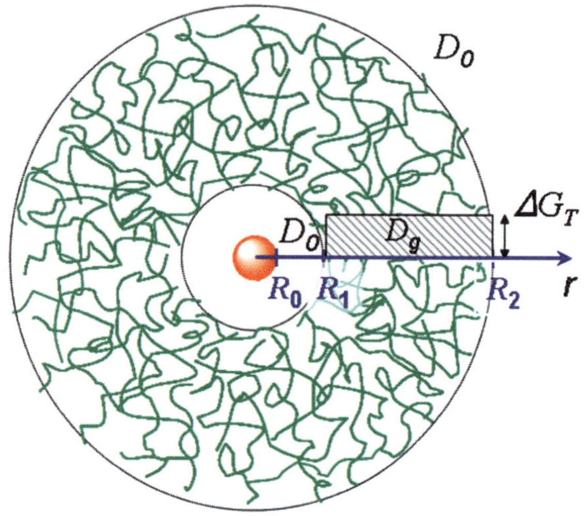

Fig. 8 Schematic presentation for the model assumption used for the analysis of the kinetic behaviour by Eqs. 2, 3, and 4. See main text for further explanation (Ref. [55])

$$k_D^{-1} = \int_{R_0}^{\infty} \frac{\exp\left[G(r)/k_B T\right]}{4\pi D(r) r^2} dr \tag{2}$$

where $D(r)$ denotes the (radial) distance-dependent diffusion constant for the reactants diffusing toward the nanoparticle, and $G(r)$ is the solvation free energy (with respect to the bulk) of the reactants along their pathway to the gold particle with radius R_0. For the sake of simplicity we assume that the reactants have a diffusion constant inside the gel D_g that is much smaller than the bulk diffusion constant D_0. Moreover, $G(r)$ is modelled by a constant transfer free energy ΔG_T from bulk solvent into the gel (see Fig. 8). Given these simplifying assumptions to Eq. 2, we obtain

$$k_D = 4\pi D_g R_1 \exp\left(-\Delta G_T/k_B T\right) \tag{3}$$

Here, terms of order $1/R_2$ have been neglected inasmuch R_0 and R_1 are typically considerably smaller than R_2. Since $\Delta G_T = \Delta H_T - T\Delta S_T$, the logarithm of the diffusion constant k_D is given by

$$\ln\left[k_D\right] = \ln\left(4\pi D_g R_1\right) + \Delta S_T/k_B - \Delta H_T/k_B T \tag{4}$$

Equation 4 shows that the diffusional barrier within such a nanoreactor consists of two terms: First, the diffusion of the reactants through the network is slowed down as compared to the bulk solution since $D_g < D_0$. Second, the transfer free energy ΔG_T enters exponentially and will have a profound influence on the measured temperature dependence. Hence, the intercept of the Arrhenius plot of k_D will be partially determined by the transfer entropy ΔS_T while the slope will be dominated by the respective enthalpy ΔH_T of the solute from the bulk into the network.

Concerning 4-NP and NB are of similar size, their diffusion constant D_g in the gel will be of same magnitude. Thus, the major change in the rate for NB when crossing the LCST must be attributed to significant differences in the affinity of NB to the microgel in its hydrophilic (swollen) state compared to the hydrophobic (collapsed) state [54]. This assumption is consistent with the more hydrophobic nature of molecule NB compared to 4-NP, and a higher number density of NB than 4-NP is expected in the hydrogel (and close to the gold particle) in the collapsed state. This demonstrates that temperature can be used as a trigger to enhance the selectivity of the catalysis for a given substrate: 4-NP reacts much faster at low temperature while the reduction of NB is preferred at higher temperature.

Binding of Proteins to Core-Shell Microgels

Introduction to Protein Adsorption on Colloidal Particles

In addition to the use as carriers for metal nanoparticles, colloidal microgels have attracted considerable attention as multifunctional systems in modern biotechnology, e.g., in drug delivery of small drugs or biomolecules [67–70], enzyme biocatalysis [71, 72] as well as tissue engineering [73, 74]. This is explained by the great mechanical flexibility as well as the low viscosity of the microgel network and the high surface area, which are essential properties for these types of applications. In this context protein adsorption has become a central problem. It is now well-recognized that upon incubation into biological media, such as blood, colloidal particles are immediately covered by biomolecules [75–77]. The control of this process requires the full understanding of the driving forces and the kinetics of the adsorption of proteins. On the one hand, the unspecific binding of proteins must be prevented to avoid adverse cellular responses and biofouling in medical applications [78–80]. On the other hand, immobilization of proteins is necessary for applications in which enzymes are used as catalysts or proteins are used as the active compounds in general [72, 81].

Due to the importance of protein adsorption to many different fields the spontaneous binding of proteins has become the subject of intense research. In many cases silica particles and bare latex particles [77, 82–86] as well as stabilized metal-nanoparticles [76, 81, 87–89] were chosen as simple model systems to study the interactions with proteins. However, these investigations need to be combined with structural studies of the adsorbed proteins since the interactions with the particles may induce misfolding and denaturation of the bound proteins [90]. This distortion of the protein conformation certainly influences the strength of binding. Moreover, structural modified proteins may trigger the immune system and may cause misfolding induced diseases [78, 79]. Additionally, the unfolding of the native proteins may also negate the use of colloidal particles as carriers for enzymes in industrial applications.

The ultimate goal of this research is the verification of the binding mechanism from a kinetic and thermodynamic point of view and eventual the semi-quantitative prediction of the tendency of a protein to adsorb to a given surface. Finally it should be possible to predict the composition of a corona of proteins around the colloidal particles as a function of the composition of the solution.

Polymeric coatings of polyethylene oxide are well known to show antifouling behaviour, i.e., they are able to suppress protein adsorption on surfaces [91]. On the other hand, layers of charged polymers, e.g., polyelectrolyte brushes [92–96] or charged microgel networks [97–106], generally cause strong adherence of proteins to the colloidal particles. The identification of the various forces resulting in attraction/repulsion of the proteins to/from the particles remains an important task for the future.

In the following section we review recent studies about the adsorption of proteins to uncharged and charged core-shell microgels. In analogy to the microgels employed for the immobilization of nanoparticles, the microgel particles consist of a polystyrene core onto which a shell of crosslinked poly(N-isopropylacrylamide) (PNIPA) is attached. Charges have been introduced by copolymerization with a certain amount of acrylic acid. The dimension of these particles ranges between 200 and 350 nm depending on the charge density, i.e., the amount of added acrylic acid. All microgels particles used are well-characterized by using a variety of methods, such as dynamic light scattering (DLS) and cryogenic transmission electron microscopy (cryo-TEM).

Strength of Interactions Between Proteins and Microgels

Uncharged microgels based on PNIPA show switchable adhesive properties because these particles exhibit a temperature-induced volume phase transition from a hydrophilic and swollen to a shrunken and mainly hydrophobic state. According to the work done by Kawaguchi et al. [107] and other studies [108–111] on protein binding to PNIPA-based microgels these systems attract proteins with higher affinity in their shrunken and hydrophobic state. Moreover, Lindman et al. investigated the thermodynamics of adsorption of human serum albumin (HSA) on a series of microgel particles which differ in their hydrophobicity by using isothermal titration calorimetry (ITC) [112]. HSA was found to bind more strongly on the microgel systems

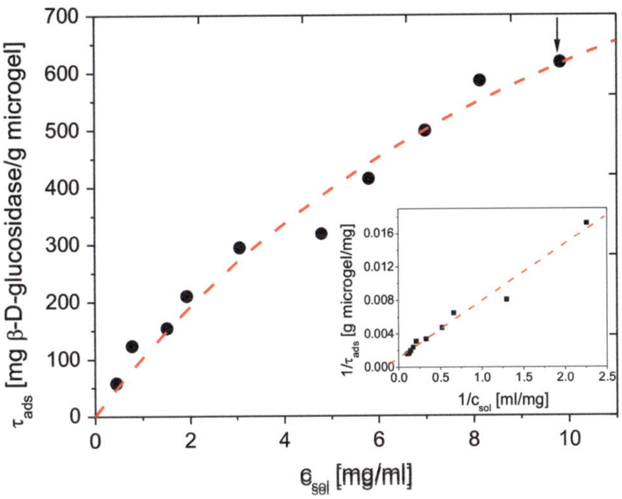

Fig. 9 Adsorption isotherm for the binding of β-D-glucosidase from *almonds* to the thermosensitive PS-PNIPA microgel. The adsorbed amount of β-D-glucosidase τ_{ads} is plotted as function of the concentration of free enzyme in solution c_{sol}. The *dashed line* represents the fit according to Eq. 5. The inset displays the data as a linearized Langmuir plot (Ref. [114])

containing a larger amount of the hydrophobic comonomer. Concluding these studies, binding of proteins on thermosensitive PNIPA-based microgels especially above the LCST is mainly caused by hydrophobic interactions. On the other hand, the total amount of protein bound to the microgel may be larger below the LCST of PNIPA, that is, in the hydrophilic and swollen state of the microgel network, due to the larger microgel volume available for the proteins [113].

In our previous study, we have shown that thermosensitive core-shell PS-PNIPA microgels act as highly efficient protein carriers in their swollen state and even for large enzymes, such as β-D-glucosidase from *almonds* [114]. The adsorption isotherm for binding of β-D-glucosidase was obtained by analyzing the efficiency of immobilization at different enzyme concentrations at 4 °C via ultrafiltration (Fig. 9). Here, the loading of the microgel τ_{ads} is plotted versus the specific protein concentration c_{sol} left free in solution. At these conditions, up to 620 mg protein per gram microgel could be bound. This amount corresponds to a total number of ~6,500 enzyme molecules incorporated into each microgel shell. The adsorption data for β-D-glucosidase can be described by the standard Langmuir isotherm which is given by

$$\frac{\tau_{ads}}{\tau_{ads,M}} = \frac{Kc_{sol}}{1 + Kc_{sol}} \tag{5}$$

where $\tau_{ads,M}$ is the maximum binding capacity and K is the binding constant. In this way, K was determined to $(1.1 \times 10^4 \pm 0.4 \times 10^4)$ M^{-1} and $\tau_{ads,M}$ to $(1{,}400 \pm 300)$ mg β-D-glucosidase per gram microgel particle. The high amount of bound protein indicates that the proteins are not concentrated on the outer shell of the microgel but rather are distributed within the total microgel network. Moreover, further structural investigations using Fourier-transform infrared (FT-IR) spectroscopy exclude large changes of the secondary structure of bound β-D-glucosidase and identified the formation of hydrogen bonds between the proteins and the amide groups of PNIPA as another driving force for protein adsorption besides hydrophobic interactions.

Incorporation of charged residues into the microgels weakens the thermosensitivity of the system and introduces attractive and repellent electrostatic interactions with proteins in solution. This leads to a complex interplay between hydrophobic and electrostatic interactions. Bysell et al. studied the effect of charge density and hydrophobicity on the interaction of negatively charged microgels to identify the role of charge distribution and apolar interactions in protein adsorption on oppositely charged particles [115, 116]. Smith et al. investigated the tunable encapsulation of cytochrome c into oppositely charged microgels [97]. The particles with the highest charge density were capable of loading extremely high amounts of protein where the protein concentration inside the microgel network exceeded the solubility concentration of cytochrome c in aqueous media. Moreover, the capacity of the microgel for protein uptake could be triggered by adjusting the salt concentration where high salt concentrations lead to weak microgel/protein complexes due to the screening of electrostatic interaction by salt while hydrophobic attraction is nearly independent of small salt ionic strength. In general it is safely to state that the presence of attractive electrostatic forces causes higher binding affinities and the salt- and pH-dependence of the adsorption process [97–99, 106].

The influence of hydrophobic and electrostatic interactions to the binding constant and to other thermodynamic parameters, e.g., the enthalpy and entropy of adsorption, can be determined by analyzing the binding of proteins as function of the salt concentration of the solution [117]. This systematic thermodynamic investigation can be done by using isothermal titration calorimetry (ITC). This method is heavily used to study the interactions between proteins and different kind of colloidal particles [83, 87, 94, 112, 117].

We recently applied ITC to the adsorption of lysozyme to negatively charged microgels (Fig. 10) [118, 119]. To evaluate the thermodynamic parameters by ITC a protein solution is injected stepwise into a solution of microgel particles and the heat change evolved upon each injection is measured as function of time. Integration of the raw output (Fig. 10a) over time gives the total heat change after each injection of lysozyme (Fig. 10b). After correcting for the protein heat of dilution, the data could be fit with an equilibrium binding model which is based on the standard

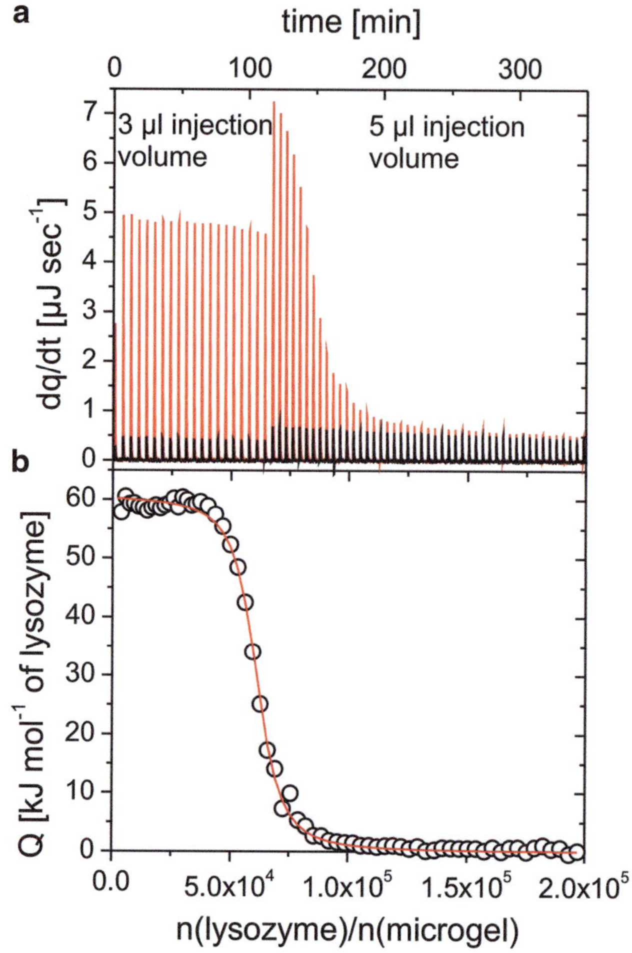

Fig. 10 (**a**) Raw ITC data of binding of lysozyme on negatively charged microgel particles at 298 K in 10 mM MOPS buffer solution pH 7.2. The *black spikes* denote the heat of dilution of lysozyme and the *colored spikes* are the change of heat corresponding to each injection of lysozyme into the microgel dispersion. (**b**) Integrated heats Q of each injection (*circles*) after correction for dilution signal of the protein. The *solid line* represents the fit of the data points according to the Langmuir binding model (Ref. [118])

Langmuir binding isotherm. The analysis of the protein binding isotherm obtained by ITC for different salt concentrations (Fig. 11) revealed that the adsorption of lysozyme to the weakly charged PNIPA-based microgels must be driven both by electrostatic interactions and specific, i.e., hydrophobic interactions and salt bridges, interactions [118].

In order to quantify the electrostatic and non-electrostatic interactions involved in the binding of proteins to charged gels a more elaborate binding model must be developed which includes the characteristic parameters of the microgel as well as of the protein. A quantitative analysis of this problem has been recently done by us using two different approaches [119]: (i) an extended Langmuir model and (ii) a binding model which is based on excluded volume packing. Both models consider electrostatic cooperativity

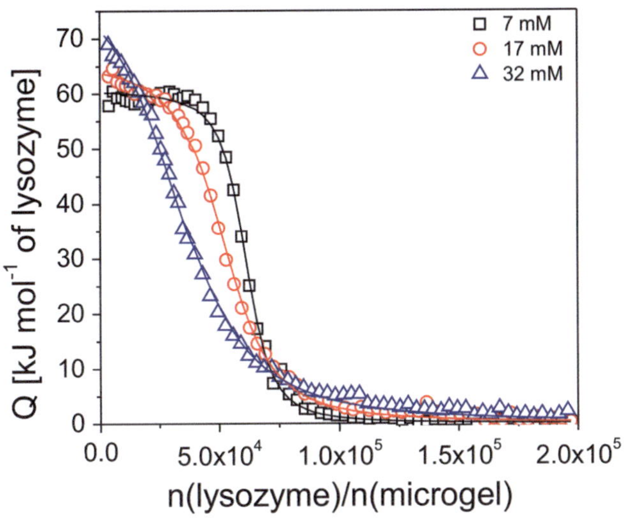

Fig. 11 ITC measurements at different salt concentrations. The integrated heats are plotted as function of the molar ratio between lysozyme and the microgel in 10 mM MOPS buffer pH 7.2 at 298 K at different ionic strength. The *solid line* represents the fit according to the Langmuir model (Ref. [118])

and any effects of microgel volume changes during protein binding [119]. In the excluded volume based model the gel is modeled as a given volume in which the proteins can move freely, that is, no immobilization on a fixed binding site is assumed. Confinement of space then leads to packing constraints that counteract the electrostatic attraction modeled in terms of a Donnan potential. In both approaches, the protein was modeled by a monopolar, homogeneously charged sphere with a given radius R_p and a net surface charge number z_p. Due to the fixed charges within the microgel network an electrostatic potential difference between bulk and the microgel is established, which is known as the Donnan potential:

$$\Delta\phi = e^{-1}k_BT\ln\left[\frac{z_gc_g}{2c_s} + \sqrt{\left(\frac{z_gc_g}{2c_s}\right)^2 + 1}\right] \quad (6)$$

Where z_g is the monomer charge valency, c_g is the concentration of charged monomers in the microgel and c_s is the salt concentration in the bulk. Further considerations given in reference [119] lead to the expression for the electrostatic contribution of the change of the free energy for protein uptake by the microgel ΔG_{el}:

$$\beta\Delta G_{el} = z_p\widetilde{\Delta\phi}(y) - \frac{z_p{}^2l_B}{2R_p}\left(\frac{\kappa_gR_p}{1+\kappa_gR_p} - \frac{\kappa_bR_p}{1+\kappa_bR_p}\right) \quad (7)$$

where $\widetilde{\Delta\phi}$ is the dimensionless Donnan potential which has been corrected for the change of the gel total net charge by the charge of bound protein. The inverse screening length in the gel κ_g is described by the salt concentration in the gel c_g via $\kappa_g = (8\pi l_B c_g)^{1/2}$ and the inverse Debye length κ_b in the bulk is defined by $\kappa_b = (8\pi l_B c_s)^{1/2}$. The first term of Eq. 7 is purely entropic and expresses the overall electrostatic interaction of a charged sphere with the Donnan potential. The second term describes the differences in the Born solvation free energies of the charged sphere in a medium characterized by the inverse screening lengths κ_b and κ_g, respectively. Since the inverse screening length is larger in the gel phase than in the bulk, the second term of ΔG_{el} is negative, i.e., favorable. The first term is always attractive, for proteins carrying an opposite net charge compared to the charge of the microgel particle. Equation 7 directly reflects the salt-dependence of electrostatic interactions. If the salt concentration in the bulk c_s is increased, both terms of ΔG_{el} diminish and the electrostatic driving force for protein binding vanishes.

Using the expression for ΔG_{el} the binding constant K can be split up into an "intrinsic" part K_0 and an electrostatic part to give the extended Langmuir binding isotherm:

$$K(x) = K_0\exp[-\beta\Delta G_{el}(x)] = \frac{\Theta(x)}{(1-\Theta(x))\,c_{sol}} \quad (8)$$

where $x = n_{\text{protein}}/n_{\text{microgel}}$ and $\theta = \tau_{\text{ads}}/\tau_{\text{ads,N}}$. The excluded volume binding model is also based on the separation of the total free energy into an electrostatic part (ΔG_{el}) and intrinsic part (ΔG_0) as it was done for the extended Langmuir model. In this work it was shown that the excluded volume based binding model is formally equivalent to the extended Langmuir approach in the low-packing regime. Both models are consistent and yield the same numbers for ΔG_0 and ΔG_{el}. The magnitude for hydrophobic (intrinsic) attraction of lysozyme to the negatively charged microgel was determined to $\Delta G_0 \approx -7k_BT$ for all salt concentrations, which constitutes roughly 1/3 to 1/2 of the total binding affinity in the small and high binding regime, respectively. The successful separation and quantification of the electrostatic and intrinsic forces by these two models was supported by the very weak dependence of ΔG_0 on the ionic strength. Figure 12 shows the binding of lysozyme to the microgel in a schematic fashion. Binding of a charged protein to an oppositely charged protein is accompanied by deswelling of the microgel. This is caused by the reduction of the Donnan potential which is followed by the partial neutralization of the total net charge of the microgel caused by the binding of oppositely charged microgel. The volume change of the microgel is of considerable magnitude and must be considered for the quantitative treatment of the protein binding to charged gels and brushes [119].

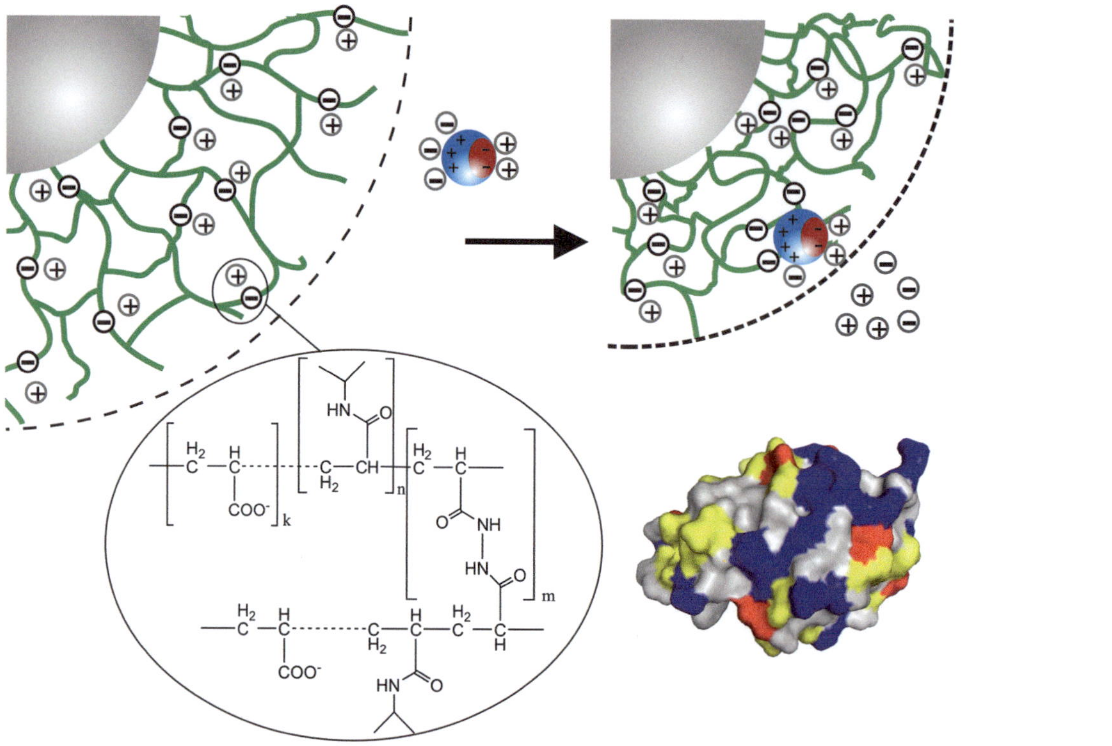

Fig. 12 *Top*: Binding of an overall positively charged protein to a weakly negatively charged microgel network. The protein is attracted by the gel through electrostatic and non-electrostatic interactions. These interactions cause the shrinking of the microgel. *Bottom left*: Chemical structure of the microgel particle. *Bottom right*: Structure of lysozyme (*chicken egg white*, PDB: 193 L). The amino acids are colored according to their polarity and charge: positive (*blue*), negative (*red*), hydrophobic (*grey*), polar (*yellow*) (Ref. [118])

Structure and Function of Adsorbed Proteins

Fourier-transform infrared (FT-IR) spectroscopy is best suited to analyze the secondary structure of proteins in aqueous solution as well as when bound to a substrate. Unlike other techniques, e.g., circular dichroism, this method can be applied to turbid samples, which strongly scatter light, such as latex particles [120–122]. The characteristic vibrational modes which are evaluated to obtain structural information are the amide I and amide II bands. These two modes appear in the IR regime between 1,500 and 1,700 cm^{-1}.

In a recent study we complemented the thermodynamic analysis of the interaction between lysozyme and a negatively charged microgel with the investigation of the protein structure after adsorption to the particle [118]. Figure 13 compares the FT-IR spectrum obtained for the bare microgel particle with the spectrum of the negatively charged microgel particle saturated with lysozyme. The PNIPA based microgel contains amide groups as well and, thus, also shows the characteristic amide I and II modes in the IR spectrum which superimpose with the signals from the bound protein. In order to analyze the secondary structure of lysozyme after immobilization the vibrational modes of the protein signal must be isolated from the signals originating from the microgel particle. This is simply done by subtraction of these two spectra. Figure 13b shows the spectrum of free lysozyme and the spectrum of adsorbed lysozyme after correction for the microgel signals and after normalization to the same enzyme concentration. For calculation of the β-sheet and α-helix content from the amide I and II bands of the FT-IR spectrum a partial least-squares (PLS) algorithm was chosen.

Using this method the β-sheet and α-helix content of the free protein was determined to (4 ± 3) % and (34 ± 4) % respectively. These results are comparable to previous values published [123]. The values for bound lysozyme were calculated to (30 ± 4) % α-helix and (2 ± 3) % β-sheet and, thus, are comparable to the values obtained for the free lysozyme. Consequently, the secondary structure of the protein was not perturbed by the strong interactions between lysozyme and the charged gel network.

One of the most sensitive methods for studying the integrity of enzymes after binding to a carrier is the analysis of the catalytic activity by suitable assays or by ITC. To avoid any misinterpretations, such an enzyme activity test must be performed in terms of the classical Michaelis-Menten kinetics. This model for enzyme kinetics is based on the following equation:

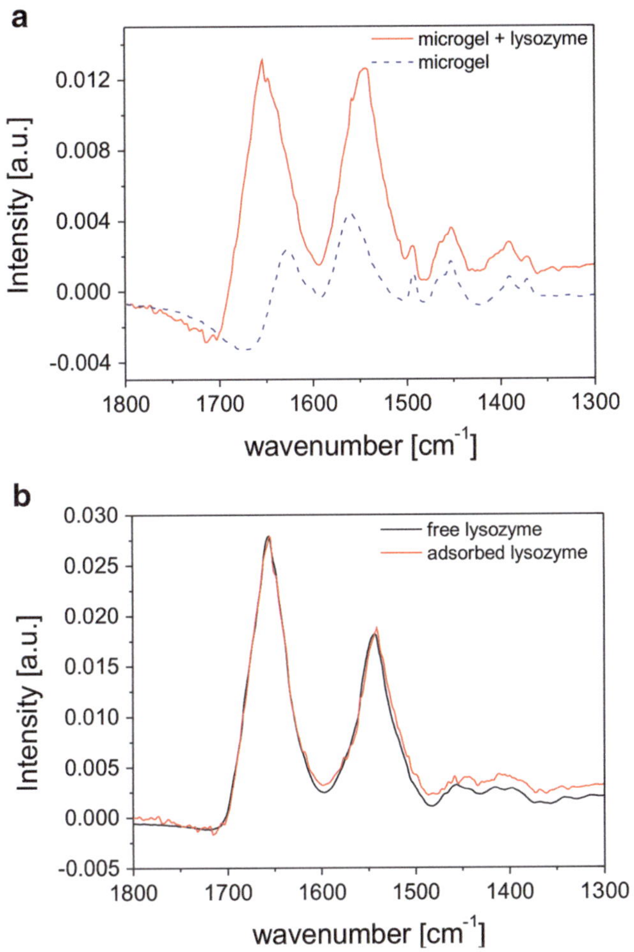

Fig. 13 (**a**) FT-IR spectra of the negatively charged microgel before and after immobilization of lysozyme. The experiment was done in 10 mM MOPS buffer pH 7.2 at 298 K. The spectra were normalized to a microgel concentration of 1 wt-%. (**b**) FT-IR spectra of free and bound lysozyme. The spectrum of the adsorbed protein was obtained by subtraction of the IR spectrum for the bare microgel particle from the spectrum of loaded microgel particles. The FT-IR spectra are normalized to the same lysozyme concentration (Ref. [118])

$$\upsilon = \frac{k_{cat}\,[E]\,[S]}{K_m + [S]} \tag{9}$$

where υ is the initial rate of the catalyzed reaction, k_{cat} is the turnover number of the enzyme, $[E]$ is the enzyme concentration, $[S]$ is the substrate concentration, and K_m is the characteristic Michaelis-Menten constant. The turnover number is independent of the enzyme and substrate concentrations and is very sensitive towards small changes of the tertiary structure. Small distortions of the protein conformation will reduce k_{cat} and, therefore, are detected by the catalytic measurements.

Recently we studied the activity of some enzymes which had been bound to microgel particles [114, 118]. Figure 14 depicts the Lineweaver-Burk plots for the hydrolytic reaction of 4-methylumbelliferyl-β-D-N,N′,N″-tiacetylchitotrioside

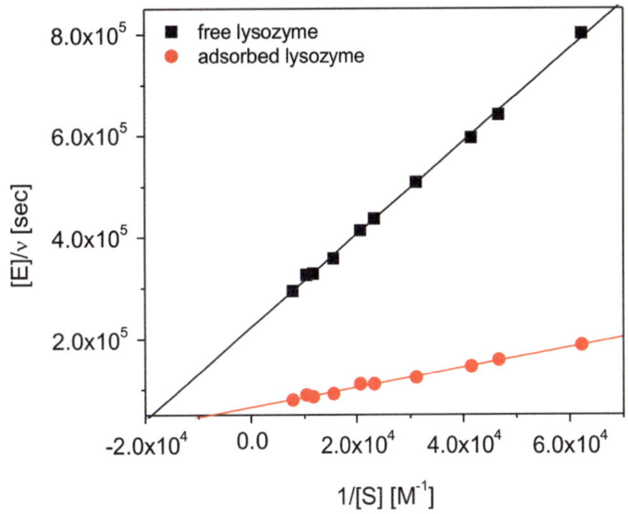

Fig. 14 Lineweaver-Burk plot for the hydrolysis of the substrate (GlcNAc)-MeU which has been catalysed by the presence of free or adsorbed lysozyme in 10 mM MOPS buffer pH 7.2 at 315 K. The *solid lines* are linear fits to the experimental data points (Ref. [118])

((GlcNA$_c$)$_3$-MeU) which was catalyzed by free and bound lysozyme at 315 K and a pH value of 7.2 [118]. The analysis of the catalytic activity of free and bound lysozyme reveals a higher activity of the immobilized enzyme. The central message from these experiments is that the function and, thus, the tertiary structure of the enzyme are persevered after immobilization to the microgel network. Actually, adsorption of lysozyme results in a pronounced enhancement of the catalytic activity. This positive change of the enzyme activity can be explained by the protonation of lysozyme after adsorption to the negatively charged microgel network [118]. The uptake of protons by lysozyme is caused by the lowered pH value inside the microgel network which is close to the pH optimum of lysozyme. Thus, the highly flexible microgel network serves as ideal carrier for the immobilization of proteins in their native conformation. In fact the catalytic activity of the bound enzyme may be positively modified by the electrostatic potential of the gel network.

Regarding thermosensitive PNIPA-based microgels we recently studied the activity of immobilized enzymes as function of temperature in order to assess the influence of the volume phase transition of the gel network on the performance of bound enzymes. The temperature-dependent analysis of the activity of β-D-glucosidase before and after immobilization to the thermosensitive carrier using the Michaelis-Menten kinetics is depicted in Fig. 15. The Lineweaver-Burk plots of the free and immobilized protein were taken at 298 K and 313 K (Fig. 15a). These plots give evidence for the integrity of the enzyme when bound to the microgel network both below and above the LCST of PNIPA.

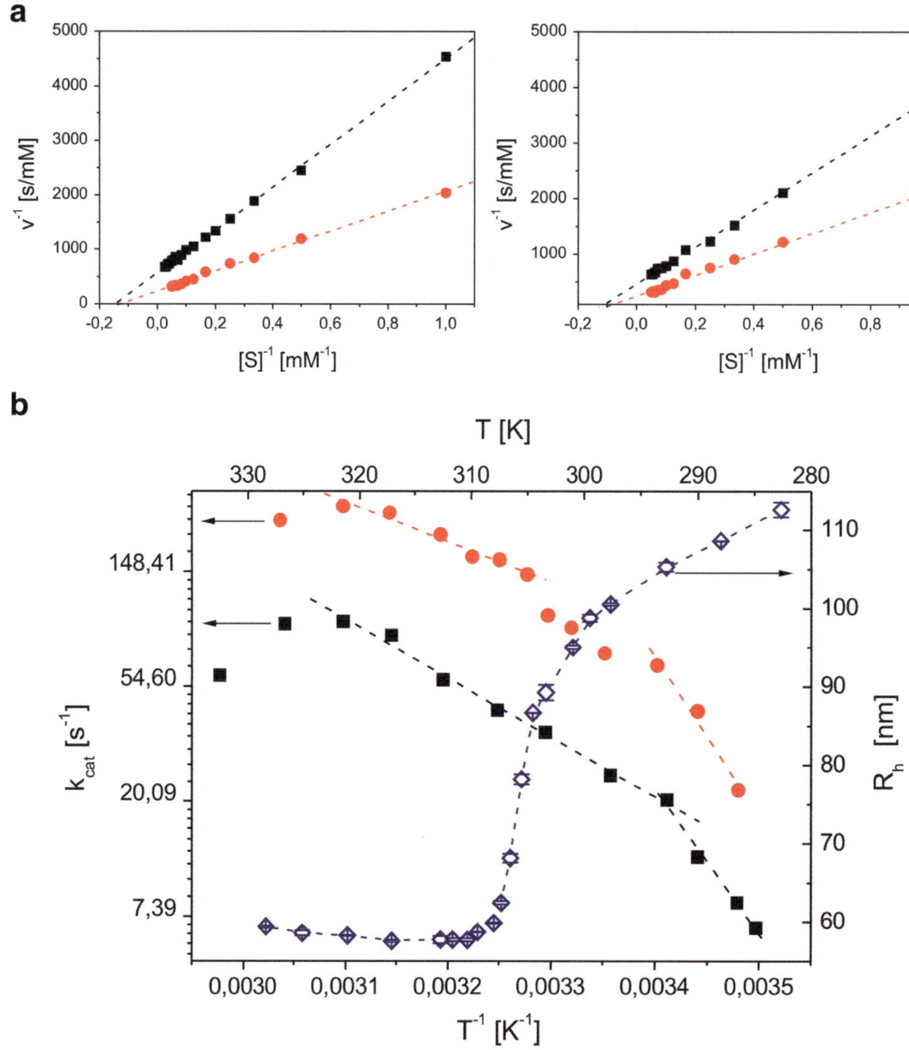

Fig. 15 (**a**) Lineweaver-Burk plots for the hydrolytic reaction of ortho-nitrophenol catalyzed by bound β-D-glucosidase (*filled circles*) and by the free enzyme (*filled squares*) at 298 K (*left*) and 313 K (*right*). The reactions were performed in 10 mM MOPS buffer pH 7.2. (**b**) Arrhenius plots of k_{cat} of the free and immobilized β-D-glucosidase. The enzyme concentration ranges between 0.005 and 0.01 g L^{-1} (free β-D-glucosidase) and between 0.0035 and 0.01 g L^{-1} (bound β-D-glucosidase). The substrate concentration was varied between 1.0 and 20.0 mM. The *dashed lines* represent linear fits to the experimental data. Additionally, the hydrodynamic radius R_h of the microgel is plotted versus the temperature (*open tilted squares*) (Ref. [114])

This is even shown more precisely in the Arrhenius plots of the turnover number for both the free and immobilized enzyme. For the whole temperature range the number of k_{cat} of the enzyme in the bound state is of some magnitudes larger than k_{cat} of β-D-glucosidase dissolved free in solution. The increase of the hydrophobicity and the drastic volume change upon temperature increase obviously do not induce denaturation of the enzyme. The weak retardation of the catalytic activity at temperatures close to the LCST indicates an emerging influence of the deswelling microgel on the free diffusion of the substrate ortho-nitrophenol. However, this influence is too weak to prevent the substrate to access the deswollen microgel and diffuse to the embedded enzymes. Thus, at higher temperatures the linear relation between ln k_{cat} and T^{-1} is recovered again. At even higher temperatures both free and immobilized β-D-glucosidase suffer from thermal denaturation and the turnover number starts to diminish.

Kinetics of Protein Adsorption

The overview so far shows that microgels provide large loading capacities for proteins. The next interesting question concerns the kinetics of this adsorption process. In case of micron-sized colloidal particles this analysis can be done by optical methods [100, 103]. The adsorption of proteins to smaller particles has to be analyzed by other techniques, such as time-resolved small angle x-ray scattering and time-resolved fluorescence spectroscopy [95, 124, 125]. In our recent work, we quantified the kinetic uptake of lysozyme by charged microgels using fluorescence spectroscopy [126]. For this purpose fluorescent labelled lysozyme was employed and the uptake of the protein by the particles with time could be monitored by the decrease of the fluorescence intensity. The drop of the fluorescence intensity of labelled lysozyme upon uptake into the gel is caused by the lowered pH value within the negatively charged network. In other words, the altered pH value within the gel was exploited to analyze the adsorption kinetics on small colloidal microgels. The kinetic data was acquired for different protein concentrations and compared to a kinetic binding model. In section "Strength of Interactions Between Proteins and Microgels" we outlined that protein adsorption on this type of microgels can be modelled by a Langmuir isotherm. Recently we demonstrated that the kinetics of the protein uptake can be described in terms of the kinetic Langmuir model which reads [127]

$$\begin{aligned}\frac{d\Theta}{dt} &= k_{on}\frac{[B][P]}{N[M]_t} - k_{off}\frac{[PB]}{N[M]_t}\\ &= k_{on}(1-\Theta)([P]_t - N\Theta[M]_t) - k_{off}\Theta \qquad (10)\end{aligned}$$

where k_{on} and k_{off} is the adsorption and desorption rate, respectively, $[M]_{\text{t}}$ is the total microgel concentration and $[P]_{\text{t}}$ is the total protein concentration in the volume. By using the relationship $K = k_{\text{on}}/k_{\text{off}}$ and by making the substitution of k_{off} for k_{on}/K Eq. 10 can be further simplified. The analytical solution of this equation was derived by Azizian [127] and can be found there. Since the binding constant K is accessible by ITC the only unknown in Eq. 10 is the constant k_{on}.

In the case that the adsorption of lysozyme is a diffusion-controlled process the motion of the protein molecules into the microgel network has to be split up into two contributions: (i) the protein diffusion in the bulk solution and (ii) the diffusion transport within the gel. The first process is characterized by the rate constant k_{D} while the motion into the gel is described by the parameter k_{i}. This is schematically shown in Fig. 16a. The characteristic constant k_{D} is identical to the rate-constant of diffusion-limited association reactions between small and large molecules with perfectly absorbing boundary conditions (Smoluchowski rate) [128]:

$$k_D = 4\pi D_0 R \qquad (11)$$

where D_0 is the diffusion coefficient of the small molecule (i.e., the protein) in solution and R is the radius of the large particle (i.e., the microgel). Consequently, k_{D} is a function of the diffusion coefficient of the protein used which is typically in the range of 10^{-10} m^2 s^{-1} and of the hydrodynamic radius of the microgel. Under these considerations the adsorption rate constant k_{on} is expressed by k_{D} and k_{i} [126].

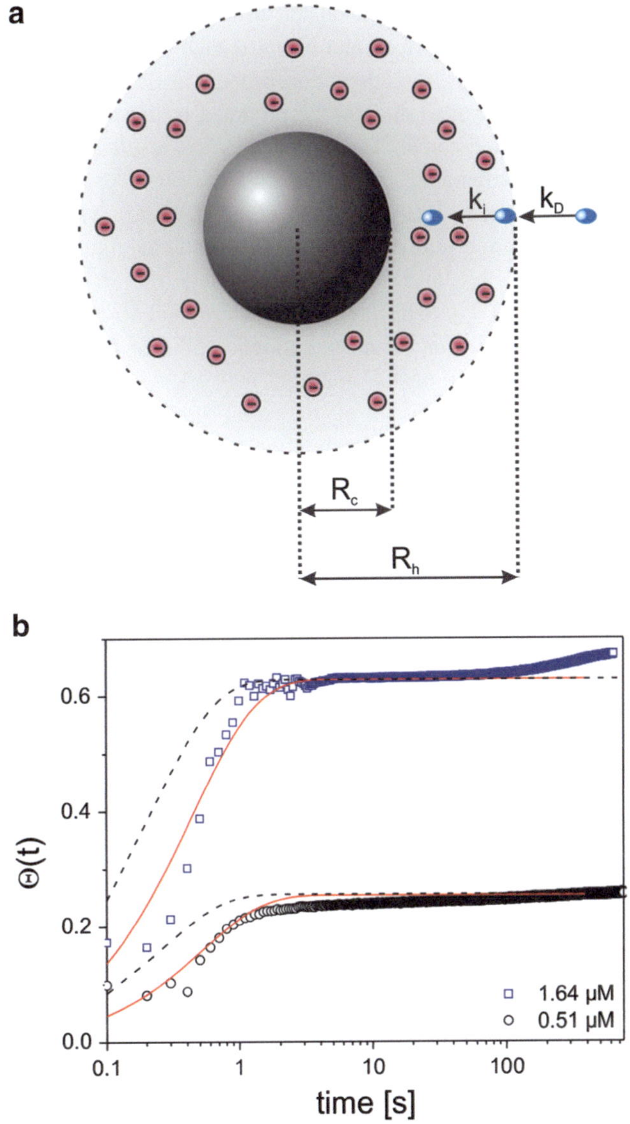

Fig. 16 (**a**) Schematic representation of the kinetic uptake of proteins by the core-shell microgel having a core of radius R_{c} and an overall hydrodynamic radius of R_{h}. The diffusion of proteins on the surface is characterized by the rate constant k_{D} while the diffusion into the shell is described by the rate constant k_{i}. (**b**) Kinetics of protein binding as measured by fluorescence spectroscopy: Fraction of occupied binding sites Θ as function of time after addition of 0.0029 g L^{-1} microgel dispersion into a solution of fluorescent labeled lysozyme of two different dilutions. The time of microgel injection was set to 0. The *solid lines* represent the fit according to Eq. 10. The *dashed lines* represent the theoretical curve when k_{i} is set to ∞ (Ref. [126])

The uptake of fluorescent labeled lysozyme by the microgel with time is shown in Fig. 16b for two different protein concentrations at low ionic strength. From this plot it is obvious that the adsorption process must be divided into two time regimes: a fast binding regime which covers the first seconds after mixing and a slow binding regime which continues over several hundred seconds. We recently demonstrated that the fast binding step must be related to diffusion of lysozyme towards the gel network and by its uptake by the microgel shell [126]. Thus, the adsorption process in the first step is diffusion controlled and can be modeled by the theoretical model outlined above. The kinetic analysis resulted in a k_D and k_i of comparable size.

The kinetic analysis was complemented with the thermodynamic study done by ITC. Knowing the values for K and k_{on} allows one to calculate the desorption rate constant k_{off}. The latter was determined to $\sim$0.5 s^{-1} which translates into a mean residence time of fluorescent labeled lysozyme on the microgel network k_{off}^{-1} of $\sim$2 s. Although about 90 % of the total protein amount is adsorbed in the first seconds after mixing, that is, in the fast binding regime, there is a significant protein fraction left which binds to the network much more slowly in a following step. The second binding step is probably caused by cooperative phenomena which originate from the protein-induced deswelling of the microgel network and the rearrangement of preadsorbed proteins within the microgel.

Conclusions

In conclusion, core-shell microgel particles consisting of a solid core and a shell of crosslinked PNIPA network present model systems of high interest, which opens the way for the fabrication of smart and adaptable functional colloidal carriers. In this review we have discussed how core-shell microgel particles are paving the way from fabrication of designed hybrid system to advanced applications. As possible applications we discussed in full detail the preparation of catalytically active metal nanoparticles and the immobilization of proteins. All results herein demonstrate that such particles may serve as "active" nanoreactor for catalytically active nanostructures, namely for metal nanoparticles and enzymes.

Acknowledgements Financial support by the Deutsche Forschungsgemeinschaft, SPP "Intelligente Hydrogele" and the Helmholtz Virtual Institute are gratefully acknowledged.

References

1. Nayak S, Lyon LA (2005) Angew Chem Int Ed 44:7686–7708
2. Soppimath K, Tan DCW, Yang Y (2005) Adv Mater 17:318–323
3. Hu Z, Chen Y, Wang C, Zheng Y, Li Y (1998) Nature 393:149–152
4. Biffis A, Orlandi N, Corain B (2003) Adv Mater 15:1551–1555
5. Ballauff M, Lu Y (2007) Polymer 48:1815–1823
6. Dusek K, Patterson D (1968) J Polym Sci Part A-2 6:1209–1216
7. Tanaka T (1978) Phys Rev Lett 40:820–823
8. Shibayama M, Tanaka K (1993) Adv Polym Sci 109:1–62
9. Heskins M, Guillet JE (1968) J Macromol Sci Chem 2:1441–1455
10. Das M, Zhang H, Kumacheva E (2006) Annu Rev Mater Res 36:117–142
11. Singh N, Lyon LA (2007) Chem Mater 19:719
12. Zhang J, Xu S, Kumacheva E (2005) Adv Mater 17:2336–2340
13. Lu Y, Ballauff M (2011) Prog Polym Sci 36:767–792
14. Welsch N, Lu Y, Dzubiella J, Ballauff M (2013) Polymer, 54: 2835–2849
15. Pelton RH (1988) J Polym Sci A 26:9–18
16. Keerl M, Pedersen JS, Richtering W (2009) J Am Chem Soc 131:3093–3097
17. Makino K, Yamanoto S, Fujimoto K, Kawaguchi H, Oshima H (1994) J Colloid Interface Sci 166:251
18. Okubo M, Ahmad H (1996) Colloid Polym Sci 274:112
19. Kim JH, Ballauff M (1999) Colloid Polym Sci 277:1210
20. Dingenouts N, Norhausen C, Ballauff M (1998) Macromolecules 31:8912
21. Lu Y, Wittemann A, Ballauff M, Drechsler M (2006) Macromol Rapid Comm 27:1137–1141
22. Stuart MAC, Huck WTS, Genzer J, Müller M, Ober C, Stamm M, Sukhorukov GB, Szleifer I, Tsukruk VV, Urban M, Winnik F, Zauscher S, Luzinov I, Minko S (2010) Nat Mater 9:101–113
23. Debord JD, Lyon LA (2003) Langmuir 19:7662–7664
24. Bhattacharya S, Eckert F, Boyko V, Pich A (2007) Small 3:650–657
25. Pinkrah VT, Snowden MJ, Mitchell JC, Seidel J, Chowdhry BZ, Fern GR (2003) Langmuir 19:585–590
26. Hoffmann M, Siebenbürger M, Harnau L, Hund M, Hanske C, Lu Y, Wagner CS, Drechsler M, Ballauff M (2010) Soft Matter 6:1125
27. Hoffmann M, Lu Y, Schrinner M, Ballauff M, Harnau L (2008) J Phys Chem B 112:14843
28. Chu F, Siebenbürger M, Polzer F, Stolze C, Kaiser J, Hoffmann M, Heptner N, Dzubiella J, Drechsler M, Lu Y, Ballauff M (2012) Macromol Rapid Comm 33:1042
29. Park JG, Forster JD, Dufresne ER (2010) J Am Chem Soc 132:5960
30. Wittemann A, Drechsler M, Talmon Y, Ballauff M (2005) J Am Chem Soc 127:9688
31. Li Z, Kesselman E, Talmon Y, Hillmyer MA, Lodge TP (2004) Science 306:98
32. Crassous JJ, Ballauff M, Drechsler M, Schmidt J, Talmon Y (2006) Langmuir 22:2403–2406
33. Dingenouts N, Seelenmeyer S, Deike I, Rosenfeld S, Ballauff M, Lindner P, Narayanan T (2001) Phys Chem Chem Phys 3:1169–1174
34. Seelenmeyer S, Deike I, Rosenfeldt S, Norhausen C, Dingenouts N, Ballauff M, Narayanan T, Lindner P (2001) J Chem Phys 114:10471
35. Álvarez-Puebla RA, Contreras-Cáceres R, Pastoriza-Santos I, Pérez-Juste J, Liz-Marzán LM (2009) Angew Chem Int Ed 48:138–143
36. Sánchez-Iglesias A, Grzelczak M, Rodríguez-González B, Guardia-Girós P, Pastoriza-Santos I, Pérez-Juste J, Prato M, Liz-Marzán LM (2009) ACS Nano 3:3184–3190
37. Welsch N, Ballauff M, Lu Y (2011) Adv Polym Sci 234:129–163
38. Lu Y, Mei Y, Drechsler M, Ballauff M (2006) Angew Chem Int Ed 45:813–816
39. Karg M, Hellweg T (2009) J Mater Chem 19:8714–8727
40. Lu Y, Wittemann A, Ballauff M (2009) Macromol Rapid Commun 30:806
41. Lu Y, Mei Y, Drechsler M, Ballauff M (2006) J Phys Chem B 110:3930–3937
42. Mei Y, Lu Y, Polzer F, Ballauff M, Drechsler M (2007) Chem Mater 19:1062–1069

43. Lu Y, Proch S, Schrinner M, Drechsler M, Kempe R, Ballauff M (2009) J Mater Chem 19:3955–3961
44. Habas S, Lee H, Radmilovic V, Somorjai GA, Yang P (2007) Nat Mater 6:692–697
45. Grzelczak M, Pérez-Juste J, Mulvaney P, Liz-Marzán LM (2008) Chem Soc Rev 37:1783–1791
46. Ajayan PM (1999) Chem Rev 99:1787
47. Yuan J, Xu Y, Walther A, Bolisetty S, Schumacher M, Schmalz H, Ballauff M, Müller AHE (2008) Nat Mater 7:718–722
48. El-Sayed M (2001) Acc Chem Res 34:257–264
49. Pérez-Juste J, Pastoriza-Santos I, Liz-Marzán LM, Mulvaney P (2005) Coord Chem Rev 249:1870–1901
50. Lu Y, Yuan J, Polzer F, Drechsler M, Preussner J (2010) ACS Nano 4:7078–7086
51. Zhang Q, Lee I, Ge J, Zaera F, Yin Y (2010) Adv Funct Mater 20:2201–2214
52. Arnal P, Comotti M, Schüth F (2006) Angew Chem Int Ed 45:8224–8227
53. Lee I, Albiter MA, Zhang Q, Ge J, Yin Y, Zaera F (2011) Phys Chem Chem Phys 13:2449–2456
54. Pelton R (2010) J Colloid Interface Sci 348:673–674
55. Wu S, Dzubiella J, Kaiser J, Drechsler M, Guo X, Ballauff M, Lu Y (2012) Angew Chem Int Ed 51:2229
56. Karg M, Pastoriza-Santos I, Liz-Marzán LM, Hellweg T (2006) Chem Phys Chem 7:2298–2301
57. Wu S, Kaiser J, Drechsler M, Ballauff M, Lu Y (2013) Colloid Polym Sci 291:231–237
58. Contreras-Cácares R, Sánchez-Iglesias A, Karg M, Pastoriza-Santos I, Pérez- Juste J, Pacifico J, Hellweg T, Fernández-Barbero A, Liz-Marzán LM (2008) Adv Mater 20:1666
59. Astruc D (2008) Nanoparticles and catalysis. WILEY-VCH Verlag GmbH & Co. KGa, Weinheim
60. Wunder S, Polzer F, Lu Y, Mei Y, Ballauff M (2010) J Phys Chem C 114:8814–8820
61. Wunder S, Lu Y, Albrecht M, Ballauff M (2011) ACS Catal 1:908–916
62. Carregal-Romero S, Pérez-Juste J, Hervés P, Liz-Marzán LM, Mulvaney P (2010) Langmuir 26:1271–1277
63. Schrinner M, Proch S, Mei Y, Kempe R, Miyajima N, Ballauff M (2008) Adv Mater 20:1928–1933
64. Yuan C, Luo W, Zhong L, Deng H, Liu J, Xu Y, Dai L (2011) Angew Chem Int Ed 50:3515–3520
65. Hervés P, Pérez-Lorenzo M, Liz-Marzán LM, Dzubiella J, Lu Y, Ballauff M (2012) Chem Soc Rev 41:5577–5587
66. Debye P (1942) Trans Electrochem Soc 82:265–273
67. Smith MH, Lyon LA (2012) Acc Chem Res 45:985–993
68. Raemdonck K, Demeester J, De Smedt S (2009) Soft Matter 5:707–715
69. Deka SR, Quarta A, Di Corato R, Riedinger A, Cingolani R, Pellegrino T (2011) Nanoscale 3:619–629
70. Murthy N, Xu M, Schuck S, Kunisawa J, Shastri N, Frechet JM (2003) Proc Natl Acad Sci U S A 100:4995–5000
71. Huang X, Yin YZ, Tang Y, Bai XL, Zhang ZM, Xu JY, Liu JQ, Shen JC (2009) Soft Matter 5:1905–1911
72. Gawlitza K, Wu C, Georgieva R, Wang D, Ansorge-Schumacher MB, von Klitzing R (2012) Phys Chem Chem Phys 14(27):9594–9600
73. Jia X, Kiick KL (2009) Macromol Biosci 9:140–156
74. Seliktar D (2012) Science 336:1124–1128
75. Mahmoudi M, Lynch I, Ejtehadi MR, Monopoli MP, Bombelli FB, Laurent S (2011) Chem Rev 111:5610–5637
76. Casals E, Pfaller T, Duschl A, Oostingh GJ, Puntes V (2010) ACS Nano 4:3623–3632
77. Monopoli MP, Walczyk D, Campbell A, Elia G, Lynch I, Bombelli FB, Dawson KA (2011) J Am Chem Soc 133:2525–2534
78. Deng ZJ, Liang MT, Monteiro M, Toth I, Minchin RF (2011) Nat Nanotechnol 6:39–44
79. Linse S, Cabaleiro-Lago C, Xue WF, Lynch I, Lindman S, Thulin E, Radford SE, Dawson KA (2007) Proc Natl Acad Sci U S A 104:8691–8696
80. Aggarwal P, Hall JB, McLeland CB, Dobrovolskaia MA, McNeil SE (2009) Adv Drug Deliver Rev 61:428–437
81. You CC, Agasti SS, De M, Knapp MJ, Rotello VM (2006) J Am Chem Soc 128:14612–14618
82. Shang W, Nuffer JH, Muñiz-Papandrea VA, Colón W, Siegel RW, Dordick JS (2009) Small 5:470–476
83. Baier G, Costa C, Zeller A, Baumann D, Sayer C, Araujo PHH, Mailander V, Musyanovych A, Landfester K (2011) Macromol Biosci 11:628–638
84. Kumar S, Aswal VK, Kohlbrecher J (2011) Langmuir 27:10167–10173
85. Koutsopoulos S, Patzsch K, Bosker WTE, Norde W (2007) Langmuir 23:2000–2006
86. Lundqvist M, Stigler J, Cedervall T, Berggård T, Flanagan MB, Lynch I, Elia G, Dawson K (2011) ACS Nano 5(9):7503–7509
87. De M, You CC, Srivastava S, Rotello VM (2007) J Am Chem Soc 129:10747–10753
88. Dobrovolskaia MA, Patri AK, Zheng J, Clogston JD, Ayub N, Aggarwal P, Neun BW, Hall JB, McNeil SE (2009) Nanomed-Nanotechnol 5:106–117
89. Xia XR, Monteiro-Riviere NA, Riviere JE (2010) Nat Nanotechnol 5:671–675
90. van der Veen M, Stuart MC, Norde W (2007) Colloids Surf B Biointerfaces 54:136–142
91. Herrwerth S, Eck W, Reinhardt S, Grunze M (2003) J Am Chem Soc 125:9359–9366
92. Wittemann A, Haupt B, Ballauff M (2003) Phys Chem Chem Phys 5:1671–1677
93. Anikin K, Rocker C, Wittemann A, Wiedenmann J, Ballauff M, Nienhaus GU (2005) J Phys Chem B 109(12):5418–5420
94. Henzler K, Haupt B, Lauterbach K, Wittemann A, Borisov O, Ballauff M (2010) J Am Chem Soc 132:3159–3163
95. Henzler K, Rosenfeldt S, Wittemann A, Harnau L, Finet S, Narayanan T, Ballauff M (2008) Phys Rev Lett 100:158301
96. Henzler K, Haupt B, Rosenfeldt S, Harnau L, Narayanan T, Ballauff M (2011) Phys Chem Chem Phys 13:17599–17605
97. Smith MH, Lyon LA (2011) Macromolecules 44:8154–8160
98. Li Y, de Vries R, Kleijn M, Slaghek T, Timmermans J, Stuart MC, Norde W (2010) Biomacromolecules 11:1754–1762
99. Li YA, Kleijn JM, Stuart MAC, Slaghek T, Timmermans J, Norde W (2011) Soft Matter 7:1926–1935
100. Li Y, Zhang Z, van Leeuwen HP, Cohen Stuart MA, Norde W, Kleijn JM (2011) Soft Matter 7:10377–10385
101. Li Y, Norde W, Kleijn JM (2012) Langmuir 28:1545–1551
102. Malmsten M, Bysell H, Hansson P (2010) Curr Opin Colloid Interface Sci 15:435–444
103. Johansson C, Hansson P, Malmsten M (2009) J Phys Chem B 113:6183 6193
104. Johansson C, Hansson P, Malmsten M (2007) J Colloid Interface Sci 316:350–359
105. Bysell H, Månsson R, Hansson P, Malmsten M (2011) Adv Drug Deliver Rev 63:1172–1185
106. Hoshino Y, Koide H, Furuya K, Haberaecker WW, Lee SH, Kodama T, Kanazawa H, Oku N, Shea KJ (2012) Proc Natl Acad Sci U S A 109:33–38
107. Kawaguchi H, Fujimoto K, Mizuhara Y (1992) Colloid Polym Sci 270:53–57
108. Shamim N, Liang H, Hidajat K, Uddin MS (2008) J Colloid Interface Sci 320:15–21
109. Halperin A, Kroger M (2011) Macromolecules 44:6986–7005

110. Cole MA, Voelcker NH, Thissen H, Horn RG, Griesser HJ (2010) Soft Matter 6:2657–2667
111. Cole MA, Jasieniak M, Thissen H, Voelcker NH, Griesser HJ (2009) Anal Chem 81:6905–6912
112. Lindman S, Lynch I, Thulin E, Nilsson H, Dawson KA, Linse S (2007) Nano Lett 7:914–920
113. Grabstain V, Bianco-Peled H (2003) Biotechnol Prog 19:1728–1733
114. Welsch N, Wittemann A, Ballauff M (2009) J Phys Chem B 113:16039–16045
115. Bysell H, Hansson P, Malmsten M (2010) J Phys Chem B 114:7207–7215
116. Bysell H, Ransson P, Schmidtchen A, Malmsten M (2010) J Phys Chem B 114:1307–1313
117. Becker AL, Welsch N, Schneider C, Ballauff M (2011) Biomacromolecules 12:3936–3944
118. Welsch N, Becker AL, Dzubiella J, Ballauff M (2012) Soft Matter 8:1428–1436
119. Yigit C, Welsch N, Ballauff M, Dzubiella J (2012) Langmuir 28:14373–14385
120. Wittemann A, Ballauff M (2004) Anal Chem 76:2813–2819
121. Jackler G, Wittemann A, Ballauff M, Czeslik C (2004) Spectroscopy 18:289–299
122. Henzler K, Wittemann A, Breininger E, Ballauff M, Rosenfeldt S (2007) Biomacromolecules 8:3674–3681
123. Dousseau F, Pezolet M (1990) Biochemistry 29:8771–8779
124. Rocker C, Potzl M, Zhang F, Parak WJ, Nienhaus GU (2009) Nat Nanotechnol 4:577–580
125. Dell'Orco D, Lundqvist M, Oslakovic C, Cedervall T, Linse S (2010) PLoS One 5:e10949
126. Welsch N, Dzubiella J, Graebert A, Ballauff M (2012) Soft Matter 8:12043–12052
127. Azizian S (2004) J Colloid Interface Sci 276:47–52
128. Jackson MB (2006) Molecular and cellular biophysics. Cambridge University Press, New York

Magnetomechanical and Magnetothermal Coupling in Ferrohydrogels

E. Roeben, L. Roeder, R. Messing, N. Frickel, G. Marten, T. Gelbrich, and A.M. Schmidt

Abstract

By merging soft, hydrogel-based matrices with nanoscopic inorganic nanoparticles to organic-inorganic hybrid materials, novel properties can arise from the unique interplay of the components´ properties. The introduction of magnetic nanoparticles of different size and shape into hydrophilic polymer network architectures leads to nano- or macroscopic hybrid gel structures that respond to magnetic fields in a predetermined way. A variety of complex gel structures are designed that allow a mutual interaction of their mechanical and thermal properties. In this review, we highlight recent accomplishments and trends in the field of magnetically active hybrid hydrogels, and conclude with an outline on future prospects in the design and application of magnetic soft matter with particle-matrix interaction.

Keywords

Particle-matrix interaction • Magnetic soft matter • Superparamagnetism • Biomedical applications • Magnetostriction • Magnetoactive materials

Introduction

The convergence of soft elastic materials with inorganic magnetic particulate matter to nanostructured hybrids has a long tradition in magnetic composites for flexible magnets, magnetic inks and for data storage. From a fundamental point of view, however, magnetic soft matter shows a great promise for more complex material design with stimuli-sensitive or remote-controlled property manipulation, including magnetoactuators, sensors, and manipulable carriers. When properly designed, the hybrids comprehend one of several different modes for magnetomechanical or magnetothermal interplay, that is, the mutual conversion of magnetic, thermal and mechanical signals within the material.

Polymer hydrogels, that are chemically or physically cross-linked networks of hydrophilic polymers swollen with aqueous media, are a natural choice for the flexible component in magnetic soft matter. They are generally supple and wet materials, and can easily be combined to hybrids using physical or chemical methods. Using standard procedures, they can further be chemically modified and functionalized. In addition, they are often biocompatible, opening the option for biomedical applications, and can be seen as biomimetic models for biological tissue. Their structure-property relationships are basically known, and they can be obtained with different water content and mesh size. Finally, some natural and synthetic hydrogels are known to be stimuli-sensitive – their swelling behavior and their mechanical performance can be triggered by a number of environmental factors. As a consequence, controllable changes in shape, volume, mesh size, permeability, and their mechanical or optical properties can be obtained by the tailored design of gel composition

E. Roeben and L. Roeder are contributed equally to the article.

E. Roeben • L. Roeder • R. Messing • N. Frickel • G. Marten • T. Gelbrich • A.M. Schmidt (✉)
Institut für Physikalische Chemie, Department Chemie, Universität zu Köln, Luxemburger Str. 116, D-50939 Köln, Germany
e-mail: annette.schmidt@uni-koeln.de

G. Sadowski and W. Richtering (eds.), *Intelligent Hydrogels*, Progress in Colloid and Polymer Science 140,
DOI 10.1007/978-3-319-01683-2_11,

and architecture. They find present and prospective applications in surgery and drug delivery [1]. However, owed to the generally weak nature of synthetic hydrogels, they are hardly considered for uses where mechanical toughness is an issue. The recent development of hybrid hydrogels by the incorporation of inorganic nanoparticles into the gel matrix currently results in new families of mechanically strong gels with novel response properties [2, 3].

When particles with dipolar nature or polarizability are combined with mechanically soft materials like gels, new possibilities are opened for the manipulations of materials by the application of external fields [4, 5]. In particular the combination of hydrogels with magnetic, nanoscopic particles offers a broad variety of interesting effects. Such materials are now known as ferrohydrogels [6]. The easy design and versatility of magnetic fields and gradients with the help of either permanent magnets or coil-based systems is one aspect – another is the compatibility and selectivity of lab-range magnetic fields that show neglectible interaction with commonly diamagnetic matrix materials as compared to the magnetic particle phase. In addition, the unique properties of magnetic nanoparticles lead to a growing interest in nanostructured magnetic materials, composites, and dispersions, with interesting perspectives in current technology and future applications [7]. Colloidal magnetic dispersions, known as magnetic fluids or ferrofluids [8, 9], usually contain magnetic particles in the range of 10–100 nm. The magnetic properties are strongly size-dependent and can be tuned in a broad range from superparamagnetic to ferromagnetic by size or composition. In general, they are single-domain, and various methods are known to prepare their stable dispersions in water, irregardless their dipolar interactions that forward their aggregation. In aqueous systems, ferrites such as magnetite (Fe_3O_4), maghemite (γ-Fe_2O_3) or mixed ferrites are mostly employed due to their excellent long-term hydrolytic stability.

At least three different particle-field interactions are of interest for the magnetomechanical manipulation of ferrohydrogels. First, the particles' magnetic susceptibility (or rather its deviation from the susceptibility of the surrounding medium) is responsible for the possibility to orient and guide superparamagnetic particles (and their attachments) in field gradients by particle-field interactions. By using homogeneous magnetic fields, the particles are rather oriented than moved, and interparticulate interactions may become significant. Third, the particles' energy dissipation observed in AC magnetic fields (magnetic heatability) is an additional feature; the ability of magnetic nanoparticles to convert magnetic energy into heat by relaxational and hysteresis effects [10] is known for a long time and has recently gained much attention for antitumor therapy [11]. In polymeric environments, it can be used to heat up the matrix and may be used to activate a thermal effect selectively at the particles location [12–20].

In current research, much respect is paid to the combination of such particles with polymers to benefit from their mechanical properties or the presence of functional groups along the polymer chain. When magnetic nanoparticles are combined with hydrophilic polymers to so-called ferrohydrogels, the latter component can take over several tasks for prospective applications: it can serve as a stabilizing or compatibilizing shell with an external matrix, e.g., in polymer stabilized magnetic dispersions, and it allows the attachment of biologically or catalytically active functional sites [21–23]. On the other hand, the magnetic properties of the nanoparticles allow the manipulation of the colloidal or composite systems. The hybrid materials, that are either of particulate submicron (chapter "Nanoscopic Ferrohydrogels") or macroscopic scale (chapter "Macroscopic Ferrohydrogels"), can either be directly magnetically activated, or they can be stimulated by magnetic heating in order to induce thermal effects in the particle environment or the polymer matrix.

This review highlights recent developments in the synthesis and investigation of ferrohydrogels with a special emphasis on the interplay between the magnetic, mechanical, and thermal properties.

Nanoscopic Ferrohydrogels

The combination of magnetic nanoparticles and polymers to nanoscopic core-shell or related structures leads to the formation of hybrid particle dispersions with a variety of interesting properties and perspectives, including instant dispensability and magnetoresponsive properties. While polymer shells based on poly(vinylpyrrolidone)s (PVP) or carboxymethyldextranes are good stabilizers in a broad range of temperature, pH and ionic strength, important for many biomedical applications, other polymers show a stimuli-responsive behavior that can be exploited for the triggering of the dispersion properties by changing the environmental conditions. The additional incorporation or attachment of biologically active components to the polymer shell provides nanocarrier systems that can be interesting in fields like biocatalysis or drug delivery.

In the biomedical field, the cores are generally intended to be superparamagnetic, allowing for only low particle interactions when external fields are absent, and an effective magnetic orientation under field influence. This behavior is useful for an easy spatial manipulation, separation, and recovery of the nanocarriers, and can act as local sources for magnetic heating. Thus, superparamagnetic nanoparticles are of concern as support for the detection, quantification,

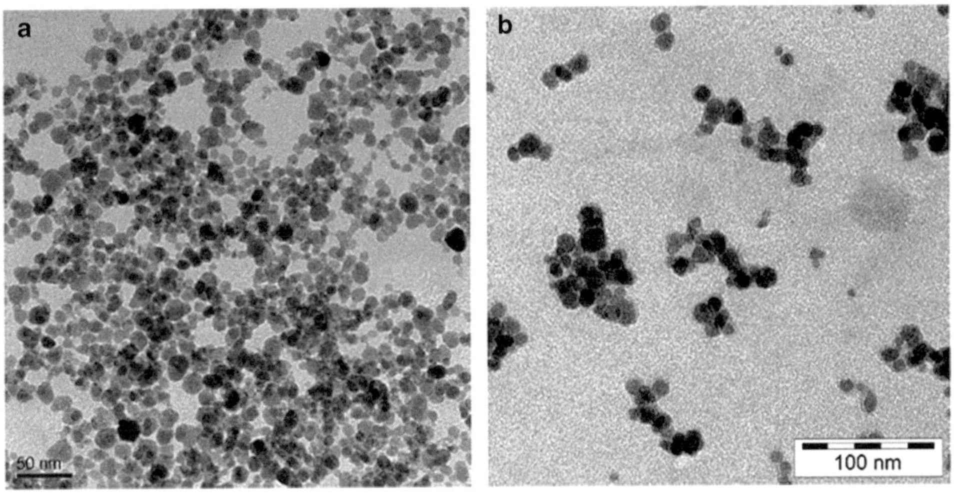

Fig. 1 TEM images of (**a**) Fe_3O_4 nanoparticles electrostatically stabilized by citric acid; (**b**) Fe_3O_4@PMEMA nanoparticles after surface-initiated ATRP (Reprinted with permission from [31]. Copyright (2010) Beilstein-Institut)

localization, separation or purification of biologically active substances, and are useful in many applications for instance fluid hyperthermia, active drug delivery, separation kits or as contrast agents for MRI [24–29].

Synthetic Strategies Towards Waterbased Magnetic Nanogels

Several routes are known for the preparation of magnetic polymer particles, including (1) the synthesis of nanocrystalline particles in the presence of preformed polymer beads (polymer-first), (2) the absorption of magnetic nanoparticles on seed or porous polymer particles followed by an encapsulation step, (3) the creation of a polymer shell in the presence of magnetic nanoparticles by emulsion/precipitation polymerization, and (4) the formation of a surface-attached polymer shell after covalent surface modification [6]. With respect to the magnetic properties, the latter two routes show clear advantages, as the particle formation can be controlled in a preceding step, and the architecture can readily be controlled. While in emulsion or precipitation polymerization single- or multicored (sub)microgels with a physical particle encapsulation are obtained, a covalent shell attachment can result in a network-like or a brush-like architecture, depending on the synthetic route.

As the magnetic core material, the iron oxides (Fe_3O_4 and γ-Fe_2O_3) are dominating, due to their biocompatibility, hydrolytic stability, acceptable saturation magnetization, and easy synthesis. A common preparation route is based on the alkaline precipitation of aqueous solutions of ferric and ferrous chloride, following Massart's method [30]. The advantages of this method combine a room temperature procedure and easy scale-up, allowing the synthesis of several hundred gram of nanoparticles with standard lab equipment. Figure 1a shows a TEM image of Fe_3O_4 nanoparticles obtained by this method. The particles are of nearly spherical shape and possess a core diameter between 10 and 15 nm [31].

For a subsequent decoration of the particles with a polymer shell, an initial stabilization is necessary, based on e.g. citric acid [30] or a poly(acrylic acid) treatment [32]. Afterwards, the particles can be employed in e.g. w/o miniemulsion polymerization [33]. For an inverse miniemulsion polymerization of *N*-isopropylacrylamide, cyclohexane serves as the continuous phase, and sorbitan monooleate as the surfactant. First, two emulsions are prepared with one of them containing the monomer and nanoparticles, and the other one the initiator. The emulsions are mixed at ambient temperature in a sonication bath, and after a short reaction time, the synthesized microgels form a suspension in cyclohexane [34]. The core-shell-structure of the gels is proven by TEM (Fig. 2). Well-defined mononuclear microgels with a spherical magnetic core consisting of agglomerated particles surrounded by a porous polymer shell are obtained.

An alternative method for the creation of a poly (*N*-isopropylacryamide) (PNiPAAm) shell is by emulsifier-free precipitation polymerization. Here, the thermoresponsive solubility of the polymer in water is exploited in the sense that the process is performed above the lower critical solution temperature (LCST), thus resulting in a precipitation polymerization [35–37]. The magnetic nanoparticles are in an aqueous dispersion with NiPAAm as monomer and *N,N'*-methylenbisacrylamide (MBA) as crosslinker. The polymerization is started through the thermal decomposition of ammonium peroxodisulfate (APDS). The magnetic nanoparticles serve as seeds for the apposition of the precipitated polymer chains, and upon reaching a critical molar mass, the oligomers precipitate on the appropriately

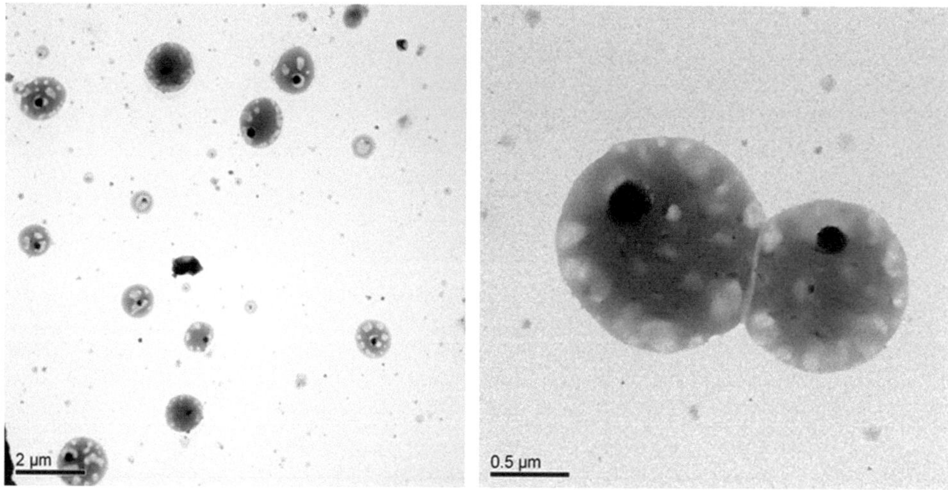

Fig. 2 TEM images of the Fe_3O_4@PNiPAAm microgels prepared by inverse miniemulsion polymerizaition [34]

primed magnetic particle surface. By this method, however, the control over the shell thickness is restricted, and the polymer content of the resulting composite is low.

In order to allow a controlled covalent attachment of polymers, the magnetic particles are directly surface-functionalized with alkoxysilanes in ethanol (Scheme 1) [38]. In an optimized procedure, no additional hydrolysis catalyst is added, and the condensation of the alkoxysilanes is therefore restricted to the vicinity of the alkaline particle surface, resulting in a chemically stable monolayer. Depending on the used alkoxysilane, the particles possess different functional groups on the particle surface.

Particles functionalized with (3-methacryloxy)propyltrimethoxysilane (MTS) can serve as macromonomers and macrocrosslinkers due to the free double bonds. Via free-radical polymerization of suitable monomers like methyl methacrylate [38] or acryl amide [39], a covalent polymer attachment to the particle surface is achieved in a grafting-through process. By adjusting the polymerization conditions accordingly, either nano-, micro- or macrogels (see chapter "Macroscopic Ferrohydrogels") are obtained.

The surface-modification with 4-(chloromethyl)phenyltrimethoxysilane (CTS) leads to particles, that can act as macroinitiators for surface-initiated atom-transfer radical polymerization (ATRP) [40, 41]. Various methacrylates, methacrylamides and vinyl monomers can be attached in a grafting-from process, resulting in brush particles with a magnetic core and polymeric arms (Fig. 1b). By using functional monomers or post-polymerization processes, the core-shell structures can be used for different application areas like biocatalysis, drug delivery or functional pickering emulsions (see section "Biomedical Applications of Functional Magnetic Nanogels").

Some approaches for the hybrid nanoobject design are successful by using a smart self-assembly process. This is the case for a system based on ionic interactions between a charged surface of superparamagnetic maghemite particles and an oppositely charged triblock terpolymer [12, 42]. The polymer is obtained by sequential anionic polymerization and is constructed of a thermoresponsive poly(glycidyl methyl ether-*co*-ethyl glycidyl ether) block, a water soluble poly(ethylene glycol) middle block and a partially quaternized poly(2-vinylpyridine) block (P[(GME-*co*-EGE)-*b*-EG-*b*-2qVP]). Caused by the attraction between the negatively charged particles and the positively charged polymer, hybrid micelles are formed in aqueous solution by spontaneous self-assembly. The micelle-like structures contain each a small cluster of about 3.5 nanoparticles, and about 30 triblock terpolymer molecules.

Thermoresponsive and Magnetothermal Effects in Nanogels

Stimuli-sensitive materials are designed to respond to an external stimulus by generating a desired effect. Possible stimuli may include temperature, light, hydrostatic pressure, mechanical stress, chemical or biological substances, pH, ionic strength, electric and magnetic fields, etc. Responds typically consist in a discontinuous change of physical properties like optical or mechanical behavior, surface properties, or a change in volume or shape.

Macromolecules offer many ways of thermoresponsive effects, including temperature-depending swelling behavior [43, 44], mechanical performance, membrane permeability [45], colour [46], and shape [47–50]. In the context of magnetic hydrogels, the thermoresponsive behavior is generally

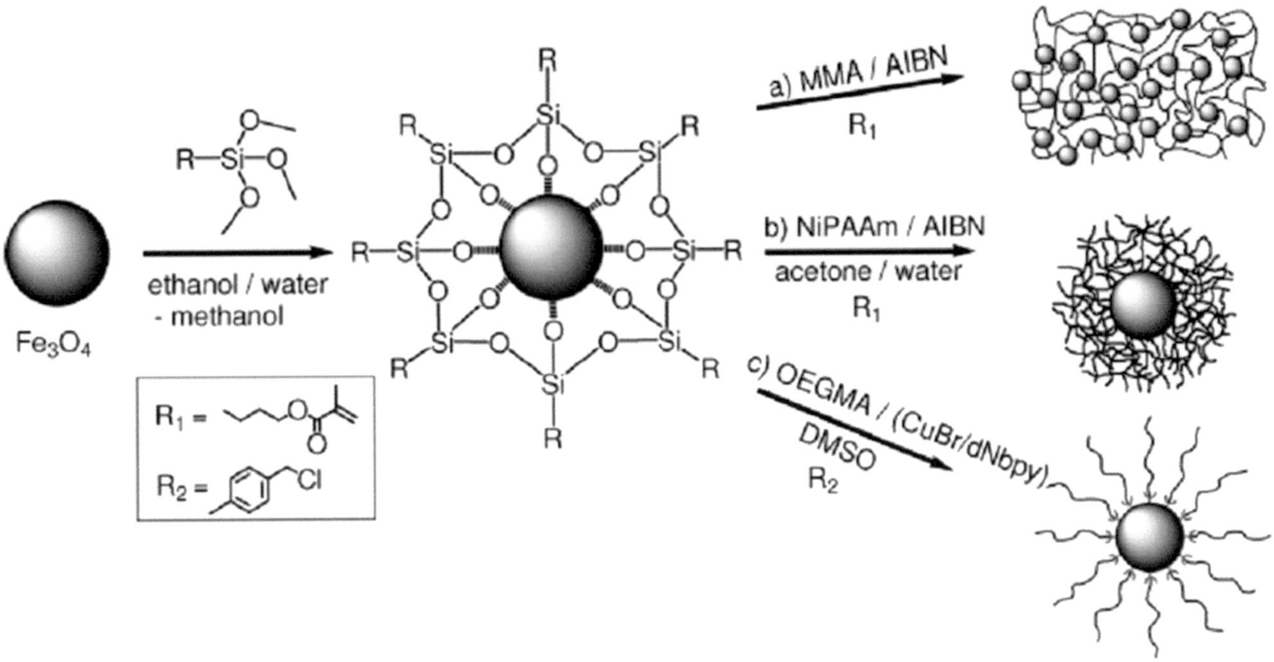

Scheme 1 Synthesis of magnetic polymer hybrid nanocomposites of different architecture by employing surface functionalized Fe_3O_4 nanoparticles as multifunctional nanosized comonomers or initiators (Reprinted with permission from [38]. Copyright (2010) American Chemical Society)

based on a transition in solubility or solvation of the polymer compound in the respective aqueous environment. While polymers with an upper critical solution behaviour at relevant temperatures (UCST) are well-known for organic systems, the special properties of water as a solvent (e.g., the strong tendency to form hydrogen bonds) results in the finding that in this solvent, a lower critical solution temperature (LCST) behavior is found for several polymers.

A well-known example is PNiPAAm, showing a coil-to-globule transition in water at about 32 °C. The transition is caused by the failure of hydrogen bonds that essentially contribute to the overall solvation energy [44]. The phase separation temperature can be varied by copolymerization with other monomers of higher or lower water solubility [51–53].

An interesting class of thermoresponsive polymers with LCST behavior in water is based on oligo(ethylene glycol) methacrylates (OEGMA) [54]. Their polymerization, preferable by controlled polymerization routes, leads to a carbon main chain with pending water-soluble oligo(ethylene glycol)-based side groups. By adapting the length of the side chains, the transition temperature can be tuned beween about 5 °C and 90 °C. A convenient solution is the copolymerization of oligo(ethylene glycol) methacrylates with different side chain length, or e.g. with methoxyethyl methacrylate (MEMA).

When thermoresponsive polymers are employed as the shell material in core-shell colloids, nanoobjects are obtained that change their size and dispersion stability with temperature. Here, the change in swelling or solvation behavior of the polymer shell triggers the solution properties of the colloidal structures. For example, the shell solubility can be used to reversibly induce a transition between flocculation and dispersion [14]. This can be of impact for the chemical activity of reactive groups within the corona (see section "Biomedical Applications of Functional Magnetic Nanogels"). While for functional groups in the polymeric corona, good binding conditions and quasi-homogeneous behaviour are observed in the well-dispersed state, a shell collapse under temperature influence causes a severe reduction of the accessibility of these groups (Scheme 4). This behaviour is useful for the activation control of immobilized enzymes, in trypsin and in β-galactosidase containing microspheres [55].

In this context, the introduction of a magnetic core can take over two functions: First, the flocculated particles can be separated easily in a magnetic field gradient, using a simple permanent magnet, and second, the cores can be used as magnetic heaters that allocate the thermal energy necessary for the phase transition [6].

For brush particles based on a superparamagnetic Fe_3O_4 core and a thermoresponsive polymer shell based on poly[oligo(ethylene glycol) methacrylate]s (POEGMA), the interplay between the composition, magnetic and thermal properties, and the dispersion behavior has been intensively studied [56–58]. In optical turbidity experiments, their aqueous dispersions show a sharp thermoflocculation

Scheme 2 Thermoflocculation of magnetic core-shell particles (here: Fe_3O_4@POEGMA) in water (Reprinted with permission from [58]. Copyright (2010) Elsevier)

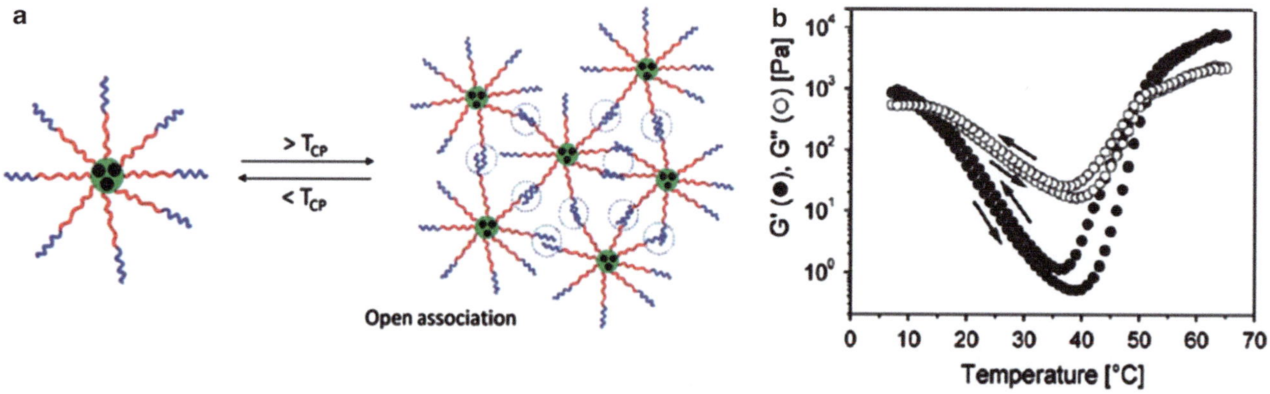

Fig. 3 (**a**) Reversible gelation via open association of hybrid micelles based on γ-Fe_2O_3 and P[(GME-*co*-EGE)-*b*-EG-*b*-2qVP] at temperatures above the cloud point (T_{cp}) of the P(GME-*co*-EGE) block; (**b**) temperature-dependent real (G') and imaginary part (G'') of the complex shear modulus ($\gamma = 0.7$ %, f = 1 Hz, 0.1 K min^{-1}) for a 36 wt% solution with a γ-Fe_2O_3/P[(GME-*co*-EGE)-*b*-EG-*b*-2qVP] ratio of 45/55 w/w (Reprinted with permission from [12]. Copyright (2010) The Royal Society of Chemistry)

temperature T_c that can be precisely adjusted by the shell composition [58]. In diluted systems, a decrease in the object size is observed by Dynamic Light Scattering (DLS) that is attributed to the shell collapse (Scheme 2). At temperatures below T_c, the hydrophilic side chains of the polymer are highly solvated, and thus the polymer shell serves as an excellent steric stabilizer for the particle dispersion. In contrast, above T_c, the polymer arms collapse due to a weaker polymer-solvent interaction, and at sufficient particle concentration, thermoflocculation is induced [59]. As a consequence of the magnetic properties, the particles can be magnetically separated from the solution. The process is fully reversible, so that cooling below T_c leads again to a stable magnetic fluid [58].

A similar behavior is observed for PNiPAAm-based core-shell nanogels obtained by precipitation polymerization. A reversible thermoflocculation is observed, however the transition regime is broad between 20 and 50 °C.

In self-assembled micelles based on γ-Fe_2O_3 and P[(GME-*co*-EGE)-*b*-EG-*b*-2qVP] (see section "Synthetic Strategies Towards Waterbased Magnetic Nanogels"), the micellar architecture involves the presence of the thermoresponsive P(GME-coEGE) block on the corona (Fig. 3a) [12]. This is connected to a reversible sol-gel transition of the micelles at elevated temperatures above a critical micelle fraction. The gelation succeeds via open association of the hybrid micelles. This self-assembly process and the size distribution of the aggregates are investigated with DLS and temperature-dependent rheology. A feature of this system is the occurrence of both an UCST and a LCST at relevant temperatures, indicated by two temperature regimes with $G' > G''$ (Fig. 3b).

The ability of magnetic nanoparticles to dissipate heat in response to a high-frequency magnetic field, known as magnetic heating [60], can be used to conveniently trigger the induction of an originally thermal response by applying an external alternating magnetic field. The heat development occurs due to relaxation processes and hysteresis losses where magnetic energy is transformed in thermal energy, and depends strongly on the particle concentration (Fig. 4) [58].

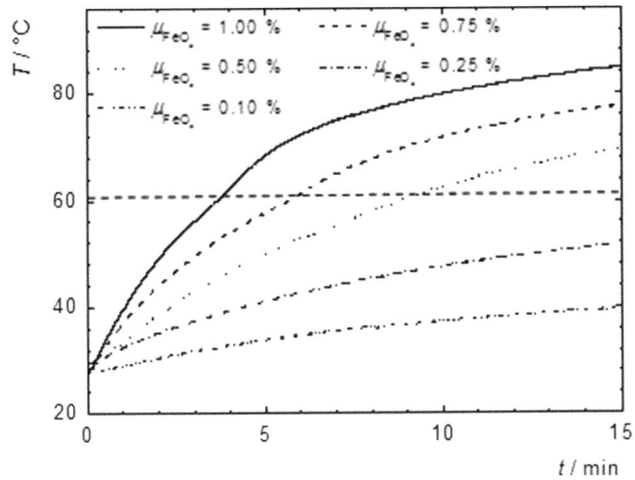

Fig. 4 Sample temperature T versus irradiation time t for Fe_3O_4@P(OEGMA) nanoparticle dispersions in an AC magnetic field (300 kHz, $H = 42.6\ kAm^{-1}$) (Reprinted with permission from [58]. Copyright (2010) Elsevier)

Accordingly, a thermosensitive hydrogel in which superparamagnetic nanoparticles are entrapped, swells or shrinks in response to magnetic heating. This behaviour may be exploited for drug release, as has recently been reported for nanoparticulate systems based on brush coated magnetic nanoparticles [15, 57], micellar solutions [12] and microgel beads [33].

When an enthalpic process of sufficient intensity is involved in the thermally induced response, the transition can be observed and analyzed directly from the time-temperature curve of the magnetic heating experiment, as shown for ice/water-embedded particles as a model system [60]. When superparamagnetic probe particles are involved, the thermal energy evolution is relatively constant over temperature. Thus, the deceleration of the temperature curve in the area of the volume transition facilitates the determination of the transition temperature and the transition enthalpy, and a quantitative calorimetric analysis is possible.

The magnetic heating is furthermore of interest for the activation of chemical or physical (thermoresponsive) processes in the particle environment – in magnetic thermoresponsive nanogels this can result in shell collapse, thermoflocculation or sol-gel transition [61–63]. Possible applications of such hydrogels systems are in the biomedical area.

Biomedical Applications of Functional Magnetic Nanogels

The magnetic properties of the core-shell structures described above involve the advantage that the particles can be transported, separated or heated up under the influence of an external magnetic field. This ability makes them very interesting for biomedical applications [6, 29, 64] like hyperthermia, drug delivery [15, 65, 66], solid-supported immunoassays [67], the temperature-controlled separation of biological compounds [68] and as contrast agents for magnetic resonance imaging (MRI) [69].

Based on their modular synthesis pathway, in particular brush-like core-shell structures show a promising versatility when bioactive species are immobilized within their shell. A flexible base for the incorporation of biological compounds in the polymer shell is the presence of reactive functional groups. This can be achieved by introducing active ester groups by using succinimidyl methacrylate (SIMA) as additional comonomer during the polymerization step (Scheme 3) [55]. Due to high affinity of the succinimidyl ester for the reaction with nucleophiles, the particles can easily be functionalized with substances bearing for instance free amino groups.

A possible application of magnetic particles with a polymer brush shell containing succinimidyl ester groups is the capture, isolation and identification of cell membrane proteins. This is demonstrated for endothelial cell surface membranes (Fig. 5) [55], and has recently been extended to vascular endothelial cells by in situ perfusion of mouse hearts [70]. Free nucleophilic, e.g. amino groups along the membrane, prevalently from proteinic complexes, are captured by attack of the active ester groups, and succinimide is eliminated as a good leaving group. The magnetic properties of the particles allow a magnetic separation of the labeled species and thus their readily isolation after cell lysis. Afterwards the species can be identified.

The possibility to functionalize active-ester functional brush particles with biological molecules can be used for the immobilization of enzymes [31, 55]. This way, nanoscopic carrier-supported biocatalysts are obtained that can serve as bioreactors for the conversion of suitable substrates. While the magnetic core allows for an easy catalyst recovery, the employment of a thermoresponsive polymer shell results in a switchable catalytic activity. Here we make use of the different accessibility of the catalytically active site depending on if the shell is swollen or collapsed. The shell collapse can directly be triggered by magnetic heating, resulting in a remotely controlled biocatalytic system.

We investigate core-shell particles with immobilized trypsin moieties as the enzyme [31]. The catalytic activity is proved by the classical N_α-benzoyl-L-arginine *p*-nitroanilide (BAPNA) method [71] which is based on the trypsin catalyzed hydrolysis of a specific peptide bond of BAPNA to N_α-benzoyl-arginine and *p*-nitroaniline in aqueous solution. The formation of *p*-nitroaniline is monitored by UV-vis spectroscopy, and the results were compared to the activity of free trypsin. The temperature dependence of the Michaelis constant K_m, a measure for the stability of the

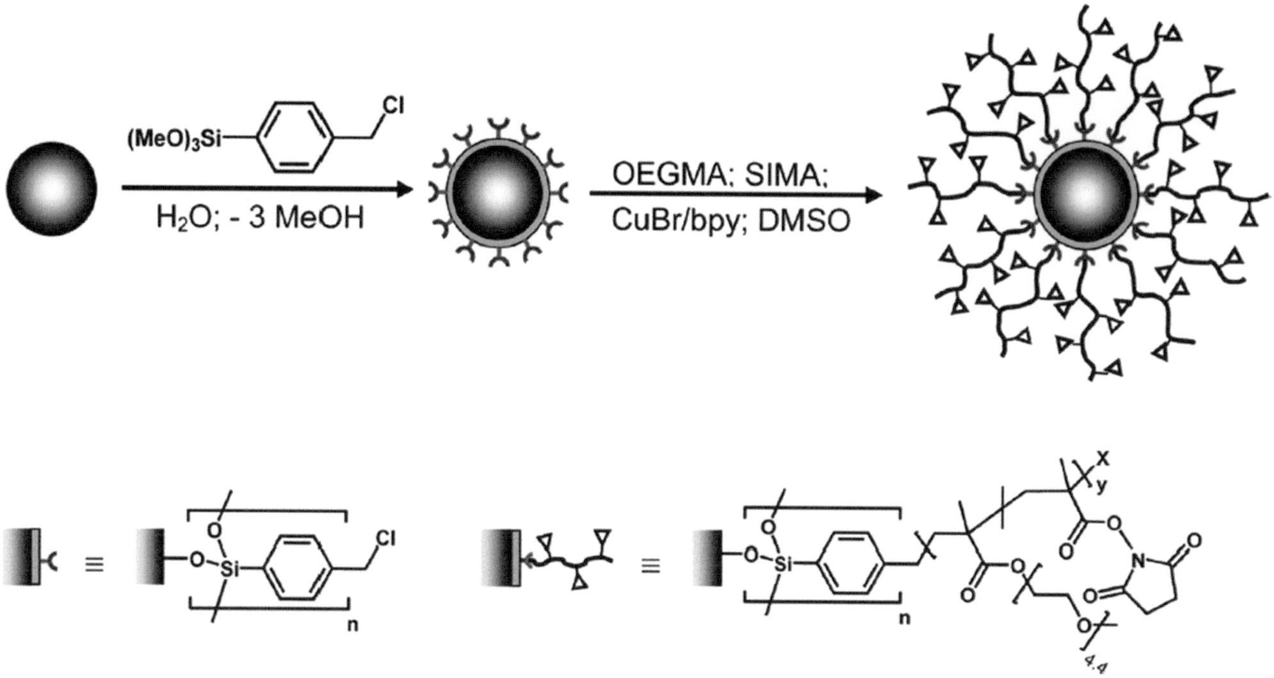

Scheme 3 Synthesis of Fe_3O_4@P(OEGMA-co-SIMA) magnetic polymer brush particles by surface-initiated ATRP (Reprinted with the permission from [55]. Copyright (2010) American Chemical Society)

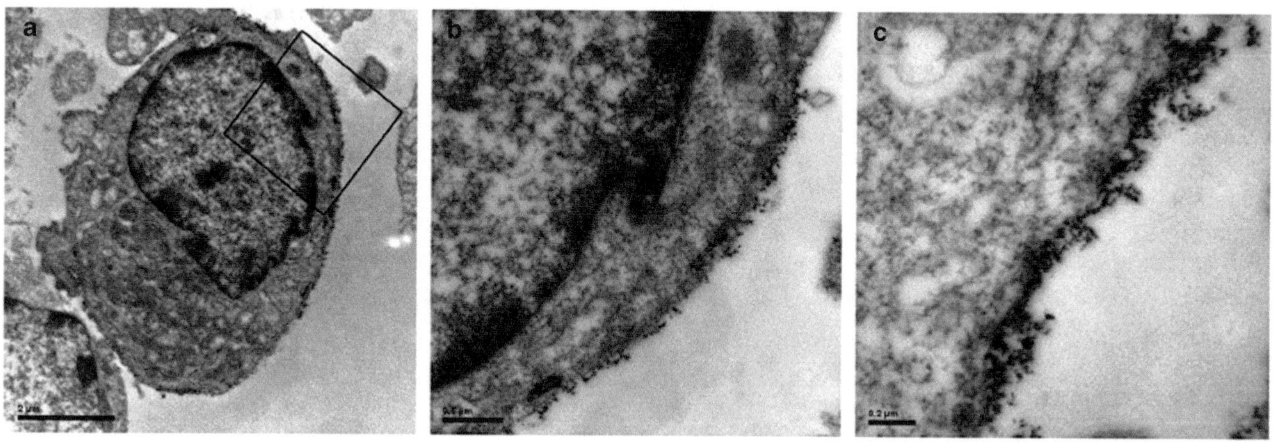

Fig. 5 (**a**) TEM image of a human endothelial cell after incubation with Fe_3O_4@P(OEGMA-co-SIMA) nanoparticles; (**b**) detail marked in (a) showing nanoparticles attached to the cell surface; (**c**) magnification of an endothelial cell surface efficiently coated with nanoparticles (Reprinted with the permission from [55]. (Copyright (2010) American Chemical Society)

enzyme-substrate complex, is shown to increase strongly for particle bound trypsin above the T_c of the particles, accompagnied with a strong acceleration of the catalytic activity, while for free trypsin only a weak temperature dependence is observed (Fig. 6). One possible interpretation is that the shell collapse results in a better accessibility of the catalytically active sites in the shell (Scheme 4).

When the polymer brush shell is instead functionalized with host molecules like cyclodextrins, thermo- and magnetoresponsive drug-delivery systems are obtained. Cyclodextrins are broadly used in medical treatment applications because of their ability to form water-soluble complexes with hydrophobic drugs, leading to a better uptake, bioavailability and transportation of the drug [72, 73, 74]. Cyclodextrin-functional copolymers are accessible, e.g. by using 3-(6-desoxy)-β-cyclodextrin-3*H*-1,2,3-triazol-4-yl) methylmethacrylate (CDMA) as a comonomer [75, 76]. In combination with a thermoresponsive main chain and a magnetic core, a magnetoresponsive drug delivery system is obtained. The host unit can reversible uptake a suitable substrate which is strongly tethered within the host molecule below the T_c of the polymer arms. Due to the

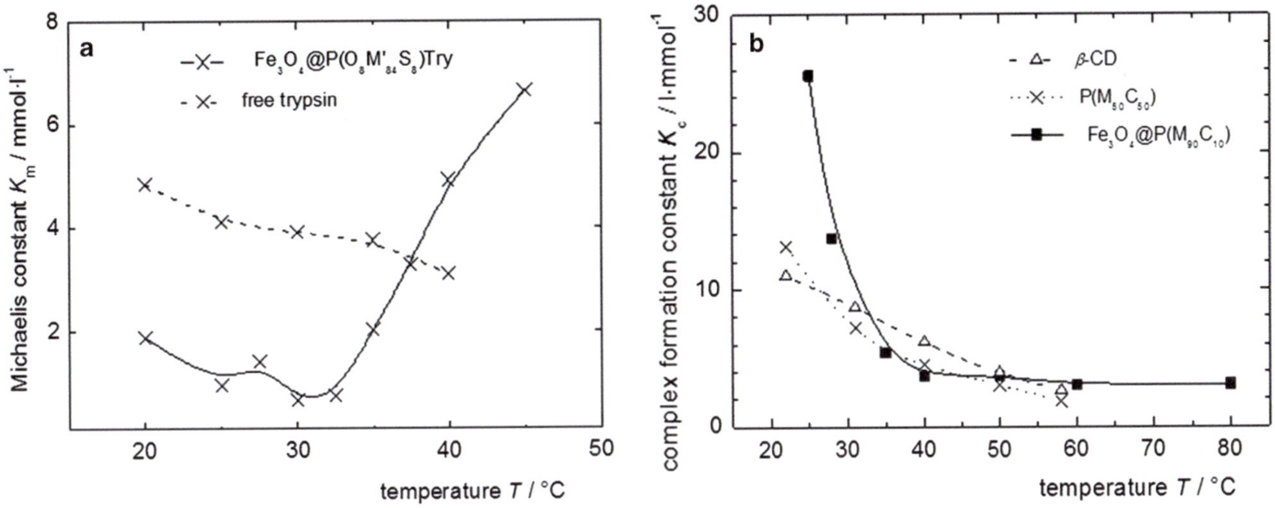

Fig. 6 (**a**) Temperature-dependent Michaelis constant K_m of particle-immobilized trypsin versus free trypsin (Reprinted with permission from [31]. Copyright (2010) Beilstein-Institut). (**b**) Temperature-dependent complex formation constant K_c of PP with free β-CD and CD-functional polymer (P[OEGMA-co-CDMA]) and brush particles (Fe_3O_4@P[OEGMA-co-CDMA]) (Adapted with permission from [72]. Copyright (2013) IEEE)

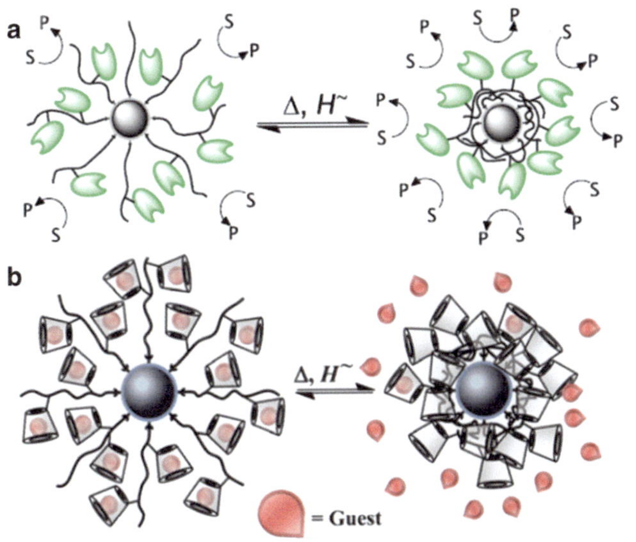

Scheme 4 (**a**) Proposed mechanism of catalytic activity after heating magnetic biocatalyst particles above T_c. Reprinted with permission from [31] (Copyright (2010) Beilstein-Institut). (**b**) On-demand drug delivery system based on Fe_3O_4@P[MOC] nanoparticles (Reprinted with permission from [72]. Copyright (2013) IEEE)

magnetic core, the particles have the potential as nanocarrier system for the transport of the substrate to the requested place, known as magnetic targeting. Heating the particles magnetically above the T_c of the polymer brushes leads to a strong decrease of the complex formation constant K_c of the host-guest-complex (Fig. 6), and to the release of the guest molecule (Scheme 4b) [72]. This can be explained on the one hand by the enhanced entropic contribution of the released molecules (and water) in the case of multivalent CD-bearing moiety, and on the other hand by the enhanced local availability of the oligoethylene glycol side chains above the T_c that can be seen as potential concurrent guest molecules. The concept is verified for β-cyclodextrin functional brush particles based on Fe_3O_4@P(OEGMAco-CDMA), and phenolphthalein (PP) as a model substance. PP is selected as the model, as its complex formation and release can easily followed by UV-Vis spectroscopy. As shown in Fig. 6b, the complex formation constants of as well free, polymer bound, and particle-immobilized β-CD are progressively temperature-dependent, indicating the increasing impact of entropy that is more pronounced in the multivalent systems. For this system, the release of the PP triggered by magnetic heating is even higher than for conventional heating.

Another interesting concept towards magnetoresponsive drug delivery systems is based on Pickering emulsions [77]. The use of magnetic nanoparticles with a thermoresponsive shell as nanostructured o/w emulsion stabilizers leads to a magnetically active transport and release system. The emulsions are prepared by sonication of the oil/water mixture in the presence of the particles, and active substances like drugs can easily be incorporated in the evolving droplets. If the emulsion stability is high enough, the particle-encapsulated, drug-containing oil droplets can directly be transported through the solution by using a static magnetic field. When the polymer-coated particles possess a UCST-related behavior in one of the two phases (water or oil), the emulsion is destabilized when heated across this temperature. Accordingly, the droplets can be destabilized by magnetic heating resulting in the release of the drug.

We realized an according model system by emulsifying a water/cyclohexane/carotenoid (β-apo-12′-carotenal) mixture using Fe_3O_4@polystyrene (PS) brush paricles as

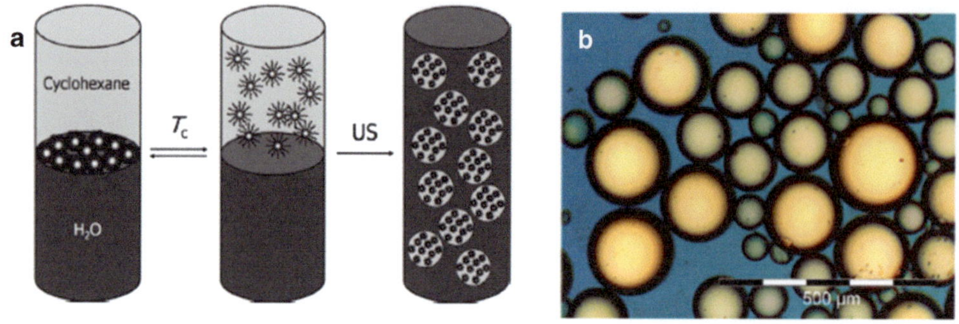

Fig. 7 (**a**) Preparation of solid-stabilized emulsions by using core-shell particles above their critical solution temperature T_c in cyclohexane; (**b**) Microscopic pictures of the colored o/w emulsions from $MFFeO_x$@PS07/H_2O/cyclohexane with acid blue 25 as water-soluble dye (Reprinted (adapted) with permission from [78]. Copyright (2009) American Chemical Society)

emulsion stabilizers (Fig. 7) [78]. Polystyrene serves as the thermoresponsive polymer in this system, as it possesses an UCST-type behaviour in cyclohexane at about 32 °C. Above T_c, the particles form a homogenous dispersion in the cyclohexane phase. After emulsification, the system is cooled down to ambient temperature, and the particles are concentrated on the interphase between the oil/carotenoid droplets and the aqueous continuous phase, forming a polymer-entangled, magnetic particle-doped shell (Fig. 7). The droplets are stable against treatment and transport with a static magnetic field. However, they show droplet evanescence and phase separation after magnetic heating above T_c. The findings indicate that such a system is suited for the transport and release of lipophilic molecules, as shown for the employed antioxidant/carotenoid.

It can be summarized that the combination of nanoscopic magnetic cores and functional soft shells leads to architectures with stimuli-responsive behavior that can directly be remote-controlled by external magnetic fields. Due to their unique properties, this class of colloidal nanostructured materials shows high potential for many applications, especially in the fields of biocatalysis and biomedicine.

Macroscopic Ferrohydrogels

Magnetostrictive polymer gels are a sub-class of soft materials containing magnetizable components, which have been embedded in the macroscopic gel matrix during the cross-linking process. When properly designed, it is possible to use the susceptibility of superparamagnetic particles, and the induction of particle interaction in static magnetic fields to achieve considerable gel bending or contraction effects. Since their first implementation by Zrìnyi et al. [79, 80] there is substantial interest in advancing the understanding of their performance. Meanwhile, magnetostrictive gels based on different matrices like poly(vinylalcohol) (PVA) [81], Pluronics [82], and poly(hydroxyethyl methacrylate) (PHEMA) [81] have been realized.

In the water-swollen state, a highly flexible environment is created for the magnetic component, and under the influence of an external magnetic field a reorientation of the magnetic moments and the development of a macroscopic magnetization is achieved, leading to the observation of a significant volume contraction or deformation under certain conditions [79]. Although the latter is of high interest for the development of magnetic actuators, e.g. as artificial muscles, the phenomenon is not finally understood at all, and occurs partly contrary to the theoretical expectations [83].

The magnetic component is mostly based on spherical superparamagnetic nanoparticles that are dispersed within the fluid phase in the gel meshes. The soft structure of the gels enables a strong displacement, shape change, or change of the mechanical properties as a reaction to applied magnetic fields, but a feasible mechanism on the basis of a microscopic model is usually not provided [84]. In most cases a macroscopic theory is used that is predicated on the elasticity theory and the Maxwell-equations, assuming a one-phase homogeneous, elastic and magnetizable material, although the gels consist of the polymer, the particles and a solvent [85–87].

On a closer look, at least two different types of interactions can be identified that are involved in the magnetostriction: [88] on the one hand the direct interaction of the single particle with the field, and the increase of interparticle interaction induced by their uniform orientation. In field gradients, the former leads to a magnetophoretic force, and the particles are attracted to regions of stronger field intensities. Because of the cross-linking bridges in the network, changes in molecular conformation due to force impact can accumulate and lead to macroscopic shape changes and/or motion. In contrast, this effect can be expected to be neglectable in uniform fields. Instead, mutual particle interactions occur if the uniformly oriented particles are so closely spaced that the

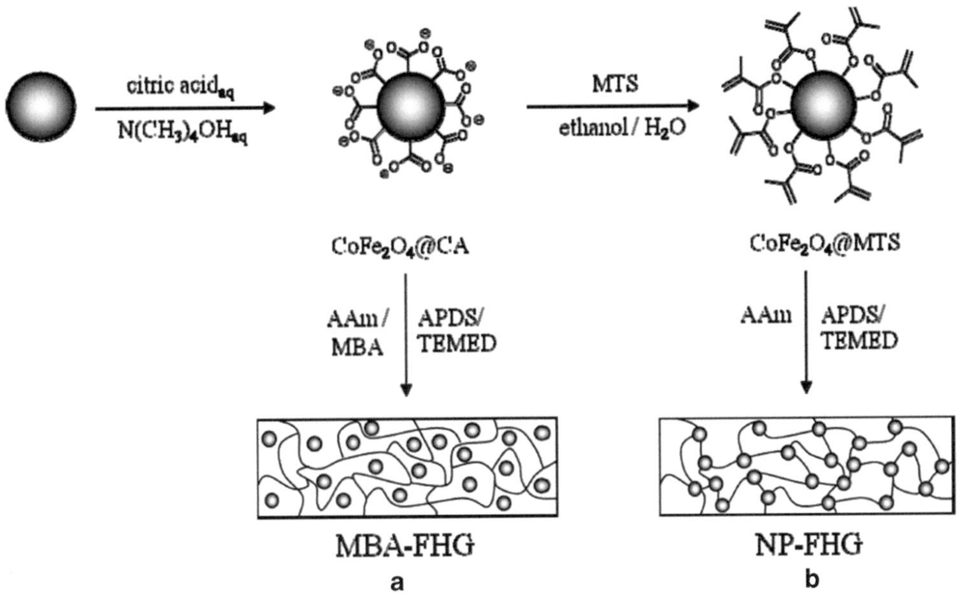

Scheme 5 Schematic illustration of the preparation of $CoFe_2O_4$@CA containing MBA-linked ferrohydrogels (**a**) and $CoFe_2O_4$@MTS-linked ferrohydrogels (**b**) (Reprinted with permission from [99]. Copyright (2011) American Chemical Society)

local field can influence their neighbours [89]. This mutual interaction can be accompagnied by a significant change in the structure of particle ensembles, e.g. chain formation. The field-induced chain formation has major implications for magnetorheological gels [90].

If already the synthesis of magnetically loaded gels is performed under the impact of an external magnetic field, magnetically and mechanically anisotropic materials are obtained with "frozen" anisotropy, manifesting itself in a direction-dependent elastic modulus and swelling [80], and in a field strength-dependent mechanical performance [80, 91]. It can be assumed that in these materials chain formation plays a significant role for the anisotropic stiffening effect.

In thermoresponsive macroscopic ferrohydrogels, magnetic heating can be used to initiate the temperature-related response. Several works have reported on the acceleration of enzymatic reactions within thermoresponsive magnetic hydrogels in suitable AC fields. Similar to conventional heated gels, where an acceleration of an enzymatic reaction is possible by up-and-down temperature change due to swelling/deswelling ("hydraulic pump" behaviour) [92], the effect is reported to be advantageously heated from incorporated magnetic nanoparticles, as the process proceeds homogeneously in the material and is therefore less limited by thermodiffusional processes [93]. The magnetic heatability of thermoresponsive gels can furthermore be used for the construction of chemo-mechanical or magneto-mechanical devices [12, 13, 94–96].

An elegant method is recently reported to minimize particle migration/agglomeration or loss in ferrogels through chemical cross-linking of nanomagnets into the organic gel matrices by employing a suitable surface modification [97–101]. Such ferrogels with modified microstructure and covalent particle-matrix interaction are expected to open perspectives for the development of soft actuators, in particular when no additional (conventional) crosslinker is employed, and the particles are ferromagnetic [99, 101].

Synthesis of Ferrohydrogels with Different Architecture

Recently it was demonstrated that under certain conditions magnetic nanoparticles can be employed as nano-probes and/or actuators to monitor the mechanical interactions between the particles and the polymer network at the length scale of the particles [102, 103]. For this purpose, the probes ought to be magnetically blocked at room temperature, so that their magnetic reorientation is dominated by Brownian rotation. Hence a direct magneto-mechanical coupling between the static or dynamic field response and the particle rotation is generated.

The magnetomechanical performance is compared for two different gel architectures with reversible and irreversible mechanical coupling, respectively. On the one hand, a conventional molecular crosslinker is used during the gel synthesis, and the particles are located in the meshes of the gel. There are no covalent bonds between the particles and the polymer network (Scheme 5a). Depending on the ratio of the hydrodynamic particle diameter and the network mesh size, the particles are more or less restricted in their mobility. When a magnetic field is applied, the dipolar particle is reoriented towards the field direction with a

b

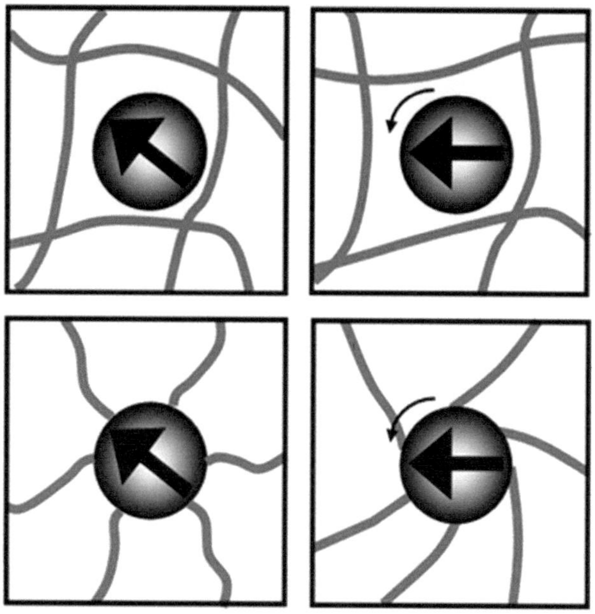

Scheme 6 Impact of matrix interaction on the remagnetization behavior of blocked magnetic particles; (**a**) conventionally crosslinked ferrohydrogels; (**b**) nanoparticle-crosslinked gels; *left*: before magnetization; *right*: during magnetization [101] (Reproduced by permission of The Royal Society of Chemistry)

irreversible (loss) effects due to field-induced changes in the gels´ microstructure (Scheme 6a).

On the other hand, gels are investigated where surface-functionalized particles serve as multifunctional crosslinker. For this purpose, the particles are equipped with a siloxane-based shell bearing methacryl groups that can participate in a copolymerization reaction (Scheme 5b). Upon a grafting-through polymerization process of (meth)acryl or vinyl monomers in the presence of the functionalized particles under appropriate conditions, covalent bonds between the particles and the polymer chains are generated. Consequently, a reorientation or displacement of the particles in an external magnetic field is directly transferred to the network. If the polymer chains are elastic the mechanical behavior is reversible (Scheme 6b). The two architectures result in different coupling mechanisms and interaction intensities between the magnetic component and the polymer matrix.

The different modes of magnetomechanical coupling are investigated by incorporating different spherical and elongated magnetic nanoprobes into poly(acrylamide) (PAAm) hydrogels. PAAm-based hydrogels are frequently used as model systems as they show a reproducible and relatively regular polymerization behavior, and as well their swelling and mechanical properties follow the basic theories within certain limits [104–108]. The synthesis of both gel types is performed in the presence of the particles so that they are homogenously embedded into the hydrogel during the radical polymerization process.

By careful analysis of the magnetization curves from Vibrating Sample Magnetometry (VSM) the mechanical properties of soft matter is investigated on the length scale of the particles. Information on the particle mobility in the network and the mechanical counter-forces of the polymer matrix/chains can be obtained by analyzing the magnetic properties based on either an extended Stoner-Wohlfarth model for systems exhibiting hysteretic loss effects (section "Magnetomechanical Properties of Conventional Ferrohydrogels"), or on the Langevin equation that can be applied for reversible magneto-mechanical effects (section "Magnetomechanical Properties of Particle-Linked Ferrohydrogels").

Properties of the Nanoscopic Ferromagnetic Probes

Before particles can be used as nanoprobes in ferrohydrogels in order to allow a semi-quantitative analysis, a detailed knowledge of characteristic particle parameters is necessary. It is essential that the particles are of rather uniform size and of ferromagnetic (blocked) nature.

Three different types of ferromagnetic particles are involved: spherical ($CoFe_2O_4$), rodlike (Ni) and ellipsoidal (α-Fe_2O_3) particles. The single domain $CoFe_2O_4$ spheres show an average core diameter of 12.1 nm [99, 101], as evidenced by TEM, XRD, and VSM, and thus they are within the ferromagnetic size range at ambient temperature [109]. The length of the Ni rods is about 250 nm with an aspect ratio of 9.4, while the ellipsoids are 500 nm in length with an aspect ratio of 4.7 (Fig. 8a, b). The latter are based on hematite and possess a weak spontaneous magnetization due to canted antiferromagnetism [110, 111]. All particles are shown to be highly crystalline, and show a hysteretic behavior in quasi-static magnetic experiments of immobilized particles, confirming their ferromagnetic nature. In contrast, water-based dispersions of all types of particles show pseudo-superparamagnetic, fully reversible behavior. The respective magnetic relaxation times τ_B of the dispersions are obtained from AC susceptometry, and are found to be in the milliseconds regime [99, 101, 111]. For each particle type, τ_B is in accordance with the calculated rotational diffusion coefficient, and thus with a magnetic reorientation dominated by particle rotation. Consequently, the selected particles are suitable for the mechanical analysis of ferrohydrogels in quasi static fields on the length scale of the probes.

Magnetomechanical Properties of Conventional Ferrohydrogels

When the magnetic probe particles are embedded in poly(acrylamide) gels that are conventionally crosslinked

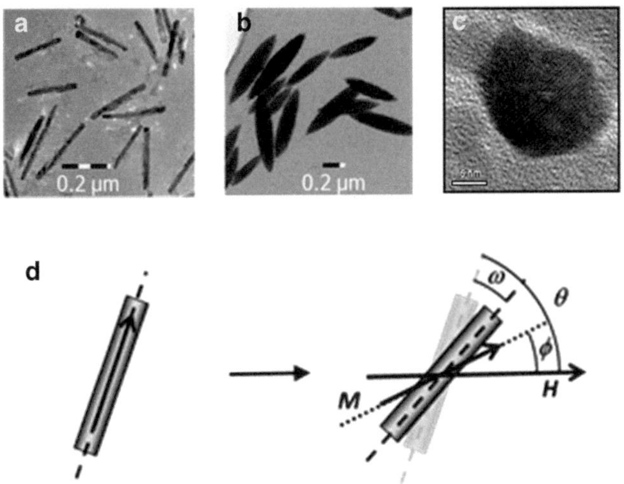

Fig. 8 TEM images of elongated magnetic probes; (**a**) Ni rods, (**b**) α-Fe_2O_3 ellipsoids; (**c**) $CoFe_2O_4$ spheres. (**d**) Schematic illustration of the field response of a cylindrical ferromagnetic particle dispersed in a soft matrix with a rotation of the magnetic moment out of the anisotropy axis and an additional rotation of the entire particle in the elastic matrix (Adapted with permission from [111]. Copyright (2012) John Wiley and Sons)

using a tetrafunctional acrylamide (*N,N*-methylene-*bis*-acrylamide, MBA), magnetic hydrogels are obtained that allow an investigation of the shear behavior of the gel at the particle level. By investigating the experimental swelling properties it is confirmed that the particle presence does not interfere significantly with the crosslinking process. Moreover, the equilibrium swelling degree of the gels shows a good correlation with the Frenkel-Flory-Rehner model [112–115]. The particles serve as nanoprobes, and are rotated by applying a quasi-static magnetic field. With this method, information about the mechanical feedback resulting from the local restoring forces on the rotational remagnetization of the probes can be received. The reversibility of this process gives information on the shear modulus, and can be compared to theoretically expected values. Resulting from the different shapes and sizes of the particles, the effects and modes of interaction with the network are found to differ.

In the case of the elongated particles, the obtained coercivity values from the magnetization measurements are analyzed based on the Stoner-Wohlfarth (SW) model [116] extended by an elastic counter-force in the works of Müller-Buschbaum [117] and Tschöpe and Birringer [118] (Fig. 8d). G can be determined by using the characteristic particle parameters (aspect ratio n, magnetic and geometric particle volume V_m and V_g, intrinsic coercivity of a SW particle $H_C{}^{SW}$, and volume saturation magnetization of the particles, M_S) and the gel factor f that is related to the experimental reduced coercivity h_c.

$$f = \frac{3\ln(n)}{4n^2}\frac{V_m}{V_g}\frac{\mu_0 H_c^{SW} M_s}{G} \qquad (1)$$

(μ_0 magnetic constant) [111].

All gels that are doped with the ferromagnetic particulate probes show a hysteretic behavior in quasi-static magnetometry that systematically increases with increasing crosslinking density, thus in accordance with a decreasing particle mobility. For $CoFe_2O_4$ spheres it was found that gels that are prepared with constant crosslinking density, but varying particle content up to 2 vol-%, show consistent swelling behavior, remanence and coercivity, indicating an absence of significant particle-particle interactions. Although the results indicate that the particles are embedded in the polymer matrix in a way that slows down the particle relaxation to a time scale that is long with respect to the experiment, this is not expected for spherical particles after the model described above, as the aspect ratio of the partices is close to 1. However, the model does not include the rheological conditions at the particle surface, i.e. if the rotation occurs under slip or no-slip conditions. It cannot be excluded that additional physical interactions between particles and matrix (e.g., hydrogen bonds) are presumably responsible for the observation of magnetic losses [101].

When elongated particles are employed, a semi-quantitative analysis according to Eq. 1 becomes possible. In the experimental series, as well the total monomer concentration ν_2 as the comonomer ratio α (molar fraction $n_{MBA}/(n_{MBA} + n_{AAM})$) in the synthesis process are systematically varied. Figure 9a shows the magnetization curves of a series of swollen hydrogel samples doped with α-Fe_2O_3 ellipsoids. A hysteretic behavior can be observed that is correlated to the comonomer ratio α (or mesh size) in the swollen gel. This is in correspondence to a reduced rotational freedom of the probes with decreasing mesh size of the swollen networks. For the determination of the relative shear modulus G/G_0, the softest ferrogel (lowest comonomer ratio α) of each series of magnetically doped gels is used as reference. The relative shear moduli estimated with Ni nanorods and α-Fe_2O_3 ellipsoids as probes show a good correlation within each experimental series, and for low comonomer ratios α, the values are in acceptable agreement with theoretical values from rubber elasticity theory (Fig. 9b). The results illustrate that in a certain modulus regime that covers e.g. the typical mechanical behavior of biological cell assemblies, the rheological characteristics of soft matter is accessible from the quasi-static performance of nanoscopic elonganted magnetic probes [111].

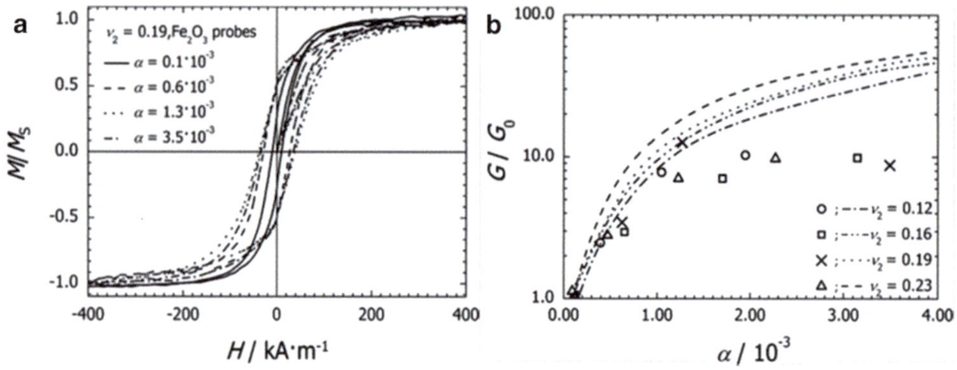

Fig. 9 (**a**) Magnetization curves of equilibrium-swollen PAAm hydrogels doped with α-Fe_2O_3 ellipsoids; (**b**) relative shear modulus G/G_0 of the Ni- and Fe_2O_3-doped hydrogels as a function of comonomer ratio α (*symbols*). The *lines* represent the expected relation from rubber elasticity theory (Reprinted with permission from [111]. Copyright (2012) John Wiley and Sons)

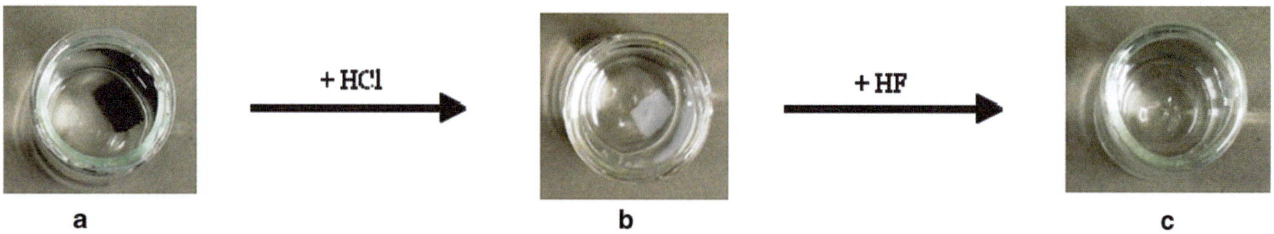

Fig. 10 Photographs of the swollen nanoparticle-linked ferrohydrogels before the dissolution experiments (**a**), the hydrogels after addition of HCl aq., and washing (**b**), and the polymer solution after addition of HF aq. (**c**) (Reprinted with permission from [99]. Copyright (2011) American Chemical Society)

Magnetomechanical Properties of Particle-Linked Ferrohydrogels

A novel class of ferrohydrogels is characterized by a direct covalent coupling between the particles and the polymer network segments. The covalent bonds are created during the synthesis of PAAm hydrogels by using surface-functionalized magnetic particles as multifunctional crosslinker. While a mechanical stability of resulting gels can be guaranteed by adding conventional crosslinkers (like MBA) [98–102], gels that are exclusively crosslinked via magnetic nanoparticles acting as the network nodes are also accessible [100, 102].

Such unique network types are achieved by functionalizing spherical ferromagnetic $CoFe_2O_4$ nanoparticles (Fig. 8c) with methacryloxy groups via the alkoxysilane route, and performing the free-radical polymerization of acrylamide in water in the presence of the functional particles in a grafting-through process [100]. In a broad compositional regime, the gels are well-defined in shape, and they swell in an affine way when immersed in water, indicating an effective crosslinking of the chains through the particle surface. A simple dissolution experiment confirms the absence of other interchain crosslinks. After treatment with HCl aq., whereby the particles are dissoluted, and addition of a few droplets of HF aq. in order to cleave the Si-O-bonds resulting from the surface coating, the gels fully disintegrate resulting in an aqueous polymer solution (Fig. 10) [99].

The microscale homogeneity of the network with respect to the particle distribution is confirmed via 3D X-ray tomography. TEM images were obtained by freeze-fracture/freeze drying direct imaging (FFFD-DI) of a representative gel. The images indicate that the particles are localized at the segment crosslinks (Fig. 11a) [99].

The equilibrium swelling degree Q_V of the particle-crosslinked gels decreases as expected with increasing particle content, confirming an increase in the overall crosslink density. Based on the Frenkel-Flory-Rehner model [112–115], an estimation of the segment length and the crosslinker functionality is possible from Q_V. The values are compared with a model based on the particle volume fraction, where the average particle distance is taken as a measure for the effective segment length. The results indicate that the network architecture can be described to be based on a fraction of comparably long polymer segments connecting two individual particles (effective network segments), and a significant fraction of segments that are looped around one particle, thus are ineffective with respect to crosslinking (Fig. 11b) [99].

The magnetic behavior of the swollen nanoparticle-crosslinked ferrohydrogels gives information on the force equilibrium in the magnetoelastic materials in

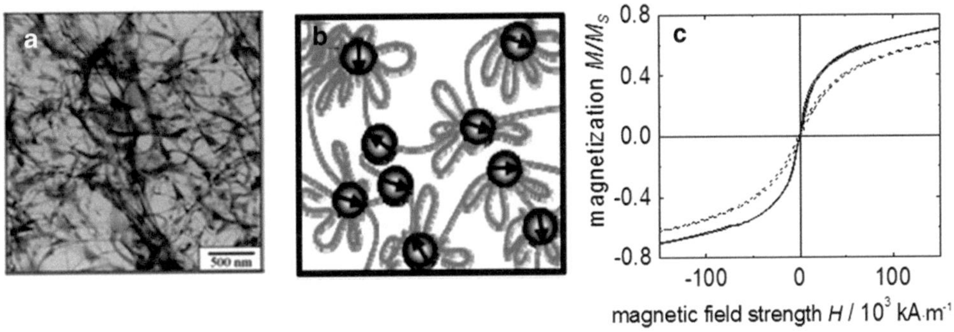

Fig. 11 (**a**) TEM; and (**b**) model of the gel architecture for the $CoFe_2O_4$@MOTS-linked ferrohydrogels; (**c**) quasi-static magnetograms of $CoFe_2O_4$@MOTS-linked ferrogel in comparison to the $CoFe_2O_4$ dispersion (Adapted with permission from [99]. Copyright (2011) American Chemical Society)

magnetic fields. In contrast to the conventionally crosslinked ferrohydrogels (section "Magnetomechanical Properties of Conventional Ferrohydrogels"), no hysteresis is observed in quasi-static magnetograms (Fig. 11c), indicating a magnetically reversible behavior that is ascribed to the covalent integration of the particles into the network architecture, resulting in elastic restoring forces from the attached polymer segments (Scheme 6b).

According to the Langevin equation [119] that readily describes the quasi-static magnetization graphs of non-hysteretic materials, the initial volume susceptibility χ_{ini} is a measure for the ratio of potential energy of the magnetized body in an external field E_{mag}, and the restoring component. In the case of the particle-linked gels, the latter is dominated by the elastic counterforce resulting from the attached polymer chains, and is of entropic nature (E_{elast}). Thus, the experimental susceptibility can be used to get quantitative information on the elastic modulus E of the attached polymer chains using

$$\chi_{ini} = \frac{E_{magn}}{E_{elast}} = \frac{\mu_0 \cdot m \cdot M_s}{E \cdot f} \cdot \frac{r_{\mathrm{cont}}}{r_{\mathrm{p}}} \tag{2}$$

(*m:* average magnetic moment of single particle, M_{s}: saturation magnetization of the sample, r_{cont}: average contour length of the effective network segments, r_{p}: particle radius, f: average particle functionality with respect to effective network segments)

Although the implemented model is rather simple, it is shown that the experimental results are in good agreement with the respective elastic moduli calculated from statistical thermodynamics [101].

Summary and Perspective

The potential of magnetic soft matter to enable access to a magnetothermal and/or magnetomechanical manipulation by means of external fields is demonstrated by the design and characterization of such materials in different structure sizes and with different architectures and modes of interaction between the materials´ components.

While superparamanetic nanoparticles can be combined with hydrophilic polymer matrices to structures that can be attracted in field gradients or selectively heated by means of dynamic magnetic fields, ferromagnetic particles reorient in homogenous fields and thus provide unusual modes of magnetomechanical manipulation.

Different synthetic strategies, involving miniemulsion-based techniques, surface-initiated polymerization, and self-assembly, lead to magnetic polymer core-shell nanostructures, allowing under more the attachment of thermoresponsive and/or biofunctional polymer shells. Consequently, these colloidal systems employ a temperature- or field-induced sol-gel transition or thermoflocculation, and are useful for improved bioseparation systems, triggered biocatalysis, and active drug delivery.

While macroscopic magnetostrictive gels that change their shape in response to external fields are known for several years, the present works demonstrate that a read-out of the mechanical characteristics of complex soft matter is possible by using ferromagnetic nanoprobes. In this respect, the presented pathways to well-defined hybrid gel matrices are promising for the development of original magneto-mechanical and magnetostrictive effects.

The course of the recent development suggests that the fundamental understanding of hybrid hydrogels and the basic particle-matrix interactions will continue, and that modern techniques for their systematic characterization will be of impact for their rational structure design. It is expected that future directions include the rational design of biomedical hybrid hydrogels that require not only the control of chemical and physical properties, but also the consideration of biological variables in order to simulate biological tissues.

The mutual interplay between particles and soft matrices and its tailored design is in all cases non-trivial, and its understanding is an important parameter for the performance of polymer composites and biological matter. Extending the investigation of magnetic particle-matrix interaction to dynamic magnetic methods and to viscoelastic matrices is consequently the next complexity step and will allow a nanorheological investigation of soft matter in the force- as well as the frequency regime on the particle scale.

Finally, it is expected that the potential impact of magnetic hybrid hydrogel components on our daily life will range from stimuli-responsive sensors and actuators to microfluidics, pharmaceutical, and biomedical devices.

Acknowledgement This work was supported by the Deutsche Forschungsgemeinschaft (DFG) within the priority programme SPP 1259 "Intelligent Hydrogels", and partly by the Emmy Noether programme (SCHM1747-4).

References

1. De Las Heras Alarcon C, Pennadam S, Alexander C (2005) Stimuli responsive polymers for biomedical applications. Chem Soc Rev 34:276–285
2. Messing R, Schmidt AM (2011) Perspectives for the mechanical manipulation of hybrid hydrogels. Polym Chem 2:18–32
3. Karg M (2012) Multifunctional inorganic/organic hybrid microgels. Colloid Polym Sci 290:673–688
4. Dagallier C, Dietsch H, Schurtenberger P, Scheffold F (2010) Thermoresponsive hybrid microgel particles with intrinsic optical and magnetic anisotropy. Soft Matter 6:2174
5. Laurenti M, Guardia P, Contreras-Cáceres R, Pérez-Juste J, Fernandez-Barbero A, Lopez-Cabarcos E, Rubio-Retama J (2011) Synthesis of thermosensitive microgels with a tunable magnetic core. Langmuir 27:10484–10491
6. Schmidt AM (2007) Thermoresponsive magnetic colloids. Colloid Polym Sci 285:953–966
7. Craig D (1995) Magnetism – principles and applications. Wiley, Chichester
8. Berkovsky BM, Bashtovoy V (1996) Magnetic fluids and applications handbook. Begell House, New York
9. Blums E, Cbers A, Mairov MM (1997) Magnetic fluids. Walter der Gruyter, Berlin
10. Brown W (1959) Relaxational behavior of fine magnetic particles. J Appl Phys 30:S130
11. Andrä W, Nowak H (1998) Magnetism in medicine. Wiley-VCH, Berlin
12. Reinicke S, Döhler S, Tea S, Krekhova M, Messing R, Schmidt AM, Schmalz H (2010) Magneto-responsive hydrogels based on maghemite/triblock terpolymer hybrid micelles. Soft Matter 6:2760–2773
13. Kaiser A, Winkler M, Krause S, Finkelmann H, Schmidt AM (2009) Magnetoactive liquid crystal elastomer nanocomposites. J Mater Chem 19:538–543
14. Kaiser A, Gelbrich T, Schmidt AM (2006) Thermosensitive magnetic fluids. J Phys Condens Matter 18:S2563–S2580
15. Schmidt A (2005) Induction heating of novel thermoresponsive ferrofluids. J Magn Magn Mater 289C:5–8
16. Schmidt AM (2005) The synthesis of magnetic core-shell nanoparticles by surface-initiated ring-opening polymerization of e-caprolactone. Macromol Rapid Comm 26:93–97
17. Feyen M, Heim E, Ludwig F, Schmidt A (2008) Magnetic nanorotors with tailored field-induced dynamics. Chem Mater 20:2942–2948
18. Sinnwell S, Ritter H (2006) Ring-opening homo- and copolymeriza tion of α-methylene-ϵ-caprolactone. Macromolecules 39:2804–2807
19. Winkler M, Kaiser A, Krause S, Finkelmann H, Schmidt AM (2010) Liquid crystal elastomers with magnetic actuation. Macromol Symp 291/2:186–192
20. Levine I, Zvi RB, Winkler M, Schmidt AM, Gottlieb M (2010) Magnetically induced heating in elastomeric nanocomposites – theory and experiment. Macromol Symp 291/2:278–286
21. Lu A, Li W, Kiefer A, Schmidt W, Bill E, Fink G, Schüth F (2004) Fabrication of magnetically separable mesostructured silica with an open pore system. J Am Chem Soc 126:8616–8617
22. Katz E, Willner I (2005) Switching of directions of bioelectrocatalytic currents and photocurrents at electrode surfaces by using hydrophobic magnetic nanoparticles. Angew Chem Int Ed 44:4791–4794
23. Lee Y, Lee J, Bae CJ, Park J, Noh H, Park J, Hyeon T (2005) Large scale synthesis of uniform and crystalline magnetite nanoparticles using reverse micelles as nanoreactors under reflux conditions. Adv Funct Mater 15:503–509
24. Franzreb M, Siemann-Herzberg M, Hobley TJ, Thomas ORT (2006) Protein purification using magnetic adsorbent particles. Appl Microbiol Biotechnol 70:505–516
25. Gu H, Xu K, Xu C, Xu B (2006) Biofunctional magnetic nanoparticles for protein separation and pathogen detection. Chem Commun. 941–949
26. Horák D, Babic M, Macková H, Benes MJ (2007) Preparation and properties of magnetic nano- and microsized particles for biological and environmental separations. J Separ Sci 30:1751–1772
27. Safarík I, Safaríková M (1999) Use of magnetic techniques for the isolation of cells. J Chromatogr B: Biomed Sci Appl 722:33–53
28. Safarik I, Safarikova M (2004) Magnetic techniques for the isolation and purification of proteins and peptides. Biomagn Res Technol 2:7
29. Pankhurst Q, Connolly J, Jones SK, Dobson J (2003) Applications of magnetic nanoparticles in biomedicine. J Phys D Appl Phys 36:R167–R181
30. Massart R, Cabuil V (1987) Effect of some parameters on the formation of magnetic nanoparticles. J Chim Phys 84:967–973
31. Marten GU, Gelbrich T, Schmidt AM (2010) Hybrid biofunctional nanostructures as stimuli-responsive catalytic systems. Beilstein J Org Chem 6:922–931
32. Guo J, Yang W, Wang C, He J, Chen J (2006) Poly (N-isopropylacrylamide) -coated luminescent//magnetic silica microspheres: preparation, characterization, and biomedical applications. Chem Mater 18:5554–5562
33. Müller-Schulte D, Schmitz-Rode T (2006) Thermosensitive magnetic polymer particles as contactless controllable drug carriers. J Magn Magn Mater 302:267–271
34. Schmidt AM, Messing R, Lu Y, Ballauff M. unpublished results
35. Kondo A, Fukuda H (1997) Preparation of thermo-sensitive magnetic hydrogel microspheres and application to enzyme immobilization. J Ferment Bioeng 84:337–341
36. Ding X, Sun Z, Zhang W (2000) Adsorption/desorption of protein on magnetic particles covered by thermosensitive polymers. J Appl Polym Sci 77:2915–2920
37. Elaissari A, Bourrel V (2001) Thermosensitive magnetic latex particles for controlling protein adsorption and desorption. J Magn Magn Mater 225:151–155
38. Frickel N, Messing R, Gelbrich T, Schmidt AM (2010) Functional silanes as surface modifying primers for the preparation of highly stable and well-defined magnetic polymer hybrids. Langmuir 26:2839–2846

39. Deng YH, Yang WL, Wang CC, Fu SK (2003) A novel approach for preparation of thermoresponsive polymer magnetic microspheres with core–shell structure. Adv Mater 15:1729–1732
40. Gürler C, Feyen M, Behrens S, Matoussevitch N, Schmidt AM (2008) One-step synthesis of functional Co nanoparticles for surface-initiated polymerization. Polymer 49:2211–2216
41. Kaiser A, Dutz S, Schmidt A.M (2009) Kinetic studies of surface-initiated atom transfer radical polymerization in the synthesis of magnetic fluids. J Polym Sci: Part A: Polymer Chemistry, 47: 7012–7020
42. Toy AA, Reinicke S, Mu AHE (2007) One-pot synthesis of polyglycidol-containing block copolymers with alkyllithium initiators using the phosphazene base t-BuP4. Macromolecules 40:5241–5244
43. Tanaka T (1978) Collapse of gels and the critical endpoint. Phys Rev Lett 40:820
44. Osada Y, Gong J-P (1998) Soft and wet materials: polymer gels. Adv Mater 10:827–837
45. Rao GVR, Krug ME, Balamurugan S, Xu H, Xu Q, López GP (2002) Synthesis and characterization of silica-poly (N-isopropylacrylamide) hybrid membranes: switchable molecular filters. Chem Mater 14:5075–5080
46. Garreau S, Leclerc M, Errien N, Louarn G (2003) Planar-to-nonplanar conformational transition in thermochromic polythiophenes: a spectroscopic study. Macromolecules 36:692–697
47. Otsuka K, Wayman CM (1998) Shape memory materials. Cambridge University Press, Cambridge
48. Lendlein A, Schmidt AM, Schroeter M, Langer R (2005) Shape-memory polymer networks from oligo(ϵ-caprolactone) dimethacrylates. J Polym Sci: Part A: Polymer Chemistry, 43:1369–1381
49. Kelch S, Steuer S, Schmidt AM, Lendlein A (2007) Shape-memory polymer networks from oligo[(epsilon-hydroxycaproate)-co-glycolate]dimethacrylates and butyl acrylate with adjustable hydrolytic degradation rate. Biomacromolecules 8:1018–1027
50. Zhou J, Schmidt AM, Ritter H (2010) Bicomponent transparent polyester networks with shape memory effect. Macromolecules 43:939–942
51. Keerl M, Smirnovas V, Winter R, Richtering W (2008) Interplay between hydrogen bonding and macromolecular architecture leading to unusual phase behavior in thermosensitive microgels. Angew Chem 120:344–347
52. Keerl M, Richtering W (2006) Synergistic depression of volume phase transition temperature in copolymer microgels. Colloid Polym Sci 285:471–474
53. Meid J, Dierkes F, Cui J, Messing R, Crosby AJ, Schmidt A, Richtering W (2012) Mechanical properties of temperature sensitive microgel/polyacrylamide composite hydrogels—from soft to hard fillers. Soft Matter 8:4254
54. Lutz J-F (2011) Thermo-switchable materials prepared using the OEGMA-platform. Adv Mater 23:2237–2243
55. Gelbrich T, Reinartz M, Schmidt AM (2010) Active ester functional single core magnetic nanostructures as a versatile immobilization matrix for effective bioseparation and catalysis. Biomacromolecules 11:635–642
56. Gelbrich T, Feyen M, Schmidt AM (2006) Magnetic thermoresponsive core-shell nanoparticles. Macromolecules 39:3469–3472
57. Gelbrich T, Feyen M, Schmidt AM (2006) Magnetic polymer brushes: towards tailor-made stabilization of magnetic fluids by surface-initiated polymerization. Z Phys Chem 220:41–49
58. Gelbrich T, Marten GU, Schmidt AM (2010) Reversible thermoflocculation of magnetic core–shell particles induced by remote magnetic heating. Polymer 51:2818–2824
59. Kaiser A, Schmidt AM (2008) Phase behavior of polystyrene-brush-coated nanoparticles in cyclohexane. J Phys Chem B 112:1894–1898
60. Messing R, Schmidt AM (2008) Heat Transfer from Nanoparticles to the Continuum Matrix. Surface and Interfacial Forces - From Fundamentals to Applications. Progr Colloid Polym Sci 134:134–140
61. Schexnailder P, Schmidt G (2008) Nanocomposite polymer hydrogels. Colloid Polym Sci 287:1–11
62. Lao LL, Ramanujan RV (2004) Magnetic and hydrogel composite materials for hyperthermia applications. J Mater Sci Mater Med 15:1061–1064
63. Ramanujan RV, Ang KL, Venkatraman S (2006) Magnet–PNIPA hydrogels for bioengineering applications. J Mater Sci 44:1381–1387
64. Pankhurst QA, Thanh NTK, Jones SK, Dobson J (2009) Progress in applications of magnetic nanoparticles in biomedicine. J Phys D Appl Phys 42:224001
65. Guowei D, Adriane K, Chen X, Jie C, Yinfeng L (2007) PVP magnetic nanospheres: biocompatibility, in vitro and in vivo bleomycin release. Int J Pharm 328:78–85
66. Liu T, Hu S, Liu T, Liu D, Chen S (2006) Magnetic-sensitive behavior of intelligent ferrogels for controlled release of drug. Langmuir 22:5974–5978
67. Elaissari A (2005) Magnetic colloids: preparation and biomedical applications. e-Polymers. 5:296–306
68. Chang Y, Su Z (2002) Preparation and characterization of thermosensitive magnetic particles. Mater Sci Eng, A 333:\break 155–159
69. Xiao L, Li J, Brougham DF, Fox EK, Feliu N, Bushmelev A, Schmidt A, Mertens N, Kiessling F, Valldor M, Fadeel B, Mathur S (2011) Water-soluble superparamagnetic magnetite nanoparticles with biocompatible coating for enhanced magnetic resonance imaging. ACS Nano 5:6315–6324
70. Schmidt AM, Gelbrich T, Marten G, Reinartz M, Schrader J unpublished results
71. Erlanger B, Kokowsky N, Cohen W (1961) The preparation and properties of two new chromogenic substrates of trypsin. Arch Biochem Biophys 95:271–278
72. Marten GU, Gelbrich T, Ritter H, Schmidt AM (2013) A magnetoresponsive drug delivery system via ß-cyclodextrin functionalized magnetic polymer brushes. IEEE Trans Magn 49:364–370
73. Van De Manakker F, Vermonden T, Van Nostrum CF, Hennink WE (2009) Cyclodextrin-based polymeric materials: synthesis, properties, and pharmaceutical/biomedical applications. Biomacromolecules 10:3157–3175
74. Loftsson T, Brewster ME (2011) Pharmaceutical applications of cyclodextrins: effects on drug permeation through biological membranes. J Pharm Pharmacol 63:1119–1135
75. Munteanu M, Choi S, Ritter H (2008) Cyclodextrin methacrylate via microwave-assisted click reaction. Macromolecules 41:9619–9623
76. Choi S, Munteanu M, Ritter H (2009) Monoacrylated cyclodextrin via "click" reaction and copolymerization with N-isopropylacrylamide: guest controlled solution properties. J Polym Res 16:389–394
77. Frelichowska J, Bolzinger M-A, Valour J-P, Mouaziz H, Pelletier J, Chevalier Y (2009) Pickering w/o emulsions: drug release and topical delivery. Int J Pharm 368:7–15
78. Kaiser A, Liu T, Richtering W, Schmidt AM (2009) Magnetic capsules and pickering emulsions stabilized by core-shell particles. Langmuir 25:7335–7341
79. Zrinyi M, Barsi L, Büki A (1997) Ferrogel: a new magneto-controlled elastic medium. Polym Gels Netw 5:415–427
80. Szabò D, Szeghy G, Zrínyi M (1998) Shape transition of magnetic field sensitive polymer gels. Macromolecules 31:6541–6548

81. Ramanujan RV, Lao LL (2006) The mechanical behavior of smart magnet – hydrogel composites. Smart Mater Struct 15:952–956
82. Qin J, Asempah I, Laurent S, Fornara A, Muller RN, Muhammed M (2009) Injectable superparamagnetic ferrogels for controlled release of hydrophobic drugs. Adv Mater 21:1354–1357
83. Raikher YL, Stolbov OV (2005) Deformation of an ellipsoidal ferrogel sample. J Appl Mech Tech Phys 46:434–443
84. Stepanov GV, Abramchuk SS, Grishin DA, Nikitin LV, Kramarenko EY, Khokhlov AR (2007) Effect of a homogeneous magnetic field on the viscoelastic behavior of magnetic elastomers. Polymer 48:488–495
85. Mauguin G (1988) Continuum mechanics of electromagnetic solids. North-Holland, Amsterdam
86. Borcea L, Bruno O (2001) On the magneto-elastic properties of elastomer–ferromagnet composites. J Mech Phys Solids 49:2877–2919
87. Bustamante R, Dorfmann A, Ogden RW (2006) A nonlinear magnetoelastic tube under extension and inflation in an axial magnetic field: numerical solution. J Eng Math 59: 139–153
88. Diguet G, Beaugnon E, Cavaillé JY (2009) From dipolar interactions of a random distribution of ferromagnetic particles to magnetostriction. J Magn Magn Mater 321:396–401
89. Zrinyi M (2000) Intelligent polymer gels controlled by magnetic fields. Colloid Polym Sci 103:98–103
90. Ginder JM, Nichols ME, Elie LD, Tardiff JL (1999) Magnetorheological elastomers: properties and applications. Proc SPIE 3675:131
91. Farshad M, Benine A (2004) Magnetoactive elastomer composites. Polym Test 23:347–353
92. Park TG, Hoffmann AS (1993) Thermal cycling effects on the bioreactor performances of immobilized β-galactosidase in temperature-sensitive hydrogel beads. Enzyme Microb Technol 15:476–482
93. Kato N, Oishi A, Takahashi F (1998) Enzyme reaction controlled by magnetic heating due to the hysteresis loss of γ-Fe_2O_3 in thermosensitive polymer gels immobilized. Mater Sci Eng C: Biol 6:291–296
94. Schmidt AM (2006) Electromagnetic activation of shape memory polymer networks containing magnetic nanoparticles. Macromol Rapid Comm 27:1168–1172
95. Kato N, Yamanobe S, Takahashi F (1997) Property of magneto-driven poly(N-isopropylacrylamide) gel containing γ-Fe_2O_3 in NaCl solution as a chemomechanical device. Mater Sci Eng C 5:141–147
96. Satarkar NS, Zhang W, Eitel RE, Hilt JZ (2009) Magnetic hydrogel nanocomposites as remote controlled microfluidic valves. Lab Chip 9:1773–1779
97. Barbucci R, Pasqui D, Giani G, De Cagna M, Fini M, Giardino R, Atrei A (2011) A novel strategy for engineering hydrogels with ferromagnetic nanoparticles as crosslinkers of the polymer chains. Potential applications as a targeted drug delivery system. Soft Matter 7:5558–5565
98. Pasqui D, Atrei A, Giani G, Cagna MD, Barbucci R (2011) Metal oxide nanoparticles as cross-linkers in polymeric hybrid hydrogels. Mater Lett 65:392–395
99. Messing R, Frickel N, Belkoura L, Strey R, Rahn H, Odenbach S, Schmidt AM (2011) Cobalt ferrite nanoparticles as multifunctional cross-linkers in PAAm ferrohydrogels. Macromolecules 44:2990–2999
100. Fuhrer R, Athanassiou EK, Luechinger NA, Stark WJ (2009) Crosslinking metal nanoparticles into the polymer backbone of hydrogels enables preparation of soft, magnetic field-driven actuators with muscle-like flexibility. Small 5:383–388
101. Frickel N, Messing R, Schmidt AM (2011) Magneto-mechanical coupling in $CoFe_2O_4$-linked PAAm ferrohydrogels. J Mater Chem 21:8466
102. Galicia JA, Cousin F, Dubois E, Sandre O, Cabuil V, Perzynski R (2009) Static and dynamic structural probing of swollen polyacrylamide ferrogels. Soft Matter 5:2614–2624
103. Barrera C, Florián-Algarin V, Acevedo A, Rinaldi C (2010) Monitoring gelation using magnetic nanoparticles. Soft Matter 6:3662–3668
104. Janas V, Rodriguez F, Cohen C (1980) Aging and thermodynamics of polyacrylamide gels. Macromolecules 13:977–983
105. Nossal R (1985) Network formation in polyacrylamide gels. Macromolecules 18:49–54
106. Richards EG, Temple CJ (1971) Some Properties of Polyacrylamide Gels. Natl Phys Sci 230:92–96
107. Zhang J, Daubert C, Foegeding E (2005) Characterization of polyacrylamide gels as an elastic model for food gels. Rheol Acta 44:622–630
108. Foegeding EA, Gonzalez C, Hamann DD, Case S (1994) Polyacrylamide gels as elastic models for food gels. Food Hydrocolloid 8:125–134
109. Veverka M, Veverka P, Kaman O, Lančok A, Závěta K, Pollert E, Knížek K, Boháček J, Beneš M, Kašpar P, Duguet E, Vasseur S (2007) Magnetic heating by cobalt ferrite nanoparticles. Nanotechnology 18:345704
110. Ozaki M, Kratohvil S, Matijević E (1984) Formation of monodispersed spindle-type hematite particles. J Colloid Interface Sci 106:146–151
111. Roeder L, Bender P, Tschöpe A, Birringer R, Schmidt AM (2012) Shear modulus determination in model hydrogels by means of elongated magnetic nanoprobes. J Polym Sci Part B: Polym Phys 50:1772–1781
112. Flory PJ, Rehner J (1943) Statistical Mechanics of Cross-Linked Polymer Networks II. Swelling. J Chem Phys 11:521–526
113. Mark JE (1982) Experimental Determinations of Crosslink Densities. Rubber Chem Technol 55:762–769
114. Frenkel J (1938) Acta Phys USSR 9:235–250
115. Frenkel J (1940) A theory of Elasticity, Viscosity and Swelling in Polymeric Rubber-Like Substances. Rubber Chem Technol 13:264–274
116. Stoner EC, Wohlfarth EP (1948) A mechanism of magnetic hysteresis in heterogeneous alloys. Philos Transact R Soc A 240:599–642
117. Schulz L, Schirmacher W, Omran A, Shah VR, Böni P, Petry W, Müller-Buschbaum P (2010) Elastic torsion effects in magnetic nanoparticle diblock-copolymer structures. J Phys Condens Matter 22:346008
118. Bender P, Günther A, Tschöpe A, Birringer R (2013) Determination of the shear modulus of gelatine hydrogels by magnetization measurements using dispersed nickel nanorods as mechanical probes. J Magn Magn Mat 346: 152–160
119. Coffey TW, Kalmykov YP, Waldron JT (2004) The Langevin equation. World Scientific Publishing, London

Hydrophobically Covered Hydrogels: Preparation Approaches and Possible Applications

Marta Horecha, Volodymyr Senkovskyy, Anton Kiriy, and Manfred Stamm

Abstract

This article overviews methods for preparation of micrometer-sized particles having hydrogel core and hydrophobic shell. It demonstrates that water-in-oil inverse suspension polymerization is a versatile method for *one-step synthesis* of either non-hollow spherical hydrogel particles or hollow capsules, depending on the nature of the gel-forming polymer. It further shows that rather thin hydrophobic shell provides a good stability of the particles in organic solvents while still does not preclude swellability of the hydrophilic core with water and small water-soluble molecules. *A stepwise approach* for fabrication of the core-shell hydrogel particles is also described which assumes preparation of hydrogel particles as the first step, their arrangement on surfaces followed by growing of the hydrophobic shell. Two methods for preparation of the hydrophobic shell were explored. The first one assumes free-radical polymerization of the polyisoprene from adsorbed hydrogel particles while the second one implies electrostatically-driven adsorption of polyisoprene latex particles onto oppositely charged hydrogels. Thus-obtained core-shell particles were used in fabrication of switchable coatings.

Keywords

Hydrophobically covered hydrogels • Core-shell particles • Switchable surfaces • Micro-capsules

Core-Shell Microgels

Hydrogel represents a very important class of "smart" materials because of its superabsorbent property, high degree of flexibility very similar to natural tissue, aqueous inner environment, porous structure, etc. To date, various hydrogels were utilized as interlaying materials in flameproof textile, in producing of implants, contact lenses and biosensors, as coatings for medical instruments and carriers for drug delivery. They are also commonly used in cosmetic products as moisture holders. With an idea to enrich application areas of hydrogels, material scientists turned their attention to the design and synthesis of micro- and nanosized hydrogel structures. Composite core-shell hydrogel particles were already proved to be extremely useful building blocks for preparation of novel functional materials. According to the nature and properties of the core and the shell, the core-shell hydrogels can be divided onto the following major types: (1) composite particles having the core made of non-hydrogel, the shell made of hydrogel materials and (2) particles having both the core and the shell composed of two different hydrogel materials [1]. The particles of the first type are usually consist of solid polymeric (e.g.,

M. Horecha • V. Senkovskyy • A.Kiriy
Department of Nanostructured Materials, Leibniz Institute of Polymer Research Dresden, Hohe Straße 6, 01069 Dresden, Germany

M. Stamm (✉)
Department of Nanostructured Materials, Leibniz Institute of Polymer Research Dresden, Hohe Straße 6, 01069 Dresden, Germany

Department of Chemistry, Technical University Dresden, 01069 Dresden, Germany
e-mail: stamm@ipfdd.de

G. Sadowski and W. Richtering (eds.), *Intelligent Hydrogels*, Progress in Colloid and Polymer Science 140,
DOI 10.1007/978-3-319-01683-2_12,

polystyrene) or inorganic (e.g., silica or metal nanoparticles) cores covered by soft, frequently stimuli-responsive hydrogel shell. Such materials already found applications in sensing and catalysis [2]. The second type of core-shell hydrogels are usually particles composed of hydrogels having different composition, structure or functionality. Such particles could be straightforwardly synthesized, for example, by two-stage precipitation polymerization [3]. In this method, the shell is added onto pre-synthesized core particles, thereby allowing the control over the radial distribution of the functional groups in the particle. Core-shell particles prepared on this fashion can exhibit very interesting properties. For example, if the core and shell are made of different thermoresponsive polymers, the particles will show multiple temperature-dependent phase transition behavior [4]. Particles with unlike sensitivity of the core and the shell were also prepared and used as detectors of different changes in environment [5]. Similar synthetic approach is also applicable for preparation of hollow hydrogel structures [6]. If degradable cross-linker was used for the core preparation, but non-degradable one was utilized for the shell preparation, then a post-synthetic degradation of the cross-linker will lead to leaching of the core and formation of hollow structures.

As follows from literature, the core-shell hydrogel structures having *hydrophilic* shells constitute the major focus of researchers as they are of a great interest for various biomedical applications. However other important applications, such as catalysis, water purification, self-healing materials, cosmetic or agrochemistry, require composite particles having *hydrophobic* outer surfaces (to provide dispersibility in non-polar environment) and hydrophilic interior (e.g. for encapsulation of polar compounds). In addition, such particles could be useful in fabrication of sensitive or self-cleaning surfaces or breath-active textiles, which became water-repellent after the contact with water. Surprisingly, this promising area of research remained relatively under-explored which motivated us to contribute there. In this article we wish to review our recent results in synthesis and application of *hydrophobically modified hydrogel particles*.

Switchable Surfaces

Recently, the progress in development and improvement of modern coatings quality have turned to the creation of adaptive or sensitive (also called "smart" or "intelligent") surfaces. Based on dynamically active systems, usually achieved by a combination of materials with opposite properties, such sensitive surfaces are able to adapt their structure by a rearrangement of interfacial components under the changing of environment in order to minimize surface energy. A number of switchable or adaptive coatings, which respond to different stimuli from environment, can be created using this principle. Surfaces with switchable wetting behaviour able to change their properties from hydrophilic to hydrophobic and vice versa, are of special interest because of their potential applications in tissue engineering and cell growth, creation of "intelligent" breath-active water-repelling textiles and self-regulating valves. The design of a new generation of such "smart" materials and surfaces is also required for new concepts of sensors [7], drug delivery systems [8], microfluidic devices [9] and colloidal assembly techniques [10].

Since the wetting behaviour of material is strongly dependent on the surface structure and chemical composition, it is possible to prepare the surface as a dynamic system that consists of two components having opposite properties and with components ratio changing with environmental stimuli. The application of such responsive materials opens the possibility to create tailored surfaces adaptive to environmental conditions. Altering the balance between two components on the surface shifts the surface properties toward an energetically favourable state that enables the temporary stabilization of the system until next stimuli is applied. Homopolymers, end-functionalised polymers, random copolymers, block copolymers, amphiphilic [11], segmented polymers, graft copolymers [12], polymer brushes [13, 14] and polyelectrolyte multilayers [15] were already employed in fabrication of switchable surfaces. Modulation of surface hydrophilicity has been triggered by light irradiation [16], temperature [17], electric field [18], solvents [19] or mechanical forces. In all these cases, natural switching behaviour of amphiphilic surfaces (so-called self-adaptation) was observed, that is, the surface became more hydrophilic in polar environment and more hydrophobic if immersed in apolar solvents. However, for many practical applications, the opposite and less natural response on the external stimuli would be desirable.

There are only a few examples in literature that reported moderate effect of increase of the water contact angle forced by the increase of the environment polarity [20]. For example, the case of increased hydrophobicity for oxidized 1,2-polybutadiene surfaces exposed to hot water was revealed recently by Ferguson et al. The "contraphilic" (or inverse) switching was also demonstrated by group of Makal and Wynne for the polyurethane coating, containing semifluorinated and 5,5-dimethylhydantoin segments in a soft block. A difference of 27° and after annealing even 38° in advancing contact angles for dry (68°) and wet (106°) states was measured. The switching effect was found completely reversible upon drying the coating in vacuum. In this system the contraphilic switching was induced by chain reorientation occurring upon a contact with water that lead to partial breaking of internal intramolecular hydrogen bonds and simultaneous release of semifluorinated alkyl chains that segregated atop of the film and contributed to increase of hydrophobicity. Although this approach is very

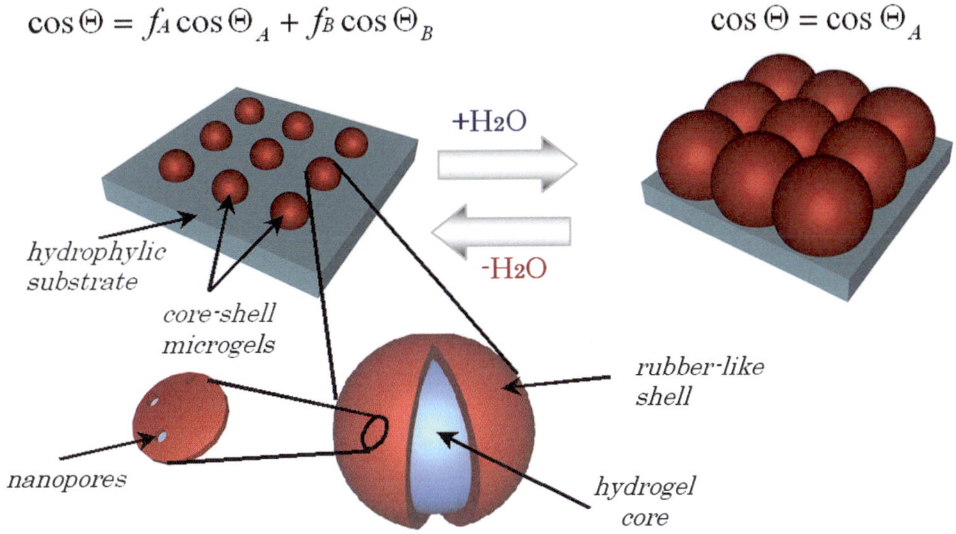

Fig. 1 Schematic representation of the core-shell microgels and the switching principle (see explanation in the main text)

elegant, the contraphilic effect was provided by a delicate balance of hydrophobic interactions and hydrogen bonds in the amphiphilic network and thus this principle might be difficult to implement in other systems with an aim to optimise the contraphilic effect.

New Design of Inversely Switchable Surfaces

We were interested to design new switchable surface coatings in which the increase of hydrophobicity upon the contact with water can be modulated in a predictable and precise way. In our opinion, the key element of the contraphilic surfaces could be hydrogel particles composed of water-swellable core and hydrophobic, elastic shell. The principle of wettability switching effect of surface modified with hydrophobically coated hydrogel particles is depicted in Fig. 1 [21]. According to the Cassie's equation [22],

$$\cos\theta = f_A \cos\theta_A + f_B \cos\theta_B; \ f_A + f_B = 1,$$

wetting behavior of the two-component heterogeneous surface depends on the coverage ratio of both components (f_A and f_B) and their own contact angles (Θ_A and Θ_B). According to our idea, an array of loosely packed core-shell hydrogel particles having a water-swellable core (e.g., cross-linked poly(N-isopropylacrylamide, PNIPAM) and a hydrophobic elastic shell (e.g., made of polyisoprene, PI) should be developed onto otherwise hydrophilic substrate. Importantly, the hydrophobic shell should not prevent the core swelling with water but has to be hydrophobic enough to provide the surface hydrophobic effect to the whole particle even in the water-swollen state. In the dry state, the hydrophobic component (hydrophobic shell) occupies a minor part of the (hydrophilic) surface, providing overall hydrophilicity to the surface (Fig. 1, left). Swelling of the particles should lead to increase of the occupation of the hydrophobic component resulting into increase of the overall hydrophobicity (Fig. 1, right). In this state, intrinsic properties (e.g., hydrophilicity) of the substrate are not important because the underlying substrate will not interact with water.

Implementation of this approach to control surface wettability requires an accurate optimization of many parameters of the very complex system. This, in turns, demands development of a number of new methods for preparation of core-shell particles and detailed investigations of their surface behaviour. Although (shell-free) dispersed hydrogel particles were studied intensively over last two decades [23], morphology, mechanical properties and swellability of hydrogel particles in the adsorbed state were investigated relatively recently [24]. The groups of Hellwig, Klitzing, Lyon and others studied immobilization of hydrogel particles on different surfaces using various deposition techniques. It was found that the behavior of hydrogel particles at the surface is affected not only by particle-particle interaction but also by their interactions with surfaces and capillary forces. It was further demonstrated that deposition conditions as well as substrate properties have a pronounced effect on packing density of particles at surfaces [25]. This knowledge was used for preparation of colloidal crystals built of hydrogel particles [26–30]. The last progress in the assembly of responsive microgel particles in two and three dimensions, highlighting their potential applications, was summarized in a comprehensive review of Lyon et al. [31]. However, most of previous studies were focused on either shell-less microgel particles or on the particles having *hydrophilic* shell [32]. Examples of hydrogel particles having *hydrophobic* shell are scarce and limited to just a few reports [32].

Hence, preparation of such particles and investigation of their surfaces behavior was an actual task. Two approaches to synthesize such core-shell particles were explored [33]. The one-step method utilized an inverse emulsion polymerization of the core-forming monomer in the presence of diblock copolymer surfactant used as stabilizer and the shell-forming material. An alternative stepwise approach implied synthesis of suitable microgels, immobilization of them on surfaces followed by preparation of the elastic hydrophobic porous shell atop of the microgels.

One-Step Preparation of Core-Shell Hydrogel Particles

Core-shell hydrogels were prepared by water-in-oil (W/O) emulsion polymerization of water-soluble monomers using polyisoprene-block-polyethylene glycol (PI_{360}-b-PEO_{130}) diblock copolymer emulsifier synthesized by living anionic polymerization (Fig. 2). Among various possible monomers suitable for fabrication of water-swellable hydrogel cores N-isopropylacrylamide (NIPAM) and acrylamide (AAm) were chosen mainly due to well-studied water-swelling behaviour of the corresponding polymers, PNIPAM and PAAm [34]. Morphology of resulting core-shell microstructures was found to be strongly dependent on the monomer nature. Non-hollow microspheres were formed only upon polymerization of highly polar AAm monomer (Fig. 2) whereas polymerization of less polar NIPAM monomer under otherwise identical conditions (above lower critical solution temperature, LCST) led to non-hollow microcapsules. The reason of the difference in the particles morphology lies in different hydration behaviour of the resulting polymers (PNIPAM versus PAAm) and this will be discussed in more detail below.

Application of Core-Shell Hydrogel Particles for Coatings

The core-shell hydrogel particles prepared by W/O emulsion polymerization in the presence of PI-b-PEO were found to be dispersible in apolar solvents (cyclohexane) but not in water suggesting the presence of hydrophobic PI shell. However, preliminary experiments revealed that these particles are not sufficiently hydrophobic to be successfully used for fabrication of coatings with well-pronounced inverse switching behaviour. Particularly, it was found that coatings prepared from these particles display rather high hydrophilicity with water contact angle about 50°. We supposed that the PI shell is too thin to ensure sufficient hydrophobic effect on the surfaces. Aiming at increase of the PI shell, isoprene

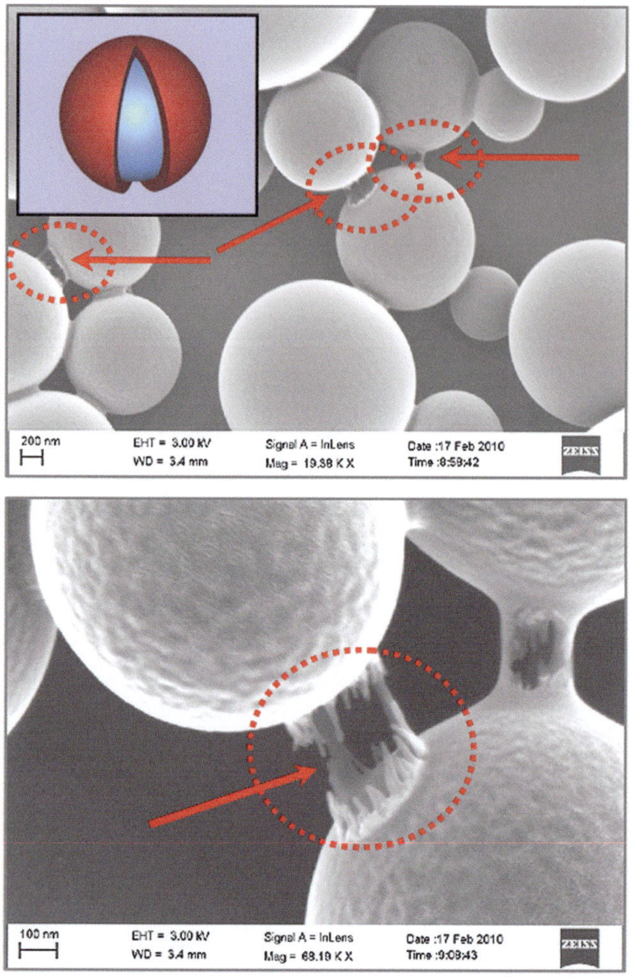

Fig. 2 SEM images of PAAm-PI core-shell particles with PI shell enhanced by SIP of isoprene from the oil phase. The presence of PI shell makes bridges between the particles (marked with *arrows*). Scale bars are 200 nm (*left*) and 100 nm (*right*)

monomer was added to the oil phase during the W/O synthesis of the core-shell hydrogel particles. According to our idea, polymerization of isoprene should not occur in the bulk solution because there is no initiator in the oil phase. However, polymerization may start in a close vicinity to the surface of the particles due to the initiator present in the water phase. We further assumed that since free-radical polymerization of AAm proceeds inside the droplets, active radicals (e.g., propagating PAAm chains) may occasionally approach to the water/oil interface initiating the polymerization of isoprene and in such a way enhancing the hydrophobic shell. We found that core-shell particles prepared by this method showed similar size and polydispersity as the particles prepared without added isoprene, but had thicker polyisoprene shell, clearly observed on SEM images (Fig. 2).

In order to characterize the ability of the core-shell particles to uptake water, dynamic light scattering (DLS) measurements were performed. Because the core-shell particles

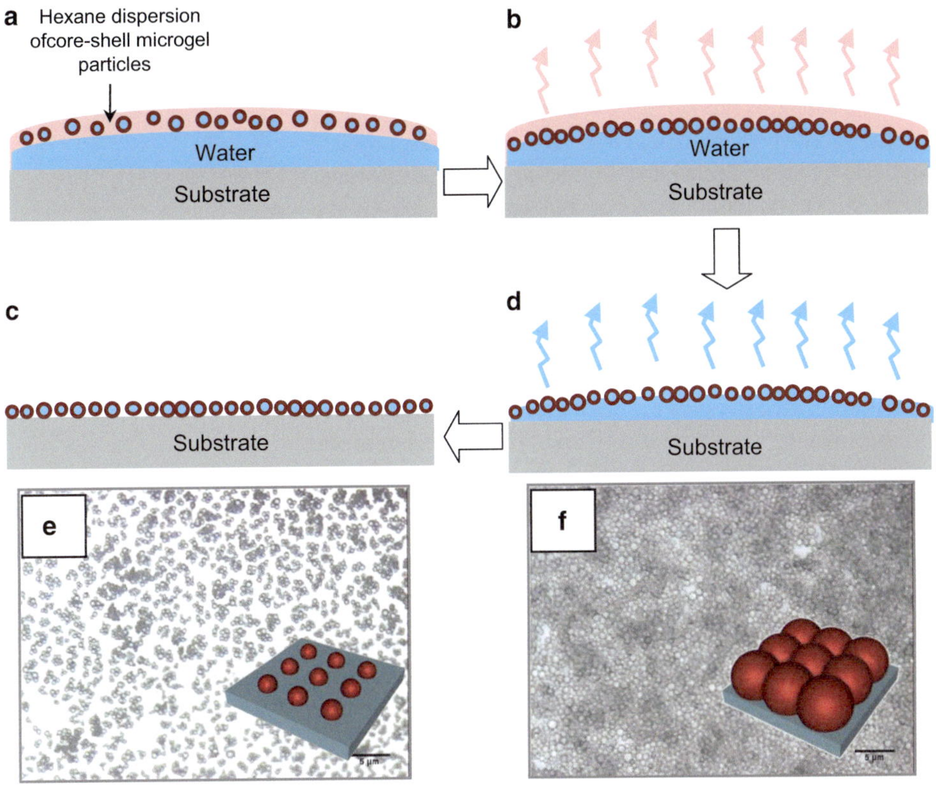

Fig. 3 Schematic representation of the two phase deposition process of PAAm-PI core-shell particles onto the hydrophilic substrate (**a–d**) and optical microscopy images of the monolayer on the core-shell particles deposited on the hydrophilic substrate before (**a**) and after (**b**) swelling. Scale bar 5 μm

were not dispersible in polar media such as water or ethanol, measurements were done in THF/water mixtures. In these experiments, a gradual increase of the particles size was observed upon a dropwise addition of water to the particles dispersed in anhydrous THF. Size of the particles reached its maximum at water content of about 10 vol%, which corresponded to a volume swelling ratio of ~6. The latter ratio was determined as the ratio of the particles volume after the swelling ($V_{swollen}$) to the particles volume before the swelling (V_{dry}). On practice, the volume swelling ratio was determined as $D_{swollen}{}^3/D_{dry}{}^3$ where $D_{swollen}/D_{dry}$ are corresponding diameters of the particles after and before the swelling, determined by DLS.

Further increase of water content led to flocculation and precipitation of the particles. It should be noted that the observed swelling ratio is probably underestimated due to the following reasons. Firstly, PAAm is partially swellable in THF as previously demonstrated by Winnik et al. [35]. Consequently, the "dry" diameter of the particles in THF is likely to be overestimated. Secondly, the volume swelling ratio of 6 was observed for THF containing only 10 % of water and higher swelling ratios might be expected for the particles placed in pure water. Since the particles flocculate in pure water, their swelling behaviour was further investigated in adsorbed state. In order to obtain the monolayer of loosely packed core-shell particles, a two-phase deposition process was developed (Fig. 3).

At first, substrates (silicon wafers or glass slides) were modified with polyacrylic acid brush to ensure their hydrophilicity. Afterwards, water was spread over the substrate and the dispersion of core-shell particles in hexane was spread atop of a water film (Fig. 3a). After evaporation of hexane, the particles appeared to be kept at the water/air interface (Fig. 3b). The amount of particles introduced onto the substrate was experimentally adjusted by variation of the concentration and quantity of the dispersion to form, after evaporation of hexane, a monolayer of closely packed particles swollen with underlying water (Fig. 3c). After the drying of water by hot air brush, the monolayer of the particles remained deposited on the substrate (Fig. 3d). As it seen on optical microscopy images, the drying process induced 3D rather than only 1D collapse of the particles. The observed lateral shrinkage of the particles upon drying is essential for successful implementation of the inverse switching. In order to attach particles to the substrate, photochemical grafting was applied. To this end, the substrate was modified by covalent immobilization of benzophenone molecules having appropriate function groups (–OH) atop of which the array of loosely packed core-shell hydrogel particles was deposited as described above. The grafting of the particles was accomplished by UV-light irradiation (254 nm, 15 min). It is known that UV-irradiation induces efficient decomposition of benzophenone and generation of radicals, which according to our idea, should react with double bonds of PI providing the covalent grafting of the particles. Similar approach was previously applied for surface-immobilization of various other polymers [35].

The change of the surface coverage during the swelling of the core-shell particles was observed using optical microscopy after immersion of the dry sample in water (Fig. 3e, f). By comparison of the particles size in the dry and swollen states, the reversible change of the surface coverage from ~40 % to ~100 % was calculated. Thereafter, water contact angle (WCA) measurements of substrates covered with surface-grafted arrays of core-shell particles were performed. The first set of WCA measurements of extensively dried samples was performed by placing of water droplets having a volume of ~10 μl. These measurements revealed WCA ranging from ~50° up to ~80° with mean value of ~65° depending on surface quality in the particular spot. Despite of a large variation of WCA, a systematic increase of advancing WCA was observed for all investigated samples after addition of new portions of water to the droplet (up to ~25 μl) and establishment of the new water contact lines (Fig. 4). The increase of the WCA upon addition of new portions of water ranged from 2° up to 19° with 2–5°increase of mean WCA.

The variation in WCA within the sample was attributed to the inhomogeneous surface coverage and structures of the coating (clustering, etc.), whereas the systematic increase of the advancing contact angle upon increase of the water droplet volume could be related to the desirable inverse switching effect. We suggest that as soon as a water droplet is reaching the sample surface, neighboring particles become exposed to saturated water vapors. This causes the swelling of the particles around the droplet which causes increase in surface hydrophobicity and increase of WCA. Although these coatings exhibited the inverse switching behaviour, small changes in contact angles indicate rather modest performance of coatings. Importantly, the only advanced WCA was increased upon the contact with water in our experiments. In contrast, the receding WCA of the modified surface decreases upon the contact with water (after water droplet was removed) and reaches very low value of 20°. Such effect was reasoned by insufficient water-repellent properties of the shell. Thus, further optimizations of the system parameters are needed and they are underway.

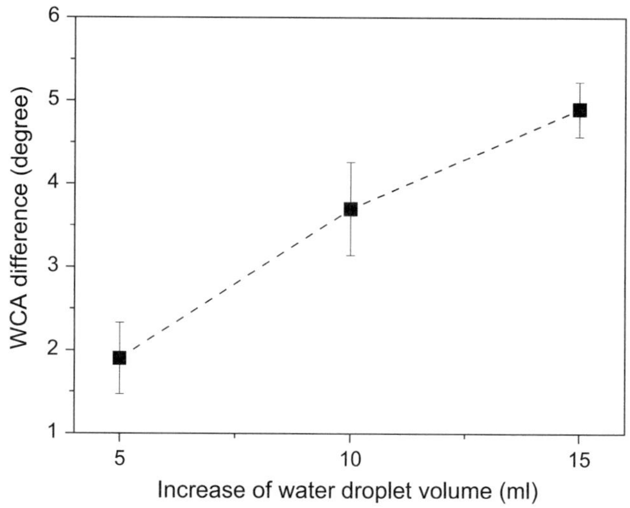

Fig. 4 Increase of advancing WCA of the hydrophilic substrate covered with PAAm-PI core-shell particles with increase of volume of water droplets as monitored by DLS measurements

Stepwise Approach for Preparation of the Core-Shell Hydrogels

Previous experiments reveal insufficient hydrophobicity of the shell of the core-shell particles. Attempts to substantially increase the shell thickness for the particles in their dispersed state failed due to aggregation of the particles during surface-initiated polymerization. To this end, a stepwise approach was developed for preparation of the core-shell particles directly "on place" (Fig. 5).

Additional advantage of the stepwise approach comes from the ability of hydrophilic microgels to form ordered 2D periodic arrays of loosely packed particles, which is favoured kind of arrangement for realization of contraphilic surfaces [36].

The stepwise approach for preparation of hydrophobically modified microgel arrays entails the following steps: (1) synthesis of suitable functional microgels; (2) adsorption of small amount of sacrificial 30 nm PS-latex particles on their surface; (3) preparation of arrays of loosely packed microgels on the hydrophilic substrates; (4) generation of hydrophobic polyisoprene shells on the surface of pre-deposited microgels; (5) dissolution of PS-latex particles to ensure porosity and water-permeability of the shell.

Monodisperse poly(N-isopropylacrylamide-co-acrylamide-co-methylene-bis-acrylamide) (PNIPAM-AAM) microgels were synthesised close to a procedure described by Pelton and Chibante [37]. Since pure PNIPAM-based particles have neither charged no reactive groups, a suitable functionality was introduced in a post-synthetic step. To this end, NIPAM was copolymerized with acrylamide to form well defined microgel particles and then acrylamide units were converted into vinylamine (VA) using oxidative rearrangement of amides [38]. Afterwards, water suspension of microgels was mixed with negatively charged PS latex in the weight ratio Wlatex/Whydrogels = 0.02 and deposited over approximately 5-nm-thick PAA brushes developed on SiO_2 wafers. After the formation of closely packed swollen microgels on the surface, the samples were rapidly heated, resulting in arrays of loosely packed dry microgels.

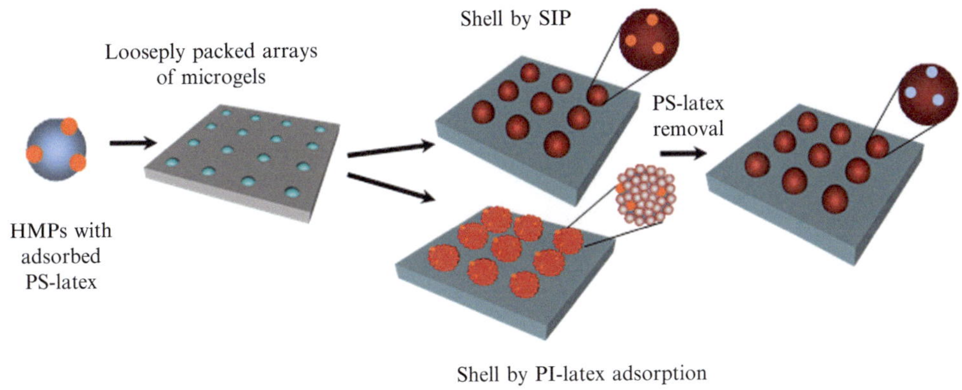

Fig. 5 Schematics of the "on-surface" preparation of hydrophobically modified microgels: PI shell microgels could be prepared either by surface-initiated radical polymerization of isoprene or by adsorption of PI-latex atop of deposited microgels

Polyisoprene shell on HMP was prepared by two different methods: "grafting-from" and "colloidosome" ones. The "grafting-from" technique is widely used for preparation core-shell particles with adjustable polymer shell thickness and grafting density [39]. This preparation procedure implies arrangement of the hydrogel particles on the surface to form loosely packed arrays [40–42], development of initiator groups on the particle surface followed by "grafting-from" polymerization. In this work, 4,4′-azobis (4-cyanovaleric acid was used as initiator for radical polymerization and isoprene as the monomer for preparation of the shell. AFM analysis of particles before and after polymerization showed their size increase in the range of ~50–60 nm (Fig. 8a, b).

A so called "colloidosome approach" was also applied for fabrication of the core-shell microgels. Similar approach has been used previously for creation a various types of core-shell structures using different particles as building blocks: "raspberry"-like particles [43], hollow spheres [44], etc. This method implies preparation of the hydrophobic shell on the oppositely charged surface of pre-synthesized hydrogel particles by adsorption of 40 nm polyisoprene latex from water suspension followed by latex coalescence. In our case, we used negatively charged 40 nm polyisoprene latex as a shell-forming material which was adsorbed atop of surface-immobilized and chemically grafted positively charged amino-functionalized microgels. A successful formation of the hydrophobic shell was confirmed by AFM (Fig. 6c). After coagulation of PI latex on the hydrogel particles surface using ethanol and drying of samples, the folded structures are clearly visible atop of each particle. We suggest that these folded structures were formed upon the drying of swollen particles already covered with polyisoprene layer. Thus, upon drying, hydrogel network collapses whereas the shell covers the shrunk microgel with folds of polyisoprene.

Thereafter, the study of changes in the surface coverage upon swelling of both types of surface-immobilized core-shell particles was performed by in-situ AFM measurements. The first conclusion which follows from AFM measurements is that the PI shell was not permeable to water if PS latex particles were not removed from the shell. Indeed, no swelling of the particles was observed upon AFM imaging of the particles performed for at least 1 h in water environment. In contrast, after removing of the PS latex and thus creating the pores in the shell, the particles became swellable in water (Fig. 7). The process was confirmed to be fully reversible: after drying of the samples, the size of the core-shell particles was the same as their initial size before the swelling. Analysis of the AFM images showed the increase of the surface coverage from 14 % to 25 % upon the swelling in water.

To verify whether the coating demonstrate the switching effect, the measurements of WCA were performed. It was found that the WCAs for the "dry" and the "wet" states reflected rather hydrophobic behaviour (around $77° \pm 1°$) with only a minor hydrophobization effect. Such a deviation from the predicted behaviour could be explained either by contamination of PAA brush by hydrophobic materials during the multistep modification procedure, or alternatively, by a fast swelling of the particles on the surface. In the latter case the swelling and, hence, the switching of the surface to the hydrophobic state occurs immediately after the water droplet is deposited onto the surface to measure WCA. To verify the latter hypothesis the swelling kinetics was studied by in-situ AFM measurements (Fig. 8a). Due to technical reasons, the first topography image of the swollen particle array was recorded about 10 min after the addition of water to the environmental cell. It was observed that during this time the surface coverage increases to about 50 % of its plateau value; the latter was reached at about 1 h. Thus, the swelling of the particles during the initial stage was proved to be a fast process, and therefore the water contact angle measurement of the "dry" state does not provide reliable result. Hence, further optimization of the system, such as decrease of the swelling rate, is necessary to record the inverse-switching behaviour of surface by WCA measurements.

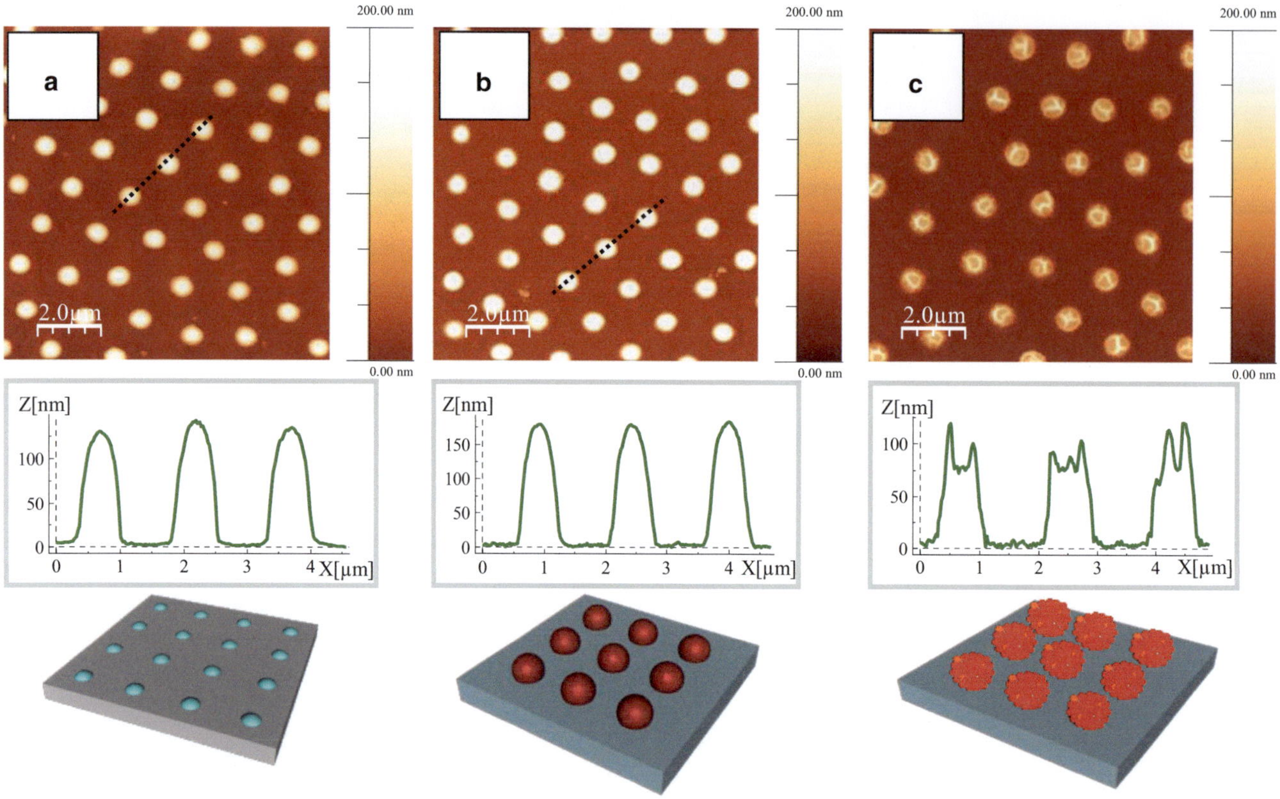

Fig. 6 AFM topography images (*top*) and corresponding cross-sectional profiles (*middle*) of surface-immobilized P(NIPAM- VA) microgels before (**a**) and after surface-initiated radical polymerization of isoprene (**b**) and after adsorption of PI-latex (**c**)

Similar results were obtained during investigation of the surfaces prepared by "colloidosome approach" for fabrication of polyisoprene shell. Exposure of the samples to water caused the swelling of hydrophobically covered microgels (Fig. 7d-f) with returning to the same dry state topology of particles upon dehydration of the samples.

Swelling kinetics showed the same trend, with the fast increase of the particles' sizes (and the surface coverage) and reaching of maximal values after 2 h. Increase of the surface coverage from 13 % to 30 % upon swelling of the particles with water was detected by in-situ AFM (Fig. 8b). The small switching effect of the water contact angle from 72° to 74° in these experiments was detected which is far below the expected switching range.

Hollow Structures

Functional hollow micro- and nanostructures, such as, hollow microcapsules have been of great research interest because of their potential applicability in chemistry, biotechnology, and pharmaceutical industry [45]. Up to date, various chemical and physicochemical methods have been explored to produce hollow structures from different materials [46]. High dispersion stability of microcapsules, easiness of their recovering, controlled permeability of their walls and etc., are properties important for various applications ranging from catalysis to drug delivery and pharmacology. From these regards, advantages of soft materials (e.g. polymers) over the solid inorganic ones are obvious. The unbreakable structure, flexibility, control over the walls mesh size and possibility to introduce various functionalities, make polymer microcapsules a promising materials for catalyst and. Polymer and hydrogel hollow structures can be prepared via layer-by-layer technique, self-assembly, vesicular polymerization or microfluidic fabrication [47]. In addition, etching of SiO_2 core from corresponding SiO_2/polymer core-shell particles also represent the way to obtain hollow structures with responsive properties [48]. In general, stimuli-responsive hollow hydrogel microcapsules offer greater versatility for encapsulation and triggered release, which can be controlled by changing temperature, pH, ionic strength, or other stimuli [49]. PNIPAM hydrogels undergo a sharp volume transition from a highly hydrated swollen state below LCST to a dehydrated collapsed state above the LCST [50]. This effect is largely exploited in "smart" PNIPAM hydrogel microcapsules, in which shell permeability is modulated by temperature. Thus, Gao et al. prepared nanocapsules by precipitation polymerization of NIPAM on silica particles templates followed by etching of

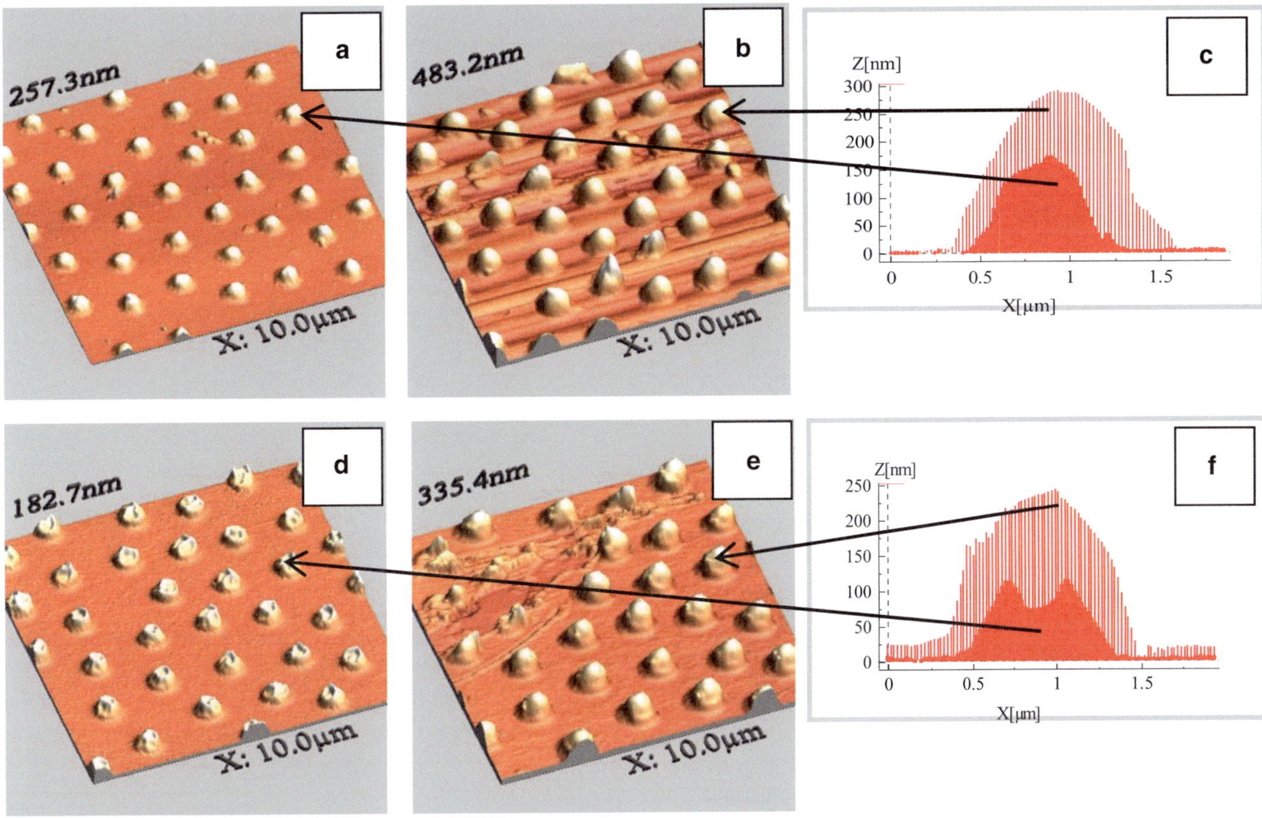

Fig. 7 3D view of AFM topography images (**a**, **b**, **d**, **e**) and corresponding cross-sectional profiles (**c**, **f**) of P(NIPAM-VA) microgels having polyisoprene shell formed by surface-initiated radical polymerization (*top*) and "colloidosome" (*bottom*) approaches in dry (**a**, **d**) and swollen (**b**, **e**) states

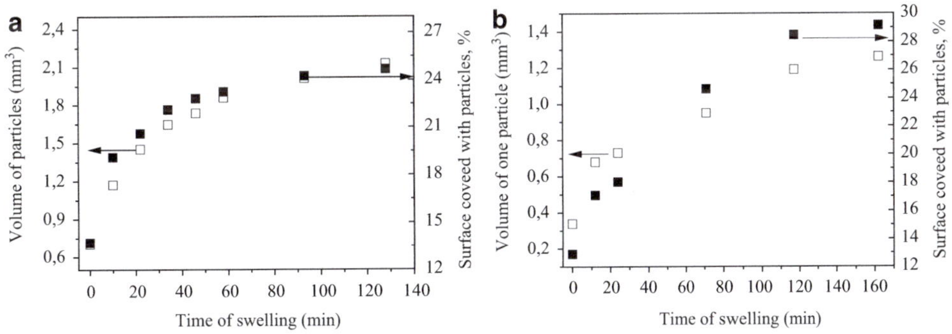

Fig. 8 Time dependence of swelling behavior of surface-immobilized P(NIPAM-VA) microgels with PI shell formed by surface-initiated radical polymerization (**a**) and "colloidosome" (**b**) approach

the SiO_2 core with hydrofluoric acid [51]. Ultrathin thermo-responsive microcapsules were fabricated through direct covalent layer-by-layer assembly of azido- and acetylene-functionalized PNIPAM onto azido-modified silica particles using click chemistry, followed by the removal of the silica templates [52].

Although template-based methods provide an excess to microcapsules of various well-defined sizes, they suffer (i) from multi-step nature of the preparation procedure and (ii) difficulties to load the microcapsules with materials possessing rather large molecular size. Choi et al. [53] reported template-less preparation of PNIAPM microcapsules by polymerization of NIPAM at temperature below LCST of PNIPAM in an oil–water (W/O) emulsion formed in a microfluidic device. To stabilize the capsulate morphology, they utilized an oil-soluble initiator that decomposed by UV-light irradiation to give radicals in the oil phase. In that case, radicals coming from the oil phase and reached the interface with water were responsible for initiation of the polymerization selectively from the interface. Sun et al. [54]

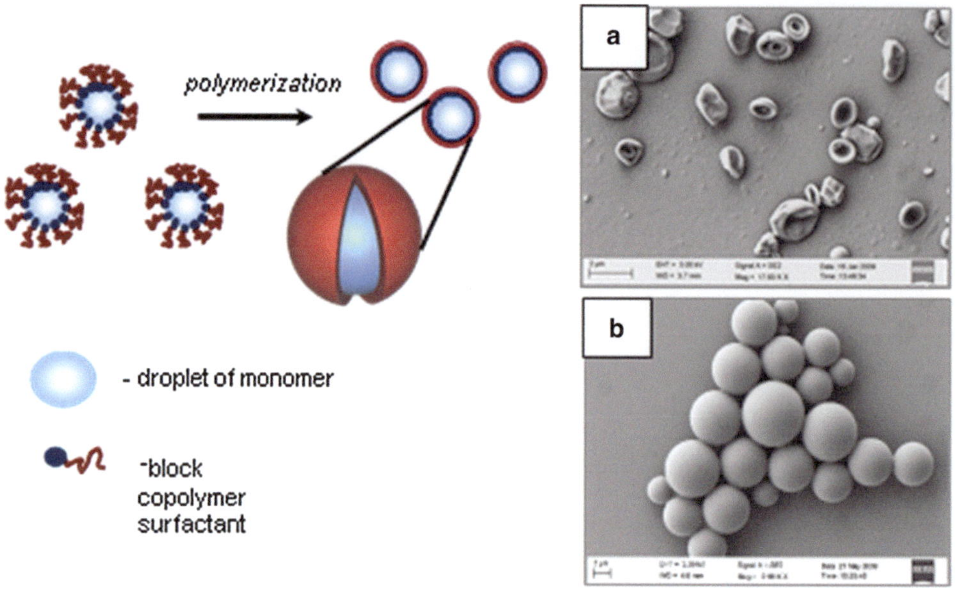

Fig. 9 Schematics of inverse W/O suspension polymerization using PI-b-PEO surfactant and SEM images of core-shell hydrophobically covered (**a**) hollow capsules and (**b**) non-hollow spheres formed upon polymerization of NIPAM (**a**) and AAm (**b**)

applied a combination of three factors to achieve the hollow morphology for PNIPAM particles. First, according to their method initiating radicals were induced selectively at the oil–water interface by diffusion and subsequent reaction of two precursors, the first one soluble in the oil, and the another one, in the water phase. Second, they used an oil-soluble cross-linker that, as expected, diffused from the oil phase into the water droplet and cross-linked NIPAM preferentially in a vicinity to the oil–water interface. Third, the polymerization was performed at temperature above LCST of PNIPAM at which formed PNIPAM is neither soluble in water nor in oil and may precipitate at the oil–water interface.

Recently, we developed a simple and universal method for preparation of PNIPAM microcapsules which does not require any special measures to ensure generation of radicals selectively at the oil–water interface [55]. Particularly, we found out that if W/O polymerization of NIPAM is carried out at temperatures above the LCST of PNIPAM phase separation of the forming cross-linked PNIPAM above the LCST is a sufficient factor to induce the formation of the microcapsules (Figs. 9 and 10).

In our case, the monomer, the cross-linker, and the initiator used for preparation of the capsules are soluble in water and insoluble in the oil phase. It can be also assumed that at relatively high temperature (80 °C) radicals are rapidly forming in the whole volume of the water droplets and thus the formation of the cross-linked PNIPAM is expected to occur in the whole volume of the droplets, and not only at the oil–water interface. Usually, such conditions lead to the formation of non-hollow particles, if water-miscible polymers are formed during the polymerization. Indeed, a control experiment demonstrated that when more polar acrylamide instead of NIPAM is polymerized under otherwise the same polymerization conditions, non-hollow polyacrylamide hard microspheres were formed. However, at temperatures above LCST of PNIPAM the polymerization of NIPAM in aqueous solutions follows a so-called precipitation-polymerization mechanism. According to it, if the polymerization is carried out in water, hydrophobic PNIPAM oligomers formed at early stages of the polymerization precipitate from the aqueous medium and aggregate to form nuclei that further grow finally resulting into well-defined PNIPAM microspheres[56]. When the polymerization of NIPAM is performed in the presence of added template particles with

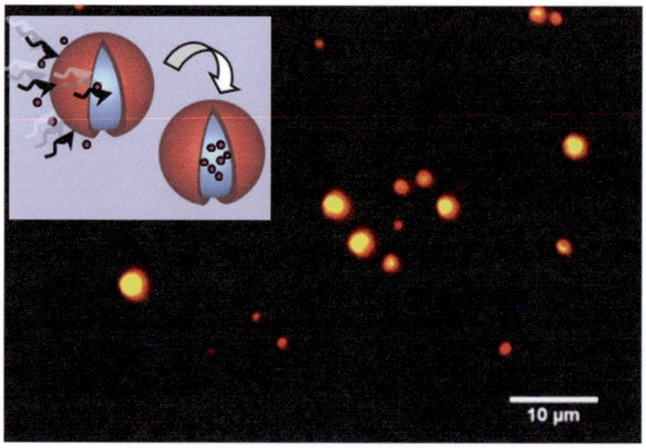

Fig. 10 Schematics of the method for loading of the microcapsules and fluorescent micrograph of the PNIPAM microcapsules loaded with Rhodamine 6G solution (Adapted from the reference [55])

an appropriate surface functionality, then the precipitation of the growing PNIPAM occurs selectively on the added particles (precipitation on convex surfaces) [57]. Similar mechanism occurs in our experiments with the only difference that the concave surfaces of the droplets featured by the macromolecular surfactant act here as seeds for selective precipitation of the phase-separated PNIPAM leading to the microcapsulate morphology which was detected by various microscopy techniques. It is likely that formed hydrophobic NIPAM oligomers preferentially accumulate at the oil–water interface rather than homogeneously nucleate in water due to a higher hydrophobicity of the continuous phase and of the diblock copolymer surfactant than of the water. Another important difference from the earlier reported protocols for preparation of PNIPAM microcapsules is that we use the *macromolecular* surfactant that provides an efficient stabilization of the microcapsules in oils even after extensive purification of the microcapsules, thus demonstrating that the macromolecular surfactant is firmly attached to the microcapsules. The later effect is, most likely, due to a covalent cross-linking of the reactive PI block by radicals formed at the oil–water interface (that came from the water phase) upon NIPAM polymerization. Due to the hydrophobic shell, the hydrogel microcapsules prepared by our methods form stable dispersions even in highly apolar organic solvents, such as cyclohexane, but rapidly flocculate in water. Most of PNIPAM hydrogel microcapsules reported so far were designed to function in water environment. However, for many other important applications, such as in catalysis, cosmetic, agrochemistry and smart coatings, microcapsules having hydrophobic outer surfaces (to provide dispersibility in organic solvents) and hydrophilic interior (for encapsulation of water-soluble compounds) are also desirable. As such, our method significantly widens application areas of PNIPAM-based microcapsules.

The size of the microcapsules dispersed in different environment and temperatures was investigated by dynamic light scattering (DLS). Microcapsules dispersed in anhydrous THF had a diameter of about 1.8–1.9 μm. Addition of small amounts of water (i.e., 3–5 % by volume) to the dispersion of the microcapsules increases their size diameter up to 2.2 μm. Even higher swelling was observed in a tertiary THF/water/cyclohexane system. In control experiments we found that addition of cyclohexane to wet THF induces phase-separation and precipitation of a water-rich phase. Particularly, the initially clear THF/water solution containing 3 % of water becomes turbid after the addition of 10 % of cyclohexane. Even higher concentration of cyclohexane derived precipitation of the water-rich phase. If similar experiments were performed in the presence of microcapsules, the phase-separating water entered the hydrophilic interior of the microcapsules, inducing an additional swelling of the microcapsules. DLS measurements showed that the hydrodynamic diameter of the microcapsules increased up to 2.8–3.5 μm upon addition of cyclohexane. Hence, it is indeed possible to reach much higher swelling degrees for the microcapsules in the tertiary THF/water/cyclohexane system than those achievable in binary THF/water mixtures.

A practically important feature of the PNIPAM-based objects is their thermoresponsiveness, the property inherently associated with the interaction of PNIPAM with water. Our experiments deal with the solvent containing minute amounts of water (i.e., 3 %) and thus, from the first sight, thermo-driven phase transitions of PNIPAM are unlikely here. Indeed, although the phase transition of PNIPAM occurs in THF/water mixtures [35] and in other mixed aqueous solutions (cononsolvency effects) [58], a substantial fraction of water in the mixed solutions is needed to ensure thermoresponsiveness. For example, in the THF/water mixtures PNIPAM exhibits the LCST if the water content exceeds a value of 35 % whereas THF-rich mixtures with water content <35 % are good solvents for PNIPAM [35]. As expected, in anhydrous THF and in THF/water mixtures the size of the microcapsules remained almost constant within the experimental error in the temperature interval from 17 °C to 55 °C. In contrast, the size of the microcapsules in the tertiary THF/water/cyclohexane system was found to be temperature-dependent. The rise of the temperature from 23 °C up to 55 °C induced continuous deswelling of the microcapsules to the diameter of 2 μm that is close to the microcapsules size in anhydrous THF. Cooling of this dispersion down to the room temperature induced reswelling of the microcapsules up to 3.5–3.7 μm.

These experiments inspired us to develop an efficient method to easy load and release the microcapsules with aqueous solutions of water-soluble compounds. To load the capsules, they should be dispersed in THF contained the aqueous solutions of the compound of interest. Consequent addition of apolar solvent induces segregation of the water solution inside the capsules. The shell of the microcapsules was found to be permeable for Rhodamine 6G and other small water-soluble molecules. Heating of the dispersion to temperatures above 40 °C leads to release of the encapsulated compound. As such, the microcapsules can be envisaged for applications that require controlled dispersions and release of hydrophilic species in an oil environment. Alternatively, the microcapsules can be loaded with materials of interest when performing W/O polymerization of NIPAM and the cross-linker in the presence of the material to be encapsulated.

The one-step preparation method of PNIPAM microcapsules successfully adopted for preparation of hybrid PNIPAM/poly(methyl methacrylate)/silica microcapsules using Pickering emulsion technique in works of Zhang and co-workers [59]. They combined both water and oil soluble monomers to obtain two-layer microcapsules with tunable wall thickness and proved their potential usefulness for controllable drug release. Narrowly size-distributed water dispersible PNIPAM nanocapsules were prepared by Landfester et al. in inverse miniemulsion [60]. These nanocapsules were found to be thermoresponsive in water environment with swelling ratio being dependent on the degree of PNIPAM cross-linking [61].

Conclusions

This paper overviewed our recent results in preparation and applications of core-shell particles having the hydrogel core and hydrophobic shell. Water-in-oil suspension polymerization was found to be a useful tool for synthesis of both core-shell particles and hollow capsules. Due to the presence of hydrophobic shell thus-prepared hydrogel particles are well-dispersible in organic solvents and, at the same time, the shell formed from the diblock copolymer surfactant was found to be permeable for water and small water-soluble molecules. The possibility to encapsulate water-soluble species into interior of the microcapsules was demonstrated that makes them promising candidates for different applications.

In addition, the method to prepare core-shell structures directly on the substrate was developed and applied in switchable coatings. Here, PNIPAM-based microgels were first self-organized into quasiperiodic loosely packed arrays and chemically grafted to the functionalized hydrophilic substrate. Two different methods were developed for preparation of the hydrophobic shell atop of the microgels: (i) surface-initiated polymerization of isoprene and (ii) electrostatic adsorption of polyisoprene latex onto oppositely charged microgels. The resulted surfaces were found to be water-sensitive and showed an increase of the fraction of the hydrophobic component from 13 % in the dry state up to 25 % in water environment due to the particles swelling, as observed by in-situ AFM measurements. However, rather small switching effect of 1–2° was recorded by water contact angle measurements because of a moderate lateral swelling of the core-shell microgels and due to a fast swelling of the microgel particles upon the contact angle measurements.

Acknowledgement Financial support by the Deutsche Forschungsgemeinschaft (SPP 1259/1 "Intelligente Hydrogele") is greatly acknowledged.

References

1. Gonsalves K, Halberstadt C, Laurencin CT, Nair LS (2007) Biomedical nanostructures. Wiley-Interscience, Hoboken
2. Lu Y, Proch S, Schrinner M, Drechsler M, Kempeb R, Ballauff M (2009) J Mater Chem 19:3955; Welsch N, Becker AL, Dzubiella J, Ballauff M (2012) Soft Matter 8:1428
3. Jones CD, Lyon LA (2000) Macromolecules 33:8301
4. Berndt I, Richtering W (2003) Macromolecules 36:8780; Berndt I, Pedersen JS, Richtering W (2005) J Am Chem Soc 127: 9372
5. Christodoulakis KE, Vamvakaki M (2010) Langmuir 26:639; Lapeyre V, Ancla C, Catargi B, Ravaine V (2008) J Colloid Interface Sci 327:316; Rodríguez-Fernández J, Fedoruk M, Hrelescu C, Lutich AA, Feldmann J (2011) Nanotechnology 22:245708
6. Nayak S, Gan D, Serpe MJ, Lyon LA (2005) Small 1:416
7. Xia F, Ge H, Hou Y, Sun TL, Chen L, Zhang GZ, Jiang L (2007) Adv Mater 19:2520
8. Jiang JQ, Tong X, Morris D, Zhao Y (2006) Macromolecules 39:4633
9. Kovarik ML, Jacobson SC (2009) Anal Chem 81:7133
10. Brown AA, Azzaroni O, Huck WTS (2009) Langmuir 25: 1744
11. Ruckenstein E, Lee SH (1987) Colloid Interface Sci 120:529, Lee SH, Ruckenstein EJ (1987) Colloid Interface Sci 117:172
12. Hopken J, Moller, M (1992) Macromolecules 25:482; Vaidya A, Chaudhury MK (2002) J Colloid Interface Sci 249:235; Senshu K, Yamashita S, Mori H, Ito M, Hirao A, Nakahama S (1999) Langmuir 15: 1754; Mori H, Hirao A, Nakahama S, Senshu K (1994) Macromolecules 27:4093; Pientka Z, Oike H, Tezuka Y (1999) Langmuir 15:3197
13. Minko S, Muller M, Motornov M, Nitschke M, Grundke K, Stamm M (2003) J Am Chem Soc 125:3896, Retsos H, Senkovskyy V, Kiriy A, Stamm M, Feldstein M, Creton C (2006) Adv Mater 18:2624
14. Retsos H, Gorodyska G, Kiriy A, Stamm M, Creton C (2005) Langmuir 21:7722
15. Hemmerle J, Roucoules V, Fleith G, Nardin M, Ball V, Lavalle P, Marie P, Voegel JC, Schaaf P (2005) Langmuir 21:10328
16. Tamber H, Johansen P, Merkle HP, Gander B (2005) Adv Drug Deliv Rev 57:376; Hammer DA, Discher DE (2001) Annu Rev Mater Sci 31:387
17. Caruso F, Caruso RA, Mohwald H (1998) Science 282:1111; Johnson SA, Ollivier PJ, Mallouk TE (1999) Science 283:963; Bronich TK, Ouyang M, Kabanov VA, Eisenberg A, Szoka FC, Kabanov AV (2002) J Am Chem Soc 124:11872
18. Dowding PJ, Atkin R, Vincent B, Bouillot P (2004) Langmuir 20:11374; Dinsmore AD, Hsu MG, Nikolaides MG, Marques M, Bausch AR, Weitz DA (2002) Science 298:1006; Qian J, Wu FP (2007) Chem Mater 19:5839; Peyratout CS, Dahne L (2004) Angew Chem Int Ed 43:3762
19. Sidorenko A, Minko S, Schenk-Meuser K, Duschner H, Stamm M (1999) Langmuir 15:8349; Tiarks F, Landfester K, Antonietti M (2001) Langmuir 17:908
20. Khongtong S, Ferguson GS (2001) J Am Chem Soc 123:3588; Makal U, Wynne KJ (2005) Langmuir 21:3742; Makal U, Uslu N, Wynne KJ (2007) Langmuir 23:209
21. Horecha M, Senkovskyy V, Schneider K, Kiriy A, Stamm M (2011) Colloid Polym Sci 289:603
22. Cassie A, Baxter S (1944) Trans Faraday Soc 40:546; Cassie A, Baxter S (1948) Discuss Faradey Soc 3:5041
23. Pelton R (2000) Adv Colloid Interface Sci 85:1; Jones CD, Lyon LA (2000) Macromolecules 33:8301; Dowding PJ, Vincent B, Williams EJ (2000) Colloid Interface Sci 221:268; Zhou S, Chu B (1998) J Phys Chem 102:1364; Saunders BR, Vincent B (1999) Adv Colloid Interface Sci 80:1
24. Nerapusri V, Keddie JL, Vincent B, Bushnak IA (2006) Langmuir 22:5036; Serpe MJ, Jones CD, Lyon LA (2003) Langmuir 19:8759; Schmidt S, Motschmann H, Hellweg T, von Klitzing R (2008) Polymer 49:749
25. Schmidt S, Hellweg T, von Klitzing R (2008) Langmuir 24: 12595; Sorrell CD, Lyon LA (2008) Langmuir 24: 7216; Lu Y, Drechsler M (2009) Langmuir 25:13100

26. Sakai T, Takeoka Y, Seki T, Yoshida R (2007) Langmuir 23: 8651; Tian E, Wang J, Zheng Y, Song Y, Jiang L, Zhu DJ (2008) Mater Chem 18:1116
27. Nakahama K, Fujimoto K (2002) Langmuir 18:10095
28. Tsuji S, Kawaguchi H (2005) Langmuir 21:8439
29. Suzuki D, McGrath J, Kawaguchi H, Lyon LA (2007) J Phys Chem 111:5667
30. Serpe MJ, Jones CD, Lyon LA (2003) Langmuir 19:8759
31. Lyon LA, Meng Z, Singh N, Sorrell CD (2009) Chem Soc Rev 38:865
32. Chen MQ, Serizava T, Li M, Wu C, Akashi M (2003) Polymer J 35:901, Chen Y, Gautrot JE, Zhu XX (2007) Langmuir 23:1047, Pabon M, Corpart JM, Selb J, Candau F (2004) J Appl Polym Sci 91:916, Paunov J (2004) Mater Chem 14:3351
33. Horecha M (2011) Coatings with inversely switching behaviour: new applications of core-shell hydrogel particles. http://tud.qucosa.de/recherche/frontdoor/cache.off?tx_slubopus4frontend[id]=6517
34. Takahashi K, Takigawa T, Masuda T (2004) J Chem Phys 120:2972; Suarez IJ, Fernandez-Nieves A, Marquez M (2006) J Chem Phys B 110:25729; Owens DE, Jian Y, Fang JE, Slaughter BV, Chen YH, Peppas NA (2007) Macromolecules 40:7306
35. Winnik FM, Ottaviani MF, Bossmann SH, Pan WS, Garciagaribay M, Turro NJ (1993) Macromolecules 26:4577
36. Horecha M, Senkovskyy V, Synytska A, Stamm M, Chervanyov AI, Kiriy A (2010) Soft Matter 6:5980
37. Pelton RH, Chibante P (1986) Colloids Surf 20:247
38. Elachari A, Coqueret X, Lablachecombier A, Loucheux C (1993) Makromolekulare Chemie-Macromol Chem Phys 194:1879
39. Edmondson S, Osborne VL, Huck WT (2004) Chem Soc Rev 33:14
40. Tsuji S, Kawaguchi H (2005) Langmuir 21:2434, Serpe MJ, Jones CD, Lyon LA (2003) Langmuir 19:8759
41. Khanduyeva N, Senkovskyy V, Beryozkina T, Horecha M, Stamm M, Uhrich C, Riede M, Leo K, Kiriy A (2009) J Am Chem Soc 131:153
42. Beryozkina T, Boyko K, Khanduyeva N, Senkovskyy V, Horecha M, Oertel U, Simon F, Komber H, Stamm M, Kiriy A (2009) Angew Chem 48:2695
43. Li R, Yang XL, Li GL, Li SN, Huang WQ (2006) Langmuir 22:8127
44. Bon SAF, Cauvin S, Colver PJ (2007) Soft Matter 3:194
45. Wang Y, Hu HP, Zhang X (2009) Adv Mater 21:2849; Lou XW, Archer LA, Yang Z (2008) Adv Mater 20:3987; Lan Y, Yang L, Zhang MC, Zhang WQ, Wang SN (2010) ACS Appl Mater Interf 2:127; Khanal A, Inoue Y, Yada M, Nakashima K (2007) J Am Chem Soc 129:1534; Li GL, Lei C, Wang C, Neoh KG, Kang ET, Yang X (2008) Macromolecules 41:9487
46. Arnal P, Comotti M, Schüth F (2006) Angew Chem Int Ed 45:8224; Kamata K, Lu Y, Xia Y (2003) J Am Chem Soc 125:2384; Lee J, Park JC, Bang JU, Song H (2008) Chem Mater 20:5839; Wu XJ, Xu D (2010) Adv Mater 22:1516; Zoldesi CI, Imhof A (2005) Adv Mater 17:924
47. Peyratout CS, Dähne L (2004) Angew Chem Int Ed 43:3762, Sukhorukov G, Fery A, Möhwald H (2005) Prog Polym Sci 30:885, Klaikherd A, Nagamani C, Thayumanavan S (2009) J Am. Chem. Soc. 131:4830, Li G, Shi L, Ma R, An Y, Huang N (2006) Angew Chem Int Ed 45:4959, Sauer M, Meier W (2001) Chem Commun 1:55, Sauer M, Streich D, Meier W (2001) Adv Mater 13:1649, Seiffert S, Weitz DA (2010) Polymer 51:5883
48. Li GL, Xu LQ, Tang XZ, Neoh KG, Kang ET (2010) Macromolecules 43:5797
49. Heskins M, Guillet JE (1968) J Macromol Sci A2:1441; Hirotsu S (1988) J Chem Phys 88:427
50. Wie W, Zhang C, Ding S, Qu X, Liu J, Yang Z (2008) Colloid Polym Sci 286:881
51. Gao H, Yang W, Min K, Zha L, Wang C, Fu S (2005) Polymer 46:1087
52. Huang CJ, Chang FC (2009) Macromolecules 42:5155
53. Choi CH, Jung JH, Kim DW, Chung YM, Lee CS (2008) Lab Chip 8:1544
54. Sun Q, Deng Y (2005) J Am Chem Soc 127:8274
55. Horecha M, Senkovskyy V, Stamm M, Kiriy A (2009) Macromolecules 42:5811
56. McPhee W, Tam KC, Pelton RH (1993) J Colloid Interface Sci 156:24
57. Zha LS, Zhang Y, Yang WL, Fu SK (2002) Adv Mater 14:1090
58. Wolf BA, Willms MM (1978) Makromol Chem 179:2265; Nandi AK, Sen UK, Bhattacharyya SN, Mandel BM (1983) Eur Polym J 19:283; Horta A, Fernandez-Pierola I (1980) Polym Bull 3:273; Schild HG, Muthukumar M, Tirrell DA (1991) Macromolecules 24:948
59. Zhang K, Wu W, Guo K, Chen J, Zhang P (2010) Langmuir 26:7971
60. Cao Z, Ziener U, Landfester K (2010) Macromolecules 43:6353
61. Kobitskaya E, Ekinci D, Manzke A, Plettl A, Wiedwald U, Ziemann P, Biskupek J, Kaiser U, Ziener U, Landfester K (2010) Macromolecules 43:3294

Influence of Salt and pH on the Swelling Equilibrium of Ionizable N-IPAAm Based Hydrogels: Experimental Results and Modeling

Luciana Ninni, Viktor Ermatchkov, Hans Hasse, and Gerd Maurer

Abstract

The influence of sodium chloride and of pH on the equilibrium swelling of ionic hydrogels consisting of N-isopropyl acrylamide (N-IPAAm), a weak cationic comonomer *N*-[3-(dimethylamino)propyl] methacrylamide (DMAPMA) and/or a strong anionic comonomer 2-acrylamido-2-methyl-1-propane sulfonic acid (AMPS) was investigated in aqueous solutions at 298 K. The mass fraction of the polymerizable substances in the aqueous solution was kept at 0.1. The sum of the mole fractions of the comonomers in the gel was 0.05 while the mole fraction of a single comonomer was varied between 0 and 0.05. The mass fraction of NaCl in the aqueous solution was varied between $3 \cdot 10^{-6}$ and 0.25. The pH-sensitive comonomer DMAPMA causes an influence of pH on the degree of swelling. The pH of the aqueous solutions was adjusted by adding HCl or NaOH. A thermodynamic framework was applied to model (i.e., correlate and predict) the experimental results. The calculations revealed that dissolved carbon dioxide (resulting from air that was in contact with the condensed phases) might have a remarkable influence on the swelling behavior.

Keywords

Ionic hydrogels • N-isopropyl acrylamide (N-IPAAm) • 2-acrylamido-2-methyl-1-propane sulfonic acid (AMPS) • N-[3-(Dimethylamino)propyl] methacrylamide (DMAPMA) • Equilibrium swelling • Experimental results • Thermodynamic modeling • pH-dependent swelling

Introduction

Hydrogels are cross-linked networks consisting of hydrophilic polymer chains. They can swell or shrink depending on external applied triggers such as temperature, pH and solvent composition of the coexisting phase, electric field, light, etc. [1–6]. The volume change of pH-sensitive hydrogels is of interest in various areas, for example, for controlled drug delivery systems [7, 8]. pH-sensitive hydrogels contain weak anionic and/or cationic comonomers. Variations in the pH of the medium can change the density of charges on the polymeric chain which influences the swelling properties of a hydrogel. The charge density of such hydrogels can also be changed by varying the amount and nature of strong electrolytes in the coexisting liquid phase. Most commonly studied anionic (cationic) hydrogels contain carboxylic or sulfonic acid groups [9, 10] (amine groups [11]). In a recent publication [12], we investigated

The online version of this chapter (doi:10.1007/978-3-319-01683-2_2) contains supplementary material, which is available to authorized users.

L. Ninni • H. Hasse • G. Maurer (✉)
Laboratory of Engineering Thermodynamics, Department of Mechanical and Process Engineering, University of Kaiserslautern, 67653 Kaiserslautern, Germany
e-mail: gerd.maurer@mv.uni-kl.de

V. Ermatchkov
Laboratory of Engineering, Thermodynamics, Department of Mechanical and Process Engineering Organization, University of Kaiserslautern, 67653 Kaiserslautern, Germany

Present Organization: Evonik Industries AG, 63457 Hanau, Germany

G. Sadowski and W. Richtering (eds.), *Intelligent Hydrogels*, Progress in Colloid and Polymer Science 140,
DOI 10.1007/978-3-319-01683-2_13, © Springer International Publishing Switzerland 2013

the swelling equilibrium behavior of synthetic ionic N-IPAAm based hydrogels in water and in aqueous solutions of NaCl at 298 K. The ionic behavior was introduced by a strong anionic comonomer (AMPS) and/or a strong cationic comonomer (MAPTAC, i.e., an ammonium salt). Here we extend these investigations to synthetic ionic hydrogels where the strong cationic electrolyte MAPTAC is replaced by the weak cationic comonomer DMAPMA (*N*-[3-(dimethylamino)propyl] methacrylamide) and study the influence of pH and NaCl on the equilibrium swelling of such hydrogels in aqueous solutions at 298 K. The new experimental data are correlated using a model that combines an expression for the Gibbs energy of aqueous solutions of (non-crosslinked) polymer chains with an expression for the Helmholtz energy of an elastic network [12, 13]. That model was extended to describe the influence of dissolved carbon dioxide on the equilibrium swelling.

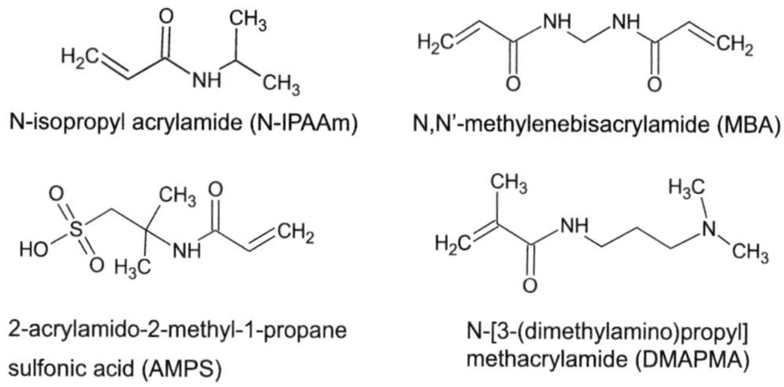

Fig. 1 Chemical structures of the hydrophilic monomers of the ionic N-IPAAm/DMAPMA/AMPS hydrogels

Experimental

Materials

The N-IPAAm based hydrogels were prepared by free-radical polymerization/crosslinking using N,N′-methylenebisacrylamide (MBA) to crosslink the polymer chains.

N-isopropyl acrylamide (N-IPAAm) (97 %, CAS 2210-25-5), the ionic comonomers 2-acrylamido-2-methyl-1-propane sulfonic acid (AMPS) (99 %, CAS 15214-89-8), *N*-[3-(dimethylamino)propyl] methacrylamide (DMAPMA) (99 %, CAS 5205-93-6) as well as the crosslinking component N,N′-methylenebisacrylamide (MBA) (≥99 %, CAS 110-26-9) are the components of the investigated ionic hydrogels. The chemical structures of these monomers are shown in Fig. 1. Ammonium peroxodisulfate ($(NH_4)_2S_2O_8$) (≥98 %, CAS 7727-54-0) and sodium disulfite ($Na_2S_2O_5$) (≥98 %, CAS 7681-57-4) were used as reaction initiator and accelerator, respectively. Sodium chloride (NaCl, ≥99 %) was used to prepare the aqueous salt solutions for the swelling experiments. Concentrated aqueous solutions of HCl and NaOH were used to prepare aqueous solutions with different pH values. All chemicals were purchased from Sigma-Aldrich, Taufkirchen, Germany and used without further purification. Oxygen-free, deionized water was used in all experimental work (for synthesis as well as for swelling experiments).

Hydrogels Production and Characterization

The structure and, consequently also the equilibrium properties of synthetic hydrogels depend on the composition of the polymerizing solutions as well as on the polymerization/crosslinking processes. A detailed description of the procedure was given in previous publications (see for example, [12, 14]). We characterize a hydrogel by its production process together with four parameters which describe the composition of the polymerizing aqueous solution using the amount of mass $\tilde{m}_i$ and the number of moles n_i of component i in the polymerizing solution:

The mass fraction of polymerizable substances in the aqueous solution:

$$\xi_{gel} = \frac{\tilde{m}_{N-IPAAm} + \tilde{m}_{MBA} + \sum_j \tilde{m}_{co_j}}{\tilde{m}_{N-IPAAm} + \tilde{m}_{MBA} + \sum_j \tilde{m}_{co_j} + \tilde{m}_{Na_2S_2O_5} + \tilde{m}_{(NH_4)_2S_2O_8} + \tilde{m}_{water}} \quad (1)$$

The mole fraction of comonomer i (co_i = AMPS, DMAPMA):

$$y_{co_i} = \frac{n_{co_i}}{n_{N-IPAAm} + \sum_j n_{co_j}} \quad (2)$$

The mole fraction of the crosslinker:

$$y_{MBA} = \frac{n_{MBA}}{n_{N-IPAAm} + n_{MBA} + \sum_j n_{co_j}} \quad (3)$$

The mass fraction of the initiator (= the mass fraction of the accelerator):

$$\xi_S = \frac{\tilde{m}_{(NH_4)_2S_2O_8}}{\tilde{m}_{N-IPAAm} + \tilde{m}_{MBA} + \sum_j \tilde{m}_{co_j} + \tilde{m}_{Na_2S_2O_5} + \tilde{m}_{(NH_4)_2S_2O_8} + \tilde{m}_{water}} \quad (4)$$

Table 1 Summary of the synthetized N-IPAAm/AMPS/DMAPMA hydrogels

Gel	ξ_{gel} $g \cdot g^{-1}$	y_{MBA} $mol \cdot mol^{-1}$	y_{AMPS} $mol \cdot mol^{-1}$	y_{DMAPMA} $mol \cdot mol^{-1}$	$10^4 \xi_s$ $g \cdot g^{-1}$	ϑ	C
1	0.1	0.02	0	0.050	2.12	1	3.12
2	0.1	0.02	0.010	0.040	2.07	0.60	2.87
3	0.1	0.02	0.015	0.035	2.10	0.40	2.77
4	0.1	0.02	0.021	0.030	1.98	0.18	2.85
5	0.1	0.02	0.032	0.020	2.05	−0.23	2.93
6	0.1	0.05	0	0.050	2.02	1	2.28
7	0.1	0.05	0.010	0.042	2.01	0.62	2.12
8	0.1	0.05	0.016	0.035	2.02	0.38	2.15
9	0.1	0.05	0.020	0.032	2.25	0.23	2.08
10	0.1	0.05	0.025	0.025	2.08	0	2.58
11	0.1	0.05	0.035	0.015	2.04	−0.40	2.24
12	0.1	0.05	0.040	0.010	1.88	−0.60	2.18
13	0.1	0.05	0.050	0	2.09	−1	2.28

The amount of comonomers is described here also by a "charge density" parameter ϑ:

$$\vartheta = \frac{y_{DMAPMA} - y_{AMPS}}{y_{DMAPMA} + y_{AMPS}} \tag{5}$$

The mass fraction ξ_{gel} was kept at 0.1. The mole fraction of a single comonomer y_{co_i} was varied between 0 and 0.05 whereas the sum of both mole fractions ($y_{AMPS} + y_{DMAPMA}$) was always 0.05. Therefore, the charge density parameter ranged from $-1 \leq \vartheta \leq 1$. The crosslinker mole fraction y_{MBA} was either 0.02 or 0.05. The mass fraction of the initiator (accelerator) was constant ($\xi_S \approx 0.0002$). Table 1 presents a summary of all investigated hydrogels.

Swelling Equilibrium Experiments

The swelling equilibrium experiments were performed as described before [12]. Therefore, only some essentials and the new features are given here.

The degree of swelling q, i.e., the ratio of the mass of an equilibrated gel $\tilde{m}_{gel}$ to the mass of the dried gel $\tilde{m}_{gel}^{(dry)}$ is calculated as:

$$q = \frac{\tilde{m}_{gel}}{\tilde{m}_{gel}^{(dry)}} \tag{6}$$

For the experiments in salt containing solutions, the mass fraction of sodium chloride (w_{NaCl}) was varied between about $3 \cdot 10^{-6}$ and 0.25. For the experiments to investigate the influence of pH on the equilibrium swelling, aqueous solutions of either HCl or NaOH were prepared by appropriately diluting those chemicals. The pH of a solution was measured using a combined electrode (glass pH electrode and Ag/AgCl electrode in a single unit from Methrom, Herisau, Switzerland, type 6.0258.010) and a pHmeter from Metrohm, Herisau, Switzerland (type 713, resolution 0.1 mV). In each experiment two pH values were determined: the pH of the solution before adding the hydrogels ($pH_{initial}$) and the pH of the aqueous solution in equilibrium with the hydrogel (pH_{final}).

Experimental Results

The experimental results for the degree of swelling of the investigated ionic hydrogels are given (as Supplementary Material to this publication) in Online resources (for the swelling in aqueous solutions of NaCl and in aqueous solutions of HCl or NaOH). Some experimental results for $pH_{initial}$ and pH_{final} and the mass fractions of HCl/NaOH (w_{HCl} or w_{NaOH}) in the fresh (i.e., without immersed hydrogels) aqueous solutions are also reported there. The arithmetic averages of the degree of swelling and of pH_{final} (from ten independent experiments) were determined and the standard deviation from the arithmetic average was calculated. That standard deviation is about 5 % (6 %) for the degree of swelling in aqueous solutions of NaCl (of HCl or NaOH). The standard deviation for pH_{final} is $\pm$ 0.2 pH units.

Swelling Equilibrium in Water and in Aqueous Solutions of NaCl

Figure 2 shows – as typical examples – the influence of NaCl on the degree of swelling of three hydrogels that differ only in the charge density parameter. One gel contains only cationic comonomers ($\vartheta = 1$) and two gels contain both types of comonomers, but a surplus of cationic comonomers ($\vartheta = 0.6$ and $\vartheta = 0.18$, respectively). In water and in highly diluted aqueous solutions of NaCl, the charge density parameter has a large influence on the degree of swelling of the ionic hydrogels and with increasing distance to the non-ionic gel (i.e., to a gel with $\vartheta = 0$) the degree of swelling increases. Increasing the mass fraction of NaCl in the liquid phase which coexists with the gel, results in a typical pattern with a maximum in the degree of swelling at around $w_{NaCl} \approx 5 \cdot 10^{-5}$ and two volume transitions. In the first transition the degree of swelling of the ionic hydrogels decreases steadily to about 13, i.e., the degree of swelling of the corresponding non-ionic gel. That decrease is completed when the mass fraction of NaCl reaches about $5 \cdot 10^{-3}$. No influence of the anionic and/or cationic comonomers on the degree of swelling could be detected at $w_{NaCl} > 5 \cdot 10^{-3}$. In that region the investigated ionic hydrogels behave just like their non-ionic counterparts. They all reveal a second transition (which occurs at $w_{NaCl} \approx 5 \cdot 10^{-2}$) where the gels collapse to a state where they have rejected nearly all water.

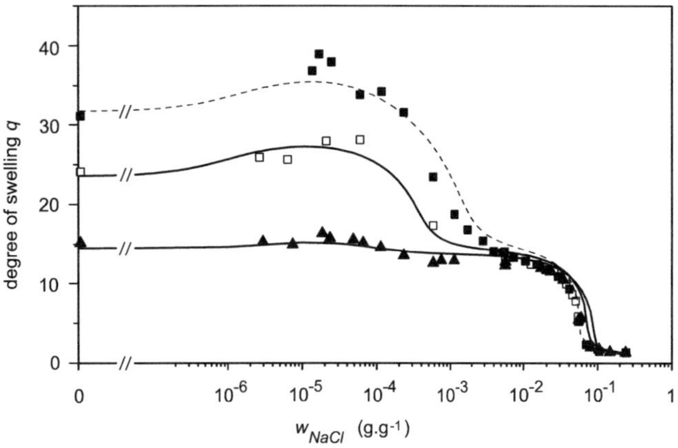

Fig. 2 Degree of swelling q of some hydrogels N-IPAAm/DMAPMA/AMPS ($y_{MBA} = 0.02$, $\xi_{gel} = 0.1$) in aqueous solutions of sodium chloride at 298 K. Experimental results: cationic gel: (■), gel 1 ($y_{AMPS} = 0$, $y_{DMAPMA} = 0.05$); anionic/cationic gels: (□), gel 2 ($y_{AMPS} = 0.01$, $y_{DMAPMA} = 0.04$); (▲), gel 4 ($y_{AMPS} = 0.021$, $y_{DMAPMA} = 0.03$). Correlation results for gel 1 (*dashed line*) and prediction results for gels 2 and 4 (*full lines*)

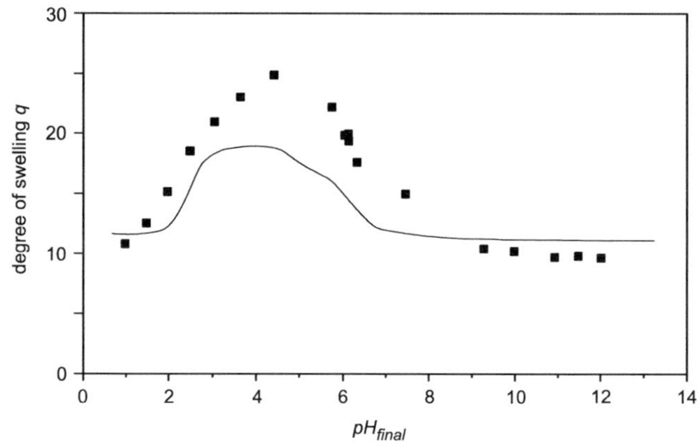

Fig. 3 Influence of pH_{final} on the swelling behavior of gel 6 ($y_{MBA} = 0.05$, $\xi_{gel} = 0.1$, $y_{AMPS} = 0$, $y_{DMAPMA} = 0.05$) at 298 K. (■) experimental results; (*full line*) prediction results

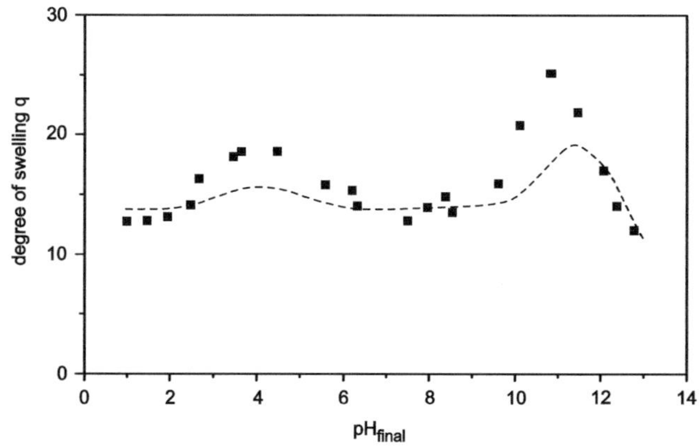

Fig. 4 Influence of pH_{final} on the swelling behavior of gel 4 ($y_{MBA} = 0.02$, $\xi_{gel} = 0.1$, $y_{AMPS} = 0.021$, $y_{DMAPMA} = 0.03$) at 298 K. (■) experimental results; (*dashed line*) correlation results

The same behavior was already observed in a prior work [12] for similar hydrogels but with the strong cationic comonomer MAPTAC (instead of the weak electrolyte DMAPMA).

Swelling Equilibrium in Aqueous Solutions at Different pH Values

Figure 3 shows experimental results for the influence of pH_{final} on the degree of swelling of an ionic N-IPAAm/DMAPMA hydrogel (gel no. 6 in Table 1) at 298 K. With increasing pH the degree of swelling increases from $q \approx 11$ at $pH \approx 1$ to $q \approx 25$ at $pH \approx 4$ and decreases to $q \approx 10$ at $pH \approx 9$. A further increase in the pH has nearly no influence on q, i.e., the ionic gel behaves like a non-ionic one. In the experiments the solutions were in contact with the atmosphere, i.e., carbon dioxide from the atmosphere might have been dissolved in the aqueous solution, resulting also in the formation of hydrogen carbonate (HCO_3^-), carbonate (CO_3^{2-}) and hydrogen (H^+) ions and in a small shift in the pH, for example, from 7 to 6.1 ±0.2.

Figure 4 shows experimental results for the influence of pH_{final} on the degree of swelling of an ionic N-IPAAm/DMAPMA/AMPS hydrogel (gel no. 4 in Table 1) at 298 K. There is also a maximum in the degree of swelling at around $pH \approx 4$, but that maximum is less pronounced. Furthermore, there is a second maximum at around $pH \approx 11$ which was not found for the cationic N-IPAAm/DMAPMA hydrogel.

Buffering of the Coexisting Liquid Phase

The experimental results for pH_{final} in the aqueous solutions that coexist with N-IPAAm/DMAPMA/AMPS gels (gels number 1, 2, 4, 5 and 6) are plotted in Fig. 5 versus the initial pH ($pH_{initial}$) of these solutions, i.e., the pH value measured at the beginning of an experiment when the gel was not yet placed into the aqueous solutions. At low as well as at high pH, the difference between pH_{final} and $pH_{initial}$ is less than the experimental uncertainty. However, in an intermediate range ($4 < pH_{initial} < 10$) there is a difference between both experimental results and $pH_{initial} > pH_{final}$. In that range the equilibration of the hydrogels reduces the pH of the solution, i.e., the hydrogel behaves like a buffer. That difference reaches up to about 2 pH units. A similar behavior was found by Horta et al. [15] and Horta and Piérola [16] when cross-linked poly(N-vinylimidazole) gels were equilibrated

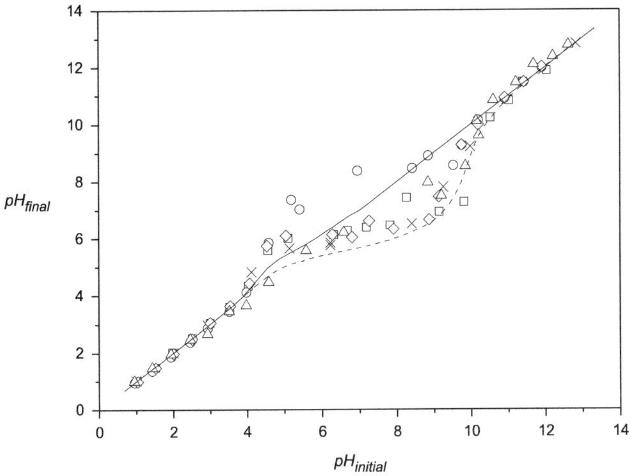

Fig. 5 Representation of the equilibrium pH of the coexisting liquid phase (pH_{final}) versus the pH of this phase at the beginning of the experiments still without a gel ($pH_{initial}$) at 298 K. Experimental results: for $y_{MBA} = 0.02$ and $\xi_{gel} = 0.1$: (◇), gel 1 ($y_{AMPS} = 0$, $y_{DMAPMA} = 0.05$); (□), gel 2 ($y_{AMPS} = 0.01$, $y_{DMAPMA} = 0.04$); (△), gel 4 ($y_{AMPS} = 0.021$, $y_{DMAPMA} = 0.03$); (○), gel 5 ($y_{AMPS} = 0.032$, $y_{DMAPMA} = 0.02$); for $y_{MBA} = 0.05$ and $\xi_{gel} = 0.1$: (×), gel 6 ($y_{AMPS} = 0$, $y_{DMAPMA} = 0.05$). Calculated results: *dashed line*: corresponds to: p_{CO_2} (initial) = 33 Pa; *continuous line*: without CO_2

in aqueous solutions under acidic conditions ($1 < pH < 6$). The authors explained that phenomena with the protonation of N-vinylimidazole that reduced the concentration of free protons in the liquid phase.

Electromotive Force Experiments

As shown before, the cationic comonomer DMAPMA has an influence on the degree of swelling of N-IPAAm-based hydrogels in aqueous solutions. DMAPMA has an amine group that can be protonated:

$$\text{DMAPMA} + \text{H}^+ \rightleftharpoons \text{DMAPMAH}^+ \qquad \text{(I)}$$

The chemical reaction equilibrium constant of that reaction is required when the experimental results for the swelling of such hydrogels are to be modeled. On the molality scale that chemical reaction equilibrium constant (protonation constant) is

$$K_I^{(m)} = \frac{a^{(m)}_{DMAPMAH^+}}{a^{(m)}_{DMAPMA}\, a^{(m)}_{H^+}} \qquad (7)$$

where $a_i^{(m)}$ is the activity of solute species i in an aqueous solution. Superscript m indicates that the reference state is based on the molality scale in water.

$K_I^{(m)}$ was determined from electromotive force (EMF) measurements at temperatures from 278 to 368 K in aqueous solutions of DMAPMA. The EMF experiments were performed with a two-cell arrangement described previously by Pérez-Salado Kamps and Maurer [17]. The experimental results can be described with an absolute mean average deviation in ln $K_I^{(m)}$ of 0.14 by:

$$\ln K_I^{(m)} = 5.299 + \frac{4739.6}{(T/K)} \qquad (278 < T/\text{K} < 368) \qquad (8)$$

These results are in good agreement with literature data by Van de Wetering et al. [18] which were determined by acid–base titrations. For 298 K, Van de Wetering et al. reported values for ln $K_I^{(m)}$ for the protonation of the monomer DMAPMA as well as of the polymer (poly(DMAPMA)) of 21.84 and 20.02, respectively while Eq. 8 results in $\ln K_I^{(m)} = 21.20$.

Modeling

As in a previous publication [12], the equilibrium swelling behavior of a hydrogel phase (II) in equilibrium with a coexisting liquid phase (I) was modeled using the thermodynamic framework derived by Maurer and Prausnitz [13]. An extension of that model is applied here. As the main features of that model were described in that previous publication, only an outline is given here and the changes/extensions are described in detail.

The hydrogel phase is treated as a liquid mixture, where the polymer network is represented by non-crosslinked chains (i.e., the network is virtually cut off at all crosslinking positions) and all contributions to the Gibbs energy which are due to crosslinking are summarized in an expression for the Helmholtz energy A of an elastic network. In equilibrium, the temperature T and the pressure p attain the same values in both coexisting phases and – when the same reference state is selected for the chemical potential of a species i that is equilibrated between both phases – the following relation holds for the activities (in the liquid phase a_i^I and in the gel phase a_i^{II}):

$$\ln a_i^I \left(T, p, n_j^I\right) = \ln a_i^{II} \left(T, p, n_j^{II}\right) + \frac{v_i^{II}}{RT}\left(\frac{\partial A}{\partial V^{II}}\right)_T + \frac{z_i F}{RT}\Delta\varphi \qquad (9)$$

Superscripts I and II characterize the phase, n_j is the number of moles of species j, v_i and z_i are the partial molar volume of species and the number of electric charges of i, respectively. V^{II}, R, F and $\Delta\varphi$ are the volume of the gel phase, the universal gas constant, Faraday's constant and the so called "electrical potential difference" between the gel and the liquid phase, respectively. Equation 9 holds for all species i that are allowed to cross the phase boundary (i.e., to equilibrate between the coexisting phases) with the exception of one of the ionic species. When there are q ionic species which partition between both phases, only $(q - 1)$ components can partition independently as the partitioning of one ionic species is restricted due to the condition of electroneutrality in both phases. If that component is designated by q, the difference in the activities of that component in both phases is used together with the expression for the Helmholtz energy to calculate the "electrical potential difference" (Maurer and Prausnitz [13]):

$$\frac{z_q F}{RT}\Delta\varphi = \ln a_q^I\left(T, p, n_j^I\right) - \ln a_q^{II}\left(T, p, n_j^{II}\right) - \frac{v_q^{II}}{RT}\left(\frac{\partial A}{\partial V^{II}}\right)_T \tag{10}$$

When hydrogels composed of N-IPAAm/DMAPMA/AMPS are equilibrated with an aqueous solution of NaCl, HCl or NaOH in the presence of carbon dioxide there are only two neutral species (H_2O and CO_2) and six ionic species (H^+, OH^-, Na^+, Cl^-, HCO_3^- and CO_3^{2-}) that can cross the phase boundary. All groups that are fixed to the polymer network are only present in the gel phase. That gel phase is considered to be an aqueous solution of non-crosslinked polymer chains. The effect of crosslinking is described by the Helmholtz energy A of the network. As in previous publications (for example, Ninni et al. [12]), a modification of the phantom network theory [19–21] is used to express the free energy that is caused by the deformation of the network from its "stress-free" state, i.e., by the volume change from V_0 (in the stress-free state, which is the state after polymerization/crosslinking) to the actual volume V of the gel:

$$\frac{A}{RT} = C\, n_{poly}\left[\left(\frac{V}{V_0}\right)^{2/3} - 1\right] \tag{11}$$

where C is an adjustable parameter which characterizes the elastic properties of the network and n_{poly} is the number of polymer chains that form that network.

The activities of species i are calculated using an expression for the Gibbs energy of the aqueous phases. The aqueous phase that coexists in equilibrium with the gel phase is described with Pitzer's equation for the Gibbs excess energy of aqueous electrolyte solutions [22]. Furthermore, and as an extension of the model described previously the presence of carbon dioxide is also taken into account. As carbon dioxide is a weak acid its presence is particularly important in the high pH-range where the reaction of carbon dioxide to bicarbonate and carbonate can result in a rather large stoichiometric concentration of CO_2. The amount of dissolved neutral CO_2 in the liquid phase that coexists with the gel phase is approximated via Henry's law:

$$p_{CO_2} = m_{CO_2} H_{CO_2,w} \tag{12}$$

where p_{CO_2} is the partial pressure of CO_2 in the atmosphere (=33 Pa), m_{CO_2} is the molality of CO_2 in water, and $H_{CO_2,w}$ is Henry's constant on the molality scale for the solubility of CO_2 in pure water ($H_{CO_2,w}$ (298 K) = 2.98 MPa kg mol^{-1}[23]). Neutral CO_2 coexists with bicarbonate and carbonate:

$$CO_2 + H_2O \rightleftharpoons HCO_3^- + H^+ \tag{II}$$

$$HCO_3^- \rightleftharpoons CO_3^{2-} + H^+ \tag{III}$$

Furthermore, the autoprotolysis of water is taken into account:

$$H_2O \rightleftharpoons OH^- + H^+ \tag{IV}$$

The chemical reaction equilibria of these reactions in the liquid phase that coexists with the gel phase are expressed on the molality scale:

$$K_{II}^{(m)} = \frac{a_{HCO_3^-}^{I,(m)} a_{H^+}^{I,(m)}}{a_{CO_2}^{I,(m)} a_w} \tag{13}$$

$$K_{III}^{(m)} = \frac{a_{CO_3^{2-}}^{I,(m)} a_{H^+}^{I,(m)}}{a_{HCO_3^-}^{I,(m)}} \tag{14}$$

$$K_{IV}^{(m)} = \frac{a_{OH^-}^{I,(m)} a_{H^+}^{I,(m)}}{a_w} \tag{15}$$

where $a_i^{I,(m)}$ is the activity of species i on the molality scale in the liquid phase I.

The corresponding chemical reaction equilibrium constants (for $T = 298$ K) $K_{II}{}^{(m)}$ ($=4.49 \cdot 10^{-7}$), $K_{III}{}^{(m)}$ ($=4.60 \cdot 10^{-11}$) and $K_{IV}{}^{(m)}$ ($=1.004 \cdot 10^{-14}$) were taken from the literature ([24–26]).

The hydrogel phase is replaced by an aqueous mixture of polymer chains (as the hydrogel is virtually cut off at all crosslinking positions), ions absorbed from the coexisting liquid and counterions provided by the comonomers of the network. As in the previous publication (with MAPTAC instead of DMAPMA) several chemical reactions are taken into account. We assume that in the polymerizing solution – but before the polymerization/crosslinking occurs – DMAPMA neutralizes AMPS resulting in an organic ammonium salt of propane sulfonic acid:

$$\text{AMPS} + \text{DMAPMA} \rightarrow \left(\text{DMAPMAH}^+/\text{AMPS}^-\right) \quad \text{(V)}$$

The neutral reaction product and the comomoner (i.e., either AMPS or DMAPMA) that is present in surplus are incorporated into the polymer chains.

The surplus of AMPS is partially dissociated:

$$\text{AMPS} \rightleftharpoons \text{AMPS}^- + \text{H}^+ \quad \text{(VI)}$$

The surplus of DMAPMA is partially protonated (cf., Eq. I)

$$\text{DMAPMA} + \text{H}^+ \rightleftharpoons \text{DMAPMAH}^+ \quad \text{(I)}$$

Reactions I and VI are assumed to be equilibrium reactions. The chemical equilibrium constant of reaction (VI)

$$K_{VI}^{(m)} = \frac{a_{AMPS^-}^{II,(m)} a_{H^+}^{II,(m)}}{a_{AMPS}^{II,(m)}} \quad (16)$$

is adopted from a previous publication [12],whereas $K_I^{(m)}$ is taken from Eq. 8.

Protonated DMAPMA can react with chloride to a neutral salt:

$$\text{DMAPMAH}^+ + \text{Cl}^- \rightleftharpoons \text{DMAPMAH}^+\text{Cl}^- \quad \text{(VII)}$$

The neutral product from reaction (V) is assumed to undergo a dissociation reaction that yields an anionic group in the network:

$$\text{DMAPMAH}^+/\text{AMPS}^- \rightleftharpoons \text{DMAPMA}/\text{AMPS}^- + \text{H}^+ \quad \text{(VIII)}$$

Furthermore, the anionic species $AMPS^-$ is assumed to react with sodium ions to a neutral salt:

$$\text{AMPS}^- + \text{Na}^+ \rightleftharpoons \text{NaAMPS} \quad \text{(IX)}$$

The chemical reaction equilibrium constant $K_{IX}^{(m)}$

$$K_{IX}^{(m)} = \frac{a_{NaAMPS}^{II,(m)}}{a_{AMPS^-}^{II,(m)} a_{Na^+}^{II,(m)}} \quad (17)$$

was also adopted from Ninni et al. [12].

Finally, it is assumed that $DMAPMA/AMPS^-$ reacts with Na^+ to a neutral species:

$$\text{DMAPMA}/\text{AMPS}^- + \text{Na}^+ \rightleftharpoons \text{DMAPMA}/\text{NaAMPS} \quad \text{(X)}$$

The chemical reaction equilibrium constant of reaction X is expressed by:

$$K_X^{(m)} = \frac{a_{DMAPMAH/NaAMPS}^{II,(m)}}{a_{DMAPMAH/AMPS^-}^{II,(m)} a_{Na^+}^{II,(m)}} \quad (18)$$

When, for example, a hydrogel with a surplus of DMAPMA (i.e., $y_{DMAPMA} > y_{AMPS}$) is equilibrated in an aqueous solution of NaCl, there are 13 species in the gel phase: (a) water, (b) non-ionic polymer chains of (N-IPAAm + MBA), (c) 4 neutral solutes ($DMAPMAH^+/AMPS^-$, CO_2, DMAPMA and $DMAPMAH^+Cl^-$), (d) 2 cationic solutes (Na^+ and H^+), (e) 5 anionic solutes (Cl^-, OH^-, HCO_3^-, CO_3^{2-} and $DMAPMA/AMPS^-$) – see also Fig. 6. When vice versa, a hydrogel with a surplus of AMPS (i.e., $y_{DMAPMA} < y_{AMPS}$) is equilibrated in an aqueous solution of NaCl, there are 14 species in the gel phase: (a) water, (b) non-ionic polymer chains of (N-IPAAm + MBA), (c) 4 neutral neutral solutes (AMPS, $DMAPMAH^+/AMPS^-$, CO_2 and NaAMPS), (d) 2 cationic solutes (Na^+ and H^+), (e) 6 anionic solutes (Cl^-, OH^-, HCO_3^-, CO_3^{2-}, $DMAPMA/AMPS^-$ and $AMPS^-$).

The chemical reaction equilibria of reactions I, VII and VIII in the gel phase (i.e., phase *II*) are expressed via the corresponding chemical reaction equilibrium constants on the molality scale:

$$K_I^{(m)} = \frac{a_{DMAPMAH^+}^{II,(m)}}{a_{DMAPMA}^{II,(m)} a_{H^+}^{II,(m)}} \quad (19)$$

$$K_{VII}^{(m)} = \frac{a_{DMAPMAH^+Cl^-}^{II,(m)}}{a_{DMAPMAH^+}^{II,(m)} a_{Cl^-}^{II,(m)}} \quad (20)$$

$$K_{VIII}^{(m)} = \frac{a_{DMAPMA/AMPS^-}^{II,(m)} a_{H^+}^{II,(m)}}{a_{DMAPMAH^+/AMPS^-}^{II,(m)}} \quad (21)$$

The chemical reaction equilibrium constants $K_{VII}^{(m)}$, $K_{VIII}^{(m)}$ and $K_X^{(m)}$ are adjustable parameters of the model. However, $K_{VIII}^{(m)}$ was approximated by using the chemical reaction equilibrium constant of reaction I:

$$K_{VIII}^{(m)} = 1/K_I^{(m)} \quad (22)$$

and $K_X^{(m)}$ was approximated by $K_{IX}^{(m)}$:

$$K_X^{(m)} = K_{IX}^{(m)} \quad (23)$$

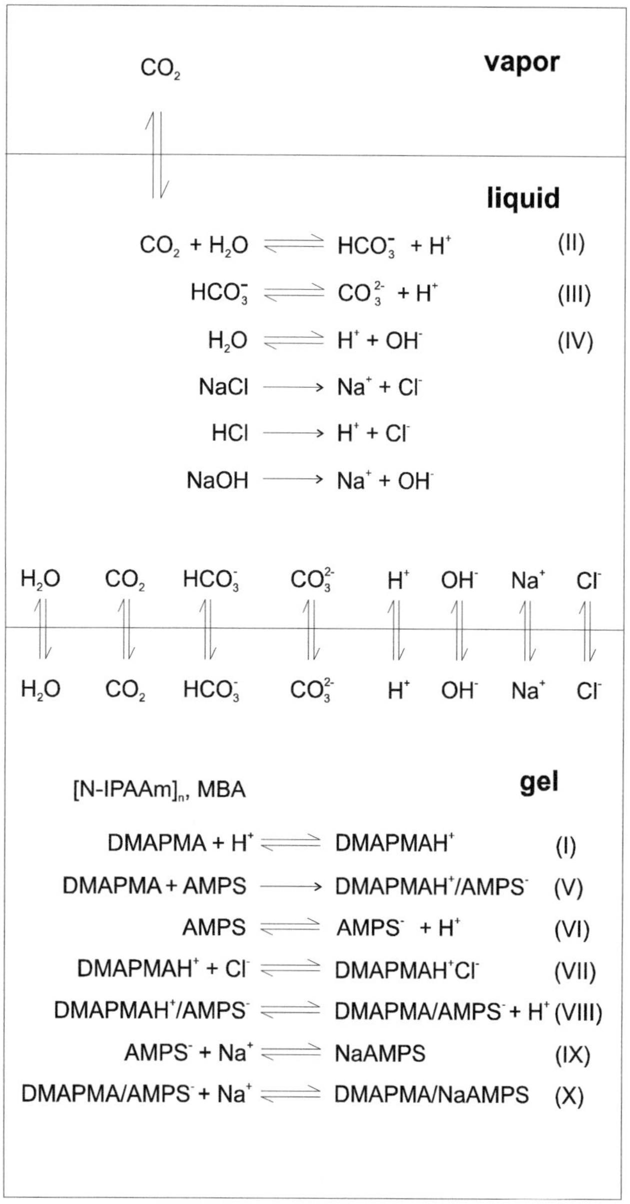

Fig. 6 Representation of the equilibrium and chemical reactions of a system containing a N-IPAAm/DMAPMA/AMPS hydrogel immersed in an aqueous coexisting liquid phase in contact with the atmosphere

i.e., the neutralization of DMAPMA/AMPS$^-$ by Na$^+$ was treated as the neutralisation of AMPS$^-$ by Na$^+$.

The activity coefficient of any solute species i that is present in the gel phase is

$$a_i^{II,(m)} = \gamma_i^{II,(m)} \frac{m_i^{II}}{m_0} \tag{24}$$

where $\gamma_i^{II,(m)}$ and m_i^{II} are the activity coefficient (on the molality scale) and the molality of solute species i in the gel phase and $m_0 = 1$ mol $\cdot$ (kg H_2O)$^{-1}$. These activity coefficients are calculated from the Virial Equation with Relative Surface fractions (VERS) model for the excess Gibbs energy of aqueous solutions of polymers and electrolytes. When no polymers are dissolved, the VERS model reduces to Pitzer's model for aqueous, low molecular electrolyte solutions [22]. Details of the VERS model were given elsewhere [12, 14] and are not repeated here. The activity of water (i.e., the solvent) follows from the activity of the solute species via the Gibbs-Duhem equation.

Correlations and Predictions

The calculations were performed assuming that one gram of a dry hydrogel is equilibrated in two kilograms of aqueous solution at 298 K. The compositions of the gel and the aqueous solution are known from the preparations. The compositions of the coexisting phases were calculated by solving the phase equilibrium conditions and the mass balances (for all components with the exception of dissolved CO_2). During equilibration it was assumed that the molality $m^I_{CO_2}$ of physically dissolved CO_2 in the aqueous solution that coexists with the gel phase is constant (and follows from Eq. 12). However, as physically dissolved CO_2 might be converted to $HCO_3{}^-$ and $CO_3{}^{2-}$ the total amount of dissolved CO_2 $m^I_{CO_2,tot}$ can be different as

$$m^I_{CO_2,tot} = m^I_{CO_2} + m^I_{HCO_3^-} + m^I_{CO_3^{2-}} \tag{25}$$

Model Parameters

Besides the parameters already introduced before, the model requires the following information:

- Volume of the hydrogel in its stress-free V_0 and swollen state V;
- Pure component parameters (molar volumes v_k and surface parameters q_k) for $k =$ water and the solute species;
- Network parameter C;
- Interaction parameters between solute species in water;
- Chemical reaction equilibrium constant for reaction VII.

The volume of the hydrogel (in the stress-free state V_0 and in the swollen state V) is calculated from the molar volumes of the components assuming an ideal liquid mixture:

$$V_0 = \sum_i n_i^0 v_i + n_w^0 v_w \tag{26}$$

where, $i = H_2O$, N-IPAAm, MBA, DMAPMAH$^+$/AMPS and either DMAPMA or AMPS and

$$V = \sum_j n_j v_j + n_w v_w \tag{27}$$

Table 2 Pure component parameters

Species	Molar mass (g mol^{-1})	Molar volume v_i (cm^3 mol^{-1})	Surface parameter q
DMAPMA	170.26	112.76	6.16
DMAPMAH$^+$	171.26	116.51	6.46
DMAPMAH$^+$Cl$^-$	206.71	131.41	6.46
DMAPMAH$^+$/AMPS$^-$	377.51	228.43	12.66
DMAPMA/AMPS$^-$	376.51	229.02	12.36
DMAPMA/NaAMPS	399.50	227.81	13.76
CO_2	44.01	24.12 [27]	1.12
HCO_3^-	61.01	23.4 [27]	1.40
CO_3^{2-}	60.01	−4.3 [27]	1.40

Table 3 Binary and ternary interaction parameters of the VERS model and chemical reaction equilibrium constants used in this work

Parameters adjusted in ionic hydrogel systems		Chemical reaction constants	
$a^{(0)}_{N-IPAAm,DMAPMAH^+}$	0.38210	$K_I{}^{(m)}$	$1.60 \cdot 10^9$
$a^{(0)}_{N-IPAAm,DMAPMAH^+/AMPS^-}$	0.02600	$K_{II}{}^{(m)}$	$4.49 \cdot 10^{-7}$ [23]
$a^{(0)}_{N-IPAAm,DMAPMA/AMPS^-}$	0.10024	$K_{III}{}^{(m)}$	$4.60 \cdot 10^{-11}$ [23]
Parameters adopted from Pitzer [22]		$K_{VII}{}^{(m)}$	7.0
$a^{(0)}_{H^+,Cl^-}$	0.17689		
$a^{(1)}_{H^+,Cl^-}$	0.29723		
b_{H^+,H^+,Cl^-}	0.00024		
$a^{(0)}_{Na^+,OH^-}$	0.08636		
$a^{(1)}_{Na^+,OH^-}$	0.25300		
b_{Na^+,Na^+,OH^-}	0.00137		

where $j = H_2O$, N-IPAAm, MBA, (either DMAPMA or AMPS), DMAPMAH$^+$/AMPS$^-$, DMAPMA/AMPS$^-$, DMAPMA/NaAMPS, AMPS$^-$, NaAMPS, DMAPMAH$^+$, DMAPMAH$^+$Cl$^-$, CO_2, HCO_3^-, CO_3^{2-}, H^+, OH^-, Na^+, and Cl^- . Most molar volumes (and surface parameters) were adopted from Ninni et al. [12]. All other pure component parameters are given in Table 2. As the surface parameter of any low-molecular ionic species (i.e., Na^+, Cl^-, H^+, OH^-, HCO_3^-, CO_3^{2-}) was set to the surface parameter q_w of water, the VERS-model for aqueous solutions of the corresponding low molecular electrolytes corresponds to Pitzer's model for such aqueous electrolyte solutions. The surface parameters for all other solute species were estimated from the method of Bondi [28].

The network parameter C (see Eq. 11) that characterizes the elastic properties of the hydrogels was adjusted to give – in combination with the equation of the excess Gibbs energy of the liquid phase – a good correlation of the degree of swelling of the ionic hydrogels in pure water. The results are summarized in Table 1.

The VERS-model requires binary ($a^{(0)}_{i,j}$, $a^{(1)}_{i,j}$) and ternary ($b_{i,j,k}$) parameters for interactions between solute species [12]. The binary and ternary parameters for interactions between sodium and chloride ions, sodium and hydroxide ions as well as between hydrogen and chloride ions were taken from Pitzer [22]. The binary interaction parameter between CO_2 and HCO_3^- was taken from Pérez-Salado et al. [23]. Binary interaction parameters between MBA and Na^+ as well as those between N-IPAAm (on one side) and either N-IPAAm, or Na^+, or NaAMPS, or AMPS$^-$ (on the other side) were adopted from a previous work [12]. The binary parameters for interactions between N-IPAAm (on one side) and DMAPMAH$^+$, DMAPMAH$^+$/AMPS$^-$ or DMAPMAH/AMPS$^-$ (on the other side) were adjusted to the experimental results for the degree of swelling of some selected hydrogels of the present work. All further binary ($a^{(0)}_{i,j}$ and $a^{(1)}_{i,j}$) and ternary ($b_{i,j,k}$) interaction parameters were set to zero. All non-neglected interaction parameters are given in Table 3.

Correlation and Prediction of the Swelling Equilibrium of N-IPAAm/DMAPMA/AMPS Hydrogels in Aqueous Solutions of NaCl

The adjustable model parameters (interaction parameters and chemical reaction equilibrium constant for reaction VII) were fitted (together with the particular network parameters) to experimental results for the equilibrium swelling of gel no. 1 in aqueous solutions of NaCl (cf. Fig. 2) and on the influence of pH on the swelling of gel 4 in water (cf. Fig. 4). The resulting numerical values for those model parameters (interaction parameters and chemical reaction equilibrium constant for reaction VII) were then used to predict the swelling behavior of the other gels. The procedure applied here was similar to that described by Ninni et al. [12]. However, in a first approach, the influence of dissolved CO_2 was neglected, i.e., it was assumed that neither CO_2, nor HCO_3^- and CO_3^{2-} were present in the solutions. However, the prediction results for the degree of swelling revealed rather large deviations from the experimental data. Considering the presence of CO_2 in the solutions resulted in a considerable improvement. The acidification of the medium caused by dissolved CO_2 shifts the chemical reaction equilibrium between DMAPMA and DMAPMAH$^+$ in favor of the protonated species (i.e., the gel carries a higher net charge and reveals a larger degree of swelling). The two reductions of the degree of swelling with increasing concentration of NaCl in the coexisting liquid (cf. Fig. 2) are caused by a) chemical reactions which transfer the previously ionic gel into a neutral gel, i.e., a hydrogel composed of N-IPAAm, DMAPMA and DMAPMAH$^+$Cl$^-$ and b) the collapse of the hydrogel at high salt concentrations which resembles the phase transition observed when a non-ionic

N-IPAAm hydrogel is dissolved in aqueous solutions of sodium chloride. Besides the network parameter C of that particular gel, three parameters were fitted: the binary parameter $a^{(0)}_{N-IPAAm,DMAPMAH^+}$ for interactions between N-IPAAm and $DMAPMAH^+$ that accounts for interactions between N-IPAAm on one side and protonated DMAPMA on the other side; the binary parameter $a^{(0)}_{N-IPAAm,DMAPMAH^+/AMPS^-}$ (that accounts for interactions between N-IPAAm and the neutral species $DMAPMAH^+/AMPS^-$) and the chemical reaction constant $K_{VII}{}^{(m)}$ which accounts for the neutralization of $DMAPMAH^+$ by chloride ions. The resulting numerical values for those parameters are given in Table 3. Figure 2 shows a comparison between the experimental data and the calculation results for 3 gels (gels no. 1, 2 and 4) with $y_{MBA} = 0.02$. Only the network parameter C was adjusted to describe the experimental results for gels 2 and 4, whereas all other model parameters were adopted as described above. The comparisons reveal the good agreement between correlation/prediction results on one side and the new experimental data on the other side. Further comparisons are shown in the Supplementary Materials.

Prediction of the Swelling Equilibrium of N-IPAAm/DMAPMA/AMPS Hydrogels in Different pH Values

The prediction of the swelling experimental data of N-IPAAm/DMAPMA hydrogels at different pH values was performed using the model (and parameters) described in the preceding section. However, for hydrogels of (N-IPAAm + DMAPMA + AMPS) a single additional binary parameter for interactions between N-IPAAm groups and $DMAPMA/AMPS^-$ was adjusted to experimental data for the influence of pH on the degree of swelling of gel 4 (shown in Fig. 4). That parameter has practically no influence on the swelling of such gels in neutral aqueous solutions of NaCl. The model is able to describe qualitatively the maxima of the degree of swelling of gel no. 4 at $pH \approx 4$ and $pH \approx 11$, but the correlation results for the degree of swelling at those pH-numbers are about 25 % below the experimental results. Figure 3 shows a comparison between the experimental results for the influence of pH on the degree of swelling q of gel no. 6 (which does not contain any AMPS) and prediction results. In accordance with the experimental data the model predicts a maximum of q at $pH \approx 4$, but again the predictions for q are about 25 % below the experimental results. We suppose that these differences are caused by the experimental conditions which did not adequately consider the influence of dissolved CO_2 on the experimental results. Further experiments (without dissolved CO_2 as well as with carefully controlled amounts of dissolved CO_2) are recommended. More comparisons between experimental results and predictions are given as Supplementary Materials (Online resource) for ionic hydrogels swollen in sodium chloride aqueous solutions as well as in aqueous solutions at different pH values. Figure 5 shows a comparison between experimental results and predictions for the pH ($pH_{initial}$ and pH_{final}) which confirm that the influence of dissolved CO_2 cannot be neglected.

Conclusions

The comparisons between calculation results and new experimental data for the degree of swelling of hydrogels composed of a neutral monomer (N-IPAAm), a strong anionic comonomer (AMPS) and a weak cationic comonomer (DMAPMA) reveal that a previously developed thermodynamic models allows to correlate (and even predict) the influence of NaCl and pH on the swelling of such ionic hydrogels in aqueous solutions. The calculation results also revealed that dissolved CO_2 might have a considerable influence on the degree of swelling. Further work on that topic as well as on the swelling behavior of N-IPAAm-hydrogels with other comonomers is recommended.

Acknowledgments The authors appreciate financial support by Deutsche Forschungsgemeinschaft (DFG), Bonn-Bad Godesberg, Germany, within the priority program "Intelligente Hydrogele" (SPP 1259).

Electronic Supplementary Material

Below is the link to the electronic supplementary material. Ninni et al suppl information.docx

References

1. Gil ES, Hudson SM (2004) Prog Polym Sci 29:1173
2. Richter A, Paschew G, Klatt S, Lienig J, Arndt K-F, Adler H-J P (2008) Sensors 8:561
3. Çaykara T, Kiper S, Demirel G (2006) J Appl Polym Sci 101:1756
4. Orlov Y, Xu X, Maurer G (2006) Fluid Phase Equilib 249:6
5. Zrinyi M (2000) Colloid Polym Sci 278:98
6. Suzuki A (1993) Adv Polym Sci 110:199
7. Qiu Y, Park K (2001) Adv Drug Deliv Rev 53:321
8. Gupta P, Vermani K, Garg S (2002) Drug Discov Today 7:569
9. Wang Y, Shen Y, Zhang Y, Yue B, Wu C (2006) J Macromol Sci Part B Phys 45:563
10. Sakiyama T, Takata H, Toga T, Nakanishi K (2001) J Appl Polym Sci 81:667
11. Vamvakaki M, Palioura D, Spyros A, Armes SP, Anastasiadis SH (2006) Macromolecules 39:5106

12. Ninni L, Ermatchkov V, Hasse H, Maurer G (2013) Fluid Phase Equilib 337:137
13. Maurer G, Prausnitz JM (1996) Fluid Phase Equilib 115:113
14. Ermatchkov V, Ninni L, Maurer G (2010) Fluid Phase Equilib 296:140
15. Horta A, Jesús Molina M, Rosa Gómez-Antón M, Piérola IF (2009) Macromolecules 42:1285
16. Horta A, Piérola IF (2009) J Phys Chem B 113:4226
17. Pérez-Salado Kamps A, Maurer G (1996) J Chem Eng Data 41:1505
18. Van de Wetering A, Moret EE, Schuurmans-Nieuwenbroek NME, van Steenbergen MJ, Hennink WE (1999) Bioconjugate Chem 10:589
19. James HM (1947) J Phys Chem 15:651
20. James HM, Guth EJ (1949) Polym Sci 4:153
21. James HM, Guth EJ (1953) Chem Phys 21:1039
22. Pitzer KS (1991) Activity coefficients in electrolyte solutions, 2nd edn. CRC Press, Boca Raton
23. Pérez-Salado Kamps A, Balaban A, Jödecke M, Kuranov G, Smirnova NA, Maurer G (2001) Ind Eng Chem Res 40:696
24. Patterson CS, Slocum GH, Busey RH, Mesmer RE (1982) Geochim Cosmochim Acta 46:1653
25. Patterson CS, Busey RH, Mesmer RE (1984) J Solution Chem 13:647
26. Edwards TJ, Maurer G, Newman J, Prausnitz JM (1978) AIChE J 24:966
27. Millero FJ (1971) Chem Rev 71:147
28. Bondi A (1964) J Phys Chem 68:441

Thermodynamic Modelling of Hydrogel Systems

Markus C. Arndt and Gabriele Sadowski

Abstract

In the present work, the PC-SAFT model was used and extended for modelling poly(N-isopropylacrylamide) PNIPAAm and its behaviour as loose polymer chains in solution and as cross-linked hydrogel. For the application of PC-SAFT to hydrogel modelling, the elastic forces which characterise swellable polymer networks were taken into account by a specific additional contribution to the Helmholtz energy. The binary system of water/PNIPAAm as polymer solution and hydrogel as well as the influence of various additives on the gel was extensively studied. These additives comprised organic solvents (methanol, ethanol, acetone), poly(ethylene glycol) as well as inorganic electrolytes (alkali halogenides and sodium nitrate). A method for predicting their influence on the re-entrant behaviour of PNIPAAm hydrogels was developed and the model capability of the extended PC-SAFT was shown for modelling of temperature, component and composition sensitivity.

Keywords

PNIPAAm • Swelling • Hydrogel • Organic solvents • Salts

Introduction

Hydrogels are networks of hydrophilic polymers which are capable of absorbing large amounts of water and often also other polar solvents. This super-absorber capability allows for a water uptake which can exceed the polymer weight by a 1,000 fold, clearly suggesting the well-established main application as drying agents. In the recent years, much research has been undertaken in the field of so-called smart hydrogels, referring to polymer networks which may consist of e.g. blended or specially functionalised polymers, copolymers or interpenetrating networks. Their characterisation as smart polymers often denotes their responsiveness to external stimuli, such as pH, temperature, or presence of certain substances. In combination with the often given biocompatibility of hydrogels a wide field in biomedical applications as e.g. drug depots or wound moisture regulators as well as micromechanical sensors and actuators is opened up.

A prominent example of these polymers is poly(N-isopropylacrylamide), PNIPAAm, which has attracted particular attraction due to its biocompatible character even in in-vivo experiments [1] and its remarkable thermodynamic behaviour in the vicinity of body temperature: for PNIPAAm chains in mixture with water, a lower critical solution temperature (LCST) is found, allowing for a homogenous mixture at low temperatures, and a phase separation into a polymer-rich phase and a nearly-pure water phase when rising the system temperature. The liquid-liquid equilibrium (LLE) of this system forms above the LCST of about 30 °C [2].

This particular behaviour can be understood as caused by preferential hydrogen bonding. Whereas the water-soluble polymer forms hydrogen bonds predominantly with its amide-structured side chain and the surrounding water molecules, the water-depleted polymer phase in the two-phase system is presumed to be characterised

M.C. Arndt • G. Sadowski (✉)
Laboratory of Thermodynamics, Department of Biochemical and Chemical Engineering, TU Dortmund, Emil-Figge-Str. 70, 44227 Dortmund, Germany
e-mail: g.sadowski@bci.tu-dortmund.de

G. Sadowski and W. Richtering (eds.), *Intelligent Hydrogels*, Progress in Colloid and Polymer Science 140, DOI 10.1007/978-3-319-01683-2_14,

by the preferential polymer-polymer annealing and intra-species hydrogen bond formation, exposing the hydrocarbon backbone of the chain. It is evident that the latter phenomenon leads to a higher hydrophobicity of the whole chain structure and thus facilitates the phase split.

Comparably to provoking the demixing behaviour by variation of the temperature, the influence of additional substances to the binary mixture of PNIPAAm and water has been examined widely. By addition of various organic and inorganic substances, the LCST of this system can be shifted to lower values and demixing behaviour can be regulated precisely [3–5]. A remarkable effect in this connection is that e.g. the not-cross-linked PNIPAAm is completely soluble in alcohols (C_1–C_4) at room temperature; same as in pure water. In water/PNIPAAm/alcohol mixtures of certain concentration, however, a phase split is observed. Winnik et al. have widely examined this behaviour, termed co-nonsolvency, in the past years in particular focussing on the influence of methanol [5–7].

Analogue to the binary system water/PNIPAAm, the explanation for the co-nonsolvency in ternary systems regards the hydrophobicity of PNIPAAm, even though the actual mechanism is not completely understood. One approach assumes a complexation of the cosolvent molecules with water. These complexes then serve as poor solvent for the polymer – in contrast to both pure solvents [8, 9]. Similarly, as causation for co-nonsolvency was proposed that the water/alcohol interactions are thermodynamically preferred to water/polymer interactions and thus not enough water molecules are available for hydrogen bonding with PNIPAAm [10]. Another explanation was provided by Tanaka et al. who assumed a cooperative hydration of the polymer. They suggested that association of one component species to the polymer energetically facilitate the formation of the same type of bonds in the direct vicinity. This behaviour would lead to a pearl-necklace type structure of the polymer chains [7].

In our recent publication we contributed another aspect by examining the extent of hydrogen bond formation of water, alcohol and PNIPAAm in systems where these three associating species were involved [11]. Calculations showed that in the polymer-rich phase a significantly lower amount of hydrogen bondsis formed than in the solvent-rich phase. Furthermore, our calculations and comparisons with experimental results showed a moderate dependence of PNIPAAm behaviour on the chains length of the alcohols. This promotes the idea of hydration of cosolvent molecules since larger alcohol species require more water molecules for a hydration sheath and thus have a stronger influence at the same molar concentration.

A direct comparison of PNIPAAm LLE systems on the one hand and of gel systems on the other hand shows that the behaviour is highly comparable in the thermodynamic point of view. Notably, in the solution the polymer is not cross-linked, whereas the swellable gel system implies interconnections of the polymer chains to a network by cross-link agents. Regardless of the number of species involved, corresponding to the LLE region exhibiting a phase split, the gel system is in a shrunken state with minimum solvent content but a very high polymer concentration. By decreasing the temperature or changing the solution composition a collapse of the phase split and obtaining a single homogenous liquid phase can be provoked. Analogue, the gel system exhibits an often dramatic change to swollen states with very little PNIPAAm concentration and a high solvent content exposing the typical super-absorber qualities. The gel PNIPAAm chains are driven to dissolve in the solution phase but are inhibited by their terminal cross-linkers. Corresponding to the co-nonsolvency effect of polymer solutions, the term of re-entrant behaviour has been established to describe high degrees of swelling in both pure solvents but gel shrinkage in mixtures of these.

Main differences in the thermodynamic evaluation of such polymer solutions and polymer gel systems are thus circumscribed by the effects of the cross-links in the polymer network. Physically, the interconnection of polymer chains has three consequences: Firstly, the actual molecular weight of the polymer takes arbitrarily high values; secondly the chemical species of the polymer must be accounted for in one phase (gel phase) only; and thirdly due to the cross-linking the interaction and configuration of the chains are hindered in their free motion, which has to be accounted for. The latter aspect can easily be epitomised by the image of solvent molecules invading a swelling polymer network and stretching the chains. Similarly to an elastic spring, a restoring force in the chains counteracts their mobility and stretching, since their ends are restrained from free spatial motion by the cross links.

Many thermodynamic theories have been developed to account for this restoring force expressed as additional energy in the system, predominantly by adding an extra elastic energy contribution to either the Gibbs or Helmholtz energy of the system. The early and ground-breaking research of Flory [12–15] as well as of Guth [16–18] on statistical thermodynamics of polymer networks gave the basis for the most-followed elasticity theories: the affine network and the so-called phantom network, which both allow for the calculation of the elastic response in a cross-linked polymer system. Both theories however differ in their view on the cross-link spots in the network. The affine network regards them as fixed and embedded in the network structure relative to each other, whereas the phantom network allows their fluctuation, so that the influence of entanglements of chains is neglected.

A significant improvement of the elastic energy has recently been given by Miao et al. [19] by accounting for the non-infinite stretchability of the polymer chains. This in contrast to the classical approach gives a physical and highly reasonable restriction to the allowed stretching of the polymer network.

Mostly based on such elasticity considerations there are various theoretical and empirical methods for a thermodynamic description of elastic polymer systems [7, 20, 21], in particular regarding the degree of swelling as response to external influences. In only few approaches though, such as in the work by Maurer and Prausnitz [22] with modifications of the Flory-Huggins equation or the UNIQUAC model, a general thermodynamic tool for the description of multicomponent phase behaviour like VLE or LLE has been taken and expanded to the descriptive capability with polymer network systems such as hydrogels. In their approach, equilibrium calculations of swelling hydrogels are performed like in LLE calculations, but with the elasticity of the gel phase regarded in a network contribution picturing the gel enclosed by an elastic membrane impermeable for the polymer species but penetrable by solvents. This approach has been used for examination of various systems and has also been followed e.g. by the group of Enders et al. [23, 24].

In our work we focus on the thermodynamic description of hydrogels in binary and multicomponent systems by the combination of an equation of state with an elasticity contribution to the Helmholtz energy. Our approach allows for a multifaceted analysis of hydrogels, regarding internal aspects such as gel structure or associative interactions as well as external factors such as temperature or the influence of organic and inorganic substances on the swelling behaviour and gel composition.

Model

Other than the prevalent approach of using an excess Gibbs energy model, we proposed the usage of the Perturbed-Chain Statistical Associating Fluid Theory (PC-SAFT), an equation of state which we combined with an explicit consideration of the elastic contribution caused by the cross links [11].

PC-SAFT has been developed by Groß and Sadowski [25–27] and comprises an approach from statistical thermodynamics to describe the residual Helmholtz energy $A^{residual}$ of a system. This quantity describing the difference between the ideal-gas state and the real system is calculated as superposed from a hard-chain term accounting for the molecular structure and repulsive molecular forces, a dispersive contribution denoting short-range molecular attraction such as Van-der-Waals forces, and an optional association contribution, which may account for molecules which are able to form hydrogen bonds. When modelling electrolyte systems, Cameretti et al. [28, 29] proposed an additional contribution based on the Debye-Hückel theory and established the ePC-SAFT model. According to the above-described elastic forces in cross-linked polymer networks, our proposal was the addition of a new energetic contribution to the Helmholtz energy [11], which is based on the elasticity theory of Miao et al. [19].

Equation 1 gives an overview over the PC-SAFT contributions applied in this work.

$$A^{residual} = \underbrace{\underbrace{A^{hard\text{-}chain} + A^{dispersion} + A^{association}}_{\text{original PC-SAFT}} + A^{Coulomb}}_{\text{ePC-SAFT}} + A^{elastic} \quad (1)$$

Advantageously, in PC-SAFT or ePC-SAFT a set of maximal five pure-component parameters (segment number m_i^{seg}, segment diameter σ_i, dispersion energy u_i/k_B, association energy ε^{AiBi}/k_B and association volume κ^{AiBi}) can be used with simple combining rules for the calculation and prediction of arbitrary and complex fluid systems. For the example of the dispersion energy, the combining rule is given in Eq. 2, viz. merely the geometric mean of the pure component parameters. In some cases it may be corrected with the only binary parameter in ePC-SAFT, a non-zero k_{ij}.

$$u_{ij} = \sqrt{u_i \cdot u_j}\left(1 - k_{ij}\right) \quad (2)$$

Conveniently, with the expression for the perturbing contribution of the association term taken from the original SAFT equation by Chapman et al. [30] the degree of hydrogen bonding is evaluable. As shown in Eq. 3, the associative Helmholtz energy contribution, conveniently divided by the molecule number N, Boltzmann constant k_B and the temperature T, depends on two vector quantities only. These are the mole fractions x_i as well as X^{Ai}, the fraction of not-bonded association sites of the bonding type A at the molecular species i.

$$\frac{A^{assoc}}{Nk_BT} = \sum_i x_i \sum_{Ai}\left(\ln\left(X^{Ai}\right) - \frac{X^{Ai}}{2} + \frac{1}{2}\right) \quad (3)$$

The quantity of interest in analysing the extent of association, X^{Ai}, can be iteratively derived from the implicit Eq. 4, which includes the number density ρ and a summation over all components and association site types for the product of X^{Bj} and the binary quantity Δ^{AiBj}, which denotes the bonding strength between each two types of association sites. Δ^{AiBj} itself depends on the radial distribution function, the two presented association parameters, the segment volume as well as on temperature and is given elsewhere in detail, same as a deduction of ρ [31].

$$X^{Ai} = \left(1 + \rho \sum_j x_j \sum_{Bj} X^{Bj} \cdot \Delta^{AiBj}\right)^{-1} \tag{4}$$

Notably, the maximum number of association sites is predetermined by the chemical structure of the molecules and not an adjustable quantity. In this work, the 2B association scheme after Huang and Radosz has been applied for modelling hydrogen bonding species [31]. In this scheme, each molecule may have two different types of association sites, viz. donator and acceptor sites for the representation of hydrogen bonds, whereas identical association-site types cannot interact with each other.

The elastic contribution to the Helmholtz energy is elementary for modelling hydrogel systems. Based on the elastic-energy contribution proposed by Miao et al. [19], we developed in our previous work the expression of Eq. 5, in which $A^{elastic}$ is given depending on the mole fraction of the polymer chains $x_{polymer}$, the network functionality Φ and three volume quantities [11].

$$\frac{A^{elastic}}{Nk_BT} = x_{polymer} \cdot \frac{\Phi - 2}{\Phi}\left[\frac{3}{2} \cdot \frac{\left(\frac{V}{V_0}\right)^{\frac{2}{3}} - 1}{1 - \left(\frac{V}{V_{max}}\right)^{\frac{2}{3}}} - \ln\left(\frac{V}{V_0}\right)\right] \tag{5}$$

The Φ parameter denotes the number of occupied cross-linker binding sites per monomer. This quantity is the one adjustable parameter that is required for quantitative modelling of network swelling. Among the network defects there are various imperfections such as varying chain lengths, entanglements, unreacted binding sites of cross-linker molecules, dangling chain ends or chain loops, which all influence the structure and elasticity of the gel. Since common elasticity approaches postulate a tetrahedral network with monodisperse chain lengths, contemplation about the experimentally-provoked non-idealities in the network makes an adjustable quantity highly plausible. For the same reason, Maurer et al. introduced a blended network parameter substituting the term (Φ-2)/Φ in their similar elasticity term [32, 33].

V/V_0 represents the volume V of the swollen gel in reference to the volume of the pure PNIPAAm V_0. The introduction of V/V_{max} in the denominator limits the swellability of the gel due to the finite stretchability of the chains, which comprises the physically reasonable innovation proposed by Miao et al.

The gel volume V is obtained directly from knowledge of the number of polymer chains between the cross-links $n_{polymer}$, its concentration and number density according to Eq. 6. The reference state V_0 is chosen as the accurately defined state of $x_{polymer} = 1$.

$$V = \frac{n_{polymer}}{x_{polymer}} \frac{N}{\rho} \tag{6}$$

The maximum gel volume can be approximated with the assumption of a monodisperse polymer with chains of equal molecular weight, a postulation which is common in both phantom and affine network theories. Further V_{max} can be obtained on the basis of an ideal tetrahedral network in which chain lengths between the cross-links equal the product of monomers per chain and the monomer diameter as given in Eq. 7. The correlation of σ_i and its temperature-dependent pendant d_i is given in Eq. 8.

$$V_{max} = n_{cross\text{-}linker} N\left(\frac{\sqrt{2}^3}{2}\sin\left(\frac{109.5°}{2}\frac{\pi}{180°}\right) \cdot m^{seg}_{polymer} d_{polymer}\right)^3 \tag{7}$$

$$d_i = \sigma_i\left[1 - 0.12 \cdot \exp\left(-\frac{3u_i}{k_BT}\right)\right] \tag{8}$$

In the case of PNIPAAm, as given in Table 1, each monomer of the chain has exactly the segment number 1, thus justifying the approximation of the maximum chain length with the product of $m_{polymer}{}^{seg}$ and $d_{polymer}$.

The molecular weight of the polymer chains between the cross-links $M_{polymer}$, their number $n_{polymer}$ as well as the amount of the linking agents $n_{cross\text{-}linker}$ depend significantly on the experimentally given relation of NIPAAm monomers and cross-linker. A reasonable quantity which can easily be obtained from experimental procedures is the mole fraction y of cross-linking agents in the polymer network as given by Eq. 9. The Eqs. 10, 11, and 12 give the other introduced relevant quantities $n_{cross\text{-}linker}$, $n_{polymer}$ and $M_{polymer}$ required for the swelling calculations.

$$y = \frac{n_{cross\text{-}linker}}{n_{cross\text{-}linker} + n_{monomer}} \tag{9}$$

$$n_{cross\text{-}linker} = \frac{m_{monomer}}{M_{monomer}} \cdot \frac{y}{1-y} \tag{10}$$

$$n_{polymer} = \frac{\Phi}{2} n_{cross\text{-}linker} \tag{11}$$

$$M_{polymer} = \frac{m_{monomer} + n_{cross\text{-}linker} \cdot M_{cross\text{-}linker}}{n_{polymer}} \tag{12}$$

Table 1 Pure-component parameters and physical data

$M_{monomer}$	Molecular weight of NIPAAm monomer				g mol^{-1}	113.15		
$M_{cross\text{-}linker}$	Molecular weight of cross-linker N, N′-methylenebisacrylamide				g mol^{-1}	154.16		
Φ	PNIPAAm network functionality				–	2.24		
	M_i	m_i^{seg}	σ_i	u_i/k_B	N_i^{assoc}	ε^{AiBi}/k_B	κ^{AiBi}	Ref.
	Molecular weight	Segment number	Segment diameter	Dispersion -energy parameter	Number of association sites	Association -energy parameter	Association -volume parameter	
	g mol^{-1}	–	Å	K	–	K	–	
Water	18.015	1.20466	2.7927 + f(T)[a]	353.9449	2	2425.67	0.04509	[29]
PNIPAAm	$M_{polymer}$ (see Eq. 12)	1 per monomer	5.38	297.343	2 per monomer	175.00	0.045	[11]
MeOH	32.042	1.52552	3.23	188.905	2	2899.49	0.03518	[27]
EtOH	46.069	2.38267	3.17706	198.237	2	2653.39	0.03238	[27]
Acetone	58.08	2.8287	3.2497	250.26	0	–	–	[36]
PEG	variable	M_i * 0.0506	2.8999	204.6	4	1799.8	0.020	[35]
Na^+	22.99	1	2.8232	230	0	–	–	[34]
K^+	39.098	1	3.3417	200	0	–	–	[34]
Cl^-	35.453	1	2.8067	200	0	–	–	[34]
Br^-	79.904	1	3.0707	190	0	–	–	[34]
I^-	126.900	1	3.6672	200	0	–	–	[34]
NO_3^-	62.005	1	3.2988	130	0	–	–	[34]

[a] $\sigma_{water} = 2.7927\ \text{Å} + 10.11\ \text{Å} * \exp(-0.01775 * T/K) - 1.1417 * \exp(-0.01146 * T/K)$

With knowledge of the residual Helmholtz energy and its derivatives with respect to the mole fractions and the volume, fugacity coefficients of all species as well as the pressure are calculable. These quantities are essential for the calculation of swelling equilibria. Postulating the balance of the chemical potential of solvents and solutes in both gel and surrounding solution, the isofugacity criterion for the equilibrium can be expressed as in Eq. 13.

$$x_i^{sol}\varphi_i^{sol}p^{sol} = x_i^{gel}\varphi_i^{gel}p^{gel} \tag{13}$$

It is noteworthy, that in swollen gels p^{gel} differs from the ambient pressure p^{sol}. This fact finds its causation in the elastic force which is induced by the solvent molecules which stretch the polymer network. It is evident, that by the contracting force within the chain network the included solvent molecules are slightly compressed and thus the macroscopic pressure inside the gel is increased. The pressure difference between two phases is exactly the elastic pressure as given in Eq. 14. It can be obtained according to one of the characteristic functions of thermodynamics by the negative differentiation of the elastic Helmholtz energy $A^{elastic}$ with respect to the volume at constant temperature and composition.

$$p^{gel} = p^{sol} - p^{elastic} \tag{14}$$

Since in a ν-component mixture in a hydrogel system only ν-1 components, viz. all solutes and solvents but not the hydrogel polymer, are distributed in both phases, the equilibrium condition from the isofugacity criterion cannot provide ν conditions. Thus an additional νth equilibrium condition to account for the hydrogel is given with Eq. 14.

Results

Swelling of PNIPAAm in Pure Water

As shown before, a set of universal pure-component parameters has to be available for equilibrium calculations with PC-SAFT.

For water, a 2B association model as used by Held et al. was utilised [29]. In this water model a temperature dependent σ_i is included for the sake of describing the water density between 273 and 373 K quantitatively with high accuracy and especially the anomaly at 277 K. Considering hydrogel applications in the biomedical sector, this temperature range is of greatest interest. The used alcohol parameters, likewise following a 2B association, were also taken from open literature [27]. The parameter sets of the ionic species is reduced to two pure-component parameters since the segment number equals 1 and no association scheme is applied; they have most recently been published by Held et al. [34]. The PEG parameters are given by Stoychev et al. [35] who introduced a 2B association scheme with one acceptor and one donator site for each chain end. For PNIPAAm as well a 2B association scheme was chosen with each one donor and acceptor site per monomeric unit, and as described in our previous work the five pure-component

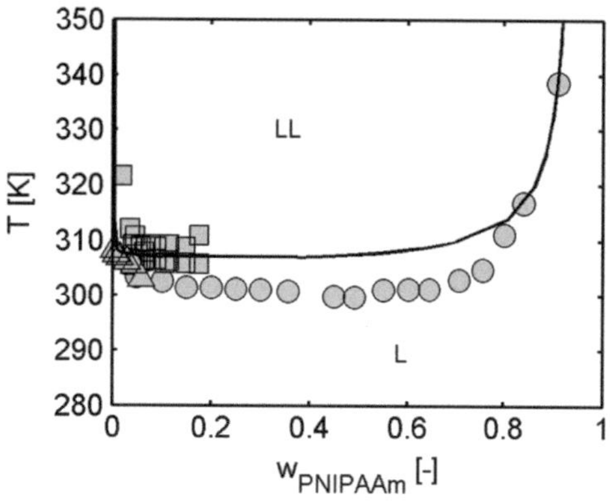

Fig. 1 LLE of the binary mixture water/PNIPAAm. *Symbols* are experimental data [2] for various molecular weights of PNIPAAm (*circles*: 53 kg mol^{-1}; *squares*: 615 kg mol^{-1}; *triangles* 2,100 kg mol^{-1}); the *line* represents the predicted binodal of a polymer with a molecular weight of 10 kg mol^{-1}, which equals $M_{polymer}$ in swelling calculations based on y = 0.01

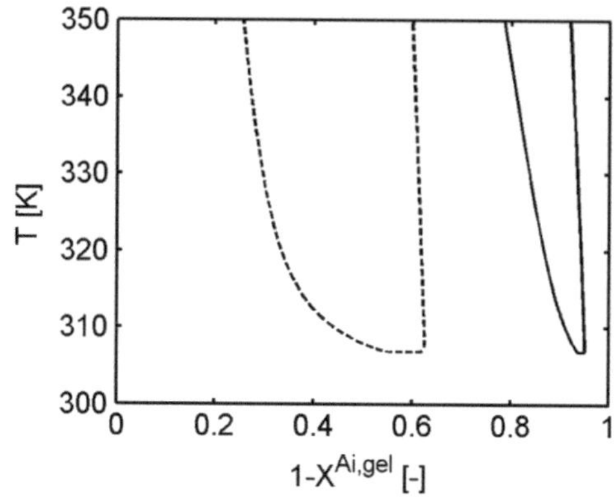

Fig. 2 Extent of association for water (*solid line*) and PNIPAAm (*dashed line*) in the two equilibrium phases. The *left* branches of the curves represent the polymer-rich phase; the *right* branches give the analogue for the polymer-depleted phase

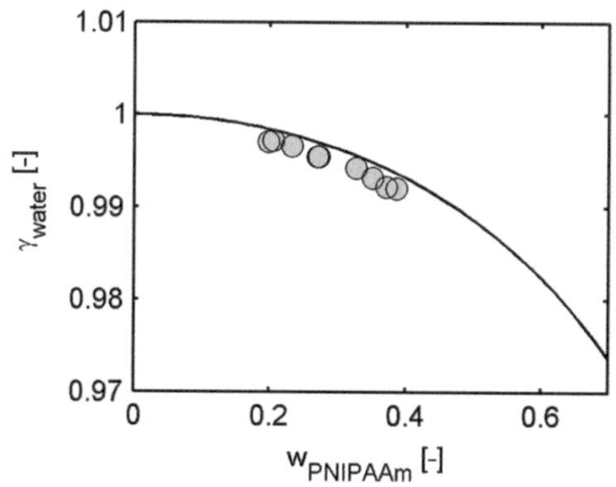

Fig. 3 Water activity coefficient at 298.15 K depending on the weight fraction of PNIPAAm with 52.3 kg mol^{-1}. *Symbols* are experimental data [32], the *line* our model prediction

parameters and the additional network parameter were fitted to aqueous mixture density data at varying temperatures and the binary swelling in water [11]. All used pure-component parameters are comprised in Table 1.

Figure 1 demonstrates the modelling results using the PNIPAAm parameters in not cross-linked polymer solutions in water, thus ignoring the subject of elasticity. It is evident, that with the parameters trained to swelling systems also polymer solutions can be modelled in good accordance to the literature data. The deviation between experimental and modelled values of the lower critical solution temperature is not negligible, but it is noteworthy that the experimental values do not follow a logical trend with their molecular weight and a general scattering of LCST values can be assumed. In the regime of the values of the experimentally examined polymer molecular mass, the calculated LCST decreases monotonically with increasing molecular weight by about 1 K, a physically reasonable and expected trend.

Since with PC-SAFT according to Eq. 4 a precise analysis of the extent of hydrogen bonding is accessible, Fig. 2 demonstrates the respective temperature dependence of the species in the above LLE system. For both, water and PNIPAAm, the average fraction of association sites involved in hydrogen bonds is given with the expression $1\text{-}X^{Ai}$. The analysis uncovers three principle facts. (I) The degree of association for the water molecules is generally higher than for the polymer species. This originates in the molecular size and structure, which suggests a comparably high sterical hindrance for the complex PNIPAAm chains. (II) There is a significant discrepancy in terms of association between the equilibrium phases, which supports the assumption that at elevated temperature the hydrophobicity of PNIPAAm is drastically increased. Particularly the PNIPAAm molecules show an up to 50 % reduced extent of association in the polymer-rich phase. The fraction of association in the solution phase, since nearly being composed solely of water, approximates the behaviour in pure water. (III) All curves follow the trend of fewer hydrogen bonds the higher the temperatures are. This is as expected due to higher kinetic energy and less close interactions of molecules.

Plausibility proof of the PNIPAAm parameters is also given by the polymer influence on the water activity. Figure 3 gives a prediction of the activity coefficient of water in the binary mixture with PNIPAAm at 298.15 K and shows a very good accordance with the literature data.

Having proven the reasonability of the parameters with the previous predictions in systems without the network contribution, in Fig. 4 the elastic contributions to Helmholtz energy and pressure are taken into account and the swelling equilibrium of PNIPAAm in pure water is given. The degree of swelling, viz. the mass of the swollen gel m in relation to the mass of the dry polymer m_0 as reference state, hereby trivially can be calculated from the reciprocal of the weight fraction of the polymer in the gel phase as

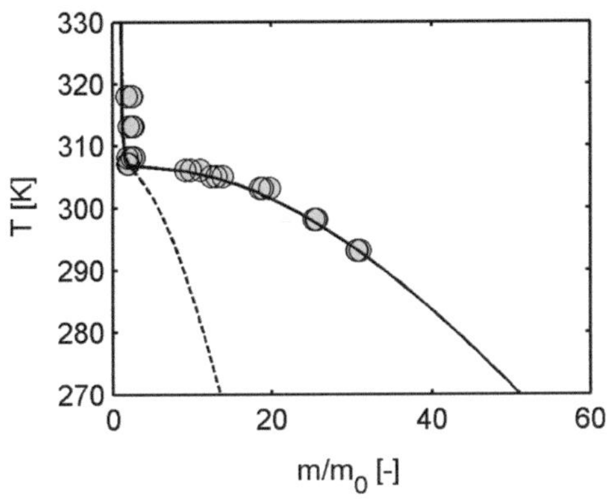

Fig. 4 Temperature dependent mass swelling of PNIPAAm gel in water with y = 0.01 and an ideal $\Phi = 4$ (*dashed line*) as well as $\Phi = 2.24$ (*solid line*), which was correlated to experimental data [37] (*symbols*)

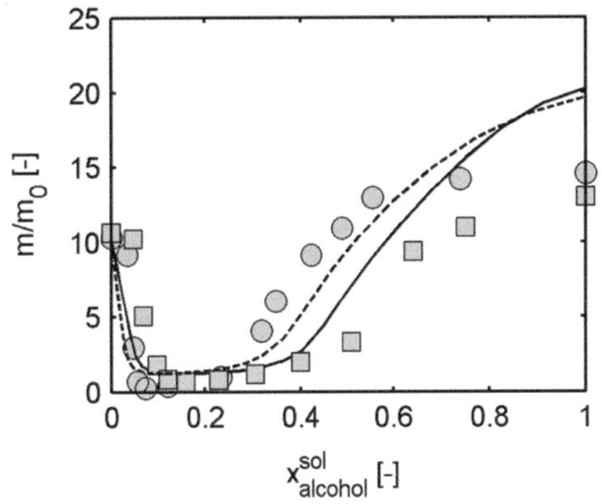

Fig. 5 Mass swelling of PNIPAAm at 298.15 K in aqueous solution of MeOH (*squares and solid line*) and EtOH (*circles and dashed line*). Lines are modelling results, symbols are experimental data [10]

$m/m_0 = w_{PNIPAAm}{}^{-1}$. The swelling prediction with an ideal network functionality of $\Phi = 4$ shows the expected swelling transition with a moderately swollen gel below the transition temperature. To match the high degrees of swelling from experimental data however, the value of Φ had to be decreased to 2.24, as adjusted parameter for the merged influences of all network errors and imperfections as well as compensation for insufficiencies of the elasticity theory. Evidently, values of Φ lower than 4 allow for greater degrees of swelling, since figuratively less than all possible cross-linker reaction sites are actually bound to polymer chains. Since clearly not a perfect reaction could be expected, such lower values of Φ had been reasonably expected. For all the subsequent modelling results the Φ value has not been changed (unless otherwise denoted), even though one could expect the requirement of a re-adjustment of Φ to correlate the experimental data of different research groups who follow different synthesis procedures. The molar fraction of cross-linker, y, however is an important input quantity from experiments for modelling the gels; the impact of its variation was examined in our previous work [11]. The reasonable correlation is that with increasing cross-linker content of the gel the polymer chains in the network must be shorter and consequently the whole network turns stiffer and cannot swell as extensively.

Influence of Alcohols on PNIPAAm Swelling

With regard to applications of hydrogels, the influence of organic compounds in aqueous mixtures is of high interest. As representative example, Fig. 5 gives the swelling dependence of PNIPAAm in alcoholic mixtures with water at a fixed temperature of 298.15 K. For two contemplated alcohols methanol and ethanol the re-entrant behaviour is clearly observable: In pure water as well as in the pure alcohol the degree of swelling is significantly high, contrasting the solvent mixture between about 5 and 40 mol% of alcohol in the solution phase where a shrunken gel can be found with hardly any solvent uptake. The experimental data by Mukae et al. [10] hereby suggest that both alcohol species behave very similarly, but that more molecules of the smaller methanol are required for both inducing the gel shrinkage and its re-swelling. This observation is fully supported by our swelling calculations. Furthermore one can see that the swelling of PNIPAAm in the pure alcohols reaches very similar degrees, valid in our model and in the experiment.

Since the PNIPAAm swelling in pure water in the experiments by Mukae at 298.15 K differs drastically from the degrees of swelling reported by Poschlad [37] at the same y = 0.01 as in Fig. 4, a re-adjustment of $\Phi = 2.88$ was required to match the average degree of swelling in water of 10.2, which is less than half the value Poschlad measured.

As shown before for the LLE system water/PNIPAAm, now the degree of association in swelling systems can be analysed. Figure 6 gives the respective modelling results in the gel for the swelling of PNIPAAm in a water/methanol solution at 298.15 K. The extent of hydrogen bond formation $1\text{-}X^{Ai}$ of all three components allows two corresponding remarks as in the analysis of the association in the LLE system. In the first place, both solvent molecules, which are significantly smaller in size than the polymer, have much higher degrees of association, an observation whose cause can be identified in the sterical hindrances. Secondly, a significant difference in the association between shrunken and swollen states is obvious: the shrunken states with the high PNIPAAm concentration show a decrease in association for all components. This is again supposed to correspond to the more hydrophobic state of the shrunken gel.

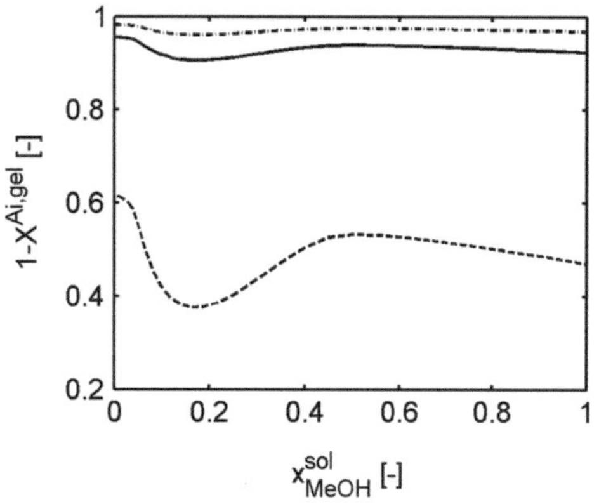

Fig. 6 Extent of association in the gel for water (*solid line*), PNIPAAm (*dashed line*) and methanol (*dash-dotted line*) depending on the molar fraction of methanol in the solution phase at 298.15 K

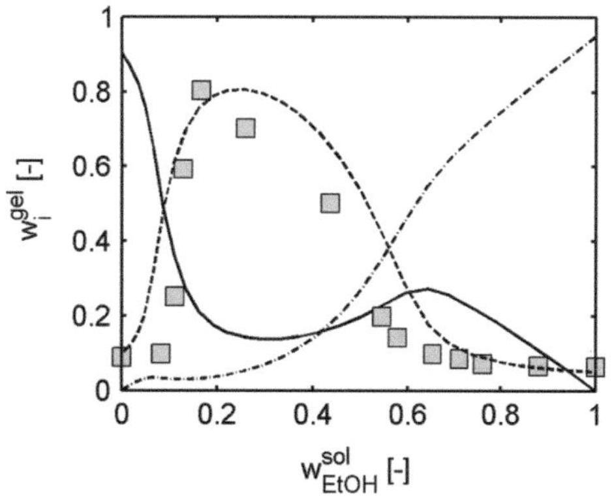

Fig. 7 Composition of the gel phase swollen in aqueous ethanol at 298.15 K. The lines give model results (*full:* water; *dashed:* PNIPAAm; *dash-dotted:* ethanol), the *symbols* are literature data [10] for PNIPAAm

It is obvious, that not only the degree of swelling can be calculated, but moreover with the equilibrium conditions of Eq. 13 we are able to calculate the gel composition. Figure 7 shows the chemical composition of a PNIPAAm gel swollen in an ethanol/water solution. The swollen states in highly aqueous or highly ethanolic solutions correspond to the rather low PNIPAAm concentrations in the gel, whereas the shrunken state comes along with high polymer concentrations. The course of the water and ethanol concentrations in the gel phase visualise the same behaviour: the sudden gel shrinkage at ethanol concentrations of around 10 wt% is caused by a dramatic loss of water whose concentration decreases from 90 to 20 wt%. A continuous increase of ethanol in the gel then is responsible for the re-swelling at higher ethanol concentrations.

For the calculation of the swelling in aqueous alcohol systems a binary k_{ij} between the two solvents had to be adjusted, which were however not fitted to the experimental data shown here; the exact approach was explained previously [11]. Our general finding was that the k_{ij} exhibits a strong

Table 2 Binary interaction parameters with water and PNIPAAm

	Water	Ref.	PNIPAAm	Ref.
PNIPAAm	$-0.17+0.0013 * (T(K)-298.15)$	[11]	0	[11]
Methanol	−0.200	[11]	0	[11]
Ethanol	−0.185	[11]	0	[11]
Acetone	−0.105	This work	0	This work
PEG	−0.137	This work	0	This work
Na^+	0.000477	[34]	−0.3	This work
K^+	0.1994	[34]	0.4	This work
Cl^-	−0.25	[34]	0.4	This work
Br^-	−0.25	[34]	0.1	This work
I^-	−0.25	[34]	−0.1	This work
NO_3^-	−0.09783	[34]	1.0	This work

influence on the extent of swelling and the location of the swelling transitions and that furthermore there is an inherent relation between the required value of k_{ij} and the dispersion energies of the components involved. As shown in Eq. 2, the greater the value of k_{ij} the weaker the attractive forces between the components i and j. Generalised statement is that the geometric mean of the interaction u_{ij} of polymer/water (u_{12}) and polymer/alcohol (u_{23}) has to be lower than the interaction strength between both solvents (u_{13}) to grant for the modelling of the re-entrant behaviour. Precisely, in the case of the alcohols the relation $(u_{13})/(u_{12}\,u_{23})^{1/2} > 1.034$ was found appropriate [11] and the respective binary parameters had to be negative, strengthening the interaction of alcohol with water. Notably, their values had to be significantly more negative than commonly used for vapour-liquid equilibria calculations. An overview of the used parameters is given in Table 2.

Influence of PEG on PNIPAAm Swelling

According to our procedure tested with the alcohol systems, a prediction for the approximate value of the k_{ij} based on the dispersion parameters can be conducted. A very different species than alcohols, poly(ethylene glycol) (PEG) is used for the prediction of its influence on the swelling behaviour of PNIPAAm. PEG as polymeric species can be expected to have a similar non-miscibility effect on PNIPAAm as can be usually observed with mixtures of polymers. Following our postulation, that in the system water (1)/PNIPAAm (2)/PEG (3) the principle $(u_{13})^2 > (u_{12}\cdot u_{23})$ as indication for the binary parameter should be fulfilled, a k_{13} should have a value at least as negative as −0.137. Making use of this value, Fig. 8 shows the modelling prediction of PEG with a molecular weight of 50,000 g/mol in terms of the calculated gel volume V referenced to the gel volume in pure water V^{ref} and in comparison experimental data from the literature

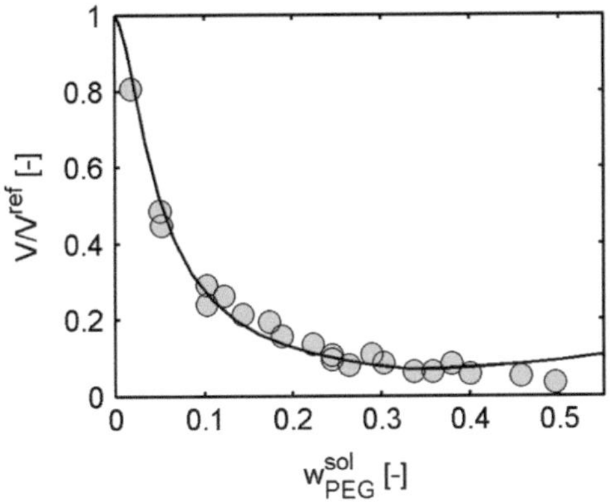

Fig. 8 Gel volume referenced to the swelling in pure water at 298.15 K depending on the influence of PEG50000. *Circles* are experimental data [38], the *line* gives model results

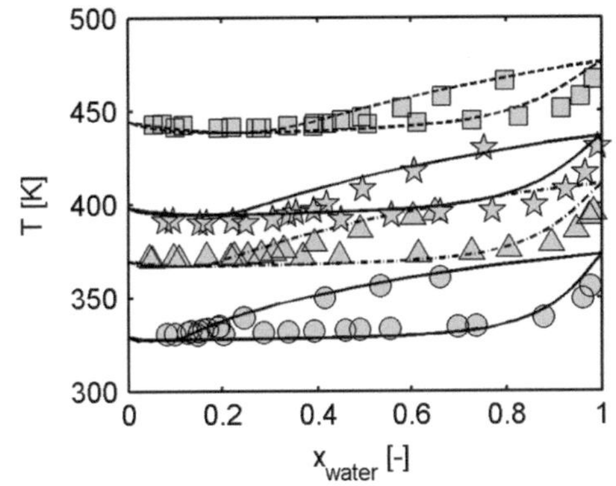

Fig. 10 Isobaric VLE of the water/acetone system at 1.013 (*circles*), 3.45 (*triangles*), 6.89 (*stars*) and 17.2 bar (*squares*). *Lines* are the modelling results matching the symbols as experimental data [39]

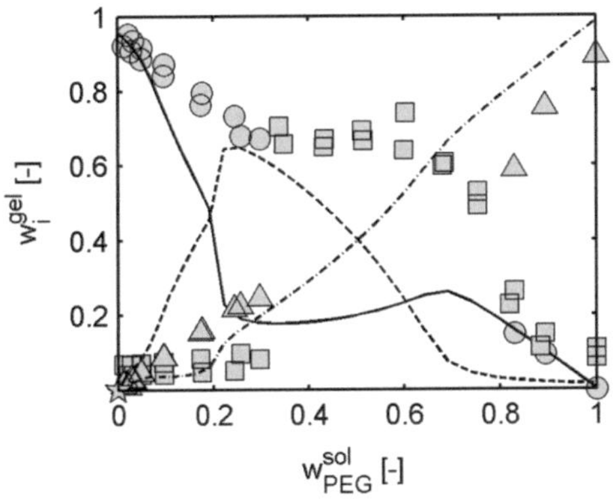

Fig. 9 Composition of the gel phase swollen in aqueous PEG400 at 298.15 K. Water: *circles and solid line;* PNIPAAm: *squares and dashed line;* PEG: *triangles and dash-dotted line*. The lines are model results, *symbols* are literature data [32]

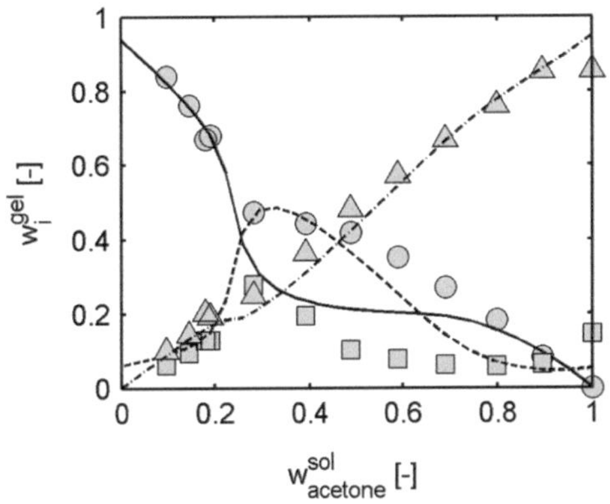

Fig. 11 Gel phase composition of PNIPAAm swollen in aqueous acetone solution at 298.15 K. Lines are the modelled results and *symbols* represent experimental data [32] for water (*circles; solid line*), PNIPAAm (*squares; dashed line*) and acetone (*triangles; dash-dotted line*)

[38] matching quantitatively. At high values of $w_{PEG}{}^{sol}$ a slightly diverging trend of experiments and our model at can be observed. With such high concentrations however, equilibrium experiments face the challenge of reaching the PEG solubility limit in water and very high viscosities may hamper accurate measurements.

Going to lower molecular weights, however, a comparison with experimental data assessing the gel composition shows that the model overestimates the influence of PEG at rather low concentrations and the whole re-entrancy effect is shifted to lower PEG solution concentrations, see Fig. 9. The principle behaviour, though, is still predicted qualitatively well, noting that only the k_{ij} estimation taken from the alcoholic solutions was applied.

Influence of Acetone on PNIPAAm Swelling

The presented prediction formula for the k_{ij} does not work perfectly in all cases as can be shown at the example of acetone, which however gives proof of the general predictive capability of PC-SAFT for thermodynamic systems. The k_{ij} value according to our presented rule yields −0.081. Taking binary vapour-liquid equilibrium (VLE) data from the literature, a more negative binary $k_{ij} = -0.105$ between water and acetone can be adjusted; see corresponding Fig. 10. Here, even though the water model is not designed for temperatures higher than 373 K, isobaric VLE data are met well with the fitted k_{ij}. Using this latter binary parameter, the ternary swelling in the water/PNIPAAm/acetone system is qualitatively well predicted when comparing with experimental data, see Fig. 11. One can see that the model predicts correctly the presence and range of a shrunken gel, even though the quantitative extent is overestimated, which however may also be ascribed to experimental uncertainties.

Most data in open literature covers experiments on PNIPAAm at 298.15 K, which is for good reason since near the thermal instability point of the binary system the influence of third components can be easily observed and

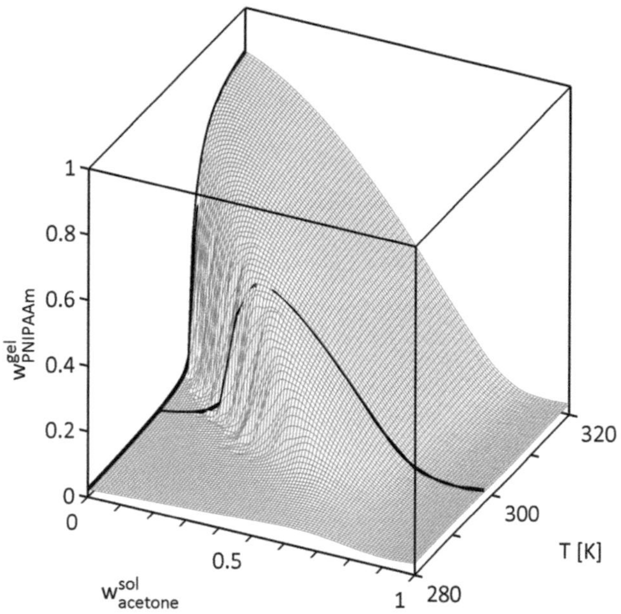

Fig. 12 PNIPAAm weight fraction in the gel phase as predicted hypersurface corresponding to the influence variables temperature and equilibrium acetone concentration in the aqueous solution phase. The *black lines* correspond to the adjusted systems of binary swelling in pure water at variable temperature as well as to the swelling at a fixed temperature of 298.15 K but with variable solution concentration

evaluated. With PC-SAFT however, having fed the model with information about the temperature dependence of the swelling in pure water as well as knowing that the prediction at 298.15 K gives good results, the predictive capability can be maxed out and the influence of both temperature and concentration in the solution phase can be evaluated. Figure 12 shows the respective hypersurface for the system water/PNIPAAm/acetone, presenting the predicted PNIPAAm concentration in the gel for arbitrary solution concentrations and in a temperature range of 280–320 K. The swelling in pure acetone is apparent to not depend on the temperature, a prediction which cannot be tested with data from the literature, but which seems probable with knowledge of the proven temperature independence of the swelling in pure ethanol [20]. The results indicate that a swollen/shrunken state transition cannot be induced by varying the concentration below a minimal temperature around 290 K.

Influence of Salts on PNIPAAm Swelling

As shown in Eq. 1, the different contributions to the residual Helmholtz energy can be superposed additively and thus depending on the component structure and properties can be switched on or off. In the case of electrolyte systems, electrostatic forces are taken into account with the Debye-Hückel approach forming the ePC-SAFT model

Table 3 Binary interaction parameters among ions according to Held et al. [34]

	Na+	K+
Cl^-	0.3166	0.0641
Br^-	0.2898	−0.1019
I^-	0.0181	−0.3116
NO_3^-	−0.300	–

[28]. For the description of salts, two basic strategies can be distinguished between: on the one hand the salt-specific approach, identifying each cation/anion combination with a specific parameter set, and on the other hand the ion specific approach, treating a salt as two separate ionic species with each an own parameter set and being dependent on each other only by the electroneutrality condition. For ePC-SAFT the second, ion-specific approach is used, drastically reducing the required amount of independent parameter sets: given m cations and n anions, a total of m*n salts can be described with $m + n$ parameter sets only, instead m*n parameter sets which would be required for the salt-specific approach. Particularly important, unknown salts can be modelled from ion parameter sets generated from known salts. Furthermore advantageously, the used theory for the Coulomb forces does not require any new adjustable parameter, and since the segment number of the ions is set to one due to their spherical shape, the overall number of adjustable pure-component parameters for ions is reduced to two, viz. diameter and dispersion energy.

Thus for modelling hydrogels in aqueous electrolyte solutions yet with the elastic contribution a total of five contributions to $A^{residual}$ is taken into account. Regarding the focus of medical applications of hydrogels, the influence of omnipresent salts in biological solutions is of particular interest, and as known from the literature there is a strong influence of these. Since salts are known to be challenging substances in thermodynamic modelling, unsurprisingly purely predictive swelling calculations showed significant deviations from experimental data. However using the binary parameters between ions and water given by Held et al. for the accurate description of various thermodynamic quantities such as osmotic coefficients or solid–liquid equilibria [34], with additionally introduced k_{ij} values between ions and PNIPAAm the electrolyte influences on the hydrogel systems below could be modelled successfully and in good agreement with experiments data. All respective binary interaction parameters are given in Tables 2 and 3.

Figure 13 shows the shifting of the temperature transition of sodium chloride. Taking the solid reference curve of PNIPAAm in pure water, the dashed and dash-dotted lines show in comparison the effect of 0.5 and 1.0 M NaCl solutions. It is obvious that by addition of salt a downshift of the whole swelling curve in terms of the temperature occurs; the 1 M NaCl evokes a shift by around 15 K. The shape of

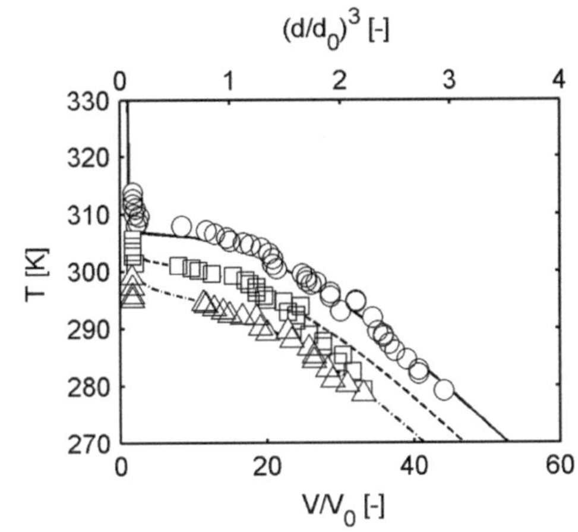

Fig. 13 Temperature-dependent influence of sodium chloride/water solutions on the PNIPAAm swelling. NaCl is present in the aqueous solution phase with 0.0 M (*circles; solid line*), 0.5 M (*squares; dashed line*) and 1.0 M (*triangles; dash-dotted line*). The *lines* are modelled results and refer to the volume swelling V/V_0, whereas symbols represent experimental data from Annaka et al. [40] giving the cubic of the diameter d of capillary gels in equilibrium over the production diameter d_0 as measure of the volume change

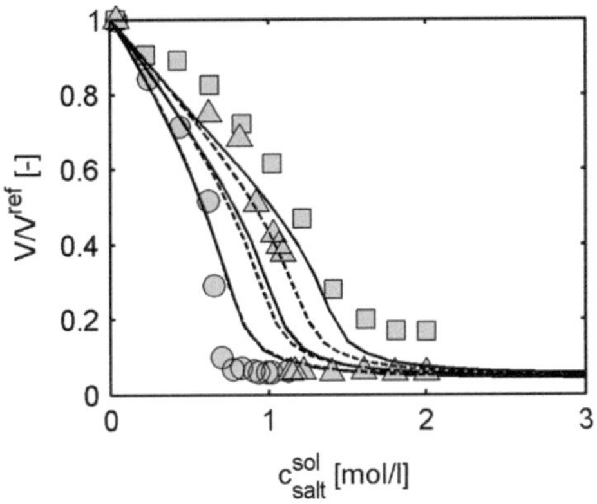

Fig. 14 Salt influence on the swelling of sodium and potassium halogenides at 298.15 K. V/V^{ref} gives the swelling of PNIPAAm gel in salt solution over the swelling of the gel in pure water. *Symbols* represent experimental data (with $V = d^3/d_0^3$ of capillary hydrogels) from the literature [40] for sodium halogenides (*circles: NaCl; triangles: NaBr; squares: NaI*), the *solid lines* give the corresponding modelling results, the *dashed lines* the analogues for the potassium halogenides. The model results for NaCl and KCl are congruent

the curve undergoes a smoothing of the previously sharper transition, a trend which has been previously observed in experiments [40].

Beneficially as explained before, with the ePC-SAFT model arbitrary cation and anion parameters can be combined to calculate various salts. In the present case, particularly alkali halogenides have been examined since playing a major role in physiological environments. Figure 14 shows the influence of sodium and potassium salts, viz. the respective chloride, bromide and iodide. Correlated to experimental data from Annaka et al. [40] and matching the experimental setup, the degree of swelling in salt solutions V is referenced to the degree of swelling in pure water V^{ref} and shown depending on the variable salt concentration in the solution. As is obvious, already small amounts of salts <1 M can induce a decrease the swelling of PNIPAAm gels. Anions hereby have a very strong influence and the series $Cl^- > Br^- > I^-$ in terms of the impact severity is identifiable. Logically, this trend follows the decrease of charge densities of the anions due to their increasing Pauling radius, which manifests in the physical parameter of the segment diameter σ_i.

Following the experimental results of Inomata et al. [41], an exchange of the cation has a far minor influence compared to the anion, as our modelling results for potassium halogenides can confirm. As found by Inomata, the behaviour curves of NaCl and KCl are practically congruent, and when going to larger anions with lower charge density the influence of the cation in scope of the comparison of sodium and potassium halogenides increases, same as in our model results.

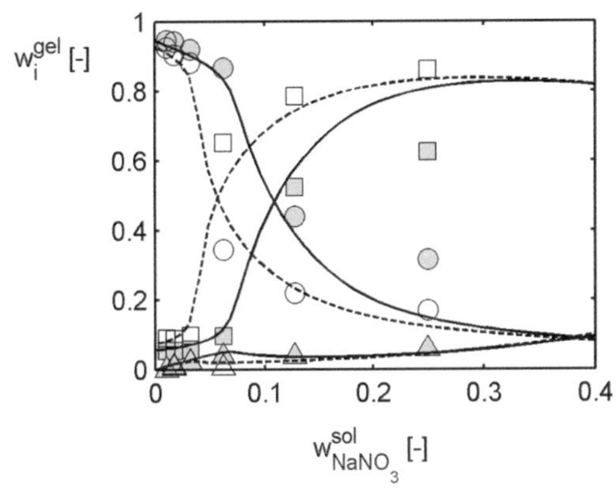

Fig. 15 PNIPAAm gel composition in aqueous sodium nitrate at 298.15 and 303.15 K. *Open symbols* represent experimental data at 303.15 K (Poggendorf S., 2008, personal communication), *full symbols* at 298.15 K (*circles:* water; *squares:* PNIPAAm; *triangles:* $NaNO_3$) and lines the corresponding modelling results (*solid: 298.15; dashed: 303.15*)

A composed view of the two possible influences temperature and salt concentration is shown in Fig. 15. Since the gel is at both described temperatures of 298.15 and 303.15 K in the swollen state, the composition of the gel in pure water at $w^{sol}{}_{NaNO3} = 0$ undergoes only minor changes; worthy of comment is that however the degree of swelling $m/m_0 = 1/w^{gel}{}_{PNIPAAm}$ does show a non-negligible decrease at risen temperatures, see Fig. 16. The temperature dependence becomes more significant in terms of the gel composition the more salt is added to the solution, which shifts the whole system nearer to the instability point of the gel. The modelled results show well the interaction of both external influence factors on the gel composition. It is important to note, that other than in the e.g. water/PNIPAAm/acetone system the salt as third component hardly diffuses into the

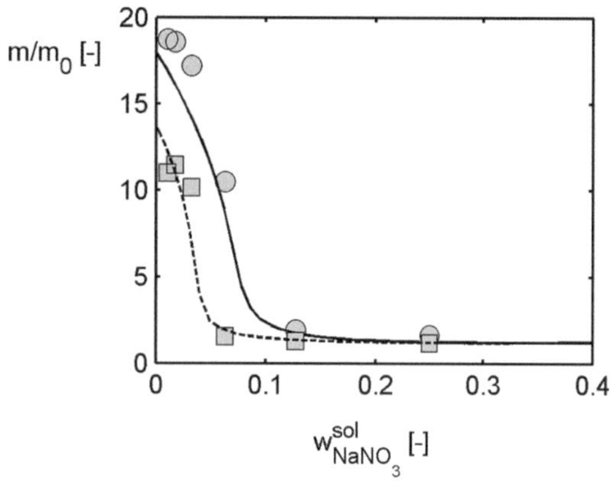

Fig. 16 Degree of swelling of water/PNIPAAm/$NaNO_3$ systems at 298.15 (*circles; solid line*) and 303.15 K (*squares; dashed line*). *Lines* are model results, *symbols* represent experimental data (Poggendorf S., 2008, personal communication)

gel. Its concentration stays distinctly low, nearly invariable over a large concentration range in the solution phase.

All binary parameters as taken from the literature or adjusted in this work are given in Tables 2 and 3.

Conclusion

In the present work, a modification of PC-SAFT was applied and used for the modelling of various PNIPAAm gel systems. The peculiarity in the calculation of swellable gels is the consideration of elastic forces introduced by the volume increase of the cross-linked polymer network. Thus we applied a new elasticity term proposed by Miao et al. [19] as superposing contribution to the residual Helmholtz energy and introduced an additional adjustable parameter, the effective network functionality, whose influence on the swelling of the prominent hydrogel polymer PNIPAAm was examined. With pure-component parameters for PNIPAAm adjusted to the swelling system in water as well as solution density data not only the gel properties, viz. the degree of swelling and the gel phase composition, could be modelled, but also thermodynamic behaviour of not cross-linked PNIPAAm chains in aqueous solution such as liquid-liquid equilibria and water activity coefficients. Of particular interest with PNIPAAm is its behaviour near a thermal instability point in the vicinity of the body temperature, at which the gel undergoes a dramatic swelling transition and analogous the polymer solution shows a lower miscibility solution temperature LLE. Our observations with PC-SAFT here allow for the analysis of hydrogen bonding and show a strong correlation of polymer behaviour and association, a phenomenon which is valid both for systems of cross-linked gel polymer and the PNIPAAm solutions.

Adding a third component, which in this work are both organic and inorganic solvents or solutes, to the water/PNIPAAm mixture discloses clearly that the position of the transition temperature can be shifted drastically. Resulting, a re-entrant behaviour of PNIPAAm gels with swollen states in pure solvent A and pure solvent B but a shrunken state in mixtures of A and B is observable. For the prediction of a potentially required binary parameter a simple correlation between the dispersion energy parameters has been identified, which allows for at least qualitative predictions of gel swelling not only regarding the influence of the solution phase composition but also of the temperature. For the first time, also the electrolyte influence on the behaviour of PNIPAAm hydrogels has been studied with a general equation of state such as PC-SAFT for various systems.

List of Symbols

Latin symbols	
A	Helmholtz energy
d	Diameter
k_{ij}	Binary interaction parameter
k_B	Boltzmann constant
M	Molar mass
m	Mass
m_i^{seg}	Segment diameter
N_i^{assoc}	Number of association sites
n	Molar amount
p	Pressure
T	Temperature
u_i	Dispersion energy
V	Volume
w_i	Weight fraction of component i
X^{Ai}	Fraction of not-bonded association sites Ai
x_i	Mole fraction of component i
y	Molar content of cross-linker
Greek symbols	
γ	Activity coefficient
Δ^{AiBj}	Association bonding strength
ε^{AiBj}	Association energy
κ^{AiBj}	Association volume
ρ	Density
σ_i	Segment diameter
φ_i	Fugacity coefficient of species i
Φ	Network functionality
Indices	
0	Reference
Ai (Bj)	Association site type A (B) of component i (j)
AiBj	Binary association between Bj and Ai

i	Component i
j	Component j
max	Maximal
ref	Reference

The authors gratefully acknowledge the support of the priority programme SPP 1259 "Intelligent Hydrogels" by the Deutsche Forschungsgemeinschaft (DFG).

References

1. Patenaude M, Hoare T (2012) Injectable, degradable thermoresponsive poly(N-isopropylacrylamide) hydrogels. ACS Macro Lett 1:409–413
2. Wohlfarth C (2000) Thermodynamic data of copolymer solutions. CRC Press, Boca Raton
3. Tao C-T, Young T-H (2005) Phase behavior of poly(N-isopropylacrylamide) in water–methanol cononsolvent mixtures and its relevance to membrane formation. Polymer 46:10077–10084
4. Winnik FM et al (1993) Cononsolvency of poly(N-isopropylacrylamide): a look at spin-labeled polymers in mixtures of water and tetrahydrofuran. Macromolecules 26:4577–4585
5. Winnik FM, Ringsdorf H, Venzmer J (1990) Methanol–water as a co-nonsolvent system for poly(N-isopropylacrylamide). Macromolecules 23:2415–2416
6. Tanaka F, Koga T, Winnik FM (2008) Temperature-responsive polymers in mixed solvents: competitive hydrogen bonds cause cononsolvency. Phys Rev Lett 101(2):028302–1–028302–4
7. Tanaka F et al (2011) Preferential adsorption and co-nonsolvency of thermoresponsive polymers in mixed solvents of water/methanol. Macromolecules 44:2978–2989
8. Costa ROR, Freitas RFS (2002) Phase behavior of poly(N-isopropylacrylamide) in binary aqueous solutions. Polymer 43:5879–5885
9. Schild HG, Muthukumar M, Tirrell DA (1991) Cononsolvency in mixed aqueous solutions of poly(N-isopropylacrylamide). Macromolecules 24:948–982
10. Mukae K et al (1993) Swelling of poly(N-isopropylacrylamide) gels in water-alcohol (C1–C4) mixed solvents. J Phys Chem B 97:737–741
11. Arndt MC, Sadowski G (2012) Modeling poly(N-isopropylacrylamide) hydrogels in water/alcohol mixtures with PC-SAFT. Macromolecules 45:6686–6696
12. Flory PJ (1953) Principles of polymer chemistry. Cornell University Press, London
13. Flory PJ (1950) Statistical mechanics of swelling of network structures. J Chem Phys 18(1):108–111
14. Flory PJ (1951) Statistical thermodynamics of rubber elasticity. J Chem Phys 19(12):1435–1439
15. Flory PJ, Rehner J (1943) Statistical mechanics of cross-linked polymer networks. J Chem Phys 11(11):512–520
16. Guth E, James HM (1941) Elastic and thermoelastic properties of rubberlike materials. Ind Eng Chem 33(5):624–629
17. James HM, Guth E (1953) Statistical thermodynamics of rubber elasticity. J Chem Phys 21(6):1039–1049
18. Peterson LE, Anthony RL, Guth E (1942) Equation of state of some synthetic rubbers. Ind Eng Chem 34(11):1349–1352
19. Miao B et al (2010) Effect on finite extensibility on the equilibrium chain size. Macromol Theory Simul 19:414–420
20. Zhi D et al (2010) A molecular thermodynamic model for temperature- and solvent-sensitive hydrogels, application to the swelling behavior of PNIPAm hydrogels in ethanol/water mixtures. Chem Eng Sci 65:3223–3230
21. Koga T et al (2008) Theoretical modeling of associated structures in aqueous solutions of hydrophobically modified telechelic PNIPAM Based on a neutron scattering study. Macromolecules 41:9413–9422
22. Maurer G, Prausnitz JM (1996) Thermodynamics of phase equilibrium for systems containing gels. Fluid Phase Equilibr 115:113–133
23. Poschlad K, Enders S (2011) Thermodynamics of aqueous solutions containing poly(N-isopropylacrylamide). Fluid Phase Equilibr 302:153–160
24. Althans D, Langenbach K, Enders S (2012) Influence of different alcohols on the swelling behaviour of hydrogels. Mol Phys 110(11–12):1391–1402
25. Gross J, Sadowski G (2001) Perturbed-chain SAFT: an equation of state based on a perturbation theory for chain molecules. Ind Eng Chem Res 40(4):1244–1260
26. Gross J, Sadowski G (2002) Modeling polymer systems using the perturbed-chain statistical associating fluid theory equation of state. Ind Eng Chem Res 41(5):1084–1093
27. Gross J, Sadowski G (2002) Application of the perturbed-chain SAFT equation of state to associating systems. Ind Eng Chem Res 41(22):5510–5515
28. Cameretti LF, Sadowski G (2005) Modeling of aqueous electrolyte solutions with perturbed-chain statistical associated fluid theory. Ind Eng Chem Res 44:3355–3362
29. Held C, Cameretti LF, Sadowski G (2008) Modeling aqueous electrolyte solutions – Part 1. Fully dissociated electrolytes. Fluid Phase Equilibr 270(1–2):87–96
30. Chapman WG et al (1990) New reference equation of state for associating liquids. Ind Eng Chem Res 29:1709–1721
31. Huang SH, Radosz M (1990) Equation of state for small, large, polydisperse, and associating molecules. Ind Eng Chem Res 29:2284–2294
32. Hüther A (2000) Experimentelle und theoretische Untersuchungen zum Quellungsgleichgewicht von Hydrogelen in wässrigen Lösungen. In: Lehrstuhl für Thermodynamik. Technische Universität Kaiserslautern, Kaiserslautern
33. Orlov Y, Xu X, Maurer G (2005) Swelling of a N-isopropyl acrylamide hydrogel in two aqueous/organic two-phase systems. Fluid Phase Equilibr 235:18–25
34. Held C et al. (2013) ePC-SAFT revised (in preparation)
35. Stoychev I et al (2009) Modeling the phase behavior of PEO-PPO-PEO surfactants in carbon dioxide using the PC-SAFT equation of state – application to dry decontamination of solid substances. J Chem Eng Data 54:1551–1559
36. Kleiner M (2008) Thermodynamic modeling of complex systems: polar and associating fluids and mixtures. In: Laboratory of Thermodynamics. Technische Universität Dortmund, Dortmund
37. Poschlad K (2011) Experimentelle und theoretische Untersuchungen der Wirkstoffaufnahme und -abgabe von Poly-(N-isopropylacrylamid)-Hydrogelen. In: Thermodynamik und Thermische Verfahrenstechnik. Technische Universität Berlin, Berlin
38. Ishidao T et al (1993) Swelling behaviors of poly(N-isopropylacrylamide) gel in poly(ethylene glycol)-water mixtures. Macromolecules 26:7361–7362
39. Othmer DF, Chudgar MM, Levy SL (1952) Binary and ternary systems of acetone, methyl ethyl ketone, and water. Ind Eng Chem 44(8):1872–1881
40. Annaka M et al (2000) Salt-induced volume phase transition of poly(N-isopropylacrylamide) gel. J Chem Phys 113(14):5980–5985
41. Inomata H et al (1992) Effect of additives on phase transition of N-isopropylacrylamide gels. Langmuir 8:687–690

Modeling and Simulation of Hydrogels for the Application as Bending Actuators

T. Wallmersperger, A. Attaran, K. Keller, J. Brummund, M. Guenther, and G. Gerlach

Abstract

Polyelectrolyte gels show a quite large swelling or bending behavior under external physical, chemical, thermal or electrical stimulation. In this paper the bending actuation of polyelectrolyte gels under applied electric fields is studied and the mechanisms occurring in polyelectrolyte gels due to the applied stimulus are investigated.

In the present research, a complete formulation for describing actuation of polyelectrolyte gels in a solution bath is presented. First, the kinematics, the balance laws of continuum chemo-electro-mechanics and the constitutive equations of the involved fields are given.

Then, in the numerical simulation part, the changes of the mobile concentrations, of the electric potential and the displacements in the gel domain are shown. It is demonstrated by a comparison between a chemo-electrical and a chemo-electro-mechanical test case, that the full coupling also leads to a change of the concentration of bound groups resulting in alterations of the local concentrations of the mobile ions, of the electric potential and again of the local gel displacement. It will be shown, that the presented coupled multi-field model is predestined for investigating electrical stimulation of hydrogels, and for simulating hydrogel bending actuation.

Keywords

Hydrogel • Chemo-electro-mechanical model • Multi-field formulation • Finite element simulation • Bending actuator

List of Nomenclature

$\mathcal{P}$	continuum domain
$\mathcal{P}_0$	continuum domain at the reference configuration
$\mathcal{G}$	gel domain
$\mathcal{G}_0$	gel domain at the reference configuration
$\mathcal{S}$	solution domain
ε	boundary layer between gel and solution
$\partial\mathcal{P}$	boundary of the continuum domain
$\partial\mathcal{G}$	gel boundary
$\partial\mathcal{S}$	solution boundary
$\partial\varepsilon$	surface of the boundary layer
$\partial\mathcal{P}_{j_\alpha}$	boundary for the prescribed flux of the ionic species α
$\partial\mathcal{P}_{\mu_\alpha}$	boundary for the prescribed electro-chemical potential
$\partial\mathcal{P}_{\omega}$	boundary for the prescribed electric charge density
$\partial\mathcal{P}_{\Psi}$	boundary for the prescribed electric potential
$\partial\mathcal{P}_{\mathbf{p}}$	boundary for the prescribed traction

T. Wallmersperger (✉) • A. Attaran • J. Brummund
Institut für Festkörpermechanik, Technische Universität Dresden, 01062 Dresden, Germany
e-mail: thomas.wallmersperger@tu-dresden.de

K. Keller
Institut für Statik und Dynamik der Luft- und Raumfahrtkonstruktionen, Universität Stuttgart, 70569 Stuttgart, Germany

M. Guenther • G. Gerlach
Institut für Festkörperelektronik, Technische Universität Dresden, 01062 Dresden, Germany

G. Sadowski and W. Richtering (eds.), *Intelligent Hydrogels*, Progress in Colloid and Polymer Science 140,
DOI 10.1007/978-3-319-01683-2_15,

$\partial\mathcal{P}_{\mathbf{u}}$	boundary for the prescribed displacement
$\mathcal{R}^3$	three dimensional Euclidean space

$\mathbf{b}$	body forces
c_α	concentration of ionic species α
c_{A-}	concentration of the fixed charges
c^0_{A-}	initial concentration of the fixed charges
c^{ref}_α	reference ionic concentration
$\hat{c}_0$	standard concentration
$\tilde{\mathbf{D}}$	dielectric displacement
D_α	diffusion coefficient of the species α
D^{ref}	diffusion coefficient of the reference ions
$\mathbf{E}$	fourth-order elasticity tensor
E	Young's modulus
$\tilde{E}$	electric field
$\mathbf{F}$	deformation gradient tensor
$\mathbf{F}^{-1}$	inverse deformation gradient tensor
F	Faraday constant
$\tilde{\mathbf{g}}$	tensor of swelling coefficients
g	swelling coefficient
$\mathbf{j}_\alpha$	ionic flux of the species α
$\bar{\mathbf{j}}_\alpha$	prescribed ionic flux
l	length
l_x	length in x-direction
l_y	length in y-direction
N_b	number of bound ions
N_f	number of (free) mobile charges
$\mathbf{n}$	outward normal unit vector
$\mathbf{p}$	traction vector
r_α	total external chemical supply
R	universal gas constant
t	time
T	temperature
$\mathbf{u}$	displacement
$\bar{\mathbf{u}}$	prescribed displacement
$\hat{\boldsymbol{u}}$	generalized displacement
u	displacement in x-direction
v	displacement in y-direction
$\mathbf{v}$	velocity
dV_0	volume element in reference state
dV	volume element in actual state
$\mathbf{X}$	reference position vector
$\mathbf{x}$	current position vector
z_α	valence of the ions

$\Delta\pi$	osmotic pressure
ε_0	vacuum permittivity
ε_r	relative permittivity
$\boldsymbol{\varepsilon}$	strain tensor
$\boldsymbol{\varepsilon}^e$	elastic strain
$\boldsymbol{\varepsilon}^\pi$	swelling strain
μ_α	electro-chemical potential of the species α
$\bar{\mu}_\alpha$	prescribed electro-chemical potential
ν	Poisson ratio
ρ_e	electric charge density
ρ	density
$\boldsymbol{\sigma}$	stress tensor
φ	motion mapping
φ^{-1}	inverse motion mapping
Ψ	electric potential
$\bar{\Psi}$	prescribed electric potential
ω_e	electric charge density per area

List of Mathematical Symbols

$u_{i,j}$	derivative of u_i with respect to x_j
$\cup$	union of ...
$\subset$	is a subset of ...
δ_{kl}	Kronecker tensor
$\mathbf{e}_i$	unit vector in x_i-direction
∂	differential operator
$\cap$	intersected with ...
$\setminus$	remainder of ...
ε_{ijk}	permutation tensor
$\emptyset$	empty set

Introduction

In the current research, the behavior of a polymer gel immersed in a solution bath with and without applied electric field is investigated. In the initial state without applied electric field the gel in the solution is considered, where the gel has reached an equilibrium swollen state compared to a dry (unswollen) state. Such a gel, which is also referred to as *hydrogel*, *polyelectrolyte gel* or *polymer gel* can be regarded as a saturated-porous medium [1].

A schematic of the gel in a solution in initial equilibrium state is shown in Fig. 1.

The polymer gel consists of a polymer network as the solid phase and a pore fluid. In the gel, bound charges are fixed to the polymer; mobile ions are present in both, the gel and the solution. Therefore, the polymer gel can be seen as a mixture of different constituents.

In the following, the different constituents of both, gel (g) and solution (s) are given. The gel (g) comprises

- Polymer (network) or solid phase (P),
- Fixed charges (A^-) bound to the polymer network,
- Pore fluid or solvent (F) and
- Mobile anions ($-$) and cations ($+$)

The solution (s) contains

- Fluid or solvent (F),
- Mobile anions ($-$) and cations ($+$).

Any external stimuli such as a change in the chemical composition of the solution, application of an electric field,

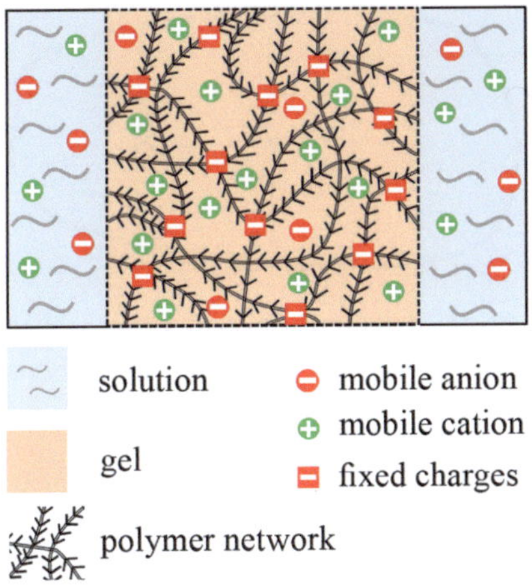

Fig. 1 Microstructure of a hydrogel in a solution bath in initial equilibrium state

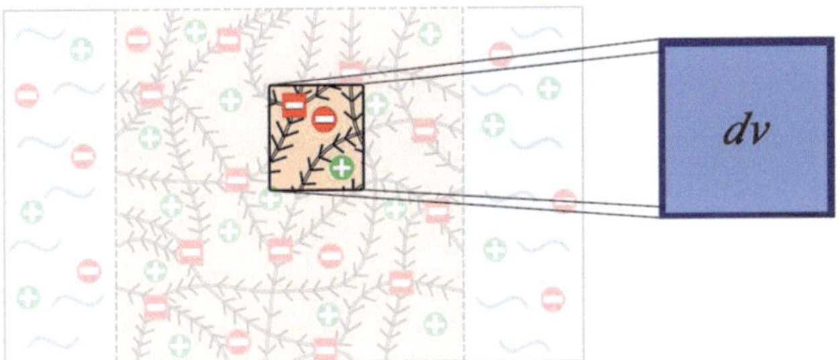

Fig. 2 Volume element of the hydrogel

change in the ambient temperature, etc. can disturb the initial equilibrium state of the hydrogel; then the hydrogel reaches a new equilibrium state. The principal objective of the current work is, therefore, to study the process from the initial equilibrium state towards a new equilibrium state.

In the present research, a continuum model of a hydrogel is formulated where all the constituents are assumed to be entirely mixed together creating an aggregate of mutually interacting constituents [1]. This is illustrated in Fig. 2.

On the basis of the considered homogenized domain $\mathcal{P}$ and for the sake of simplification we further impose the following assumptions in our model:

- In a first approximation, the fluid (F) stays at rest, i.e. $v_l^F = 0$ where v_l^F is the velocity of the fluid, and all state variables in the solvent remain unchanged after external stimulation. Therefore the state changes of the solvent are not considered.
- Only small deformations of the polymer occur.

The present paper is structured as follows:

In section "Literature Review", a brief overview of the existing modeling approaches of polyelectrolyte gels available in literature is given. The kinematics of the considered continuum body (hydrogel), the balance laws of continuum chemo-electro-mechanics and the constitutive equations of the involved fields are all presented in section "Continuum Chemo-Electro-Mechanics". A detailed description of the governing equations, as well as the initial and boundary conditions are depicted in section "Governing Equations, Boundary Conditions and Initial Conditions".

The modeling and discretization of the coupled multi-field formulation – applicable for both, chemical and electrical stimulation, is given in a summarized form in section "Discretization".

In section "Numerical Simulation", a test case for a bending actuator by performing electrical stimulation is presented. A comparison between the results of a chemo-electrical and a full chemo-electro-mechanical coupling is performed. The coupled chemo-electro-mechanical test case shows the capabilities of the developed model.

The last section contains conclusion and outlook.

Literature Review

In the 1950s, Flory and Rehner [2, 3] were the first to formulate a macroscopic theory for investigating the swelling of network structures. One of the first groups who quasi-statically investigated the dynamic behavior of ionic polymer gels in electric fields were Doi et al. [4], Shiga and Kurauchi [5] and Shahinpoor [6]. Brock et al. [7] have given a coupled formulation for the ion-dynamics and a theory for large displacements.

Grimshaw et al. [8] have presented a coupled formulation for chemically and electrically induced swelling in polyelectrolyte gels. In de Gennes et al. [9], the mechano-electric effects in ionic gels are described, while Tamagawa and Taya [10] and Li et al. [11] consider the ion distribution and the ion transport by applying a chemo-electric model. De and Aluru [12] proposed a chemo-electro-mechanical model for the simulation of pH-sensitive hydrogel while using a multiphysical modeling approach. Li and Lai [13] simulated ionic-strength-sensitive hydrogels using a meshless simulation technique. Wallmersperger et al. [14, 15] have published electro-chemical and electro-mechanical formulations for ionic gels under chemical and electrical stimulation. The modeling of temperature-sensitive polyelectrolyte gels has been performed by Keller et al. [16].

Gerlach et al. [17] developed chemical and pH-sensors based on the swelling of hydrogels.

Equilibrium swelling of hydrogels in aqueous solutions based on thermodynamics has been investigated by Orlov et al. [18, 19] and Ermatchkov et al. [20]. The mass transport within membranes as well as the swelling behavior of gels have been described by Kind et al. The mass transport measurements in membranes were performed by means of in situ Raman spectroscopy [21], for the modeling, a multi-component transport model was used [22]. Yoon et al. [23]

used the fluorescence microscopy to track the kinetics of micron-size hydrogel layers and verified the linear theory of poroelasticity. Recently, Prudnikova and Utz [24] have experimentally investigated the electrochemical equilibrium properties of poly(acrylic acid/acrylamide) hydrogels and compared the results to the one predicted by theory with an overall satisfactory outcome. In Mann et al. [25], the swelling of charged hydrogels is discussed. A review article on gel swelling theories has been published by Quesada-Perez et al. [26].

Sun et al. [27] and van Loon et al. [28] presented a triphasic and quadriphasic mixture theory for the simulation of hydrated tissues, respectively. The swelling of charged tissues and gels in the framework of the Theory of Porous Media has been investigated by Ehlers et al. [29] and Acartürk [30]. On the atomistic scale, the conformation change of hydrogels in water has been studied by Walter et al. [31]. An overview over modeling on different scales has been given in [32]. Experiments and modeling of non-linear effects in hydrogel-based chemical sensors have been performed by Guenther et al. [33].

Recently, Bouklas and Huang [34] compared the linear poroelasticity theory and a nonlinear theory to study the kinetics of swelling hydrogels. They have demonstrated that within the linear regime, both theories have yielded consistent results. Lucantonio and Nardinocchi [35] also investigated the bending deformation of gel bars through a stress-diffusion mechanism in the absence of the electric field. Their model has been developed based on a thermodynamically consistent theory of poroelasticity.

In the special issue on "responsive gels" [36], e.g. phase field model simulations of hydrogel dynamics under chemical stimulation [37] and simulation results of tracer diffusion in hydrogels [38] were published. Also, the modeling and simulation of pH-sensitive hydrogels [39] and experimental results on diffusion of polyethylene glycol in N-isopropylacrylamide (NIPAAm) hydrogels with Raman spectroscopy [40] were released within this special issue.

Due to the complexity of the governing physics and the coupling nature of the problem, the available modeling approaches for hydrogels still present a wide gap between theory and practice. This work, therefore, intends to bridge this gap and contribute to the modeling and simulation of hydrogels.

Continuum Chemo-Electro-Mechanics

In this section, the continuum chemo-electro-mechanical formulation is presented. First, general definitions are given, then the kinematics are discussed. After showing the balance laws of the different constituents for the involved fields, the mechanical laws for the chemical, electrical and mechanical fields are given.

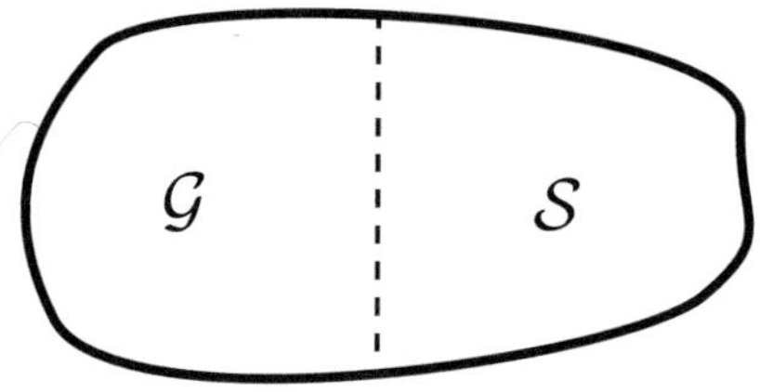

Fig. 3 Decomposition of the whole domain $\mathcal{P}$ into the hydrogel $\mathcal{G}$ and the solution domain $\mathcal{S}$

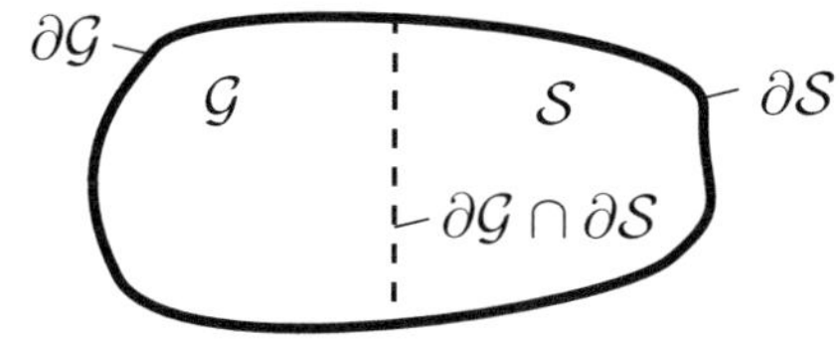

Fig. 4 Gel-solution boundary

General Definitions

A hydrogel, as described in the previous section, placed in a solution bath is considered. The whole domain $\mathcal{P}$ is decomposed into the hydrogel domain $\mathcal{G}$ and the solution domain $\mathcal{S}$, see Fig. 3. This may be given by

$$\mathcal{P} = \mathcal{G} \cup \mathcal{S} \qquad \text{and} \qquad \emptyset = \mathcal{G} \cap \mathcal{S} \, .$$

The hydrogel body $\mathcal{G}$ is enclosed by a surface $\partial\mathcal{G}$ and the solution domain $\mathcal{S}$ is enclosed by a related surface $\partial\mathcal{S}$. The surfaces (boundaries) between gel and solution are given by $\partial\mathcal{G} \cap \partial\mathcal{S}$, see Fig. 4.

The surface of the complete (gel-solution) domain $\partial\mathcal{P}$ is given by

$$\partial\mathcal{P} = (\partial\mathcal{G} \cup \partial\mathcal{S}) \setminus (\partial\mathcal{G} \cap \partial\mathcal{S}) \, .$$

It is assumed that at each instance of time t, the hydrogel body $\mathcal{G}$, can occupy a different region in three dimensional Euclidean space, $\mathcal{R}^3$ ($\mathcal{G} \subset \mathcal{R}^3$) enclosed by the surface $\partial\mathcal{G}$.

It is necessary to specify the configuration of the hydrogel at any instance of time. The configuration $\mathcal{G}_0$ at $t = 0$ is referred to as the reference or initial configuration, while any other configurations at time $t > 0$ are called the current or deformed configurations. The hydrogel at reference and current configuration is shown in Fig. 5.

In the present research, the problem is formulated in a Cartesian coordinate system. For this, the index notation is used. A general vector $\mathbf{u}$ is defined by

$$\mathbf{u} = \sum_i u_i \, \mathbf{e}_i = u_i \, \mathbf{e}_i \qquad ; i = 1, 2, 3 \, , \tag{1}$$

where $\mathbf{e}_i$ are the unit vectors in x_i-direction and u_i are the coordinates or components in x_i-direction. The Einstein

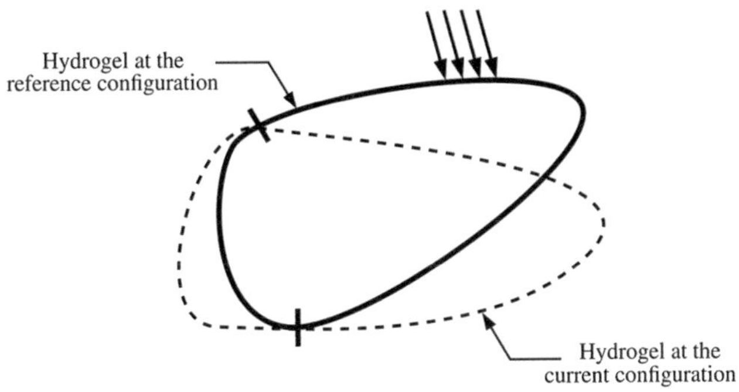

Fig. 5 Body of the hydrogel at the reference and current configurations

summation convention, i.e. the summation over identical indices in a term is performed. The spatial derivative $u_{i,j}$ of the vector $\mathbf{u}$ is given as follows:

$$u_{i,j} = \frac{\partial u_i}{\partial x_j} \quad ; i = 1,2,3 \quad \text{and} \quad j = 1,2,3 . \qquad (2)$$

Kinematics

As discussed in section "Introduction", the state change of the solvent is not considered. Due to this, only the deformation of the polymer gel is investigated.

For finite deformations, initial (reference) and current (deformed) configuration of the gel geometry have to be distinguished.

Displacements and Strains

Current and reference configurations can be converted to each other through the motion mapping and the inverse motion mapping. Denoting $\mathbf{X}$ as the position vector of an arbitrary material point of the gel domain in the reference configuration and $\mathbf{x}$ as the position vector of the same point in the current configuration, the motion mapping is defined as:

$$x_i = \varphi_i(X_J, t) . \qquad (3)$$

The displacement $\mathbf{u}$ is given by the difference of the current and initial position:

$$u_i(X_J, t) = x_i(X_J, t) - X_J\, \delta_{Ji} , \qquad (4)$$

where δ_{kl} is the Kronecker delta.

The deformation gradient $\mathbf{F}$ is defined as the partial differentiation of $\mathbf{x}$, Eq. (3), with respect to $\mathbf{X}$:

$$F_{iJ} = \frac{\partial x_i}{\partial X_J} = x_{i,J} = \delta_{iJ} + u_{i,J} . \qquad (5)$$

Conversely, the inverse motion mapping is defined as:

$$X_J = \varphi_J^{-1}(x_i, t) \qquad (6)$$

with the inverse deformation gradient defined as:

$$F_{Jk}^{-1} = \frac{\partial X_J}{\partial x_k} = X_{J,k} . \qquad (7)$$

It is assumed that such mappings are one-to-one and continuous, with continuous partial derivatives to whatever order which is required [41]. The description of motion or deformation expressed by Eq. (3) is known as material or Lagrangian description of motion, whereas the description of motion expressed by Eq. (6) is commonly referred to as spatial or Eulerian description of motion [41].

Since the analyses associated with the current study are based on the assumption of infinitesimal displacements, the distinction between the current (deformed) and the reference configurations for the gel can be neglected here, i.e.:

$$\mathcal{G} \approx \mathcal{G}_0 .$$

Due to small deformations, where the diplacement gradients are quite small, i.e. $||u_{i,j}|| \ll 1$, the strain $\boldsymbol{\varepsilon}$ of the polymer gel may be given by

$$\varepsilon_{lm} = \frac{1}{2}(u_{l,m} + u_{m,l}) . \qquad (8)$$

From this, the volume change of the gel from the initial volume element dV^0 to the current volume element dV may be expressed as

$$\frac{dV}{dV^0} \approx 1 + \varepsilon_{ll} \quad \text{or} \quad \frac{dV^0}{dV} \approx 1 - \varepsilon_{ll} . \qquad (9)$$

Please note that more precisely, the deformation is the one of the polymer gel network. As the interstitial solution is directly related to the network, the displacement can also be considered as the one of the polymer gel.

Velocities

The velocities, i.e. the partial derivatives of the displacements with respect to time, of the different constituents are specified as follows:

- The mobile species α have a velocity of v_l^α.
- As the state change of the solvent is not considered, i.e. the fluid (F) is immobile, $v_l^F = 0$.
- Also the velocities of polymer (P) and bound anions (A^-) are quite small due to quasi-static processes, i.e. $v_l^P = v_l^{A^-} \approx 0$.

Balance Laws of Continuum Chemo-Electro-Mechanics

In this section, the balance laws for the involved fields are given.

Balance Law for Number of Particles (Mass Balance)

Balance Law of Mobile Ionic Species

The equation for the numbers of mobile particles (mole numbers) can be formulated by

$$\frac{\partial c_\alpha}{\partial t} + j^\alpha_{i,i} = r_\alpha, \ in \ \mathcal{P} \tag{10}$$

where the concentration c_α is the molar number per volume and $\mathbf{j}_\alpha$ is the total ionic flux consisting of conductive (cond), i.e. diffusive and migrative, as well as convective (conv) fluxes

$$j^\alpha_i = j^{\alpha\, cond}_i + j^{\alpha\, conv}_i \tag{11}$$

and r_α is the external chemical supply.

In the present research, the convective flux is given by

$$j^{\alpha\, conv}_i = c_\alpha\, v^F_i = 0 \tag{12}$$

and the chemical supply $r_\alpha = 0$.

Balance Law of Bound Species

The equation for the numbers of bound species in the gel domain ($\mathcal{G}$) may be formulated by

$$\frac{\partial c_{A^-}}{\partial t} + \left(c_{A^-}\, v^{A^-}_i\right)_{,i} = 0, \ in \ \mathcal{G} \tag{13}$$

where c_{A^-} and $\mathbf{v}^{A^-}$ are the concentration and the velocity of the bound anions, respectively.

Equation (13) may be formulated by

$$c_{A^-}\, dV = c^0_{A^-}\, dV^0 . \tag{14}$$

By setting

$$c_{A^-} = c^0_{A^-} + \Delta c_{A^-} , \tag{15}$$

we obtain for the change of the bound concentrations Δc_{A^-}

$$\Delta c_{A^-} = -c^0_{A^-}\, \varepsilon_{ll} \tag{16}$$

or for the actual concentration of bound charges c_{A^-}

$$c_{A^-} = c^0_{A^-}\, (1 - \varepsilon_{ii}) . \tag{17}$$

Maxwell Equations

In electrostatics, the Maxwell equations for the complete domain $\mathcal{P}$ are given by

$$\tilde{D}_{i,i} = \rho_e, \ in \ \mathcal{P} \tag{18}$$

where $\tilde{\mathbf{D}}$ is the dielectric displacement and ρ_e is the charge density of both, the bound and mobile charges as well as by

$$\varepsilon_{ijk}\, \tilde{E}_{k,j} = 0 \ in \ \mathcal{P} \tag{19}$$

where $\tilde{\mathbf{E}}$ is the electric field and ε_{ijk} is the permutation tensor.

The relationship between the electric field and the electric potential Ψ is given by:

$$\tilde{E}_i = -\Psi_{,i} \ in \ \mathcal{P} \tag{20}$$

in order to fulfill Eq. (19). The charge density is given by

$$\rho_e = F \sum_{\alpha=1}^{N_f+N_b} (z_\alpha\, c_\alpha) \ in \ \mathcal{P} \tag{21}$$

where z_α is the valence of the species α, F is the Faraday constant and N_f and N_b are the numbers of mobile (free (f)) and bound (b) species.

Equation (21) can be formulated for the gel domain, where we have mobile anions and cations as well as bound charges, by

$$\rho_e = F\, \left(z^+ c^+ + z^- c^- + z_{A^-}\, c_{A^-}\right) \ in \ \mathcal{G} \tag{22}$$

and for the solution domain where only mobile charges are present by

$$\rho_e = F\, \left(z^+ c^+ + z^- c^-\right) \ in \ \mathcal{S} . \tag{23}$$

Momentum Balance

The balance of linear momentum for the gel is expressed mathematically as:

$$\rho \frac{D v_j}{D t} = \sigma_{ij,i} + \rho\, b_j \ in \ \mathcal{G} \tag{24}$$

with the material time derivative of the velocity $\mathbf{v}$ defined by

$$\frac{D v_j}{D t} = \frac{\partial v_j}{\partial t} + \frac{\partial v_j}{\partial x_k}\frac{\partial x_k}{\partial t} = \dot{v}_j + v_{j,k}\, v_k ,$$

where $\mathbf{b}$ are the body forces acting on $\mathcal{G}$ and $\boldsymbol{\sigma}$ are the total stresses. In the absence of body forces and by neglecting the inertia terms, the balance law of linear momentum can be written as:

$$\sigma_{ij,i} = 0 . \tag{25}$$

From the balance of angular momentum follows that the stresses are symmetric, i.e.

$$\sigma_{kl} = \sigma_{lk} \,. \tag{26}$$

Material Laws

For finalizing the system equations, the constitutive equations, i.e. the material laws, between the different field variables also have to be incorporated.

Material Laws for the Chemical Field

As given in Eq. (11), the total flux is the sum of the conductive and the convective flux. The conductive flux contributions are given by

$$j_l^{\alpha\,cond} = -\frac{D_\alpha\, c_\alpha}{R\,T}\, \mu_{\alpha,l} \tag{27}$$

where R is the universal gas constant, T the temperature, D_α is the diffusion constant and μ_α is the electrochemical potential of the species α. The electrochemical potential is defined (for dilute solutions) as

$$\mu_\alpha = RT \ln \frac{c_\alpha}{\hat{c}_0} + F\, z_\alpha \Psi \tag{28}$$

where $\hat{c}_0 = 1\,\frac{\text{mol}}{\text{l}}$ is the standard concentration.

Inserting Eqs. (12), (27) and (28) into Eq. (11), we obtain:

$$j_i^\alpha = \underbrace{-D_\alpha\, c_{\alpha\,,i}}_{diffusion} \;-\; \underbrace{z_\alpha\, c_\alpha\, D_\alpha \frac{F}{R\,T}\, \Psi_{,i}}_{migration} \,. \tag{29}$$

Inserting Eq. (29) in Eq. (10) results in

$$\frac{\partial c_\alpha}{\partial t} = \left(D_\alpha\, c_{\alpha,i} + z_\alpha\, c_\alpha \frac{F}{R\,T}\, \Psi_{,i} \right)_{,i} \,. \tag{30}$$

Between gel and solution there exists an osmotic pressure difference $\Delta\pi$, resulting from the different concentration of the mobile ions in the different regions [42, 43]. The normalized osmotic pressure in the gel can be calculated as follows:

$$\Delta\pi = RT \sum_{\alpha=1}^{N_f} \left(\Delta c_\alpha - \Delta c_\alpha^0 \right) \;\; in \;\; \mathcal{G} \tag{31}$$

where $\Delta c_\alpha = (c_\alpha - c_\alpha^{ref})$ and $\Delta c_\alpha^0 = (c_{\alpha 0} - c_{\alpha 0}^{ref})$.

In the case of chemical stimulation, the concentration of mobile ion species at time t is nearly constant in space in both, gel and solution, resulting in a homogeneous osmotic pressure. Applying the electrical field, however, results in the concentration variation in space (and time) of the ions leading to a space dependent osmotic pressure difference.

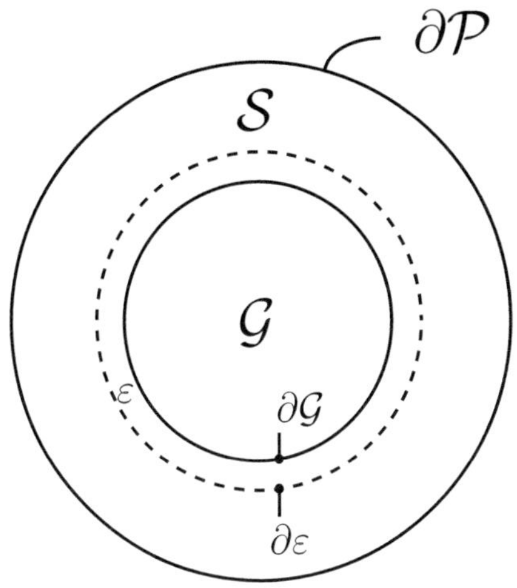

Fig. 6 Decomposition of the whole domain $\mathcal{P}$ into the hydrogel domain $\mathcal{G}$ and the solution domain $\mathcal{S}$, the boundary layer part within the solution domain is called ε; the reference layer between the boundary layer part and the undisturbed solution is $\partial\varepsilon$

To calculate the instantaneous osmotic pressure, therefore, we need to have access to the reference concentration of the mobile ion species at any instance of time. To this end, the field equation governing the reference concentration of the mobile ions has to be introduced. The cost of such a consideration is two additional "artificial" degrees of freedom. The reference concentrations are obtained as follows: First the concentrations in the solution directly outside the gel-solution interface – at the boundary $\partial\varepsilon$ have to be computed, see Fig. 6. These concentrations are set as boundary conditions for the reference concentrations. The field equation of the reference concentration can be expressed as:

$$\left(D^{ref}\, c_{\alpha\,,i}^{\;ref} \right)_{,i} = 0 \;\; in \;\; \mathcal{G} \cup \varepsilon \tag{32}$$

where D^{ref} is the diffusion coefficient of the reference ionic species.

Material Laws for the Electrical Field

The constitutive equation between the electric field $\tilde{\mathbf{E}}$ and the dielectric displacement $\tilde{\mathbf{D}}$ may be given by

$$\tilde{D}_i = \varepsilon_0\, \varepsilon_r\, \tilde{E}_i \tag{33}$$

where ε_r is the relative permittivity and ε_0 is the vacuum permittivity.

Equations (18)–(21) and (33), together construct the governing equations of the electric field as:

$$\Psi_{,ii} = -\frac{F}{\varepsilon_0\, \varepsilon_r} \sum_{\alpha=1}^{N_f+N_b} (z_\alpha\, c_\alpha) \tag{34}$$

This results in

$$\Psi_{,ii} = -\frac{F}{\varepsilon_0\,\varepsilon_r} \sum_{\alpha=1}^{N_f+N_b} (z_\alpha\, c_\alpha) \;\; in \;\; \mathcal{G} \quad \text{and} \tag{35}$$

$$\Psi_{,ii} = -\frac{F}{\varepsilon_0\,\varepsilon_r} \sum_{\alpha=1}^{N_f} (z_\alpha\, c_\alpha) \quad in \;\; \mathcal{S}\,. \tag{36}$$

Material Laws for the Mechanical Field

It is assumed that the gel shows a linear elastic behavior, therefore:

$$\sigma_{ij} = E_{ijkl}\,\varepsilon^e_{kl} \;\; in \;\; \mathcal{G} \tag{37}$$

where $\boldsymbol{\varepsilon}^e$ is the elastic strain and $\mathbf{E}$ is the fourth-order tensor of elasticity.

The total strain $\boldsymbol{\varepsilon}$ is the sum of the elastic strain $\boldsymbol{\varepsilon}^e$ and the swelling strain $\boldsymbol{\varepsilon}^\pi$:

$$\varepsilon_{kl} = \varepsilon^e_{kl} + \varepsilon^\pi_{kl}\,. \tag{38}$$

The swelling strain may be obtained by

$$\varepsilon^\pi_{ij} = \Delta\pi\,\tilde{g}_{ij} \tag{39}$$

where $\tilde{\mathbf{g}}$ is the tensor of the swelling coefficients with:

$$\tilde{g}_{kl} = g\,\delta_{kl} \tag{40}$$

Therefore the stress tensor can be calculated as:

$$\sigma_{ij} = E_{ijkl}\,\varepsilon^e_{kl} = E_{ijkl}\,(\varepsilon_{kl} - \varepsilon^\pi_{kl})\,. \tag{41}$$

Inserting Eq. (41) into Eq. (25), we finally obtain the governing equation of the mechanical field

$$\sigma_{ij,i} = \left(E_{ijkl}\left(\varepsilon_{kl} - \varepsilon^\pi_{kl}\right)\right)_{,i} = 0 \;\; in \;\; \mathcal{G}\,. \tag{42}$$

Please note, that in the isotropic case the elasticity tensor may be given by

$$E_{lmnp} = \frac{E}{2(1+\nu)}\left[\delta_{ln}\,\delta_{mp} + \delta_{lp}\,\delta_{mn} + \frac{2\,\nu}{1-2\,\nu}\,\delta_{lm}\,\delta_{np}\right] \tag{43}$$

where E is the Young's modulus and ν the Poisson ratio.

Governing Equations, Boundary Conditions and Initial Conditions

In this section, all the equations and the corresponding boundary conditions (BCs) and initial conditions (ICs) which are necessary to completely describe the chemo-electro-mechanical problem, are given.

Governing Equations

For the considered chemo-electro-mechanical multi-field problem, the governing equations, presented in detail in section "Continuum Chemo-Electro-Mechanics", will be given in summary in Box 1.

1. For the mobile ions

$$\frac{\partial c_\alpha}{\partial t} = \left(D_\alpha\, c_{\alpha,i} + z_\alpha\, c_\alpha\,\frac{F}{R\,T}\,\Psi_{,i}\right)_{,i} \quad in \;\; \mathcal{P}$$

2. For the bound anions

$$c_{A^-} = c^0_{A^-}\,(1 - \varepsilon_{ii}) \;\; in \;\; \mathcal{G}$$

3. For the reference concentrations

$$\left(D^{ref}\,c^{ref}_{\alpha,i}\right)_{,i} = 0 \;\; in \;\; \mathcal{G} \cup \varepsilon$$

4. For the electric potential

$$\Psi_{,ii} = -\frac{F}{\varepsilon_0\,\varepsilon_r}\left(z^+\,c^+ + z^-\,c^- + z_{A^-}\,c_{A^-}\right) \;\; in \;\; \mathcal{G}$$

$$\Psi_{,ii} = -\frac{F}{\varepsilon_0\,\varepsilon_r}\left(z^+\,c^+ + z^-\,c^-\right) \;\; in \;\; \mathcal{S}$$

5. For the displacements

$$\left[E_{ijkl}\left(\varepsilon_{kl} - g\,\delta_{kl}\,R\,T\sum_\alpha\left(\Delta c_\alpha - \Delta c^0_\alpha\right)\right)\right]_{,i} = 0 \;\; in \;\; \mathcal{G}$$

where

$$\varepsilon_{ij} = \frac{1}{2}\left(u_{i,j} + u_{j,i}\right) \;\; in \;\; \mathcal{G}$$

and

$$\Delta c_\alpha = (c_\alpha - c^{ref}_\alpha) \quad \text{and} \quad \Delta c^0_\alpha = (c_{\alpha 0} - c^{ref}_{\alpha 0}) \;\; in \;\; \mathcal{G}$$

Box 1: Field equations of the chemo-electro-mechanical field

Boundary Conditions and Transition Conditions

In this section, both, the boundary and transition conditions which are necessary to solve the equations given in the section "Governing Equations" are formulated in Boxes 2 and 3.

As [| |] in Box 3 equals zero for the electric potential, the electro-chemical potential, and the normal components of the dielectric displacement and the fluxes, this means that these quantities are all continuous at the interface.

The definitions of the domains and the boundaries are as follows:

$$\mathcal{P} = \mathcal{G} \cup \mathcal{S}\,,$$

$$\varnothing = \mathcal{G} \cap \mathcal{S}\,,$$

$$\partial\mathcal{P} = (\partial\mathcal{G} \cup \partial\mathcal{S}) \setminus (\partial\mathcal{G} \cap \partial\mathcal{S})\,,$$

$$\partial\mathcal{P} = \partial\mathcal{P}_\Psi \cup \partial\mathcal{P}_\omega = \partial\mathcal{P}_j \cup \partial\mathcal{P}_\mu\,,$$

$$\varnothing = \partial\mathcal{P}_\Psi \cap \partial\mathcal{P}_\omega = \partial\mathcal{P}_j \cap \partial\mathcal{P}_\mu\,,$$

$$\partial\mathcal{G} = \partial\mathcal{G}_{u_l} \cup \partial\mathcal{G}_{p_l}\,,$$

$$\varnothing = \partial\mathcal{G}_{u_l} \cap \partial\mathcal{G}_{p_l}\,,$$

see also Figs. 3 and 4.

1. For the flux, the electro-chemical potential and the reference concentrations

$$j_l^\alpha\, n_l = \bar{j}^\alpha \ \text{ on } \ \partial\mathcal{P}_j$$

$$\mu_\alpha = \bar{\mu}_\alpha \ \text{ on } \ \partial\mathcal{P}_\mu$$

$$c_\alpha^{ref} = c_\alpha \ \text{ on } \ \partial\varepsilon$$

where **n** is the unit normal vector at the boundary, defined positive in outwards direction.

2. For the electric potential and the dielectric displacement

$$\Psi = \bar{\Psi} \ \text{ on } \ \partial\mathcal{P}_\Psi$$

$$\tilde{D}_l\, n_l = -\omega_e \ \text{ on } \ \partial\mathcal{P}_\omega$$

where ω_e is the charge density per area.

3. For the displacements and stresses

$$u_l = \bar{u}_l \ \text{ on } \ \partial\mathcal{G}_{u_l}$$

$$\sigma_{ml}\, n_m = \bar{p}_l \ \text{ on } \ \partial\mathcal{G}_{p_l}$$

where **p** is the traction vector.

Box 2: Boundary conditions for the chemo-electro-mechanical problem

$$[|\Psi|] = 0 \ \text{ on } \ \partial\mathcal{G} \cap \partial\mathcal{S}$$

$$[|D_l|]\, n_l^{\mathcal{G}} = 0 \ \text{ on } \ \partial\mathcal{G} \cap \partial\mathcal{S}$$

$$[|j_l^\alpha|]\, n_l^{\mathcal{G}} = 0 \ \text{ on } \ \partial\mathcal{G} \cap \partial\mathcal{S}$$

$$[|\mu_\alpha|] = 0 \ \text{ on } \ \partial\mathcal{G} \cap \partial\mathcal{S}$$

where $[|\text{variable}|] = \text{variable}^{\mathcal{G}} - \text{variable}^{\mathcal{S}}$ and

$\mathbf{n}^{\mathcal{G}}$ is the unit normal vector at the gel boundary $\mathcal{G}$, defined positive in outwards direction.

Box 3: Transition conditions for the chemo-electro-mechanical problem

Initial Conditions

The initial conditions (at the initial state equilibrium) for the given equations are formulated in Box 4.

$$c_\alpha(x_i, t = t_0) = c_\alpha^0(x_i) \ \text{ in } \ \mathcal{P}$$

$$c_{A^-}(x_i, t = t_0) = c_{A^-}^0(x_i) \ \text{ in } \ \mathcal{G}$$

Box 4: Initial Conditions for the chemo-electro-mechanical problem

Discretization

In this paper a full coupling between the chemical, electrical and mechanical field is applied. For the solution of the non-linear system of partial differential equations (PDEs), a Newton-Raphson scheme is used. The discretization in space is realized by the Finite Element Method while the time discretization is conducted by the implicit Euler-backward algorithm. The 2D finite element consists of degrees of freedom for the concentrations of mobile ions c_α and bound charged groups c_{A^-} in the chemical field, the electric potential Ψ in the electric field and the displacement $u = u_x$ and $v = u_y$ in the mechanical field. Additional degrees of freedom are used for the reference concentrations $\tilde{c}_\alpha = c_\alpha^{ref}$. The generalized displacement vector then reads

$$\hat{\boldsymbol{u}} = \left[u\ v\ c^+\ c^-\ c_{A^-}\ \tilde{c}^+\ \tilde{c}^-\ \Psi\right]^T\,. \tag{44}$$

In the Newton-Raphson scheme for the solution of the non-linear system the increment of the generalized displacement $\boldsymbol{\Delta u}_{i+1}$ (between the old iteration i and the actual iteration $i+1$) with

$$\boldsymbol{\Delta u}_{i+1} = \boldsymbol{u}_{i+1} - \boldsymbol{u}_i \tag{45}$$

is calculated – as given in detail in previous publications [42, 44] for 1D problems – from

$$\boldsymbol{K}(\boldsymbol{u}_i)\ \boldsymbol{\Delta u}_{i+1} = \boldsymbol{R}(\boldsymbol{u}_i)\,. \tag{46}$$

where the Jacobian $\boldsymbol{K}(\boldsymbol{u}_i)$ reads

$$\boldsymbol{K}(\boldsymbol{u}_i) = \begin{bmatrix} K_{uu} & K_{uv} & 0 & 0 & 0 & K_{u\tilde{c}^+} & K_{u\tilde{c}^-} & 0 \\ K_{vu} & K_{vv} & 0 & 0 & 0 & K_{v\tilde{c}^+} & K_{v\tilde{c}^-} & 0 \\ 0 & 0 & K_{c^+c^+} & 0 & 0 & 0 & 0 & K_{c^+\Psi} \\ 0 & 0 & 0 & K_{c^-c^-} & 0 & 0 & 0 & K_{c^-\Psi} \\ K_{c_{A^-}u} & K_{c_{A^-}v} & 0 & 0 & K_{c_{A^-}c_{A^-}} & 0 & 0 & 0 \\ 0 & 0 & 0 & 0 & 0 & K_{\tilde{c}^+\tilde{c}^+} & 0 & 0 \\ 0 & 0 & 0 & 0 & 0 & 0 & K_{\tilde{c}^-\tilde{c}^-} & 0 \\ 0 & 0 & K_{\Psi c^+} & K_{\Psi c^-} & K_{\Psi c_{A^-}} & 0 & 0 & K_{\Psi\Psi} \end{bmatrix}. \tag{47}$$

The vector $\boldsymbol{R}(\boldsymbol{u}_i)$ contains the residuals of the applied differential equation system.

The reference concentrations $\tilde{c}_\alpha$ are coupled only with themselves and via the osmotic pressure difference with the displacements u and v. In the model setup a local constraint is defined in the solution at the boundary of the gel-solution interface in order to tie the reference concentrations $\tilde{c}_\alpha$ to their respective mobile ion concentrations c_α.

Between these constraints the reference solution is obtained by solving the diffusion equation (32) in the 2D domain.

The resulting four-noded element with bi-linear shape functions now contains 32 degrees of freedom. This allows the simulation of both, chemical and electrical stimulation, as the osmotic pressure difference for the swelling strain ε_π of the gel is calculated locally.

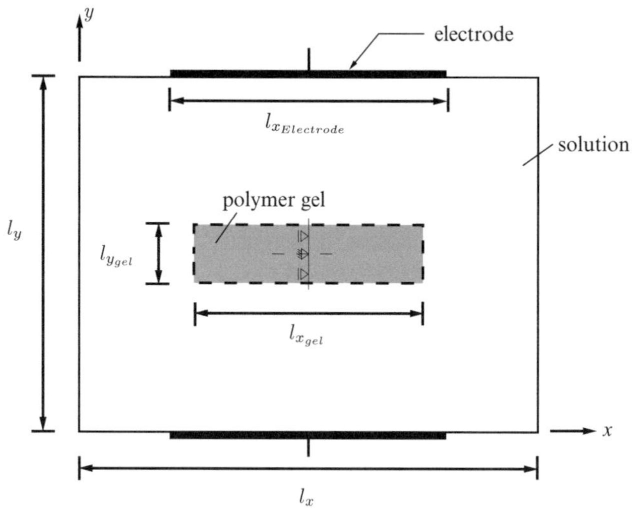

Fig. 7 Polymer gel in solution bath under electrical stimulation

Numerical Simulation

In this section the application of hydrogels as bending actuators is investigated. For this reason, a hydrogel (5×1 mm) is placed in a solution bath (15×11 mm). In the numerical test cases, an electrical stimulation is performed, i.e. in both test cases, an electric potential difference is applied between the electrodes which are placed at the top and bottom of the solution bath, see Fig. 7. In test case 1, only the chemo-electrical coupling, while in test case 2, the full coupling is applied.

The initial concentrations in the NaCl-solution bath are $c^0_{Na^+} = 1.0$ mM and $c^0_{Cl^-} = 1.0$ mM. In the gel, the initial concentration of bound charges in equilibrium state is $c^0_{A^-} = 1.0$ mM. The electric potential as well as the mobile and bound concentrations are depicted in Fig. 8. The unknowns are just shown versus y at $x = \frac{l_x}{2} = 7.5$ mm (in the middle of the gel and the solution domain), as there occur no gradients in space outside the regions of the boundary layer.

The gel-solution domain is discretized with 400 by 400 elements in space (using 200×200 elements for the gel). In order to capture the high gradients of the concentrations and of the electric potential in the gel-solution interface, an adaptive mesh refinement is applied, see Fig. 9.

The parameters of the numerical simulation are given in Table 1.

The numerical simulation is carried out as follows: The simulation is performed for the whole domain, (i) gel, i.e. polymer network and solvent, and (ii) solution. For computing the mechanical deformation, we then have (i) the polymer domain and (ii) a polymer free domain.

In the test cases, at $t = 0$ s, an electric field between the upper ($y = y_{top} = l_y = 11$ mm) and the lower electrode ($y = y_{bottom} = 0$ mm) is applied by imposing different electric potentials. The electric potential is prescribed versus time by a ramp function reaching the final values

$$\Psi(y = y_{bottom}) = 0.05\,\text{V} \quad \text{and} \quad \Psi(y = y_{top}) = -0.05\,\text{V}\,. \tag{48}$$

The mobile concentrations at the boundaries of the solution bath are prescribed $c_{Na^+} = c_{Cl^-} = 1$ mM.

Test Case 1: Chemo-Electrical Simulation

In the first test case, only the chemo-electrical coupling is investigated by setting the swelling coefficient $g = g_1$ equal to zero.

The steady-state solution of the electric potential versus the gel-solution domain is depicted in Fig. 10.

In Fig. 11a, b, the electric potential versus y at $x = \frac{l_x}{2} = 7.5$ mm and versus x at $y = \frac{l_y}{2} = 5.5$ mm is given. It can be seen that the decrease over y of the electric potential is larger in the solution than in the gel. This is compensated by a larger jump at the anode side of the gel (at $y = 5$ mm) than at the cathode one ($y = 10$ mm), see Fig. 11a. The electric potential versus x is nearly constant in both, gel and solution. The jump at the interface which is solved by five elements is $\Delta\Psi \approx -12.5$ mV, see Fig. 11b. This jump is equivalent to the one at initial equilibrium state which may be obtained by solving the Donnan equation between gel (g) and solution (s)

$$\Delta\Psi = \Psi^{(g)} - \Psi^{(s)} = -\frac{1}{z_\alpha}\frac{RT}{F}\ln\left(\frac{c_\alpha^{(g)}}{c_\alpha^{(s)}}\right)\,. \tag{49}$$

In Fig. 11c, d, the concentrations of the mobile and bound charges versus y at $x = 7.5$ mm and versus x at $y = 5.5$ mm

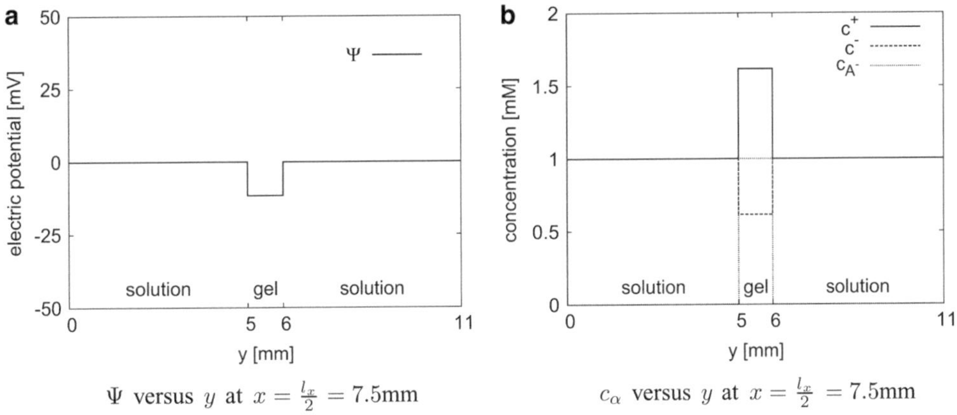

Fig. 8 Initial equilibrium state: electric potential and concentrations versus y. (**a**) Ψ versus y at $x = \frac{l_x}{2} = 7.5\,\text{mm}$. (**b**) c_α versus y at $x = \frac{l_x}{2} = 7.5\,\text{mm}$

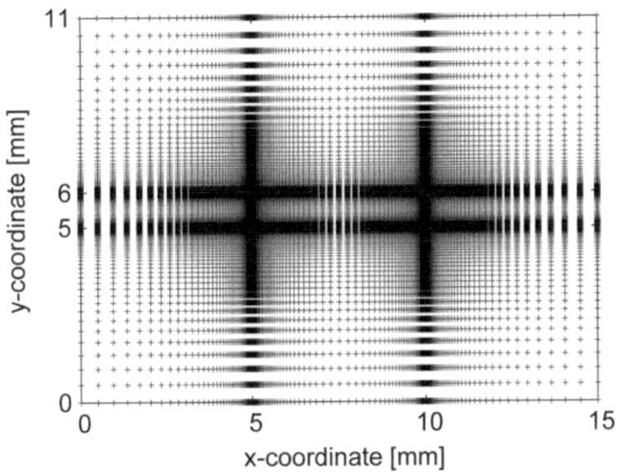

Fig. 9 Adaptive mesh applied for the numerical simulation

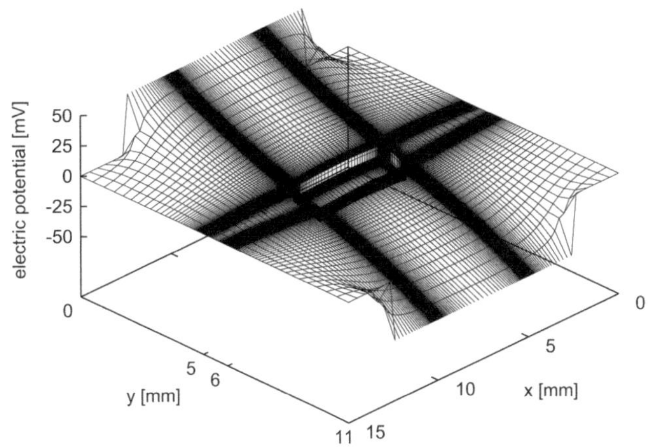

Fig. 10 Electric potential Ψ versus length x and width y

Table 1 Parameters for the numerical simulation

Parameter		Value	Dimension
Domain length	l_x	15	mm
Domain width	l_y	11	mm
Gel length	$l_{x_{gel}}$	5	mm
Gel width	$l_{y_{gel}}$	1	mm
Electrode length	$l_{x_{Electrode}}$	10	mm
Diffusion coefficient Na^+	D_{Na^+}	$1.0 \cdot 10^{-7}$	$m^2\,s^{-1}$
Diffusion coefficient Cl^-	D_{Cl^-}	$1.0 \cdot 10^{-7}$	$m^2\,s^{-1}$
Cation valence	z^+	1.0	–
Anion valence	z^-	−1.0	–
Bound charges valence	z_{A^-}	−1.0	–
Temperature	T	293	K
Gas constant	R	8.3143	$J\,mol^{-1}\,K^{-1}$
Faraday constant	F	96,487	$C\,mol^{-1}$
Permittivity	$\varepsilon_0\,\varepsilon_r$	$8.854 \cdot 10^{-10}$	$A\,s\,V^{-1}\,m^{-1}$
Young's modulus (of polymer)	E	1.0	$N\,m^{-2}$
Poisson ratio	ν	0.37	–
Test case 1: Swelling coefficient	$g = g_1$	0	$m^3\,J^{-1}$
Test case 2: Swelling coefficient	$g = g_2$	$1.0 \cdot 10^{-3}$	$m^3\,J^{-1}$

are given. As the gel remains undeformed, see Fig. 12, the concentration of the bound charges in the gel is $c_{A^-} = c^0_{A^-}$ throughout the gel.

It can be seen that the concentrations versus x remain the same as prescribed, see Fig. 11d. The mobile concentrations show a decrease versus y in the solution for both cations and anions satisfying the boundary conditions at the bottom $c_{Cl^-}(y = 0\,\text{mm}) = c_{Na^+}(y = 0\,\text{mm})$ and top boundary $c_{Cl^-}(y = 11\,\text{mm}) = c_{Na^+}(y = 11\,\text{mm})$. Inside the gel, the concentration of bound charged groups is constant and the mobile concentration increases versus y satisfying the electro neutrality condition.

Normally, the larger jump in the electric potential and in the quotient of the concentrations would lead to an increase of the gel geometry on the anode side and a decrease at the cathode side resulting in a bending of the gel towards the cathode. As the swelling coefficient $g = g_1 = 0$, there is no coupling between the osmotic pressure difference (between gel and solution) and the swelling strain ε^π, the gel does not swell and remains straight.

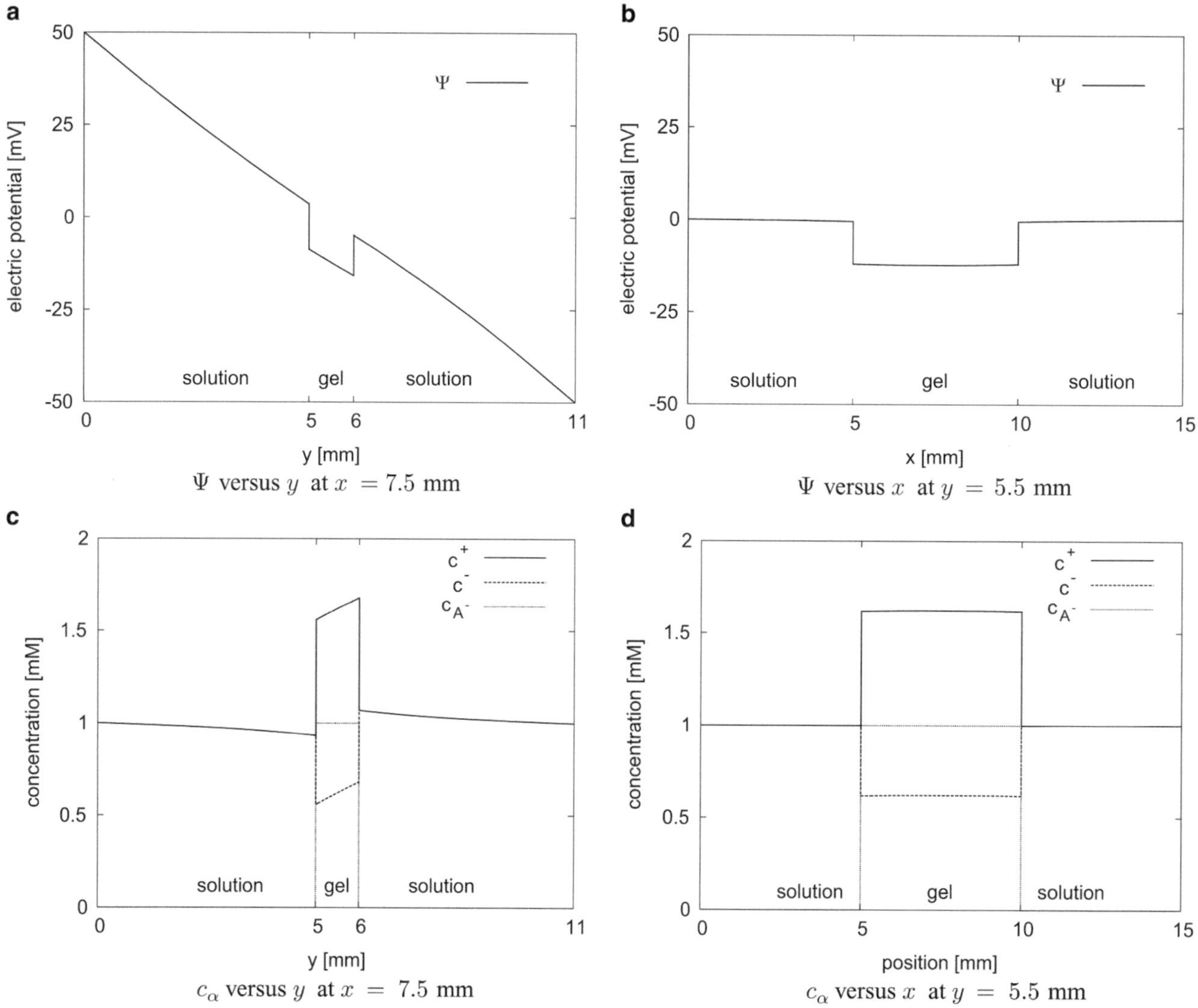

Fig. 11 Test case 1: Electric potential Ψ and concentrations c_α versus x and versus y. (**a**) Ψ versus y at $x = 7.5$ mm. (**b**) Ψ versus x at $y = 5.5$ mm. (**c**) c_α versus y at $x = 7.5$ mm. (**d**) c_α versus x at $y = 5.5$ mm

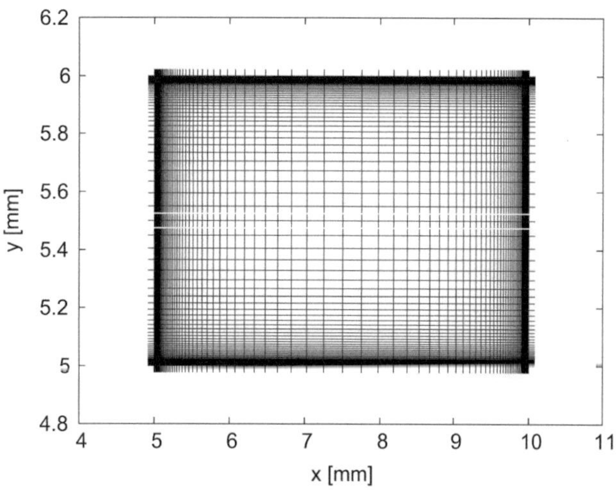

Fig. 12 Test case 1: Undeformed gel

Test Case 2: Electrical Stimulation for the Application of Hydrogels as Bending Actuators

In the second test case, the full chemo-electro-mechanical coupling is investigated; in this test case, the swelling coefficient is set to $g = g_2 = 1 \cdot 10^{-3}\ \mathrm{m^3\ J^{-1}}$. As in this test case the displacements and the change of the concentrations due to this are quite small, the reference concentrations have been prescribed the steady-state value and are not updated each time step.

The steady-state result of the bended gel in the solution bath is depicted, see Fig. 13 and in detail in Fig. 14. It can be seen that the gel is now bended towards the cathode as predicted.

In Fig. 15a, c, the electric potential and the concentrations versus y at the nodes which were initially at $X = 7.5$ mm

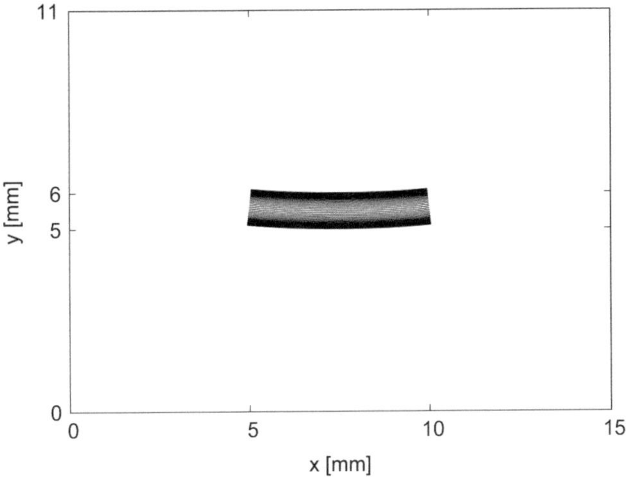

Fig. 13 Test case 2: Bended polymer gel in solution bath

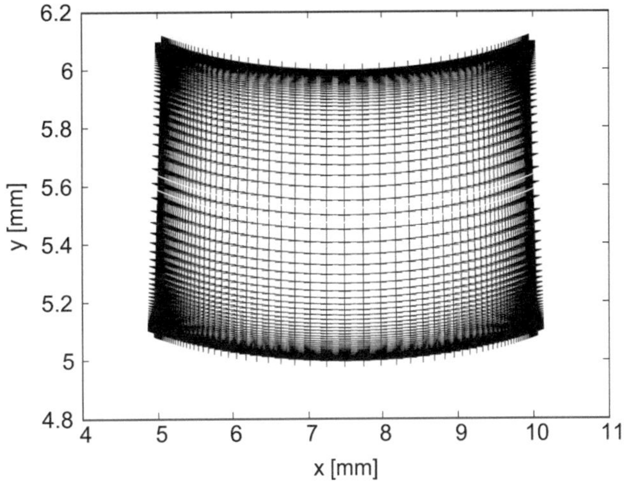

Fig. 14 Test case 2: Bended polymer gel

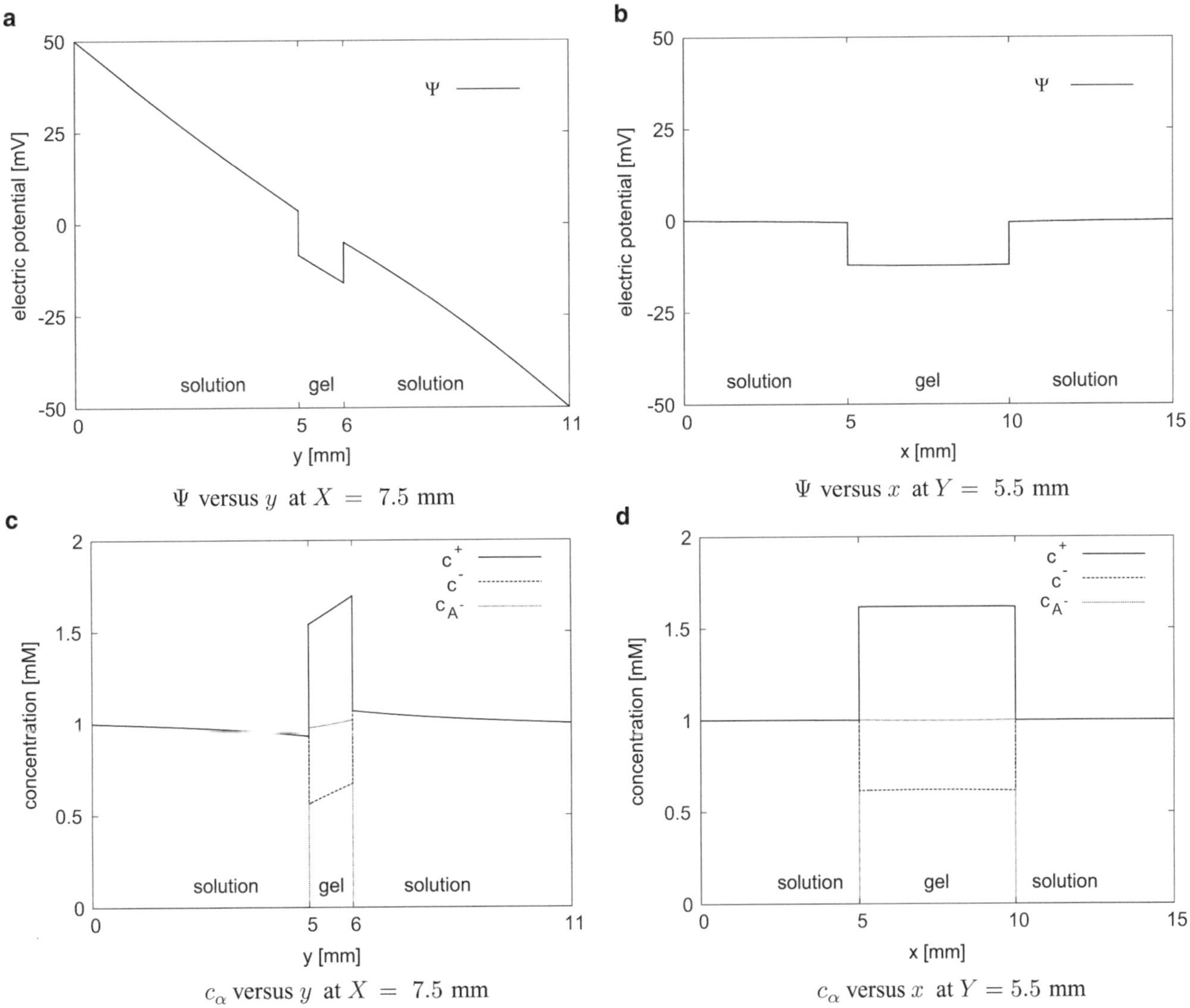

Fig. 15 Test case 2: Electric potential Ψ and concentrations c_α versus x and versus y. (**a**) Ψ versus y at $X = 7.5$ mm. (**b**) Ψ versus x at $Y = 5.5$ mm. (**c**) c_α versus y at $X = 7.5$ mm. (**d**) c_α versus x at $Y = 5.5$ mm

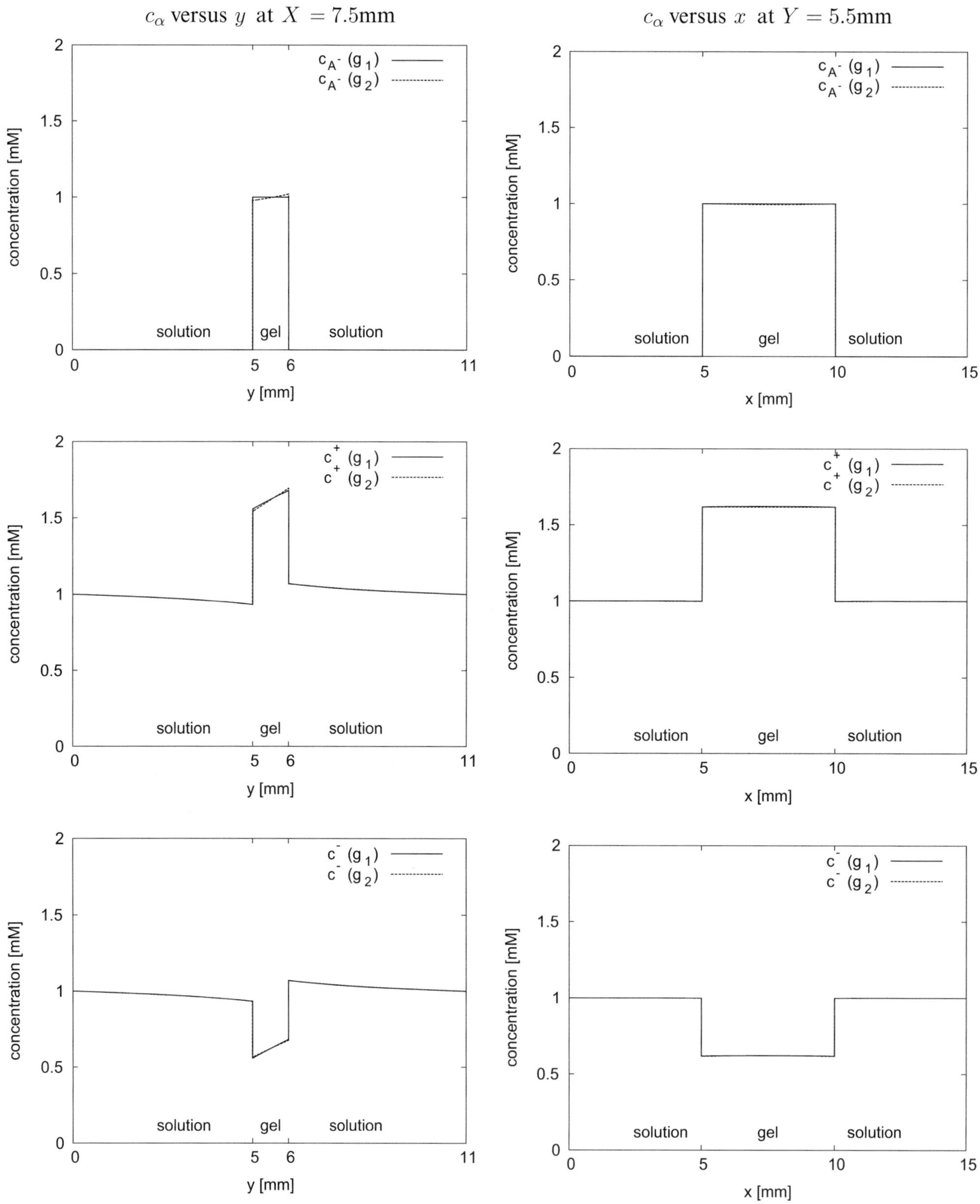

Fig. 16 Comparison of concentrations c_α between test case 1 and test case 2

are depicted, while in Fig. 15b, d, the unknowns versus x at the initial coordinate $Y = 5.5$ mm are given. Please note that the steady-state electric potential and the concentrations are plotted versus the actual coordinates x and y.

Looking at Fig. 15a, c, they are quite similar to Fig. 11a, c. However, it can be seen that the concentration profile and the run of the curve of the electric potential is no longer 100 % straight. This is due to the bending towards the cathode.

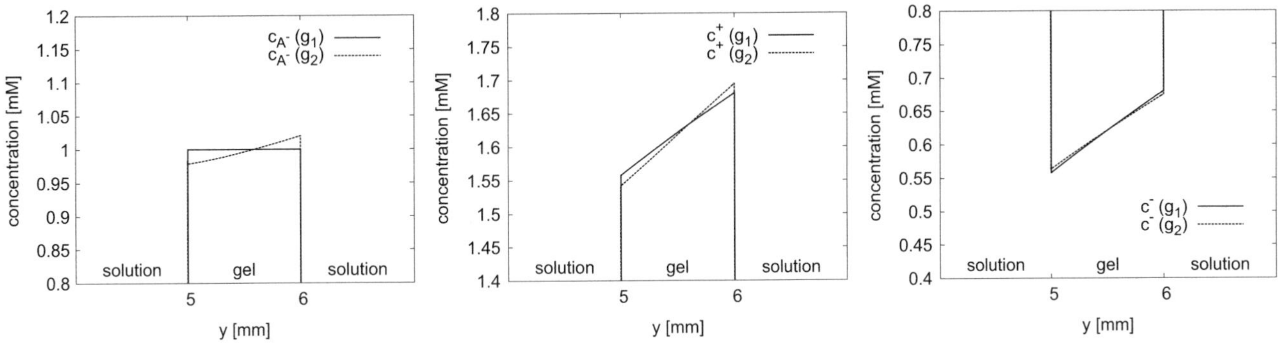

Fig. 17 Comparison of concentrations c_α between test case 1 and test case 2 (detail)

Comparison of the Test Cases

In Fig. 16, a comparison between test cases 1 (electrical stimulation, $g = g_1 = 0$) and 2 (electrical stimulation, $g = g_2$) at $x = 7.5$ mm versus y and at $y = 5.5$ m versus x is shown.

As explained in section "Test Case 1: Chemo-Electrical Simulation", there is a large change in the run of the curve of both, the concentrations and the electric potential between the initial state and the equilibrium states after electrical stimulation.

Comparing the results of the two test cases, it can be seen that – due to the small deformations – the differences are quite small: In the solution there is no modification between the results of both test cases. In the gel domain (see detailed plots in Fig. 17), the increase of the mobile anions versus y is slightly smaller for test case 2 than for test case 1, see Fig. 17 (right). Due to the increase of the polymer gel domain at the anode side and its decrease at the cathode side, there results a decrease of the bound concentration on the anode side and an increase of c_{A^-} at the cathode side, Fig. 17 (left). This leads to a larger increase of the mobile cations versus y for the chemo-electro-mechanical case (test case 2) than for the chemo-electrical case (test case 1) in the gel as well as a reduced jump in the bound concentrations at the anode side and a larger one at the cathode side, see Fig. 17 (left) and (middle).

Concluding, the presented coupled multi-field formulation is predestined for modeling the chemo-electro-mechanical coupling and simulating hydrogel actuation.

Conclusion and Outlook

In the present paper, the complete formulation for describing actuation of polyelectrolyte gels in a solution bath has been presented. After giving the kinematics of the considered continuum body (hydrogel), the balance laws of continuum chemo-electro-mechanics and the constitutive equations of the involved fields were depicted.

In this research, the mechanism of bending actuation – by inducing electric fields – of polyelectrolyte gels has been studied. When varying the electrical field applied to the solution bath surrounding the hydrogel, the employed model is capable of showing the changes of the mobile concentrations, of the electric potential and the locally different swelling in the gel domain. The full chemo-electro-mechanical coupling also leads to a change of the concentration of bound groups resulting in alterations of the local concentrations of the mobile ions, of the electric potential and again of the local gel displacement.

Concluding, it has been shown that the presented coupled multi-field model can excellently describe electrical stimulation of polyelectrolyte gels, and is predestined for simulating hydrogel bending actuators. So, the derived model can also be applied to gel grippers consisting of anionic and cationic polymer gels.

Acknowledgements This research has been financially supported by the Deutsche Forschungsgemeinschaft (DFG) in the framework of Priority Programme 1259 (SPP) "Intelligente Hydrogele" under the grant number Wa 2323/2.

References

1. Ehlers W (2009) Challenges of porous media models in geo- and biomechanical engineering including electro-chemically active polymers and gels. Int J Adv Eng Sci Appl Math 1(1):1–24
2. Flory J, Rehner J (1943) Statistical mechanics of cross-linked polymer networks i. Rubberlike elasticity. J Chem Phys 11(11):512–520
3. Flory J, Rehner J (1943) Statistical mechanics of cross-linked polymer networks ii. Swelling. J Chem Phys 11(11):521–526
4. Doi M, Matsumoto M, Hirose Y (1992) Deformation of ionic polymer gels by electric fields. Macromolecules 25:5504–5511
5. Shiga T, Kurauchi T (1990) Deformation of polyelectrolyte gels under the influence of electric field. J Appl Polym Sci 39:2305–2320
6. Shahinpoor M (1995) Micro-electro-mechanics of ionic polymeric gels as electrically controllable artificial muscles. J Intell Mater Syst Struct 6:307–314
7. Brock D, Lee W, Segalman D, Witkowski W (1994) A dynamic model of a linear actuator based on polymer hydrogel. J Intell Mater Syst Struct 5:764–771

8. Grimshaw PE, Nussbaum JH, Grodzinsky AJ, Yarmush ML (1990) Kinetics of electrically and chemically induced swelling in polyelectrolyte gels. J Chem Phys 93(6):4462–4472
9. de Gennes PG, Okumura K, Shahinpoor M, Kim KJ (2000) Mechanoelectric effects in ionic gels. Europhys Lett 50(4):513–518
10. Tamagawa H, Taya M (2000) A theoretical prediction of the ions distribution in an amphoteric polymer gel. Mater Sci Eng A 285 (1–2):314–325
11. Li H, Luo R, Lam KY (2007) Modeling of ionic transport in electric-stimulus-responsive hydrogels. J Membr Sci 289(1–2): 284–296
12. De SK, Aluru NR (2004) A chemo-electro-mechanical mathematical model for simulation of pH sensitive hydrogels. Mech Mater 36(5–6):395–410
13. Li H, Lai F (2011) Transient analysis of the effect of the initial fixed charge density on the kinetic characteristics of the ionic-strength-sensitive hydrogel by a multi-effect-coupling model. Anal Bioanal Chem 399(3):1233–1243
14. Wallmersperger T, Kröplin B, Holdenried J, Gülch RW (2001) A coupled multi-field formulation for ionic polymer gels in electric fields. In: Bar-Cohen Y (ed) 8th international symposium on smart structures and materials, Newport Beach. Electroactive polymer actuators and devices, vol 4329. SPIE, pp 264–275
15. Wallmersperger T, Kröplin B, Gülch R (2004) Coupled chemo-electro-mechanical formulation for ionic polymer gels – numerical and experimental investigations. Mech Mater 36(5–6):411–420
16. Keller K, Wallmersperger T, Kröplin B, Günther M, Gerlach G (2011) Modelling of temperature-sensitive polyelectrolyte gels by the use of the coupled chemo-electro-mechanical formulation. Mech Mater 18(7):511–523
17. Gerlach G, Guenther M, Sorber J, Suchaneck G, Arndt K-F, Richter A (2005) Chemical and pH sensors based on the swelling behavior of hydrogels. Sens Actuators B Chem 111–112:555–561
18. Orlov Y, Xu X, Maurer G (2006) Equilibrium swelling of N-isopropylacrylamide based ionic hydrogels in aqueous solutions of organic solvents: comparison of experiment with theory. Fluid Phase Equilibria 249(1–2):6–16
19. Orlov Y, Xu X, Maurer G (2007) An experimental and theoretical investigation on the swelling of N-isopropylacrylamide based ionic hydrogels in aqueous solutions of (sodium chloride or di-sodium hydrogen phosphate). Fluid Phase Equilibria 254(1–2):1–10
20. Ermatchkov V, Ninni L, Maurer G (2010) Thermodynamics of phase equilibrium for systems containing N-isopropylacrylamide hydrogels. Fluid Phase Equilibria 296:140–148
21. Scharfer P, Schabel W, Kind M (2007) Mass transport measurements in membranes by means of in situ Raman spetroscopy – first results of methanol and water profils in fuel cell membranes. J Membr Sci 303:37–42
22. Scharfer P, Schabel W, Kind M (2008) Modelling of alcohol and water diffusion in fuel cell membranes. Chem Eng Sci 63(19):4676–4684
23. Yoon J, Cai S, Suo Z, Hayward RC (2010) Poroelastic swelling kinetics of thin hydrogel layers: comparison of theory and experiment. Soft Matter 6(23):6004–6012
24. Prudnikova K, Utz M (2012) Electromechanical equilibrium properties of poly(acrylic acid/acrylamide) hydrogels. Macromolecules 45(2):1041–1045
25. Mann BA, Kremer K, Holm C (2006) The swelling behavior of charged hydrogels. Macromol Symp 237(1):90–107
26. Quesada-Perez M, Maroto-Centeno J, Forcada J, Hidalgo-Alvarez R (2011) Gel swelling theories: the classical formalism and recent approaches. Soft Matter 7:10536
27. Sun DN, Gu WY, Guo XE, Lai WM, Mow VC (1999) A mixed finite element formulation of triphasic mechano-electrochemical theory for charged, hydrated biological soft tissues. Int J Numer Methods Eng 45(10):1375–1402
28. van Loon R, Huyghe JM, Wijlaars MW, Baaijens FPT (2003) 3D FE implementation of an incompressible quadriphasic mixture model. Int J Numer Methods Eng 57(9):1243–1258
29. Ehlers W, Markert B, Acartürk A (2002) A continuum approach for 3-d finite viscoelastic swelling of charged tissues and gels. In: Eberhardsteiner J, Mang HA, Rammerstorfer FG (eds) Proceedings of the fifth world congress on computational mechanics, Vienna, 7–12 July 2002
30. Acartürk A (2009) Simulation of charged hydrated porous materials. PhD thesis, Universität Stuttgart, Stuttgart
31. Walter J, Ermatchkov V, Vrabec J, Hasse H (2010) Molecular dynamics and experimental study of conformation change of poly(N-isopropylacrylamide) hydrogels in water Fluid Phase Equilibria 296:164–172
32. Wallmersperger T, Wittel FK, D'Ottavio M, Kröplin B (2008) Multiscale modeling of polymer gels – chemo-electric model versus discrete element model. Mech Adv Mater Struct 15(3–4):228–234
33. Guenther M, Gerlach G, Wallmersperger T (2009) Non-linear effects in hydrogel-based chemical sensors: experiment and modeling. J Intell Mater Syst Struct 20(8):949–961
34. Bouklas N, Huang R (2012) Swelling kinetics of polymer gels: comparison of linear and nonlinear theories. Soft Matter 8(31):8194–8203
35. Lucantonio A, Nardinocchi P (2012) Reduced models of swelling-induced bending of gel bars. Int J Solids Struct 49(11–12):1399–1405
36. Sadowski G (2011) Special themed issue on "responsive gels". Colloid Polym Sci 289:453
37. Li D, Yang H, Emmerich H (2011) Phase field model simulations of hydrogel dynamics under chemical stimulation. Colloid Polym Sci 289(5–6):513–521
38. Tabatabaei F, Lenz O, Holm C (2011) Simulational study of anomalous tracer diffusion in hydrogels. Colloid Polym Sci 289 (5–6):523–534
39. Wallmersperger T, Keller K, Kröplin B, Günther M, Gerlach G (2011) Modeling and simulation of pH-sensitive hydrogels. Colloid Polym Sci 289(5–6):535–544
40. Poggendorf S, Mba GA, Engel D, Sadowski G (2011) Diffusion of poly(ethylene glycol) and ectoine in NIPAAm hydrogels with confocal Raman spectroscopy. Colloid Polym Sci 289(5–6):545–559
41. Mase G (1969) Schaum's outline of continuum mechanics. McGraw-Hill, New York
42. Wallmersperger T, Ballhause D (2008) Coupled chemo-electro-mechanical FE-simulation of hydrogels – Part II: electrical stimulation. Smart Mater Struct 17(4):045012
43. Wallmersperger T, Ballhause D, Kröplin B, Günther M, Gerlach G (2009) Coupled multi-field formulation in space and time for the simulation of intelligent hydrogels. J Intell Mater Syst Struct 20(12):1483–1492
44. Ballhause D, Wallmersperger T (2008) Coupled chemo-electro-mechanical FE-simulation of hydrogels – Part I: chemical stimulation. Smart Mater Struct 17(4):045011

Molecular Simulations of Hydrogels

Peter Košovan, Tobias Richter, and Christian Holm

Abstract

We review the state-of-the-art of simulational approaches to understand the complex behavior of swollen hydrogels. We concentrate on molecular simulation approaches that use coarse-grained polymer models, and atomistic simulations. Continuum, finite-element, and other modeling approaches are briefly mentioned. The current understanding of the topic, as well as open problems are highlighted.

Keywords

Hydrogel • Swelling • Polymer • Polyelectrolyte • Simulation • Atomistic • Coarse-grained

Introduction

Most synthetic polymers are carbon based, and have a hydrophobic backbone. Water is therefore a poor solvent for them. An important exception are polymers with a large polarity of the backbone, and charged polymers, also called polyelectrolytes (PEs) [1], for which the polar solvent water is normally a good solvent. If one crosslinks these polymers one obtains polymer networks that can swell in aqueous solutions, therefore the name hydrogel. The swelling comes from the tendency of the network chains to adopt a coil-like conformation similar to a free chain in good solvent. The latter is a loose structure, containing much more solvent than polymer. The enormous swelling capacity of charged or polyelectrolyte hydrogels is a consequence of the high osmotic pressure of counterions entrapped in the gel network in order to preserve electroneutrality. To reduce the pressure, the gel swells (the network expands) and absorbs the surrounding solvent, which is counteracted by the elastic response of the polymer chains. The detailed physical picture is rather complicated and involves the electrostatic interactions, polymer-solvent interactions, counterion condensation, etc. [1–4]. The swelling equilibrium is achieved when all the counteracting forces are balanced. Scientific literature on hydrogels is enormous, comprising thousands of papers and several monographs or book chapters, e.g. [2, 4–6].

Dry hydrogels immersed in water can take up in special cases up to 1,000 times water volume compared to their dry volume. The swelling behavior depends on several factors. On one hand there are the parameters relevant for the electrostatic interactions like salt concentration of the swelling agent, counterion valency, solution pH, degree of charging of the polyelectrolyte network. On the other hand it also depends on the network topology, content of non-crosslinked polymers, backbone stiffness and temperature. Extraordinary swelling capabilities of hydrogels and their responses to external stimuli make them useful for a number of applications [5, 7]. Typical examples of commonplace items containing hydrogels are baby diapers or hygiene products. Apart from such large-scale material applications, micro- or nanogels are intensively studied as carriers for

T. Richter • C. Holm (✉)
ICP University Stuttgart, Allmandring 3, 70569 Stuttgart, Germany
e-mail: holm@icp.uni-stuttgart.de; richter@icp.uni-stuttgart.de

P. Košovan
ICP University Stuttgart, Allmandring 3, 70569 Stuttgart, Germany

Faculty of Science, Department of Physical and Macromolecular Chemistry, Charles University in Prague, 120 00 Prague, Czech Republic
e-mail: peter.kosovan@natur.cuni.cz

G. Sadowski and W. Richtering (eds.), *Intelligent Hydrogels*, Progress in Colloid and Polymer Science 140,
DOI 10.1007/978-3-319-01683-2_16,

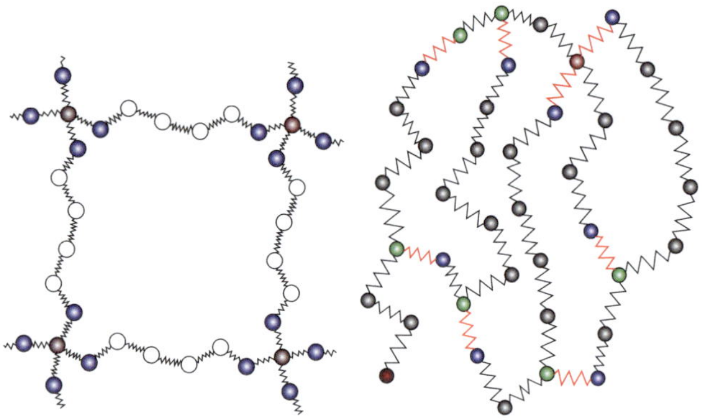

Fig. 1 Illustration of a tetra-functional regular end-linked network (*left*) and a tri-functional randomly crosslinked one (*right*)

controlled release and drug delivery, actuators in microfluidic devices, but also in the pharmaceutical industry, in agriculture or in the food industry [8–13]. The extracellular matrix of living organisms also consists of hydrogels and synthetic hydrogels mimicking its structure have been exploited as smart biomaterials [14–16].

Hydrogels can comprise either physical polymer networks (i.e. a polymer solution above overlap concentration) that are held together by entanglements, or, more typically, the network can be crosslinked. The most straightforward methods to produce a crosslinked polymer network involve incorporating a small fraction of crosslinkers as comonomers. Upon activation by elevated temperature or UV radiation the crosslinkers chemically react with each other producing tetra-functional crosslinked networks. Alternatively, crosslinkers at the end of the chains can react with the main monomers of the polymer, producing tri-functional end-linked networks. In either case, unreacted crosslinker and dangling chain ends can be left in the network structure. Overall, the length of segments in the networks prepared as described above covers a broad range, yielding also irregular mesh size distributions. Consequently, both structure and topology of such network are irregular and random. It is possible to prepare more regularly structured hydrogel networks but these require special synthetic techniques [17]. An alternative route is to crosslink biopolymers which can be produced and controlled on the monomer level. Figure 1 provides a schematic illustration of a regular tetra-functional network and a randomly crosslinked tri-functional one. Note, however, that the network functionality and regularity of its topology are independent parameters and the current selection has been made just for illustration purposes.

In the current review we focus on molecular simulations of polymers which are relevant for understanding the behavior of hydrogels. These include interactions of polymers with water as well as swelling of generic model polymer networks, and in particular charged ones. We focus on progress in the field since the last review on this topic by Escobedo and de Pablo in 1999 [18].

The review is structured as follows: Introduction is followed by a section dedicated to various approaches to modeling of hydrogels, apart from molecular simulations. This is followed by a section about molecular simulations with atomistic resolution. The subsequent section on corase-grained simulations is the main focus of our review. The last section presents some concluding remarks and a brief outlook for future studies.

Approaches to Modeling of Hydrogels

In this section we present a brief overview of other modeling approaches which are closely related to the topic of our article but are outside its scope or have been reviewed by other authors. A detailed overview over earlier theoretical works can be found in the first two chapters of Vol. 109 of Advances in Polymer Science [19, 20].

Phenomenological Approaches

A concise summary of the foundations of phenomenological approaches to hydrogel swelling can be found in the article by Maurer and Prausnitz [21], where the equilibrium conditions for the swelling of a gel are derived from the first two laws of thermodynamics. The gel system consists of two phases: the hydrogel swollen by the solvent and the pure solvent. Both phases are in equilibrium if they are in chemical and mechanical equilibrium. The mechanical contribution originates from the fact that the osmotic pressure inside and outside can be different. Additionally the hydrogel is an elastic object and not a pure fluid.

There are different approximations that can be used for the elastic properties of the network, which were used by different groups. In the article by Ballhause and Wallmersperger [22] the gel elasticity is characterized by a constant Young's modulus. They propose a set of equations for treating the chemical, electrical and mechanical fields. Furthermore, they couple the mechanical and chemical fields via a strain caused by the osmotic pressure difference. With this model they can quantitatively reproduce the swelling induced by a chemical stimulus, namely variation of the electrolyte concentration outside of the hydrogel. In a following publication [23] they demonstrated that this model is also capable of qualitatively describing the response of the gel to an electric field. In contrast to the stimulation by different salt concentrations, the electrical stimulus leads to an asymmetric swelling of the gel. This effect explains the experimentally observed asymmetric bending of a hydrogel in an electric field. In a further concurrent article [24] they also take pH into account. In this way gels with pH-dependent dissociation can be modeled in various environments.

Flory-Rehner type theories [25] use a modification of the Flory-Huggins theory [26] to describe swelling of polymer gels. Different variations of this type of theories have been reviewed recently by Quesada-Pèrez and we refer a potentially interested reader to this review, and references therein [27]. In recent years, several works appeared which use field-theoretical approaches to model the swelling of charged polymer gels [28–30].

Sadowski and coworkers developed an approach for modeling complex mixtures of fluids [31] which is based on the work of Chapman [32]. In the original PC-SAFT approach the Helmholtz free energy is approximated by independent terms that handle the different contributions due to hard-chain interactions, dispersion, and association. Arndt and Sadowski [33] extended it to enable the modeling of hydrogels by including the elastic response of the network. Their approach successfully describes a hydrogel in pure solvents like water, methanol and ethanol as well as mixtures of water and ethanol. Here, the cononsolvency effect is captured quite well.

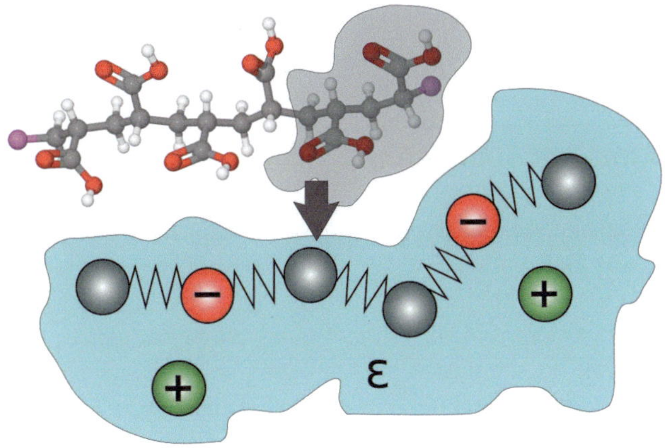

Fig. 2 Schematic illustration of the coarse-graining procedure. One or several monomer units of the polymer are combined to one effective monomer unit in the coarse-grained system. The coarse-grained units interact via effective potentials, usually of the Lennard-Jones type. Charges on monomers and small ions are modeled explicitly. Usually the solvent is integrated out and represented by a dielectric medium with dielectric constant ε

Molecular Simulations

Molecular simulation methods [34, 35] use interaction potentials and topological information about bonding as direct input. With these input parameters system configurations in various thermodynamic ensembles can be generated either using Monte Carlo (MC) or Molecular Dynamics (MD) methods. The MD method is based on integration of Newton's equations of motion and besides system configurations it provides also information on the dynamical evolution of the system. The MC method is based on generating system configurations according to Markov chain theory, and accepting them with some probability [36]. In most of its versions, MC cannot be used to study temporal evolution. However, it can be used to generate new system configurations in an unphysical way which is a great advantage e.g. if high energy barriers occur between different states of the system or in simulations of phase equilibria. The obtained configurations are analyzed using statistical physics in order to obtain thermodynamic properties of the investigated systems. Both MC and MD methods can in principle provide all kinds of thermodynamic data but their efficiency can largely differ, depending on the problem under consideration. For more information we refer the reader to textbooks dedicated to the topic [34–36].

The key input of a molecular simulation is a model of the system under investigation, in particular the interaction potentials (the force-field) between the studied molecules. It is convenient to classify the commonly used models in two groups: atomistic and coarse-grained. In atomistic models each atom and each covalent bond is modeled as a separate entity. They preserve all molecular information down to the scale on which classical physics is still applicable. Computations using atomistic models are typically expensive and limited to small systems and short time scales. In order to simulate bigger systems on longer time scales, one reduces the resolution and is simultaneously gaining CPU time through a coarse-graining procedure, where the degrees of freedom that are considered "unimportant" are being integrated out. The first coarse-graining level comprises united atom models, in which several small atoms (e.g. a methyl group) are modeled as a single entity. Their resolution is close to that of the atomistic models and in this review we do not make a distinction between full atomistic and united atom models.

Further steps in coarse-graining include uniting one or several monomer units into one coarse-grained particle which results in the so-called bead-spring polymer model. Furthermore, solvent in the coarse-grained models is often treated implicitly as a dielectric continuum with a given permittivity ε while charges are typically included explicitly in the model. The concept of coarse-graining is schematically illustrated in Fig. 2. The coarse-grained models lose most of the chemical specificity inherent to atomistic models. They often use generic interaction potentials to account for connectivity, excluded volume and polymer-solvent interactions. On the other hand, they enable simulations of much larger systems over a longer time-span (measured in experimental time units). With these models it is possible to study the effect of chain length, crosslinking density, fraction of charged monomers or added salt on the swelling behavior of the hydrogel.

Atomistic Simulations

Atomistic simulations are currently restricted to about 10^6 atoms and runtimes of 10^{-7} s. In practical terms, it means that a single polymer chain composed of tens of monomer units immersed in solvent molecules can be simulated. It is possible to simulate bigger systems on massively parallel computer architectures but the real bottleneck is the slow dynamics of long polymer chains. Parallelization can reduce the CPU time per time step only down to a certain level. Thus longer simulations can only be achieved by longer simulation runtimes. Typical runtime is several days and cannot be easily extended by several orders of magnitude which would be required for larger polymeric systems. Since CPU power in the last years increases through parallelization rather than clock frequency, this unfavorable situation is not expected to change in near future, unless an algorithmic breakthrough appears.

With the above limitations it becomes clear that atomistic simulations of hydrogels are restricted to single chains. Here they provide an invaluable tool for explanations of specific effects which are sensitive to details of the molecular structure. In such cases only the atomistic simulations can provide molecular interpretation of the effects in terms of e.g. change in the solvation shell around the polymer, hydrogen bonding, cononsolvency or similar. These are especially important for solubility of uncharged polymers in water, where the relation between the polymer molecular structure and the structure of water around it plays an important role.

Poly(Vinyl Alcohol)

Several simulation studies investigated the behavior of poly(vinyl alcohol) (PVA) which is a typical uncharged water-soluble polymer forming hydrogels. Tamai and Tanaka [37] investigated the influence of the polymer on the behavior of water. They found that the presence of PVA prevents water from a subtle phase transition between the strong and fragile character liquid [38]. A similar result was found by Chiessi and coworkers [39] where the presence of PVA alters the water-water correlation function up to 2 nm away from the hydrogel.

A series of studies from Müller-Plathe and van Gunsteren [40] and Müller-Plathe [41, 42] also deal with PVA 15-mer in water and in water/ethanol mixtures. They have used the SPC water model in combination with force field for PVA obtained from ab initio calculations. The first article [40] compared solvent structure and dynamics of water, ethanol and 1:1 water/ethanol mixture solvating PVA. It has revealed the clear hydrophilic character of PVA, manifested in the formation of PVA-water hydrogen bonds and in preferential solvation by water. The analysis of hydrogen bond lifetimes has shown that the solvent dynamics near the polymer is slower than in the bulk. In [41] and [42] the origin of different dynamic properties of water is tracked down to confinement and the different correlations of the rotational reorientation times. These are slowed down close to the polymer and are faster when the water is surrounded by other water molecules. This is correlated to the number of hydrogen bonds, which typically increases in the vicinity of PVA. All articles from Müller-Plathe indicate that the slowdown of solute diffusion in PVA-water system can be described treating the polymer as simple geometric obstacles, leading to the relation $D/D_0 = [(1-\phi)/(1+\phi)]^2$. At several spots Müller-Plathe in his articles pointed out that his simulations could not capture all system-relevant time scales, particularly those related to relaxation of the whole chain. In addition, he emphasized that confidence in validity of simulation results can only be gained by quantitatively comparing properties which can be calculated as well as measured experimentally.

Poly(*N*-Isopropyl Acrylamide)

The behavior of poly(*N*-isopropyl acrylamide) (PNiPAAm) in aqueous solutions has become a subject of numerous simulation works. Upon an increase of temperature, the polymer exhibits a collapse transition (Lower Critical Solution Temperature, LCST), experimentally observed around 32 °C (305 K). The phenomenon of cononsolvency refers to PNiPAAm being soluble both in pure water and methanol but poorly soluble in a composition window of the solvent mixture at the side of low methanol content. The simulation studies are aimed at molecular-level explanation of this behavior.

Tönsing and Oldiges [43] were in 2001 presumably first who studied structure of water around PNiPAAm oligomers composed of 15, 23, and 33 units combined with 2, 3, and 4 crosslinker molecules at three temperatures of 285, 300, and 325 K. They used the GROMOS force field for the polymer and the SPC/E water model. Initial equilibration of 200 ps followed by a production run of 250 ps revealed preferential presence of the water hydrogens around the oxygen atoms of PNiPAAm and of the water oxygens around the hydrogen of the N–H group of the polymer. These findings corroborate the notion of strong hydrogen bonding between PNiPAAm and water.

A similar study, focusing on the interactions of PNiPAAm with water molecules was conducted by Longhi et al. in 2004 [44]. They used a 50-mer of PNiPAAm pre-equilibrated in vacuo for 50 ps and subsequently simulated for 4.4 ns in water at temperatures of 300 and 310 K, i.e. slightly below and above its LCST. The polymer was modeled using the Amber force field and the partial charges were assigned based on a custom quantum-mechanical calculation. The authors presented an analysis of the number of molecules in the first and second hydration shell and observed a reduction in the number of hydration waters as the temperature was increased from 300 to 310 K. At the same time they observed 10–15 % increase in the mobility of water molecules in the vicinity of the polymer. A single hairpin-like structure was found to persist from the initial conformation throughout the whole simulation at either temperature, indicating strong intramolecular hydrogen bonding but also the fact that the polymer did not explore the whole conformational space within the simulation runtime. Nevertheless, the main message is clear and consistent with other simulation studies: hydrogen bonding and water structure around PNiPAAm are related to the conformational transition around its LCST.

Considerably longer simulations of 75 ns of a 26-mer of PNiPAAm and of the same polymer "doped" by amino acid units have been performed by the same group in 2008 [45]. Similarly to the previous study, the authors concentrated on hydrogen bonding between the polymer and water molecules and showed how the presence of comonomers alters its conformation around LCST. In addition they examined how the presence of amino acid comonomers alters the polymer-water interactions.

Hasse and coworkers [46] investigated conformational transitions of PNiPAAm in aqueous solutions. They used combinations of several atomistic force fields and water models to see if they can reproduce the experimentally observed LCST of PNiPAAm in water. In the preliminary work they used the all-atom force field OPLS-AA and the united atom force fields GROMOS-96 53A6 (G53A6) and GROMOS-87 (G87) in all possible combinations with the SPC/E and TIP4P water models. It turned out that only OPLS + SPC/E and G53A6 + TIP4P combinations yield both a stretched PNiPAAm conformation at 280 K (well below LCST) and a collapsed conformation at 360 K (well above LCST). Compared to experiments with the transition temperature of about 305 K and the width of the transition of 20 K, the OPLS + SPC/E combination yields a gradual conformational transition of 40 K width with LCST about 320 K. The G53A6 combination yields a sharp transition at 353 ± 3 K. The work of Walter et al. [46] has clearly shown that a simulation of such a subtle thing as the conformational transition is very sensitive to the choice of the force field and in many cases can result in a quantitative failure. The OPLS + SPC/E and G53A6 + TIP4P combinations appear robust in this respect, even though the force fields have not been trained on such data and they have been recommended for use with other water models.

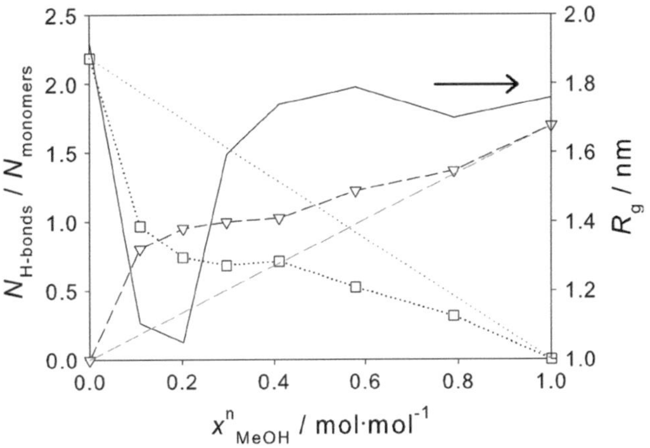

Fig. 3 Average number of hydrogen bonds between a monomer unit of the PNiPAAm chain and water (□) or methanol (∇) and as a function of methanol mole fraction. Changes in the hydrogen bond numbers correlate with changes in the Radius of Gyration (*solid line*) (Adapted from [47]. Copyright (2012) American Chemical Society)

In a subsequent study in 2012 Walter et al. [47] studied the cononsolvency of PNiPAAm in water/methanol mixture. Experimental literature [48–50] was suggesting preferential adsorption of methanol on the polymer but a clear molecular picture was missing. In this study, the authors used the OPLS + SPC/E force-field combination, which they had validated in the previous work [46], supplemented with the GROMOS-96 force field for methanol. The simulations could very well capture the experimentally observed composition dependence of cononsolvency. Furthermore, they could also capture the effect of temperature change upon crossing the LCST of PNiPAAm in water. Based on the analysis of the number of hydrogen bonds between the polymer and water or methanol shown in Fig. 3, the authors concluded that methanol binds to the polymer much stronger than water. Especially at low methanol content its concentration around the polymer significantly differs from that in the bulk. Figure 3 also shows that the change in the number of hydrogen bonds correlates with the conformational change of the polymer quantified by its radius of gyration. Preferential orientation of methanol, exposing the hydrophobic methyl groups to the outside results in poor solubility of such methanol-solvated polymer in the bulk solvent composed predominantly of water.

It is instructive to follow the chronological development of simulations of aqueous PNiPAAm solutions. Tönsing and Oldiges first concluded from a 250 ps simulation about the presence of hydrogen bonding between water and the polymer. Later, Longhi et al. from a 4 ns simulation arrived at similar conclusions but also showed that even though water equilibrates quickly around the polymer, the whole chain needs much longer time to explore its conformational space. Finally, Walter et al. showed that only certain force field combinations can reproduce the experimentally observed conformational transition, even though they considered much broader temperature window than in previous studies. In addition they showed that a shorter polymer than those studied previously needs about 10 ns to reach equilibrium conformation and the temperature at which the transition occurs in the simulation can notably differ from the experimentally observed one. Even though the computing power increases rapidly and so does also the runtime of the simulations, still the maximum achievable runtime combined with the limitations of the available force fields restrict the use of full atomistic simulations of hydrogels to studies of conformational changes of oligomers or restructuring of small molecules around the polymer.

Coarse-Grained Simulations

In the following, we review the coarse-grained simulations of hydrogels. In contrast with the atomistic simulations, the coarse-grained models are not capable of handling the molecular interactions in detail. On the other hand, the neglect of the details enables studies of significantly bigger systems over longer time spans which are necessary to examine the effects of topology, crosslink density, polydispersity and similar effects.

Neutral Hydrogels

In this section we include models of hydrogels which do not contain bare charges. Typical real-world examples of such polymers include PNiPAAm or PVA hydrogels.

Models of neutral hydrogels could be further classified as athermal, i.e. with pure excluded-volume interactions, and thermal which in addition to excluded volume contain an enthalpic term in the interaction, conventionally denoted by ϵ as the depth of the potential well.

In 1999 Escobedo and de Pablo published a review [18] which summarizes simulation works on polymer networks and gels up to that time. In addition, it also covers simulations of Everaers and Kremer on elasticity and kinetics of crosslinking of rubber-like polymer networks [51, 52]. Although the latter is a closely related topic, it is outside the scope of the current review since rubber-like polymers are typically much denser and solvent-free polymer networks. Therefore we omit them here and rather refer a potentially interested reader to the above-mentioned review [18] and references therein.

In the 1990s Escobedo and de Pablo published a series of simulation studies of hydrogels [53–55] accompanied by methodological developments to enable MC simulations of hard particle networks at constant pressure and fixed chemical potential of the solvent [56–60]. A thermal polymer was modeled by a square-well potential. In the first article [53] they studied the compressibility behavior of linear and branched polymers which they compared to the behavior of ideal diamond-like networks with different lengths of strands connecting the nodes. They have shown that at vanishing pressure the network exists at a finite packing fraction which is actually lower than what would correspond to full stretching. This is in strong contrast with uncrosslinked polymers in which the packing fraction at vanishing pressure approaches zero. Further they conclude that at a semi-quantitative level the *PVT* behavior of the studied systems is well described by the Flory-Rehner theory [61]. In a later article [55] they studied the swelling of both athermal and thermal gels by monomeric solvent, oligomeric solvent and solvent mixtures. The chemical potential of the solvent in the gel phase was set equal to that in the pure solvent via the particle insertion method [62]. In pure solvent cases the authors observed a higher uptake of the smaller solvent molecules as compared to bigger ones. In the mixed solvent they observe that the smaller of the solvents preferentially enters the gel which was the first direct evidence of entropically driven segregation upon swelling of athermal gels. Additionally they introduced the connectivity ratio as

$$\nu = \frac{\text{Total number of bonds in the system}}{\text{Total number of sites}} \tag{1}$$

which distinguishes crosslinked networks ($\nu > 1$) from uncrosslinked branched polymers ($\nu < 1$). They proposed a simple mean-field theory which uses a reference polymer to predict swelling behavior of the networks. They have shown that with a suitable reference system the theory very well predicts almost all of their simulation data. Furthermore, with the proper reference system they were able to separate the deformation contribution of the network from the additional contribution to pressure due to the crowding around the branching points. Finally they pointed out that the elastic component of gel pressure is typically very small. Therefore it might be difficult to critically examine the applicability of the questioned assumption of additivity of the elastic and mixing contributions to the swelling free energy, which is inherent to the Flory-Rehner theory. In a later article in 1997 [55], Escobedo and de Pablo studied *PVT* behavior of athermal and thermal networks. They observed that upon increasing pressure the solvent uptake in the gel also increases. They also reported on a volume phase transition in thermal (square-well) networks upon change in pressure.

A few years later, a series of papers from the Hentschke group appeared, extending the early work of Escobedo and de Pablo. The first paper by Aydt and Hentschke [63] was soon followed by one from Lu and Hentschke [64] which was very similar in spirit. In both above-mentioned articles, the authors investigated the swelling of a hydrogel upon addition of a monomeric or oligomeric solvent over a range of temperatures and pressures. The first one of the two articles [63] used the Gibbs Ensemble molecular dynamics [65, 66] while the latter [64] used a combination of Widom particle insertion [62] with molecular dynamics. In addition, both articles present a Flory-Huggins-like theory which was used to obtain the equation of state of the hydrogel swollen by the solvent. In the latter article [64], the authors showed that their original derivation of the theory [63] was erroneous. Nevertheless, in both cases good qualitative agreement between the simulations and theory was observed. In contrast with the classical Flory-Huggins formalism, they considered the presence of voids which was necessary to account for the pressure dependence of the swelling. Their network model had a cubic topology (six chains connected to one network node) with $N = 1$ segment in between two network sites. The main observation was that the behavior of the swelling ratio as a function of pressure and temperature is complex and non-monotonous. Under sub-critical conditions for the pure solvent the gel shrinks with increasing temperature or pressure. At intermediate super-critical conditions both pressure and temperature dependence of the swelling ratio exhibit a maximum which shifts to higher super-critical pressures or temperatures with increase of either pressure or temperature. At high super-critical conditions the network monotonously swells upon increase in pressure or temperature. The presented theoretical treatment provided predictions in good qualitative agreement with the simulation data, being able to predict all the above-mentioned trends. The first two articles were followed by a study from Lu and Hentschke of hydrogel swelling by a six-site chain-like solvent [67] using the same methodology as earlier [64] but with Rosenbluth sampling [68] to transfer the chain-like solvent between the two simulation boxes. The results for the chain-like solvent resembled those of the one-site solvent simulations, but the non-monotonous behavior was observed both at sub-critical and super-critical conditions. In addition to that, the maximum in the swelling ratio was much sharper and more asymmetric as compared to the one-site solvent case. In the case of chain-like solvent the agreement between the theory and simulations was much worse. In particular the theory could not capture the non-monotonous behavior at sub-critical conditions and predicted much milder changes in the swelling ratio as a function of pressure. Finally, the study also revealed a considerable slowdown of the solvent diffusion in the network as compared to self diffusion of pure solvent at the same conditions. In the final paper of the series, Lu and Hentschke [69] studied the swelling of hydrogels of different crosslink densities, ranging from $N = 1$ to $N = 5$ segments between the nodes of the cubic network. Both one-site and six-site solvents were used. Similar to the preceding study, they found fair agreement between their theory [64] and the simulation data in the case of one-site solvent. In the case of the six-site solvent only some features observed in the simulation data could be reproduced by the theory. They observed that with decreasing crosslink density (increasing N), the maximum in the swelling ratio as a function of pressure is getting more pronounced and sharper and shifts towards lower pressures. The diffusion slowdown in the network with respect to free solvent at the same conditions is much less pronounced in the less dense networks. Several years later, two more articles by Oyen and Hentschke further extended the previous series. The first of them [70] examined finite size effects in tightly meshed networks. The authors have shown that considerable deviations from the limiting value expected at infinite system size are present when the number of particles constituting the network is $M \lesssim 10^3$. They also derived an equation to describe the finite-size scaling of the density of the swollen network. The second article [71] has shown that the same kind of finite-size scaling relation of the form

$$Y = Y_\infty - C_Y \exp(a_Y M^{-1/3}), \tag{2}$$

where C_Y and a_Y are adjustable parameters and Y is the quantity of interest, can be applied also to other quantities obtained from the simulation. Of particular interest is $M^{-1/3}$ in the argument of the exponential. This relation allows for extrapolation of simulation results on just a few small systems to the thermodynamic limit. Next, the authors studied swelling of the gel by a mixture of two single-site solvents, one of them being of the same size as the polymer beads, the other one having a smaller radius by 20 %. They observed that the smaller solvent is preferentially taken up by the network. The preference as well as the total swelling ratio are relatively insensitive to solvent composition at high temperatures while at low temperatures both effects are more pronounced.

Coarse-Grained Simulations of Charged Hydrogels

Coarse-grained simulations of charged hydrogels comprise contributions from four research groups. Historically the first to simulate charged hydrogels has been the Lund group (Linse and coworkers) in 2002. In 2003 first contribution of the Madison group (de Pablo, Olvera de la Cruz, and coworkers) appeared, followed a year later by the first contribution from the Mainz group (Kremer, Holm, and coworkers), and finally in 2011 the Spanish group in Linares/Granada (Quesada-Pérez, Martín-Molina) started to publish some works.

In all concerned publications, a diamond-like network was used as the hydrogel model with N being the length

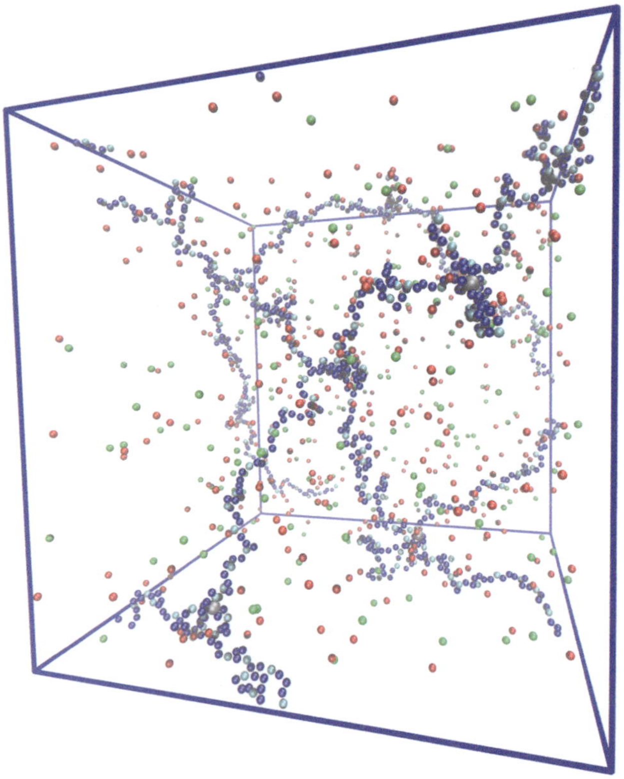

Fig. 4 Snapshot of a model hydrogel consisting of chains of length N = 59, charge fraction $f = 0.25$, external salt concentration $c_s = 0.2\,\text{mol}/\text{l}$. *Grey* spheres are network nodes, *blue* uncharged monomers, *cyan* charged monomers, the salt ions are colored *red* and *green*

between the network nodes, $1/N$ being the crosslink density and $f = 1/n$ the fraction of charged groups in the chains where every n-th monomer unit carries a dissociated charge. A simulation snapshot of such a model hydrogel is shown in Fig. 4. The default model was further modified in order to study some specific issues as described below. The Lund group used hard sphere repulsion to account for the excluded volume in the athermal case and square-well potential to account for poor solvent conditions. The other groups used the WCA interaction potential for the athermal case and the full Lennard-Jones potential for the poor solvent case. In all the reviewed simulations the solvent was modeled as dielectric continuum with dielectric constant introduced through the Bjerrum length, λ_B. All ions were treated explicitly using the Ewald summation [72] or one of its mesh modifications [73–76].

Solution Properties of Single-Chain Polyelectrolytes

Most effects observed in the behavior of polyelectrolyte gels have a close analogy to properties of single polyelectrolyte chains in dilute solutions. Therefore, before reviewing the simulations of charged hydrogels, we briefly summarize the known properties of single-chain polyelectrolytes and their microscopic explanations. Since simulations and theory related to single-chain polyelectrolytes have been reviewed recently in several articles [77–80], we merely summarize here the main results.

Dilute solution properties of single-chain polyelectrolytes are determined by the interplay of a number of entropic and enthalpic effects. A polymer is said to be in a theta solvent, if its conformation is that of a Gaussian coil. The short range repulsion between monomers due to excluded volume slightly swells the coil. If the polymer is charged, the repulsion due to long-ranged electrostatic interactions favors further swelling at the expense of entropy loss. The polyelectrolyte conformation is a result of balance of the two counteracting effects. In the first approximation, an increase in the linear charge density on the chain (fb/λ_B, where b is the monomer size) leads to a more stretched conformation which is simultaneously manifested in an increased persistence length.

The electrostatic self-repulsion of the chain is further screened by the counterions – oppositely charged small ions which compensate the polymer charge. They are electrostatically attracted to the polymer (multivalent macroion) which favors their localization near the chain, while entropy favors their homogeneous distribution. The balance between the entropy and the strength of electrostatic coupling results in a monotonously decaying concentration profile of counterions around the polymer. At a certain threshold charge density, $fb/\lambda_B \approx 1$, the enthalpy gain due to electrostatic interactions between the chain and the free ions exceed the entropy gained when the ions escape. In such situations, a fraction of counterions condense on the chain (i.e. are located in its vicinity), compensating its excess charge such that the resultant effective charge is kept at the threshold value, commonly called Manning condensation [81]. The preferential localization and condensation are much more pronounced in the case of multivalent counterions with valence z. Their entropy loss is the same as for the monovalent ions but their electrostatic interactions with the polymer are z times stronger. Relatively early, simulations with explicit ions have shown that mean-field approximations, e.g. Debye-Hückel, fail to describe interactions in charged polymers due to significant coupling between the macroion and the small ions [82–84]. The behavior of counterions and the chain cannot be effectively separated and treated independently. Therefore counterions need to be included explicitly in the simulations. Correlations are particularly strong at high charge densities or in the presence of multivalent ions. As an extreme case, stiff macroions in the presence of multivalent counterions tend to bundle, as has been observed in simulations [85, 86] or experimentally for DNA [87]. This effect is known as correlation-induced attraction of like charges. In general, addition of salt suppresses the ion correlations and all electrostatic effects.

Water molecules in contact with a carbon-based polymer exhibit preferential orientation which results in a loss of entropy. The entropy can be re-gained when several host molecules are clustered, so that the number of structured molecules of water is reduced. It results in a net attractive force between the host molecules, commonly known as hydrophobic interaction, which refers to poor solubility in water. Despite the fact that it is predominantly entropic in origin, in the implicit solvent models it is conveniently introduced as an additional short-ranged attractive potential. The interplay of short-ranged hydrophobic attraction and long-ranged electrostatic repulsion in hydrophobic (poor solvent) polyelectrolytes results in the formation of pearl-necklace structures, predicted theoretically and found by Monte Carlo simulations by Dobrynin et al. [88] for polyelectrolytes at infinite dilution. Collapsed domains (pearls) are formed on the chain, connected by extended parts of the same chain (strings). Upon an increase in the strength of the hydrophobic attraction or a decrease in Bjerrum length, the polymer conformation goes through a sequence of transitions involving pearls of increasing size. This behavior has also been observed in simulations of single-chain polyelectrolytes at finite concentrations [89–93]. Later simulation studies of polyelectrolyte brushes [94] and stars [95] in poor solvents have shown that introduction of topological constraints add to the complexity of conformational structures. Also experimentally pearl-necklace structures have been meanwhile observed in [96–98], and indirectly in the experiments of Combet et al. [99].

Of specific relation to charged hydrogel networks are star-branched polyelectrolytes which possess a single branching point in the centre and a spherical symmetry (to first-order approximation). Simulations and theories of star polyelectrolytes have been reviewed recently by Borisov et al. [80]. They exhibit a variation of conformational properties and counterion distribution as a function of distance from the central branching point (radial direction). Near the centre of a star, the arms are stretched and counterions are more strongly localized. Further towards the periphery, the increase in available volume allows for more coiled conformations and weaker localization of counterions. Specifically in the poor solvent regime, formation of bundles of chains is preferred near the centre of the star while pearl-necklace structures similar to linear polyelectrolytes appear further away [95].

Salt-Free Athermal Gels

In the very first article which deals with swelling of charged hydrogels, Schneider and Linse [100] have studied the pressure as a function of density (network packing fraction). The swelling equilibrium of a hydrogel is attained when the

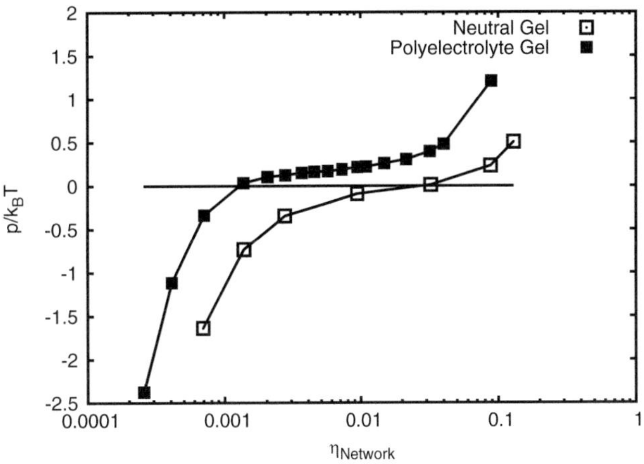

Fig. 5 Pressure in the simulation box as a function of network packing fraction for a polyelectrolyte gel and a neutral gel of the same structure (Adapted from the article by Schneider and Linse [100])

pressure in the gel is equal to that of the surrounding solution. In the case of implicit salt-free solvent this is realized at zero pressure in the simulation box. Negative pressures then correspond to an applied external stretching force while positive pressures indicate gel under compression. Schneider and Linse [100] observed that both for charged and neutral gels of the same structure, the pressure-density relations exhibit qualitatively the same shape as shown in Fig. 5. Starting at low packing fractions, the pressure is negative and steeply rises through zero (swelling equilibrium) towards positive values as the packing fraction increases. This rise is followed by a plateau at intermediate densities and a further steep rise at high densities when the excluded volume becomes dominant. However, clearly the equilibrium ($P = 0$) between the charged hydrogel and a pure solvent is reached at much lower packing fraction in comparison with the neutral gel, in qualitative agreement with theoretical predictions. Additionally, they have shown that the assumption of affine deformation is well fulfilled for the charged network while it fails for the neutral one. Finally, they compared the reduced pressure of polymer networks to polyelectrolyte and simple electrolyte solutions and to the same systems without electrostatic interactions. Only in the gels the pressure attains negative values. In all other systems, it approaches a fixed positive value at low densities, corresponding to the ideal gas, in which each polymer chain is equivalent to a single particle. In all neutral systems the pressure decreases monotonously with decreasing density. In solutions with charged species it exhibits a minimum at intermediate densities and approaches the dilute limit from below which is a consequence of the net cohesive effect of electrostatics.

The subsequent article by Schneider and Linse [101] presents much more extensive results on similar systems as above, examining the influence of electrostatics, crosslinking density and chain stiffness on swelling of polyelectrolyte gels in equilibrium with pure solvent. The results are analyzed in a similar manner as in the previous article [100]. They reveal that the PE gel swells more as (I) the charge density is increased; (II) crosslinking density is decreased; (III) chain stiffness is increased; (IV) valence of counterions is decreased. In agreement with the previous work, it turns out that the assumption of affine deformation is quite well fulfilled in swollen polyelectrolyte gels near equilibrium while in neutral gels or compressed PE gels it becomes questionable. It is noteworthy that both above-mentioned articles [100, 101] emphasize the non-trivial contribution due to electrostatic interaction which results in an inhomogeneous distribution of free ions in the gel and in a negative pressure contribution.

Edgecombe and Linse [102] addressed the influence of polydispersity and topological defects on the swelling behavior of neutral and PE gels under salt-free conditions. To introduce polydispersity, they modified the default model such that two chains were made shorter and two longer than the remaining ones. Similarly, topological defects were introduced by removing bonds between network nodes and some of the chains, resulting in dangling (severed) chains. The authors have shown that polydispersity noticeably decreases the swelling capacity of PE gels while it has a much weaker effect on the neutral gels. On the contrary, the presence of dangling ends notably increases the swelling of flexible PE gels while it has weaker effect on neutral or stiff PE gels.

In a later work Edgecombe and Linse [103] studied swelling and mechanical properties of interpenetrating polymer networks using the same diamond-like network model as in all previous studies. They observed that interpenetration leads to a slight decrease in swelling when two PE networks of the same crosslink density are combined. In contrast, two interpenetrating neutral networks swell slightly more than the corresponding single network due to enhanced excluded volume effects. Interpenetrating networks consisting of one network with high and one with low crosslink density exhibit intermediate behavior, swelling more than the highly crosslinked component but less than the weakly crosslinked one. Applied uni-axial stress results in reduction of the swelling volume.

Electrostatic Coupling and Counterion Valence

Yan and de Pablo [104] simulated athermal charged gels with different strength of electrostatic interactions, expressed through the ratio of Bjerrum length to segment size, $\lambda = \lambda_B/\sigma$. They observed that under conditions of sufficiently high λ (low dielectric constant) the osmotic pressure as a function of packing fraction exhibits the characteristic behavior of a van der Waals loop. These are a sign of first-order transition between the swollen and collapsed state of the gel which is also corroborated by the structure factor, $S(\mathbf{k})$, and the pair correlation function, $g(r)$, of the simulated systems. Similar observations of signs of a first-order transition in an athermal gel upon a change of λ_B have been obtained by Schneider and Linse [105] in an article which focused on the swelling of charged gels in poor solvents. The results of this work are discussed in more detail in section "Charged Gels in Poor Solvent".

These findings were further extended in a subsequent study [106] which investigated the pressure-density relations in a range of different Bjerrum lengths and for mono- di- and tri-valent counterions. The authors found out that an increase in counterion valency and in Bjerrum length have qualitatively similar effect: above a certain critical value the first-order swelling transition occurs while below this value the osmotic pressure in the gel increases monotonously with network density. In addition, the authors examined in detail the different contributions to pressure, namely the excluded volume, network elasticity and electrostatics. They observed that although both excluded-volume and electrostatic contribution quite strongly depend on Bjerrum length as well as on the counterion valency, they approximately cancel each other, irrespective of the interaction parameters. This leaves the network elasticity as the dominant contribution which justifies the use of models for neutral networks for the description of swelling of charged gels.

In their second study, Yin et al. [107] investigated gel swelling in mixtures of monovalent and divalent counterions of varying compositions. Experimentally this could be realized as swelling in an excess salt solution of the given composition. However, in the simulation the excess salt was not treated explicitly. It was observed that an addition of multivalent ions shifts the onset of the first-order collapse transition towards lower Bjerrum lengths. However, for all studied compositions the transition occurs only for values of $\lambda_B/\sigma > 2$, i.e. the model fails to reproduce the transition observed experimentally in swelling of ploy(acrylic acid) gels with a mixture of NaCl and $CaCl_2$ in water at nearly physiological conditions. Similar to the previous study [106], the cancellation of the excluded-volume and electrostatic contributions to pressure was observed also here, irrespective of interaction parameters or the ratio of mono- and di-valent ions. However, the relative strength of non-ideal contribution of the divalent ions is much stronger than that of the monovalent ones, which was attributed to the accumulation of ions of higher valency around the chains and network nodes, as revealed by $g(r)$. Last but not least, the article examined the modified Flory-Huggins theory with χ parameter expressed as:

$$\chi = \chi_0 + \chi_1 \phi \tag{3}$$

where χ_0 is the usual Flory-Huggins parameter and ϕ is the polymer volume fraction. The theory could match the simulation results with nearly ideal dependence of χ_0 and χ_1 on the ratio λ/σ. This was contrasted with the experiments of Horkay et al. [108, 109] which could be described by the same theory with strong dependence of χ_1 on $CaCl_2$ concentration, suggesting that further improvements to the simulation model are necessary to capture the experimentally observed effects.

Mann et al. [3, 110–112] studied in several publications the swelling of hydrogels under athermal and near theta conditions in a broad range of parameters such as chain length, $N \in [39 : 239]$, fraction of charged monomers, $f \in [0.0625 : 1.0]$, and Bjerrum length, $\lambda_B \in [1 : 5]$. In the first article they developed a scaling theory for swelling of polyelectrolyte gels [110] which takes into account the counterion condensation. The condensation occurs when the charge density on the polymer exceeds a critical value defined by the Manning parameter, $\xi_M^{\max} = f_{\max} l_B / b = 1/\nu_{\mathrm{CI}}$ [81]. Mann et al. proposed to renormalize the polymer charge fraction and the number of osmotically active counterions by the number of condensed counterions as

$$f_{\mathrm{eff}} = \begin{cases} f & \text{for } \frac{l_B f}{b} < \frac{1}{\nu_{\mathrm{CI}}} \\ f_{\max} = \frac{b}{\nu_{\mathrm{CI}} l_B} & \text{otherwise}. \end{cases} \quad (4)$$

With the renormalized charge fraction, the end-to-end distance of the charged gel at swelling equilibrium follows a simple scaling relation

$$R_{\mathrm{e}} = N_m b f_{\mathrm{eff}}^{1-\nu}, \quad (5)$$

where ν is the universal scaling exponent ($\nu = 1/2$ for theta conditions and $\nu \approx 0.588$ for the athermal solvent [113]). This is demonstrated by the fact that in the plot of $\langle R_{\mathrm{e}}^2 \rangle^{1/2}$ against $N_m b f_{\mathrm{eff}}^{1-\nu}$ data from many simulations at various values of λ_B, f and N collapse on a single line as shown in Fig. 6. This result is in line with the observations of Schneider and Linse [101] concerning the dependence of swelling ratio on crosslink density and charge fraction but in addition it puts their observations to a more general framework.

The later article of Mann et al. [111] focused on deviations from ideal behavior and on the cancellation of the non-ideal contributions which leads to the success of the simple scaling descriptions. Their approach was different from that of Yin et al. [106, 107]. By comparing the same system with and without electrostatic interactions, Mann et al. [111] have shown that a significant contribution to non-ideality is of electrostatic origin. An extensive discussion of all results presented in the articles of Mann et al. which goes even beyond what has been published in journals can be found in the dissertation of Mann [3].

PE gels swollen by macroions of different size and charge under salt-free conditions have been studied by Edgecombe and Linse [114]. They observed that with increasing number of macroions, the gel considerably deswells, reaching swelling ratios even below those of the corresponding uncharged gel. In excess of macroions relative to the amount of network charges the gel slightly swells again. The strongest deswelling effect was observed for small macroions with high charge. In such case the exchange of counterions by macroions is entropically most favorable. The macroions are found in close contact with the polyelectrolyte chains of the network with preferential localization near the nodes.

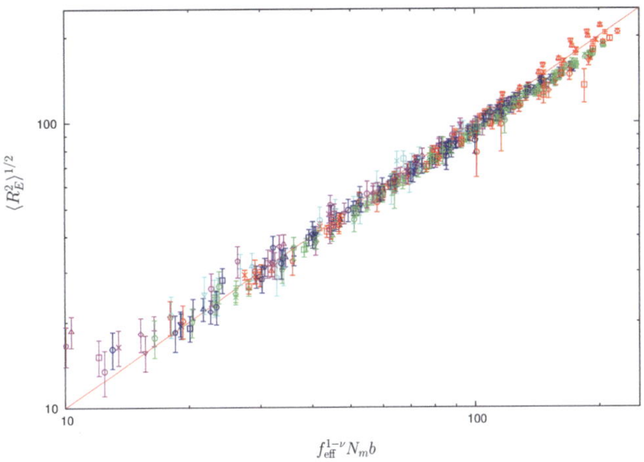

Fig. 6 Root-mean-squared end-to-end distance $\langle R_{\mathrm{e}}^2 \rangle^{1/2}$ for polyelectrolyte gels of various N, f and λ_B in good solvent and close to the Θ-point, plotted as a function of effective charge fraction, f_{eff} and chain length, N, according to Eq. (5)

Effect of Added Salt

Edgecombe, Schneider and Linse published a study focused on swelling of hydrogels in the presence of salt [115]. They used a both *NPT* and *NVT* simulations in combination with Widom's particle insertion method [62] to measure the chemical potential of the salt ions. In a separate simulation they measured the chemical potential of the corresponding model salt solution and determined the swelling equilibrium as the point where osmotic pressures and chemical potentials of the salt on both sides of the gel-solution interface were equal. In qualitative agreement with the classical Flory and Rehner [25] prediction they observed that the salt decreases the hydrogel swelling and has a more pronounced effect on gels with high charge density. Nevertheless, even at high salt (1 M) the swelling ratio is a factor of 2–3 greater than that of the corresponding neutral gel. For the salt content inside a hydrogel they found a reasonable qualitative agreement between the Flory-Rehner theory combined with the Donnan equation for partitioning of ions between the two phases, showing that the salt content in the gel is significantly lower than in the surrounding solution. However, a more detailed analysis reveals that while the above theory provides a correct direction of the change with parameters such as crosslinking degree or charge density, in quantitative terms the theory fails. This failure has been attributed to improper description of the cohesive effect of electrostatics which was further analyzed in detail.

Yin et al. [116] used grand-canonical simulations in the osmotic ensemble to study the swelling of a charged hydrogel in equilibrium with a reservoir of salt solution containing either 1:1 or 1:2 electrolyte. They observed a pronounced decrease of the gel swelling upon an exchange of 1:1 electrolyte for 1:2 electrolyte. Formation of nano-heterogeneous structures in a partly collapsed gel was observed with the divalent counterions located inside the collapsed domains near the polymer chains and mono-valent co-ions scattered randomly in the structure. This lead to a non-monotonous dependence of salt concentration inside the hydrogel, c_{in}, on the reservoir salt concentration, c_{s}. In any case $c_{\mathrm{in}}/c_{\mathrm{s}} < 1$, however this ratio is always lower for the 1:1 electrolyte as compared to 1:2 electrolyte and it exhibits a non-monotonous dependence on c_{s}: at very low c_{s}, the value of $c_{\mathrm{in}}/c_{\mathrm{s}}$ is at its minimum; it has a maximum at intermediate values of c_{s} and decreases again at high c_{s} around 10^{-1} M. While for 1:1 electrolyte in most cases $c_{\mathrm{in}}/c_{\mathrm{s}} \ll 1$, for 1:2 electrolyte, $c_{\mathrm{in}}/c_{\mathrm{s}} \approx 0.9$ in a rather broad range of c_{s}.

Charged Gels in Poor Solvent

The third article in the series by Schneider and Linse [105] deals with charged hydrogels in the presence of short-ranged polymer-polymer attractions and/or variation of permittivity and hence the strength of electrostatic coupling. Upon incorporation of short-range attraction, they observed that beyond $\epsilon \approx 0.3\,k_{\mathrm{B}}T$ a transition from a swollen to a collapsed gel occurs with the possible presence of metastable states. A decrease of permittivity from $\varepsilon_{\mathrm{r}} = 80$ (water) to $\varepsilon_{\mathrm{r}} = 15$ (organic solvent) leads to a collapse of the gel, even in the absence of short-range attraction. In the purely athermal system it is a consequence of correlations in the localization of ions which result in the net attractive interaction. In the presence of short-range attraction, the collapse transition upon variation of ε_{r} occurs at higher values of the permittivity. In either case, the collapse results in an about 100-fold decrease of the gel volume, i.e. the collapsed state has a significantly smaller volume than the corresponding uncharged athermal gel.

Qeusada-Pérez et al. introduced a phenomenological model for temperature-dependence of the solvent-mediated hydrophobic interactions [117] which includes enthalpy and entropy contributions. They tuned the potential parameters such that their model uncharged hydrogel exhibits collapse transition upon temperature increase around the known LCST of PNiPAAm $T \approx 298\,\mathrm{K}$. At the same time they emphasize that the shape and interaction strength of the potential is quite sensitive to the chosen values for the enthalpic and entropic contributions and hence the absolute values of transition temperatures reported in the simulations should not be taken too literally. Nevertheless, the main feature of the used potential is that its attraction strength (hydrophobicity), ϵ, quite abruptly increases with temperature, which makes the ratio $\epsilon/k_{\mathrm{B}}T$ a non-monotonous function of temperature. This enables reproducing the features of temperature-induced collapse transition of PNiPAAm which could not be done with T-independent interaction potential. Using this model, they investigated the T-dependent collapse transition of charged hydrogels with a varying content of charged groups and chain length. In accordance with experiments, they observed that an increase in charge fraction shifts the collapse transition to higher temperatures and makes it more abrupt. At higher charge fractions, they observed a coexistence between a swollen (metastable) and collapsed (stable) states, similar to Schneider and Linse [105]. In a subsequent study, Quesada-Pérez et al. [118] simplified their model for temperature-dependent hydrophobic interactions to an analytically tractable form. Using estimates of the interaction parameters such that they approximately capture the transition of PNiPAAm gels copolymerized with low fractions of acrylic acid (AA) as an ionizable comonomer, their model (simulations) reasonably capture the experimentally observed collapse transition of gels with higher AA contents. In contrast, a Flory-Rehner-like theory with the usual assumption $\chi(T) \approx (\Delta H - T\Delta S)/2k_{\mathrm{B}}T$ could only produce a comparable agreement if both H and S were considered as T-dependent adjustable parameters. Furthermore, the authors illustrated on their simulation results how the T-dependent transition varies with crosslink density for an uncharged and a charged gel: in both cases the transition is more abrupt and slightly shifted to higher temperatures for longer chains.

In a third study in this series, Quesada-Pérez et al. [119], examined T-dependent collapse transition of similar gels as earlier in the presence of multivalent counterions. They observed in simulations that multivalent ions induce lower swelling ratios both at high and low temperatures (collapsed and swollen gels, respectively) and that the collapse transition with multivalent ions becomes more abrupt. Finally they compared gels with high and low charge fractions, having the same number of multivalent and monovalent counterions, respectively. They have shown that the osmotic pressure contribution of multivalent ions is highly non-ideal, which is attributed to their preferential localization in the vicinity of polymer chains, as revealed by radial distribution functions. The simulations show that the gel with multivalent counterions swells less than the gel with the same number of monovalent ones, while the simple theory based on ideal osmotic pressure of small ions would predict the same swelling for both types of gels.

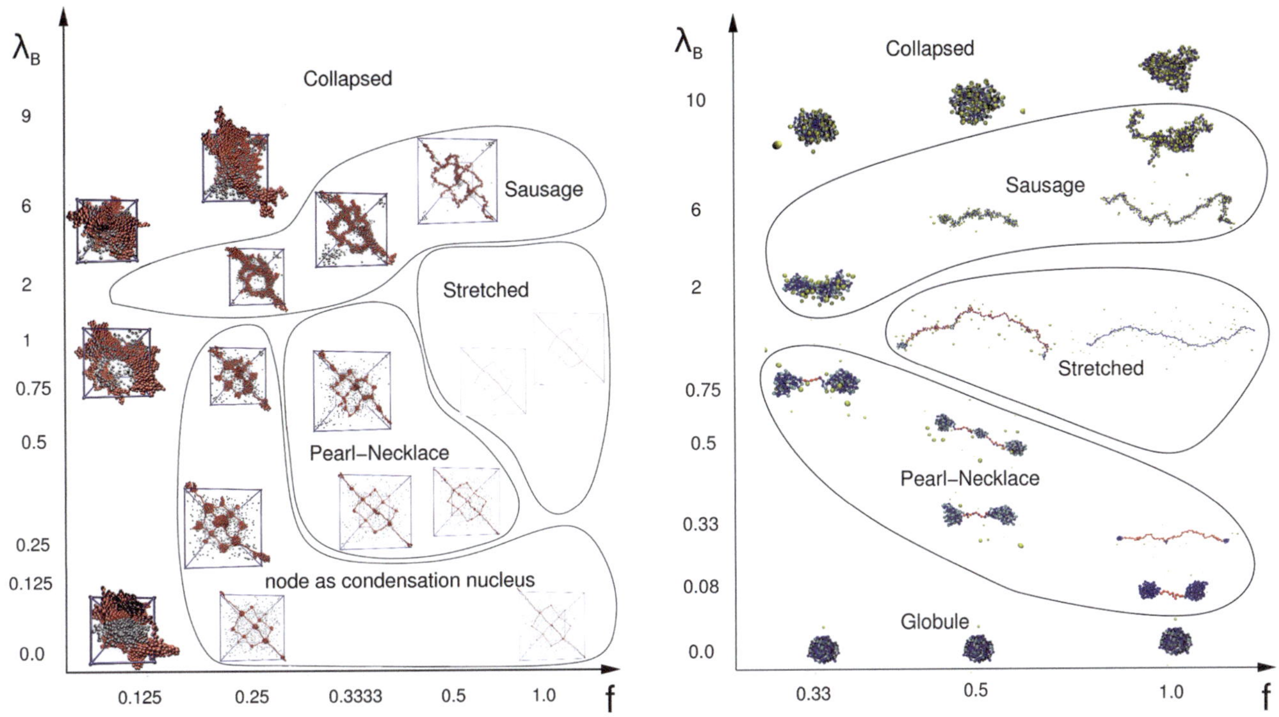

Fig. 7 Structure diagram of a hydrogel under poor solvent conditions showing the variations of structures as a function of backbone charge fraction and Bjerrum length. The behavior of hydrogels (*left*) is compared to that of single-chain polyelectrolytes (*right*) (Figure adapted from Ref. [120])

Structural properties of poor solvent hydrogels were investigated in simulations of Mann [3] and Mann et al. [112, 120]. The zoo of conformational structures observed at different values of Bjerrum length and fraction of charged monomers is summarized and compared to the behavior of single chains in Fig. 7. The simulated hydrogels exhibit conformational structures analogous to those of the single chains, namely collapsed, sausage-like, stretched and pearl-necklace conformations. The formation of pearls defines an additional characteristic length scale which is reflected as a new peak in the structure factor. This is found to be in agreement with experimental SAXS study of PSS polyelectrolytes with Cs counterions. In addition, in a certain range of parameters the nodes of the network serve as condensation nuclei for pearls while the central portions of the chains still remain stretched. This is in line with the observation for star polyelectrolytes in poor solvents [95] that formation of collapsed domains is promoted near a topological junction. At low charge fractions, phase separation occurs between the hydrophobic monomers and the purely repulsive counterions which form separate domains in the periodic network structure.

Weak Polyelectrolyte Gels

Weak polyelectrolytes are polymers containing charged groups that are weak acids or weak bases. We will restrict ourselves to weak acids, keeping in mind that the behavior of weak bases is analogous. The degree of dissociation of weak polyelectrolytes depends on the surrounding pH which adds another dimension to the complexity of their swelling behavior. This can be understood in analogy with single-chain weak polyelectrolytes, for which the coupling between the pH, degree of charging and the conformation results in a rich behavior [121].

Longo et al. [122] presented a study of the swelling behavior of weak polyelectrolyte gels. They presented a molecular theory in which they explicitly took into account the spatial variation of the electrostatic potential, which in turn influences locally the dissociation of the acidic monomers. This spatial variation of the potential was obtained from a set of conformations generated by a simulation of a coarse-grained polyelectrolyte gel of simple cubic topology. The simulations were performed for several polymer volume fractions, ranging from a collapsed to a highly swollen gel. The article presents an extensive set of results, showing non-ideal behavior of the hydrogel swelling as a function of pH. In particular, the charge fraction f exhibits a smoother variation with pH than what would be expected in an ideal case without explicit spatial coupling. Next, the pH inside the gel can be noticeably lower than that of the surrounding solution. With increasing external pH, the gel exhibits a sharp swelling transition around pH $=$ pK_A. Finally, added salt may play two opposing roles: increasing degree of charging of the acid groups and at the same time screening the electrostatic interactions.

Nanogels and Microgels

Nanogels and microgels are hydrogel particles with the size of several nm or μm. While their internal structure is similar to that of macroscopic hydrogels, the effect of the gel-solution interface cannot be neglected in this case. Therefore such object needs to be simulated as a whole, including the interface and a portion of solvent which surrounds it. Counterions can escape nanogels to distances comparable to their own size which means that the gel alone need not be electrically neutral. The size of a nanogel or a microgel is at the threshold of what can be handled by coarse-grained simulations. Therefore it is desirable to employ mean-field descriptions which can be validated by explicit simulations.

Claudio et al. [123] have compared the Poisson-Boltzmann cell model of a microgel to the explicit coarse-grained simulations. They constructed a series of intermediate models removing the differences between the two extreme cases in a stepwise fashion. In particular they focused on distribution of ions in the gel particle. They concluded that despite the differences between the two methods the resulting ion distributions are quite close to one another. Hence the Poisson-Boltzmann model of a charge-penetrable colloid can be used to predict the thermodynamic properties of charged microgels under good solvent conditions in a dilute solution.

The work of Jha et al. [124] deals with collapse transitions in charged polymeric nanogels. They refer to their modeling method as "Theoretically informed coarse-grained simulations" which is a combination of a mean-field model with explicit simulation of bonded interactions. The authors first extensively analyzed the possible influence of discretization of their mean-field model on the obtained results, concluding that with a judicious choice of discretization parameters the method is robust. They observe that the nanogels exhibit a sharp collapse transition around Bjerrum length $\lambda_B \approx 1$ which is accompanied by release of counterions into the surroundings of the gel. This behavior is different from that of the infinite networks studied in previous works: while the infinite network has to be macroscopically neutral, this is not true for a nanogel.

Conclusion

The swelling of hydrogels has been successfully modeled using molecular simulations. The atomistic simulations provide valuable insights into the details of polymer-solvent interactions which are crucial to understand phenomena such as the solubility of PVA in water, the LCST transition of PNiPAAm or cononsolvency of PNiPAAm in water-methanol mixtures. However, simulations with atomistic resolution are limited by the size of the molecule as well as by the timescales which do not allow them to be used for studying molecules bigger than about 50 monomer units, which is too small to make quantitative comparisons to experiments. Coarse-grained simulation models trade the atomistic resolution for much longer time scales and bigger system sizes. They enabled studies of the swelling behavior of both neutral and charged hydrogels as a function of crosslinking density, network imperfections, solvent type (monomeric, oligomeric), charge fraction, counterion valence, added salt or solvent quality. Such simulations provided insights into the structure-property relations of polymeric gels which would be difficult to obtain from experiments. In addition, the coarse-grained simulations provided data on systems with well-defined structure which could be used for validation of theoretical predictions. Several independent studies of charged gels under athermal and theta solvent conditions have shown that cancellation of non-ideal contributions leads to a relatively simple relation between the equilibrium swelling of the gel and its structural parameters. Several recent studies have successfully employed various combinations of coarse-grained simulation models with field-theoretical approaches which enabled studies of microgels or weak polyelectrolyte gels. However, even today, we have only understood a small portion of the complexity real hydrogels possess. The polydispersity issue and composition problems remain still a big hurdle for simulations since even for coarse grained simulations the time and length scales are beyond a realistic modeling. We should take these facts as an incentive to keep up our efforts in understanding this interesting class of materials.

Acknowledgements The authors would like to thank DFG for financial support with the project SPP 1259 "Intelligente Hydrogele". PK would like to acknowledge financial support from the grant "LK 21302 Navrat" by the Ministry of Education, Youth and Sports of the Czech Republic.

References

1. Barrat J-L, Joanny J-F (1996) Theory of polyelectrolyte solutions. Adv Chem Phys 94:1–66
2. Kokufuta E (2001) Phase transitions in polyelectrolyte gels. In: Radeva T (ed) Physical chemistry of polyelectolytes. Marcel Dekker, New York, pp 591–664
3. Mann BA (2005) The swelling behaviour of polyelectrolyte networks. PhD thesis, Johannes Gutenberg-University, Mainz
4. Hua L (2009) Smart hydrogel modelling. Springer, Berlin
5. Harland R, Prud'homme R (1992) Polyelectolyte gels. American Chemical Society, Washington, DC
6. Khokhlov AR, Starodubtzev SG, Vasilevskaya VV (1993) Responsive gels: volume transitions I. In: Dušek K (ed) Conformational transitions in polymer gels: theory and experiment. Volume 109 of advances in polymer science. Springer, New York, p 123

7. Kazanskii KS, Dubrovskii SA (1992) Polyelectrolytes hydrogels chromatographic materials, vol 104. Springer, Berlin/Heidelberg
8. Buchholz FL, Peppas NA (1994) —. In: Comstock MJ (ed) ACS symposium series, vol 573. American Chemical Society, Washington, DC
9. Eichenbaum GM, Kiser PF, Dobrynin AV, Simon SA, Needham D (1999) Investigation of the swelling response and loading of ionic microgels with drugs and proteins: the dependence on cross-link density. Macromolecules 32(15):4867–4878
10. Peppas NA, Bures P, Leobandung W, Ichikawa H (2000) Hydrogels in pharmaceutical formulations. Biopharmaceutics 50:27
11. Radeva T (2001) Physical chemistry of polyelectrolytes. Volume 99 of surfactant science series. Marcel Dekker Inc., New York
12. Rudzinski WE, Dave AM, Vaishnav UH, Kumbar S, Kulkami AR, Aminabhavi TM (2002) Des Monomers Polym 5:39
13. Hamidi M, Azadi A, Rafiei P (2008) Hydrogel nanoparticles in drug delivery. Adv Drug Deliv Rev 60:1638–1649
14. Samsonov GV, Kuznetsova NP (1992) Crosslinked polyelectolytes in biology. Adv Polym Sci 104:1–50
15. Jia X, Kiick KL (2009) Hybrid multicomponent hydrogels for tissue engineering. Macromol Biosci 9:140–156
16. Jagur-Grodzinski J (2009) Polymeric gels and hydrogels for biomedical and pharmaceutical applications. Polym Adv Technol 21:27–47
17. Mengel C, Meyer WH, Wegner G (2001) Photocrosslinkable star polymers: precursors for model polyelectrolyte networks. Macromol Chem Phys 202(7):1138–1149
18. Escobedo FA, de Pablo JJ (1999) Molecular simulation of polymeric networks and gels: phase behavior and swelling. Phys Rep Rev Sect Phys Lett 318(3):86–112
19. Shibayama M, Tanaka T (1993) Phase transitions and related phenomena of polymer gels. In: Dušek K (ed) Volume phase transition and related phenomena of polymer gels. Springer, Berlin/Heidelberg, pp 1–62
20. Onuki A (1993) Theory of phase transition in polymer gels. In: Dušek K (ed) Volume phase transition and related phenomena of polymer gels. Springer, Berlin/Heidelberg, pp 63–122
21. Maurer G, Prausnitz JM (1996) Thermodynamics of phase equilibrium for systems containing gels. Fluid Phase Equilib 115(1–2):113–133
22. Ballhause D, Wallmersperger T (2008) Coupled chemo-electro-mechanical FE-simulation of hydrogels-part I: chemical stimulation, smart materials and structures. Smart Mater Struct 17:045011
23. Wallmersperger T, Ballhause D (2008) Coupled chemo-electro-mechanical FE-simulation of hydrogels-part II: electrical stimulation. Smart Mater Struct 17:045012
24. Wallmersperger T, Keller K, Kroeplin B, Guenther M, Gerlach G (2011) Modeling and simulation of pH-sensitive hydrogels. Colloid Polym Sci 289(5–6):535–544
25. Flory PJ, Rehner J (1943) Statistical mechanics of crosslinked polymer networks i. Rubberlike elasticity. J Chem Phys 11:512
26. Flory PJ (1953) Principles of polymer chemistry. Cornell University Press, Ithaca
27. Quesada-Perez M, Maroto-Centeno JA, Forcada J, Hidalgo-Alvarez R (2011) Gel swelling theories: the classical formalism and recent approaches. Soft Matter 7: 10536–10547
28. Jha PK, Solis FJ, de Pablo JJ, de la Cruz MO (2009) Nonlinear effects in the nanophase segregation of polyelectrolyte gels. Macromolecules 42(16):6284–6289
29. Wu K-A, Jha PK, de la Cruz MO (2010) Control of nanophases in polyelectrolyte gels by salt addition. Macromolecules 43(21):9160–9167
30. Jha PK, Zwanikken JW, Detcheverry FA, de Pablo JJ, de la Cruz MO (2011) Study of volume phase transitions in polymeric nanogels by theoretically informed coarse-grained simulations. Soft Matter 7:5965–5975
31. Gross J, Sadowski G (2001) Perturbed-chain saft: an equation of state based on a perturbation theory for chain molecules. Ind Eng Chem Res 40(4):1244–1260
32. Chapman WG, Gubbins KE, Jackson G, Radosz M (1989) Saft: equation-of-state solution model for associating fluids. Fluid Phase Equilib 52(0):31–38
33. Arndt MC, Sadowski G (2012) Modeling poly (N-isopropylacrylamide) hydrogels in water/alcohol mixtures with PC-SAFT. Macromolecules 45(16):6686–6696
34. Frenkel D, Smit B (2002) Understanding molecular simulation, 2nd edn. Academic, San Diego
35. Allen MP, Tildesley DJ (1987) Computer simulation of liquids. Oxford science publications, 1st edn. Clarendon Press, Oxford
36. Binder K, Heermann DW (2010) Monte Carlo simulation in statistical physics: an introduction, 5th edn. Springer, Berlin/Heidelberg
37. Tamai Y, Tanaka H (1998) Dynamic properties of supercooled water in poly(vinyl alcohol) hydrogel. Chem Phys Lett 285(1–2):127–132
38. Angell CA (1993) Water-II is a strong liquid. J Phys Chem 97(24):6339–6341
39. Chiessi E, Cavalieri F, Paradossi G (2007) Water and polymer dynamics in chemically cross-linked hydrogels of poly(vinyl alcohol): a molecular dynamics simulation study. J Phys Chem B 111(11):2820–2827. PMID: 17388423
40. Müller-Plathe F, van Gunsteren WF (1997) Solvation of poly(vinyl alcohol) in water, ethanol and an equimolar water-ethanol mixture: structure and dynamics studied by molecular dynamics simulation. Polymer 38(9):2259–2268
41. Müller-Plathe F (1998) Molecular simulation: understanding polymer gels at the molecular level. Ber Bunsen-Ges Phys Chem Chem Phys 102(11):1679–1682. 97th annual meeting of the Deutschen-Bunsen-Gesellschaft-fur-Physikalische-Chemie, Munster, 21–23 May 1998
42. Müller-Plathe F (1998) Microscopic dynamics in water-swollen poly(vinyl alcohol). J Chem Phys 108(19):8252–8263
43. Tönsing T, Oldiges C (2001) Molecular dynamic simulation study on structure of water in crosslinked poly(-isopropylacrylamide) hydrogels. Phys Chem Chem Phys 3:5542–5549
44. Longhi G, Lebon F, Abbate S, Fornili SL (2004) Molecular dynamics simulation of a model oligomer for poly(n-isopropylamide) in water. Chem Phys Lett 386(1–3):123–127
45. Gangemi F, Longhi G, Abbate S, Lebon F, Cordone R, Ghilardi GP, Fornili SL (2008) Molecular dynamics simulation of aqueous solutions of 26-unit segments of p(NIPAAm) and of p(NIPAAm) "Doped" with amino acid based comonomers. J Phys Chem B 112(38):11896–11906
46. Walter J, Ermatchkov V, Vrabec J, Hasse H (2010) Molecular dynamics and experimental study of conformation change of poly(n-isopropylacrylamide) hydrogels in water. Fluid Phase Equilib 296(2):164–172. VIII Ibero-American conference on phase equilibria and fluid properties for process design
47. Walter J, Sehrt J, Vrabec J, Hasse H (2012) Molecular dynamics and experimental study of conformation change of poly(n-isopropylacrylamide) hydrogels in mixtures of water and methanol. J Phys Chem B 116(17):5251–5259
48. Amiya T, Tanaka T (1987) Phase-transitions in cross-linked gels of natural polymers. Macromolecules 20(5):1162–1164
49. Mukae K, Sakurai M, Sawamura S, Makino K, Kim SW, Ueda I, Shirahama K (1993) Swelling af poly(N-Isopropylacrylamide)

gels an water-alcohol (C1-C4) mixed-solvents. J Phys Chem 97(3):737–741
50. Althans D, Langenbach K, Enders S (2012) Influence of different alcohols on the swelling behaviour of hydrogels. Mol Phys 110(11–12, SI):1391–1402
51. Everaers R, Kremer K (1996) Topological interactions in model polymer networks. Phys Rev E 53:R37–R40
52. Everaers R (1999) Entanglement effects in defect-free model polymer networks. New J Phys 1:1–54
53. Escobedo FA, de Pablo JJ (1996) Monte Carlo simulation of branched and crosslinked polymers. J Chem Phys 104(12):4788–4801
54. Escobedo FA, de Pablo JJ (1997) Phase behaviour of model polymeric networks and gels. Mol Phys 90(3):437–444
55. Escobedo FA, de Pablo JJ (1997) Simulation and theory of the swelling of athermal gels. J Chem Phys 106(2):793–810
56. Escobedo FA, de Pablo JJ (1995) Monte Carlo simulation of the chemical potential of polymers in an expanded ensemble. J Chem Phys 103(7):2703–2710
57. Escobedo FA, de Pablo JJ (1995) Chemical potential and equations of state of hard core chain molecules. J Chem Phys 103(5):1946–1956
58. Escobedo FA, de Pablo JJ (1995) Extended continuum configurational bias Monte Carlo methods for simulation of flexible molecules. J Chem Phys 102(6):2636–2652
59. Escobedo FA, de Pablo JJ (1996) Expanded grand canonical and gibbs ensemble Monte Carlo simulation of polymers. J Chem Phys 105(10):4391–4394
60. Escobedo FA, de Pablo JJ (1997) Pseudo-ensemble simulations and gibbs–duhem integrations for polymers. J Chem Phys 106(7):2911–2923
61. Flory PJ, Rehner J Jr (1943) Statistical mechanics of cross-linked polymer networks ii. Swelling. J Chem Phys 11(11):521–526
62. Widom B (1963) Some topics in the theory of fluids. J Chem Phys 39:2802–2812
63. Aydt EM, Hentschke R (2000) Swelling of a model network: a gibbs-ensemble molecular dynamics study. J Chem Phys 112(12):5480–5487
64. Lu Z-Y, Hentschke R (2001) Swelling of a model polymer network by a one-site solvent: computer simulation and flory-huggins-like theory. Phys Rev E 63:051801
65. Kotelyanskii MJ, Hentschke R (1995) Gibbs-ensemble molecular dynamics: liquid-gas equilibrium in a lennard-jones system. Phys Rev E 51:5116–5119
66. Kotelyanskii MJ, Hentschke R (1996) Gibbs-ensemble molecular dynamics: liquid-gas equilibria for lennard-jones spheres and n-hexane. Mol Simul 17(2):95–112
67. Lu Z-Y, Hentschke R (2002) Computer simulation study on the swelling of a model polymer network by a chainlike solvent. Phys Rev E 65:041807
68. Rosenbluth MN, Rosenbluth AW (1955) Monte Carlo calculation of the average extension of molecular chains. J Chem Phys 23(2):356–359
69. Lu Z-Y, Hentschke R (2002) Swelling of model polymer networks with different cross-link densities: a computer simulation study. Phys Rev E 66:041803
70. Hentschke R, Oyen E (2005) Finite size effects in tightly meshed polymer networks. J Chem Phys 122(10):104904
71. Oyen E, Hentschke R (2005) Computer simulation of polymer networks: swelling by binary lennard-jones mixtures. J Chem Phys 123(5):054902
72. Ewald PP (1921) Die Berechnung optischer und elektrostatischer Gitterpotentiale. Ann Phys 369(3):253–287
73. Eastwood JW, Hockney RW, Lawrence DN (1980) P3M3DP–the three-dimensional periodic particle-particle/particle-mesh program. Comput Phys Commun 19(2):215–261
74. Deserno M, Holm C (1998) How to mesh up Ewald sums. I. A theoretical and numerical comparison of various particle mesh routines. J Chem Phys 109:7678
75. Essmann U, Perera L, Berkowitz ML, Darden T, Lee H, Pedersen L (1995) A smooth particle mesh Ewald method. J Chem Phys 103:8577
76. Arnold A, Holm C (2005) Efficient methods to compute long range interactions for soft matter systems. In: Holm C, Kremer K (eds) Advanced computer simulation approaches for soft matter sciences II. Volume II of advances in polymer sciences. Springer, Berlin, pp 59–109
77. Holm C, Hofmann T, Joanny JF, Kremer K, Netz RR, Reineker P, Seidel C, Vilgis TA, Winkler RG (2004) Polyelectrolyte theory. Adv Polym Sci 166:67–112
78. Dobrynin AV, Rubinstein M (2005) Theory of polyelectrolytes in solutions and at surfaces. Prog Polym Sci 30(11):1049–1118
79. Dobrynin AV (2008) Theory and simulartions of charged polymers: from solution properties to polymer nanomaterials. Curr Opin Colloid Interface Sci 13:376–388
80. Borisov OV, Zhulina EB, Leermakers FAM, Ballauff M, Müller AHE (2011) Conformations and solution properties of star-branched polyelectrolytes. In: Müller AHE, Borisov O (eds) Self organized nanostructures of amphiphilic block copolymers I. Volume 241 of advances in polymer science. Springer, Berlin/Heidelberg, pp 1–55
81. Manning G (1969) Limiting laws and counterion condensation in polyelectrolyte solutions I. Colligative properties. J Chem Phys 51:924–933
82. Stevens MJ, Kremer K (1993) Form factor of salt-free linear polyelectrolytes. Macromolecules 26:4717
83. Stevens MJ, Kremer K (1993) Structure of of salt-free linear polyelectrolytes. Phys Rev Lett 71:2228
84. Stevens MJ, Kremer K (1996) Structure of salt-free linear polyelectrolytes in the Debey-Hückel approximation. J Phys II 6:1607–1613
85. Nilsson LG, Guldbrand L, Nordenskiöld L (1991) Evaluation of the electrostatic osmotic pressure in an infinite system of hexagonally oriented DNA molecules. A Monte Carlo simulation study. Mol Phys 72(1):177–192
86. Grønbech-Jensen N, Mashl RJ, Bruinsma RF, Gelbart WM (1997) Counterion-induced attraction between rigid polyelectrolytes. Phys Rev Lett 78:2477–2480
87. Bloomfield V (1991) Condensation of DNA by multivalent cations: considerations on mechanism. Biopolymers 31:1471
88. Dobrynin AV, Rubinstein M, Obukhov SP (1996) Cascade of transitions of polyelectrolytes in poor solvents. Macromolecules 29(8):2974
89. Micka U, Kremer K (2000) Strongly charged flexible polyelectrolytes in poor solvents-from stable spheres to necklace chains. Europhys Lett 49(2):189–195
90. Micka U, Holm C, Kremer K (1999) Strongly charged, flexible polyelectrolytes in poor solvents – a molecular dynamics study. Langmuir 15:4033
91. Limbach HJ, Holm C (2002) Conformational properties of poor solvent polyelectrolytes. Comput Phys Commun 147:321–324
92. Limbach HJ, Holm C (2003) Single-chain properties of polyelectrolytes in poor solvent. J Phys Chem B 107(32):8041–8055
93. Košovan P, Limpouchová Z, Procházka K (2006) Molecular dynamics simulation of time-resolved fluorescence anisotropy decays from labeled polyelectrolyte chains. Macromolecules 39(9):3458–3465
94. Carrillo J-MY, Dobrynin AV (2007) Molecular dynamics simulations of polyelectrolyte adsorption. Langmuir 23(5):2472–2482
95. Košovan P, Kuldová J, Limpouchová Z, Procházka K, Zhulina EB, Borisov OV (2010) Molecular dynamics simulations of a polyelectrolyte star in poor solvent. Soft Matter 6:1872–1874

96. Minko S, Kiriy A, Gorodyska G, Stamm M (2002) Single flexible hydrophobic polyelectrolyte molecules adsorbed on solid substrate: transition between a stretched chain, necklace-like conformation and a globule. J Am Chem Soc 124(13): 3218–3219
97. Baigl D, Sferrazza M, Williams CE (2003) On the pearl size of hydrophobic polyelectrolytes. Europhys Lett 62(1): 110–116
98. Kirwan LJ, Papastavrou G, Borkovec M, Behrens SH (2004) Imaging the coil-to-globule conformational transition of a weak polyelectrolyte by tuning the polyelectrolyte charge density. Nano Lett 4(1):149–152
99. Combet J, Rawiso M, Boué F (2006) To be published, and poster presented at: WE-Heraeus-Seminar Understanding the self organization of charged polymers, Physikzentrum Bad Honnef Germany, 4–6 Apr 2005; XX congress of the international union of crystallography IUCr 2005, Florence, 23–31 Aug 2005; 6th international symposium on polyelectrolytes Polyelectrolytes 2006, Dresden 4–8 Sept 2006
100. Schneider S, Linse P (2002) Swelling of cross-linked polyelectrolyte gels. Eur Phys J E 8:457–460
101. Schneider S, Linse P (2003) Monte Carlo simulation of defect-free cross-linked polyelectrolyte gels. J Phys Chem B 107:8030–8040
102. Edgecombe S, Linse P (2007) Monte Carlo simulation of polyelectrolyte gels: effects of polydispersity and topological defect. Macromolecules 40(10):3868–3875
103. Edgecombe S, Linse P (2008) Monte Carlo simulation of two interpenetrating polymer networks: structure, swelling, and mechanical properties. Polymer 49(7):1981–1992
104. Yan Q, de Pablo JJ (2003) Monte Carlo simulation of a coarse-grained model of polyelectrolyte networks. Phys Rev Lett 91(1):018301
105. Schneider S, Linse P (2004) Discontinuous volume transitions in cross-linked polyelectrolyte gels induced by short-range attractions and strong electrostatic coupling. Macromolecules 37: 3850–3856
106. Yin D-W, Yan Q, de Pablo JJ (2005) Molecular dynamics simulation of discontinuous volume phase transitions in highly-charged crosslinked polyelectrolyte networks with explicit counterions in good solvent. J Chem Phys 123(17): 174909
107. Yin D-W, Horkay F, Douglas JF, de Pablo JJ (2008) Molecular simulation of the swelling of polyelectrolyte gels by monovalent and divalent counterions. J Chem Phys 129(15):154902
108. Horkay F, Tasaki I, Basser PJ (2000) Osmotic swelling of polyacrylate hydrogels in physiological salt solutions. Biomacromolecules 1(1):84–90
109. Horkay F, Basser PJ, Hecht A-M, Geissler E (2000) Osmotic and sans observations on sodium polyacrylate hydrogels in physiological salt solutions. Macromolecules 33(22):8329–8333
110. Mann BA, Everaers R, Holm C, Kremer K (2004) Scaling in polyelectrolyte networks. Europhys Lett 67(5):786–792
111. Mann BA, Holm C, Kremer K (2005) Swelling of polyelectrolyte networks. J Chem Phys 122(15):154903
112. Mann BA, Holm C, Kremer K (2006) The swelling behaviour of charged hydrogels. Macromol Symp 237:90–107
113. Rubinstein M, Colby RH (2003) Polymer physics. Oxford University Press, Oxford
114. Edgecombe S, Linse P (2006) Monte Carlo simulations of cross-linked polyelectrolyte gels with oppositely charged macroions. Langmuir 22(8):3836–3843
115. Edgecombe S, Schneider S, Linse P (2004) Monte Carlo simulations of defect-free cross-linked gels in the presence of salt. Macromolecules 37(26):10089–10100
116. Yin D-W, de la Cruz MO, de Pablo JJ (2009) Swelling and collapse of polyelectrolyte gels in equilibrium with monovalent and divalent electrolyte solutions. J Chem Phys 131(19):194907–6
117. Quesada-Perez M, Ibarra-Armenta JG, Martin-Molina A (2011) Computer simulations of thermo-shrinking polyelectrolyte gels. J Chem Phys 135(9):094109
118. Quesada-Perez M, Ramos J, Forcada J, Martin-Molina A (2012) Computer simulations of thermo-sensitive microgels: quantitative comparison with experimental swelling data. J Chem Phys 136(24):244903
119. Quesada-Pérez M, Maroto-Centeno JA, Martin-Molina A (2012) Effect of the counterion valence on the behavior of thermo-sensitive gels and microgels: a Monte Carlo simulation study. Macromolecules 45(21):8872–8879
120. Mann BA, Lenz O, Kremer K, Holm C (2011) Hydrogels in poor solvents: a molecular dynamics study. Macromol Theory Simul 20:721–734. Cover issue
121. Panagiotopoulos AZ (2009) Charge correlation effects on ionization of weak polyelectrolytes. J Phys Condens Matter 21:424113
122. Longo GS, de la Cruz MO, Szleifer I (2011) Molecular theory of weak polyelectrolyte gels: the role of pH and salt concentration. Macromolecules 44:147–158
123. Claudio GC, Kremer K, Holm C (2009) Comparison of a hydrogel model to the Poisson-Boltzmann cell model. J Chem Phys 131(09):094903
124. Jha PK, Zwanikken JW, de Pablo JJ, de la Cruz MO (2011) Electrostatic control of nanoscale phase behavior of polyelectrolyte networks. Curr Opin Solid State Mater Sci 15(6):271–276. Functional gels and membranes

Relaxation Mechanisms of Physical Hydrogels Networks

Jan Zidek, Andrey Milchev, Josef Jancar, and Thomas A. Vilgis

Abstract

We study the dynamic mechanical behavior of a model hydrogel subject to deformation by means of Molecular Dynamics simulation. The model has a predefined invariable chemical composition and secondary structure of entangled network, but its macroscopic response to tensile deformation varies depending on external conditions. Our model has temperature-, *pH*, swelling degree-, and deformation rate-responsive behavior. The model is found to respond to changes in external conditions qualitatively in the same way as real hydrogels, which serve as reference in our study. The model is focused on physical hydrogels, where the self-assembly of interacting acrylic acid (*AA*)-groups plays essential role in the formation of the network. In particular, as a modeled material we have chosen the *poly-(lactide-glycolide)-acrylic acid PLGA* $-$ *AA* hydrogel, where *AA*-groups are believed to play the role of quasi-mobile nodes in the formation of a network. One output of the model is the change of energy density during tensile deformation which is then used to calculate the stress–strain relationship.

Structural changes were investigated both during stretching and relaxation of the system. Three different structural factors pertaining to *AA*-groups were monitored regarding possible changes during deformation and also when the deformation was stopped: detachment from clusters, hops between clusters, and reorientation of *AA*-clusters. Our results suggest that the proposed model provides a qualitatively faithful description of tensile deformation in physical hydrogels.

Keywords

Hydrogel • Molecular dynamics • Relaxation • Tensile deformation • Macromolecular networks

Introduction

Hydrogels, in contrast to organic gels, display a large variety of properties which can be determined by the nature of the networkforming chains, by *pH*-value, by the ionic strength, the solvent composition, and the degree of swelling. The stretching and swelling behavior is thus adjustable through the conformation of the mesh between two crosslinks. Hydrogels are therefore best examples for intelligent materials whose conformation can be adjusted in a definite way. In this paper we present evidence how some of these

J. Zidek (✉) • J. Jancar
Central European Institute of Technology, BUT, Technicka 10, 612 00 Brno, Czech Republic
e-mail: jan.zidek@ceitec.vutbr.cz

A. Milchev
Bulgarian Academy of Sciences, 1113 Sofia, Bulgaria

T.A. Vilgis
Max-Planck Institute for Polymer Research, Ackermannweg 10, 55128 Mainz, Germany

G. Sadowski and W. Richtering (eds.), *Intelligent Hydrogels*, Progress in Colloid and Polymer Science 140, DOI 10.1007/978-3-319-01683-2_17, © Springer International Publishing Switzerland 2013 factors affect the properties of hydrogels whose formation occurs as a result of competition between hydrophobicity and hydrophilicity of the different composition units along the network chains. In fact, the heterogeneous clusters form a kind of crosslinked regions. The crosslinks are spatially extended and can be dissolved by large strains/tensile forces.

Generally, properties of hydrogels are determined by network structure. Hydrogels may vary in their chemical structure, from crosslinked networks up to self-assembled micellar structures. Even if the chemical composition is exactly defined, the properties of a hydrogel are not strictly determined but rather change, depending on external conditions such as *pH*, or temperature, *T*. The resulting macroscopic properties may be thus ascribed to many factors. Therefore, the description of hydrogel properties is, in general, rather complex and laborious.

The subject of this paper is focused on a particular structure and type of properties. Within the framework of hydrogel structure, a particular type of a physical gel was investigated. Thus, the cohesion of the macromolecular network is produced by self-assembly of physically interacting groups. As a result, such hydrogels behave as transient networks. Their nodes are not permanent, but are rather formed by clusters (micelles) of interacting groups. Regarding strain–stress relationship, our model analysis targets the dynamic behavior during deformation whereas the chemical composition and the building units, i.e., the secondary structure of the system, remain unchanged. The model, in particular, has been designed to mimic the structure of $PLGA - AA$, presented in paper of Park et al. [1].

We study the effect of four external factors on the deformation behavior of the hydrogel: temperature, deformation rate, *pH*, and swelling degree. We find that deformation properties may change reversibly up to a certain deformation threshold only. This is typical for self-assembled networks where network rearrangements proceed by surmounting of an energy barrier. Such properties have been experimentally studied recently [2, 3]. The elastic response of macromolecular networks has also been described theoretically [4–6]. An overview of the models of Kuhn, James, Guth, Flory, and Treloar can be found in James and Guth [7]. Arruda and Boyce [8] presented a constitutive model of macromolecular networks as hyper-elastic materials. All the above stated models, however, describe elastic deformation. The visco-elastic behavior was considered by means of the theory of reptation [9–11].

Except in gels, permanent deformation may be observed also in glassy amorphous materials. The theoretical background of such deformation is more frequently described in glassy polymers than in the hydrogels [12]. Chen et al. described the relationship between segmental dynamics on a molecular scale, and dynamic mechanical properties. The deformation in a glassy state is presented as a series of discontinuous hops of building units (segments) in the locally perturbed domain structure [13]. Anticipating, we would like to note that similar effects were found to accompany also the deformation of hydrogels.

In classical rubbers and gels with entropy-based physical background the deformation is to great extent elastic. Glasses, partially crystalline materials, and other imperfect crosslinked materials show plastic and thus irreversible deformations at large stretching ratios. Upon exceeding some threshold value, the deformation tends to be permanent. The reconstruction of the network structure during large deformations is significant. This makes the theoretical description of such processes difficult and, as far as we are aware, there still does not exist a generally acknowledged theory of plastic deformation, in contrast to the reptation theory of visco-elastic deformation. Thus, with the lack of sufficiently comprehensive analytical description available at present, Molecular Dynamics simulations can highlight a number of interesting and important effects, which elucidate structural changes in a network during permanent deformation.

Model and Methods

Molecular Dynamics Simulation

The present model is designed to describe the behavior of a *Poly-lactide/glycolide* (*PLGA*) hydrogel crosslinked by *Acrylic Acid* [14]. The chemical structure of such gel has already been presented in the literature [1] (cf. Fig. 1a). The polymer is known to form complex structures such as micelles. The micelles are ordered into a macromolecular network (Fig. 1b).

Our model represents a partially coarse-grained system, which mimics the structure of hydrogels. Regarding this model, many properties are very similar to those of the real material. The system comprises a network composed from semi-flexible macromolecular chains (Fig. 1c). The flexibility is governed by partial charges on the individual *PEG* atoms. The charge on the macromolecular chain affects the distribution of water molecules in the vicinity of chains. The formation of clusters in the system is due to hydrogen bonds. The hydrogen bonds play significant role as in the real structure material, e.g., $-COOH$ groups are attracted by hydrogen bonds. In order to mimic the polymer entanglements in the real system, we take the polymer chains as cyclic and artificially entangled so that the originally created entanglement cannot be destroyed. The cyclic molecule is effected by the simulation box periodicity, i.e., the last macromolecular segment near the boundary of the box is connected to the first segment from the opposite boundary. This mimics a situation, when the length of macromolecular chain is significantly larger than the length of the simulation box.

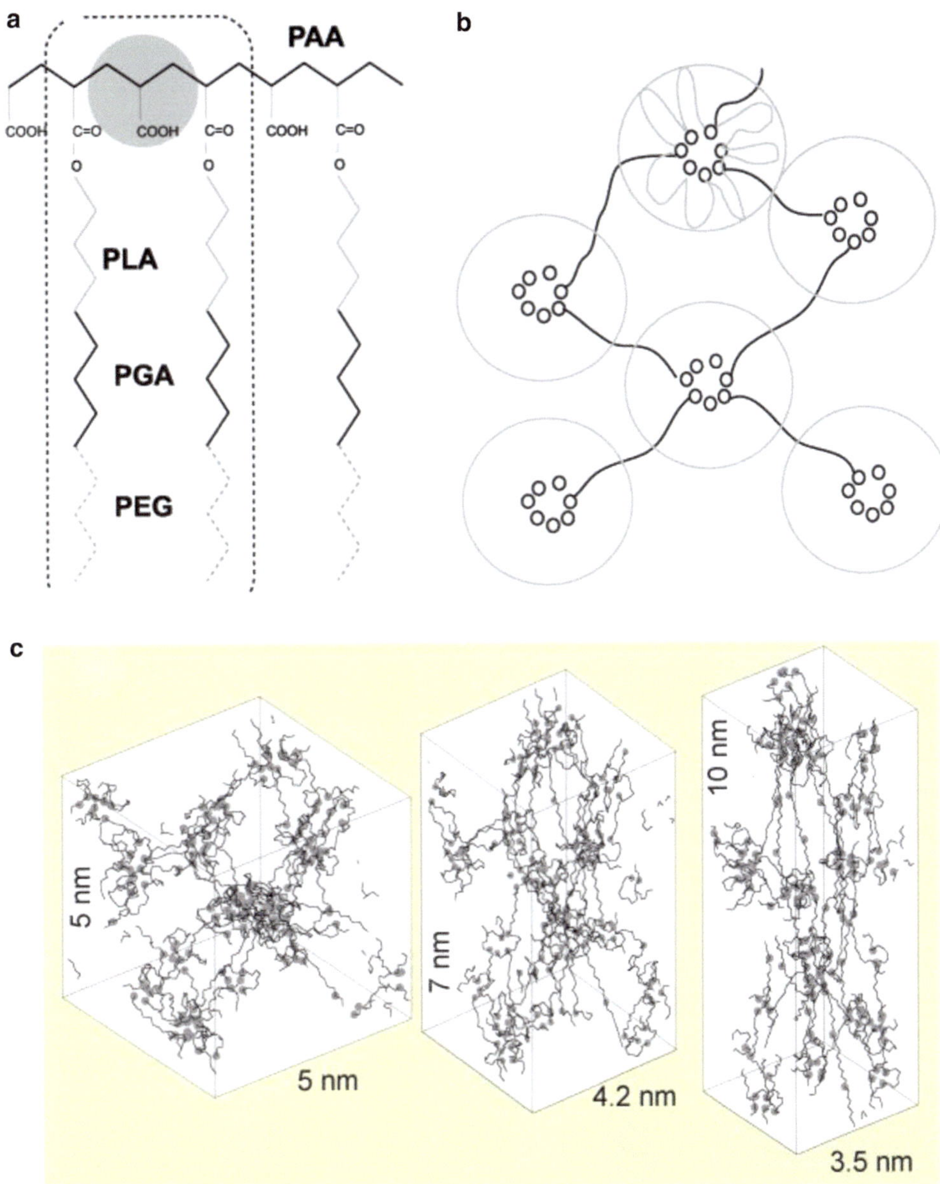

Fig. 1 Model structure of hydrogels; (**a**) chemical structure of PLGA hydrogels according to literature; (**b**) presumptive structure: *grey circles* – micelles; *black part* – simplified network structure used in the model; *circles* – interacting acrylic acid groups; *lines* – hydrophilic polyethylene glycol chains; (**c**) snapshots of hydrogel networks; blue points-*AA*-clusters; *black lines* – connecting chains

For the sake of simplicity and tractability, some structural properties of the model are significantly simplified in comparison to those of the real material. The real micellar structure is significantly more complex. Thus, micelle cores are created not only by acrylic acid (*AA*)-groups but there are also hydrophobic chains of poly-lactic-and poly-glycolic acid. Our clusters, however, are formed by strongly interacting *AA*-groups only (Fig. 1b). In contrast to the real material, which contains chains of poly-*AA* segments, we consider single *AA* – groups that are uniformly distributed in the simulation box. Thus, the number of molecules in the model is essentially reduced. As a matter of fact, each *AA*-group is taken as a coarse grained single bead.

The simulation box contains *8* chains, each composed of *200* atomic groups corresponding to *60* monomer units of *PEG* and *20 Acrylic Acid* groups. Each polymer chain is composed of alternating blocks of *3 PEG* monomer units and one *AA* group.

The simulation box with the dry phase is filled by water molecules whereby the single point charge (*SPC*) model of water is applied as a solvent [15].

The Molecular Dynamics simulation was carried out by means of the GROMACS software [16] package. Deformation of the simulation box was realized by shifting the boundary of the box. The box was stretched in one direction and compressed in transverse direction so that the volume remains constant during deformation. The stretching rate was chosen as *0.1* ns^{-1} or *0.01*ns^{-1}. The box was stretched twice its original length during *10 ns* (*100ns*). The relaxation phase was effected by *10 ns* simulation without stretching.

The main output of the model is the energy density, *W*, which is related to the stress–strain ($\sigma - \lambda$) dependence. In the case of uniaxial stretching, the true stress in stretching direction is calculated from the derivative of energy density:

$$\sigma = \lambda_z \partial W / \partial \lambda_z - \lambda_{xy} \partial W / \partial \lambda_{xy} \tag{1}$$

The stretching is performed in *z*-direction while in *x, y*-directions one has transversal shrinking.

Analysis of Clusters

Cluster fraction The cluster fraction, which eventually determines the gel behavior, was analyzed by Delaunay triangulation (*DT*) of *AA*-groups within the volume of the box. The idea of the method is to learn something about the network structure by counting the nearest neighbors and the corresponding coordination number. As a result of *DT*, one gets a set of lines connecting the pairs of neighboring acrylic acid groups. The volume fraction is calculated from the histogram of lengths of connecting lines. The histogram takes into account the different weight of lines in the statistical analysis. Several short lines occupy the same space as one long line; therefore, distances have to be normalized by the line length:

$$F\left(l_1, l_2, ..., l_n\right) = \left\{ \frac{N_1 l_1}{\sum_{i=1}^{n} N_i l_i}, \frac{N_2 l_2}{\sum_{i=1}^{n} N_i l_i}, ..., \frac{N_n l_n}{\sum_{i=1}^{n} N_i l_i} \right\}, \tag{2}$$

where F is a function of the set of distances of *AA*-groups l_1, l_2, ... l_n, and N_1, N_2, ..., N_n are the frequencies (N) of such distances (l) in the simulation box.

Another method for characterization of the nodes that we used is based on a cluster-counting algorithm. As a result of this cluster counting, one may sort *AA*-groups by their presence in particular clusters. This analysis is a prerequisite to the study of hops of *AA*-groups and their spatial reorientation during the stretching process.

Analysis of segmental hops Hops of *AA*-groups between clusters are frequently observed during the stretching process. The process of hopping does not change the overall cluster fraction yet it modifies the original composition of *AA*-clusters during the process. At the end of the simulation, the existing clusters are composed from different *AA*-groups.

A pair of two *AA*-groups is defined when both *AA*-groups occupy the same cluster. All possible pairs in a cluster can be found as a combination of two *AA*-groups from the set of all *AA*-groups in the cluster. One may thus define and calculate a correlation function of the *AA*-pairs. In case, that a selected pair is present in the initial configuration and also persists at time $t > 0$, it is considered as persistent so its coefficient of cluster cohesion is 1. Otherwise, the coefficient is *0*. The average coefficient of all pairs $c_{cohesion}$ can thus have a value between *1* and *0* and change as time goes by.

The reorientation of *AA*-clusters reflects a change in the main-axis direction of clusters, which can be caused either by rotation in space, or by deformation of clusters. The individual components of clusters (the *AA*-pairs from previous paragraph) are re-oriented in space in the same way as the clusters.

The vector of each *AA*-pair in the cluster is computed from the coordinates (**X**) of both groups AA_1 and AA_2, i.e., $\mathbf{V} = \mathbf{X}_{AA1} - \mathbf{V}_{AA2}$. The vector changes during simulation, and/or deformation of the sample V(t, λ). The re-orientation of a pair is equivalent to the change of pair vector. The cosine of the angle between a vector at zero-time and at time t was calculated from scalar product:

$$\cos\theta = \frac{\vec{V}_P \cdot \vec{V}_{P,0}}{\left|\vec{V}_P\right| \left|\vec{V}_{P,0}\right|} \tag{3}$$

where $\mathbf{V}_{P,0}$ is a vector of a pair at time $t = 0$, and $\mathbf{V}_P$ is the vector at *t*.

Another (similar) indicator of orientation is the second Legendre polynomial P_2:

$$P_2 = 0.5\left(3\langle\cos^2\theta\rangle - 1\right) \tag{4}$$

where θ is measured as the angle of the current vector with respect to the vector direction before simulation, and the average is taken over the different pairs. The value of P_2 can attain values from *–0.5*, when the direction of the *AA*-pair becomes perpendicular to the previous pair orientation, *0*, when both vectors are oriented randomly, up to *1*, when these vectors of the pair remain parallel.

Results and Discussion

Our results show that, in general, the behavior of the model system during tensile deformation exhibits a number of common features with the behavior of a real material.

One can detect a region of small deformation, where the deformation of the network is almost affine and reversible, and then observe the softening at large deformation where

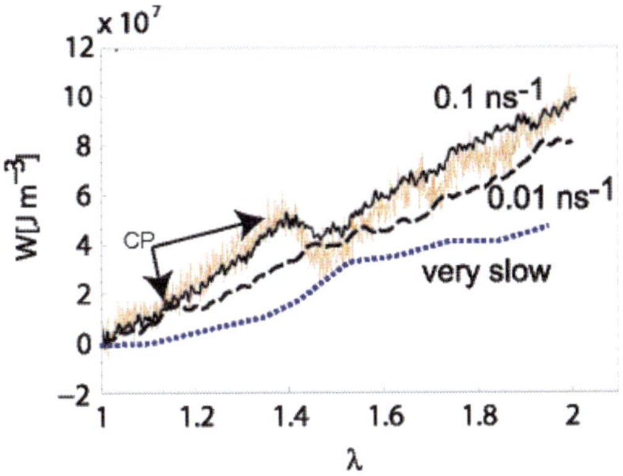

Fig. 2 Increase of energy density during stretching for different strain rates; number at lines – strain rate; CP – critical point (onset of plastic deformation)

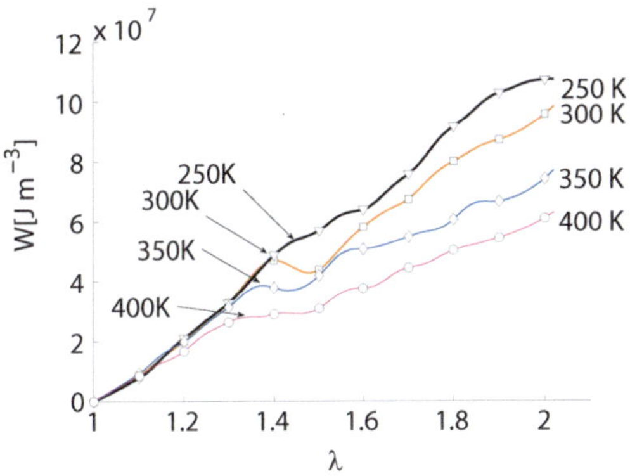

Fig. 3 Increase of energy density during stretching at different temperatures; *Arrows* – critical points

plastic flow of the network structure is observed. The same effect is also found in the experimental tensile curve, when the deformation goes beyond the so called yield point [17].

Tensile Behavior of Hydrogel Models Under Different External Conditions

In what follows we briefly present the observed mechanical response of our model with respect to *temperature*, *pH*, *strain rate*, and *swelling ratio*.

Strain rate The dependence of deformation properties on strain rate is characteristic for hydrogels. The increase of mechanical energy density as a function of strain rate is shown in Fig. 2. At large strain rates, the hydrogels have a critical point (CP) at relatively high stretching ratio. The slowing down of deformation rate leads to a shift of the critical point to lower deformation thresholds, i.e., the CP is observed at lower stretching ratio λ. When the deformation becomes very slow, the yield point disappears altogether, and the deformation of the model leads to ductile flow just from the very beginning.

In our interpretation we see the yield point as a consequence of the dissociation of *AA*-groups from the interacting clusters (which serve as network nodes). For quick deformation, the dissociation rate is much slower than the deformation itself, and the network responds elastically. Therefore, the network behaves initially as affine. When the stretching ratio exceeds a critical threshold, the polymer chains stretch and the dissociation of *AA*-groups is facilitated. In contrast, when the deformation is very slow, both dissociation and network rearrangement occur simultaneously, and creep is observed from the very onset of stretching.

One should point out that the increase of stiffness with increased shear rate is typical property of hydrogels and has been frequently reported in literature (for example, see, [18, 19]).

Effect of Temperature The dependence of the mechanical response on temperature is shown in Fig. 3. The energy density was investigated in the range of 250 ÷ 400K. Apparently, the energy density function changes with temperature. The threshold stretching ratio, where the onset of softening phase is observed, is found to decrease with increasing temperature. Such a behavior may be explained by noting that the increase of temperature facilitates the dissociation of the *AA*-groups. Unfortunately, we were unable to find corresponding $\sigma - \lambda$ curves for different temperature in the literature. However, the deformation of our model system can be compared to the deformation properties of real materials. Thus, self-assembled physical gels become more liquid-like, when the temperature in increased [20, 21]. This resembles very much the behavior of our model. On the other hand, the chemically crosslinked gels exhibit the opposite trend, their stiffness increases with increasing temperature.

Effect of pH The variation of the total energy density at different *pH* is shown in Fig. 4. Two limiting cases were investigated. The first one is the acidic *pH* to the effect that all *COOH*-groups of the acrylic acid become neutral. Another limiting case is our model network at extremely basic *pH*, whereby all the groups are dissociated. The real hydrogel is kind of intermediate network, where the *COOH*-groups are partially dissociated and partially neutral.

In the dissociated state, the interaction between *COO*–-groups is electrostatic and repulsive. Therefore, the acrylic acid *AA*-groups do not tend to aggregate. On the other hand, the neutralized *COOH*-groups are attracted by hydrogen bonds, and *AA*-clusters are formed.

Our model describes a specific case, where the micellar structure of the nodes depends on the attraction between *COOH*-groups. One example of such model is the *poly-lactide/glycolide* hydrogel, which is crosslinked by acrylic acid groups. Park and Kim [1] investigated the *sol-gel*

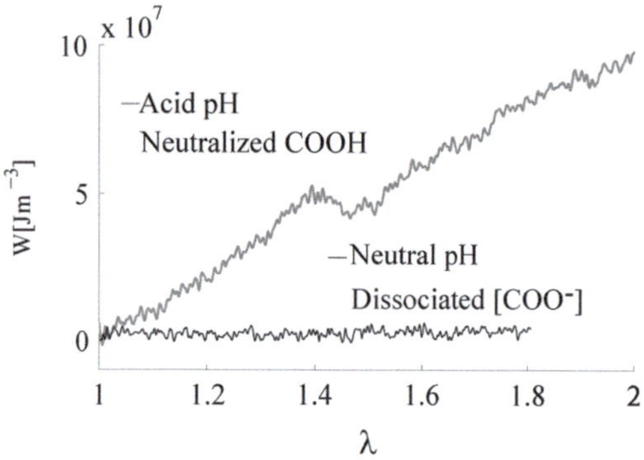

Fig. 4 Increase of energy density during the stretching at *neutral pH* (dissociated COO– in acrylic acid), and *acid pH* – (neutralized COOH groups)

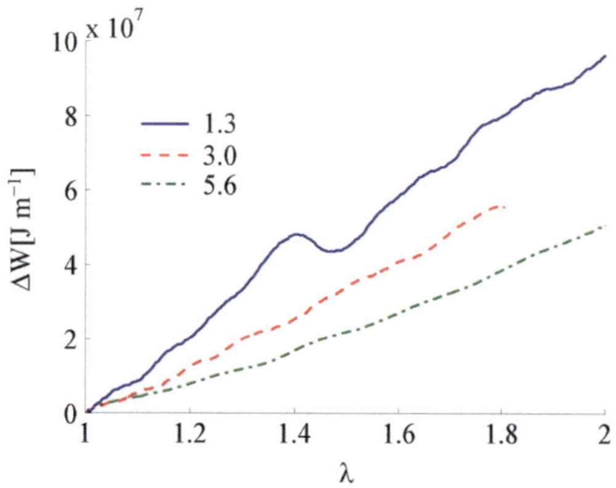

Fig. 5 Increase of energy density during stretching at different swelling ratio; legend – swelling ratio

transitions regarding its dependence on *pH*. They found that the gel is more easily created at $pH = 4$, where the *COOH*-groups are neutralized, rather than at $pH = 7$, when some *AA*-groups are dissociated. Bossard at al. [22] presented the properties of a hydrogel from *poly(methyl methacrylate)-poly(dimethyl amino ethyl methacrylate)-poly(methyl methacrylate)*. The structure was found to be in the gel phase at $pH = 4$, and to exist as a solution at $pH = 8$.

Effect of swelling ratio The tensile deformation of our model at different swelling ratio was also investigated. The results are shown in Fig. 5. The increased amount of solvent leads to inflation of the network. During this inflation again the dissociation of *AA*-clusters is facilitated. Therefore, the concentration of *AA* cluster decreases with increasing swelling ratio. The slope of the energy density against strain λ is proportional to the stiffness of underlying network. We find therefore that the stiffness decreases with increasing swelling ratio.

The same effect is also observed experimentally on sol-gel phase diagrams [1]. One finds that both phases are present at weakly swollen networks. When the swelling degree increases, however, the gel phase becomes less pronounced. When the swelling degree exceeds some limit, the gel phase disappears altogether owing to destruction of the micellar structure.

Therefore, one may conclude that the present model mimics qualitatively properly the behavior of physical (self-assembled) hydrogels although the model network density is higher than the network density of real hydrogels. The latter has a very large number of atoms and would require a prohibitively large computational effort even for a modern CPU power. Nonetheless, the trends in the responsive behavior of our model follow those in the experimental behavior of reference materials.

On the other hand, one should be aware that this model does not agree with the respective experimental behavior in the case of crosslinked gels. The properties of chemically crosslinked hydrogels depend predominantly on the concentration of elastically active chains, rather than on the propertied of nodes. Yet, the present model can be modified by additional crosslinking in order to allow for the behavior of crosslinked gels. Thus, a mixture of physical and chemical links is expected to reproduce more complex deformation behavior like in additionally treated hydrogels. We plan to report on the new results in near future.

Mechanism of Relaxation

As mentioned above, in the course of the *MD* simulation one observes a softening phase at large deformation. This phase is found also in real hydrogels, and the deformation during this phase does not vanish. In the softening phase, the material becomes ductile. This is accompanied by structural changes, which are dynamic in nature. The following changes have been taken into consideration:

- Dissociation of *AA*-groups from cluster
- Reorientation of *AA*-clusters
- Hops of *AA*-groups from one cluster to another one.

The dynamic character of structure variation with stretching can be detected also in the behavior of our model system. One such process is the dissociation of *AA*-groups. Changes in the volume fraction of *AA*-clusters can be detected by Delaunay triangulation. The cluster fraction as a function of stretching ratio is shown in Fig. 6 (see right axis). Evidently, the cluster fraction decreases during the stretching of the network. Related type of processes is hops of *AA*-groups. They are observed from the very beginning of sample deformation. Therefore, the dissociation of *AA* clusters has dynamic character. Dissociation of *AA* clusters may partially be compensated by re-attachment of free *AA*-groups, but as a whole the dissociation prevails.

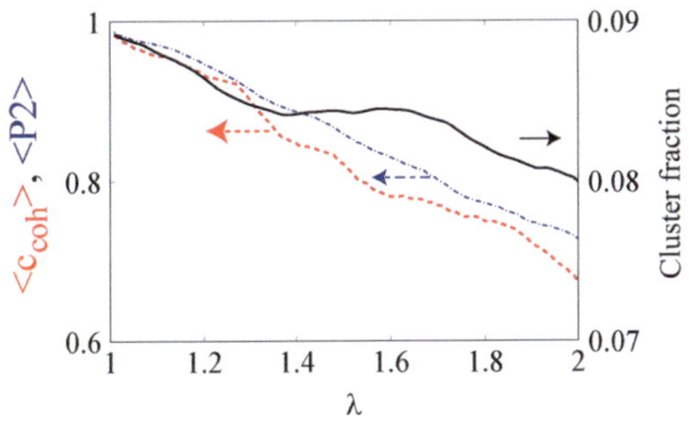

Fig. 6 Cluster fraction, (full line, right y-axis), coefficient of cohesion (left axis, *dash-dot*), and reorientation parameter (P2, *dashed*, left axis) as function of stretching ratio

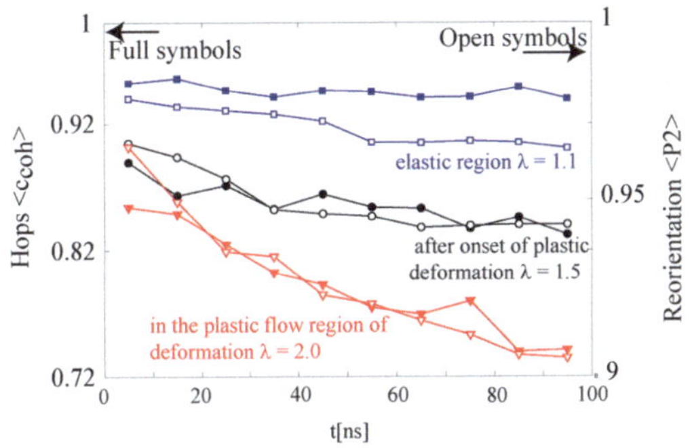

Fig. 7 Time variation of the intensity of two principal factors that affect the network structure during the relaxation process; Segmental hops (*full symbols*, *left* y-axis); reorientation of clusters (*open symbols*, *right* y-axis); The relative intensities of both processes were calculated for different degrees of tensile stretching: elastic deformation $\lambda = 1.1$; after the onset of plastic flow at $\lambda = 1.5$, and at strong plastic deformation $\lambda = 2.0$

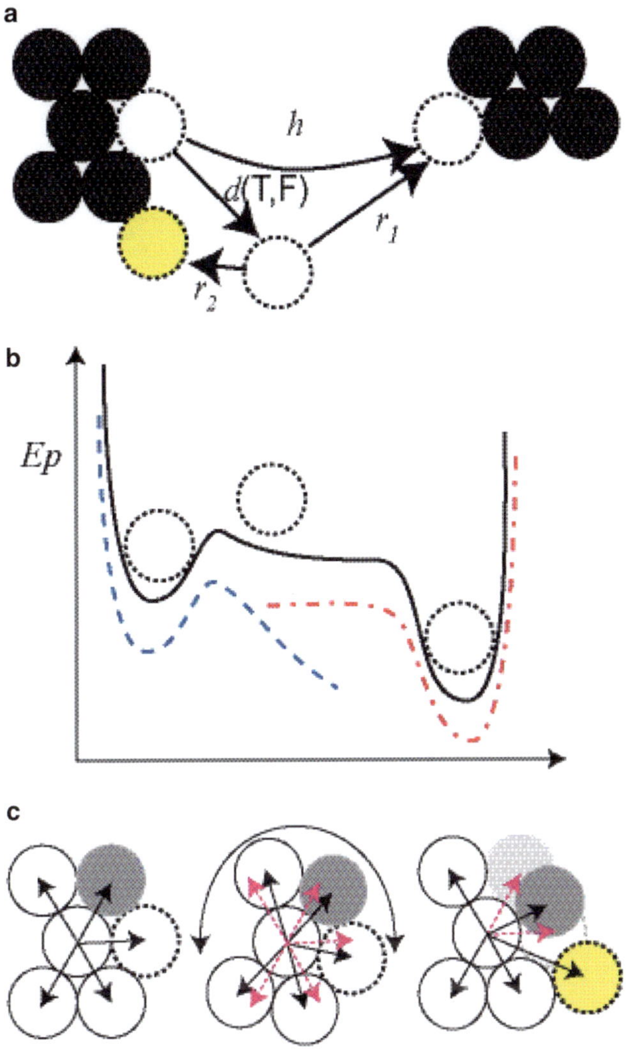

Fig. 8 Mechanisms of structural changes: (**a**) dissociation of an *AA*-group as function of temperature and force acting on the group d(T,F), and re-attachment to another cluster r1, or back to the same cluster r2; h-segmental hop; (**b**)-potential energy of an *AA*-group during the hop; *blue*(*dashed*) Lennard-Jones potential of atoms pulled by external force, *red* (*dash-dot*) LJ potential with relaxed force; (**c**) mechanism of cluster re-orientation; *left* – original cluster, *arrows* – pair vectors; *middle* – cluster after rotation; *magenta* – original pair vectors; *right* – shift of *AA*-groups; *yellow circle AA*-group, which could be relocated by dissociation d(T,F) and re-attachment r2, see (**a**)

In the interval between $\lambda = 1.3$ and 1.6 the cluster fraction is found to exhibit a plateau. This plateau corresponds to the onset of the softening phase. At the onset of softening phase at $\lambda = 1.4$, the hops intensity is increased.

The softening phase is due also to the reorientation of clusters. The intensity of reorientation and hops is shown in Fig. 6 (see left axis).

All dynamic processes take place when the network stretching proceeds at some rate (Fig. 6). However, only the segmental hops and cluster reorientation continue also when the sample is held under constant strain. Their variation with time is shown in Fig. 7. Evidently, the relaxation process depends on the degree of applied strain, i.e., on the strain ratio (λ). When the deformation is stopped during the elastic phase ($\lambda = 1.1$), segmental hops are not observed whatsoever. Only reorientation of clusters proceeds for a short while before coming to an end too.

The other two cases concern the plastic flow just after the onset of softening ($\lambda = 1.5$), and the flow well in the softening phase ($\lambda = 2.0$). They look similar whereby both reorientations and hops take place. The reorientation is more intense than in the elastic phase, because the connections between two nodes are not so strong as in the case of affine network.

The mechanism of all possible structural changes is shown in Fig. 8. Part (a) describes the basic mechanism. The acrylic acid *AA*-groups dissociate from a cluster. This process depends on temperature as far as the *AA*-group must overcome some energy barrier, which is determined by interactions within the cluster. The dissociation depends also on the force which acts on the *AA*-groups. A detached *AA*-group can re-attach to a neighboring cluster. Then, a

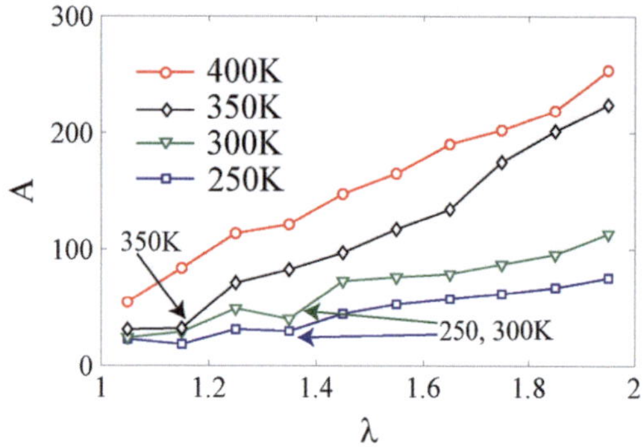

Fig. 9 Non-affinity parameter as function of the stretching ratio of the simulation box at different: (**a**) temperatures; (**b**) swelling ratio

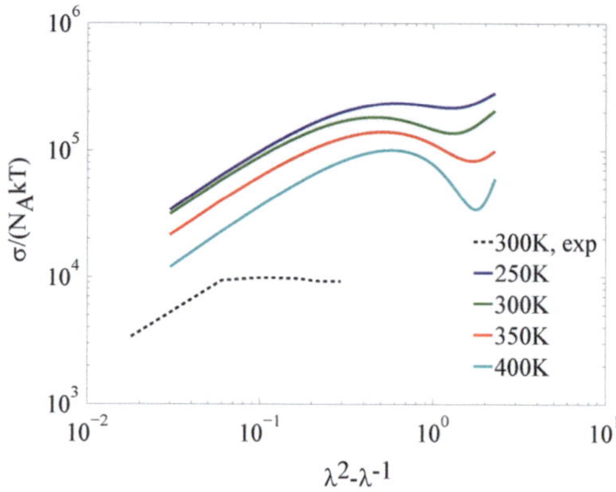

Fig. 10 Network theory plot: reduced stress as function of $\lambda^2 - \lambda^{-1}$; Simulation data is compared to experimental results

segmental hop (h) may occur. Alternatively, a detached *AA*-group may re-attach to another place in the original cluster (yellow point).

Part (b) describes the energy profile during a hop (black curve). The energy barrier in the original cluster is decreased due to the applied external force. The Lennard-Jones potential of a pair, pulled by external force, is shown as blue dashed curve. The hop of an *AA* leads to relaxation of the external force on the *AA*-group. Therefore, the change of potential energy during re-attachment corresponds to the Lennard Jones potential – red dash-dotted curve.

Part (c) describes possible mechanisms of re-orientation of clusters. During this phase, pair vectors (arrows) change their orientation, manifested by the coefficient P_2 which decreases below the value of *1*. The change of such a vector can be achieved in two possible ways. The first one is rotation of the entire cluster (middle picture); The second is a shift of an *AA*-group in the cluster (right picture). The shift of an *AA*-group can also exhibit dynamic mechanism (yellow point), i.e., it can take place by dissociation of an *AA* – group and reattachment back to the original cluster.

We analyzed in more detail which is the genuine mechanism of reorientation actually is. If cluster rotation occurs, the value of P_2 should be similar for all pairs in this cluster. If, on the contrary, reorientation is due to a shift, differences in P_2 of the pairs in a single cluster should be observed. The rotation (middle picture) was almost not observed. Only the shift of *AA* group (right picture) was detected. It was observed in any case, that the vectors do not reorient to the stretching direction during the deformation of network.

The information about the nature of deformation is supplemented by the analysis of affinity. The coefficient of non affinity, (*A*), was calculated as function of the network degree of stretching λ. Values close to zero mean affine deformation. The higher the value of *A* is, the lower is the affinity of the observed deformation. The non-affinity parameter *A* for different temperatures is shown in Fig. 9.

The deformation has usually a first (elastic) phase, where the non-affinity parameter is relatively low. When the stretching ratio exceeds some critical deformation, the non-affinity increases. The increase of non-affinity is indicated in Fig. 9 by an arrow. It happens close to the critical point in the energy density function (Fig. 3). Only at 400K the function of non-affinity does not exhibit a critical point. The deformation is non-affine in the entire range of deformation.

Conclusion

In this paper we presented a model of a hydrogel which is examined by Molecular Dynamics simulation. The model describes physically crosslinked hydrogels whereby the molecular composition of the polymer phase, and the secondary structure (an entangled network) were kept unchanged. The model turns to be *temperature-*, *pH-*, *swelling degree-*, and *deformation rate*-sensitive. It was found that the model behavior is qualitatively in agreement with that of hydrogels, whose properties are driven by self-assembly of interacting groups, as e.g., in physical gels with micellar structure. The calculated tensile curves are qualitatively similar to the experimental tensile deformation of *poly-(lactide glycolide) hydrogels* (Fig. 10).

The mechanism of relaxation was analyzed. Three different mechanisms of structure variation were studied with respect to their changes during network deformation, including changes after deformation is stopped. The first process, detachment of *AA*-clusters, appears related solely to deformation. It seizes after the deformation is halted. The second process comprises hops of *AA*-groups between clusters. This is a dynamic process, which is observed also during the relaxation phase, when the deformation is stopped. The segmental hops seem to be a persistent process. The third process is that of cluster reorientation. This process continues also during the relaxation phase. It is relatively fast in comparison to the segmental hops. It acts as quick reduction of forces acting on the cluster, after the cluster was instantaneously put out of equilibrium.

Acknowledgements This work was supported by DFG under research project SPP 1259 Intelligent hydrogels. J. Zidek, and J. Jancar acknowledge European Regional Development Fund (CEITEC, CZ.1.05/1.1.00/02.0068). A. Milchev is indebted to the Max-Planck Institute for Polymer Research in Mainz, Germany, for hospitality during his visit and to CECAM nano SMSM for financial support. ***Notice:*** Figs. 1a, c, 2, 3, and 10 were reprinted with permission from ref [14], Copyright (2012), American Institute of Physics.

References

1. Park C, Kim S (2010) Sol–gel transition behavior of amphiphilic comb-like poly[(*PEG*- b-*PLGA*)acrylate] block copolymers. J Polym Sci Part A Pol Chem 48:1287–1297. doi:10.1002/pola.23877
2. MacKintosh F, Kas J, Janmey P (1995) Elasticity of semi-flexible biopolymer networks. Phys Rev Lett 75:4425–4428. doi:10.1103/PhysRevLett.75.4425
3. Djonlagic J, Zugic D, Petrovic Z (2012) High strength thermoresponsive semi-IPN hydrogels reinforced with nanoclays. J Appl Polym Sci 124:3024–3036. doi:10.1002/app.35334
4. Miao B, Vilgis T, Poggendorf S, Sadowski G (2010) Effect of finite extensibility on the equilibrium chain size. Macromol Theor Simul 19:414–420. doi:10.1002/mats.201000009
5. Tamai Y, Tanaka H, Nakanishi K (1996) Molecular dynamics study of polymer-water interaction in hydrogels. 1. Hydrogen-bond structure. Macromolecules 29:6750–6760. doi:10.1021/ma951635z
6. He Y, Shao Q, Tsao H, Chen S, Goddard W, Jiang S (2011) Understanding three hydration-dependent transitions of zwitterionic carboxybetaine hydrogel by molecular dynamics simulations. J Phys Chem B 115:11575–11580. doi:10.1021/jp204682x
7. James H, Guth E (1996) Simple presentation of network theory of rubber, with a discussion of other theories. J Polym Sci B 34:7–36. doi:10.1002/polb.1996.883
8. Arruda E, Boyce M (1993) A three-dimensional constitutive model for the large stretch behavior of rubber elastic materials. J Mech Phys Solids 41:389–412. doi:10.1016/0022-5096(93)90013-6
9. De Gennes P (1971) Reptation of a polymer chain in the presence of fixed obstacles. J Chem Phys 55:572. doi:10.1063/1.1675789
10. Doi M, Edwards S (1978) Dynamics of concentrated polymer systems. Part 1."Brownian motion in the equilibrium state". J Chem Soc Farad T 2 74:1789. doi:10.1039/F29787401789
11. Doi M, Edwards S (1978) Dynamics of concentrated polymer systems. Part 2."Molecular motion under flow". J Chem Soc Farad T 2 74:1802. doi:10.1039/F29787401802
12. Chen K, Saltzman E, Schweizer K (2009) Segmental dynamics in polymers: from cold melts to ageing and stressed glasses. J Phys Condens Matter 21:503101. doi:10.1088/0953-8984/21/50/503101
13. Warren M, Rottler J (2010) Deformation-induced accelerated dynamics in polymer glasses. J Chem Phys 133:164513. doi:10.1063/1.3505149
14. Zidek J, Milchev A, Vilgis T (2012) Dynamic behavior of acrylic acid clusters as quasi-mobile nodes in a model of hydrogel network. J Chem Phys 137:244908. doi:10.1063/1.4769833
15. Miyamoto S, Kollman P (1992) Molecular dynamics studies of calixspherand complexes with alkali metal cations: calculation of the absolute and relative free energies of binding of cations to a calixspherand. J Am Chem Soc 114:3668–3674. doi:10.1021/ja00036a015
16. Hess B, Kutzner C, Van der Spoel D, Lindahl E (2008) GROMACS 4: algorithms for highly efficient, load-balanced, and scalable molecular simulation. J Chem Theor Comput 4:435–447. doi:10.1021/ct700301q
17. Krok M, Pamula E (2012) Poly(L-lactide-co-glycolide) microporous membranes for medical applications produced with the use of polyethylene glycol as a pore former. J Appl Polym Sci 125:E187–E199. doi:10.1002/app.36697
18. Kwon J, Subhash G (2010) Compressive strain rate sensitivity of ballistic gelatin. J Biomech 43:420–425. doi:10.1016/j.jbiomech.2009.10.008
19. Urayama K, Taoka Y, Nakamura K, Takigawa T (2008) Markedly compressible behaviors of gellan hydrogels in a constrained geometry at ultraslow strain rates. Polymer 49:3295–3300. doi:10.1016/j.polymer.2008.05.045
20. Saravanakumar K, Tata B, Aswal V (2012) Thermoreversible viscoelastic to weak gel transition in a micellar ionic liquid with salt. Colloids Surf A Physicochem Eng Asp 414:359–365. doi:10.1016/j.colsurfa.2012.08.061
21. Fernández E, Mijangos C, Guenet J, Cuberes M, López D (2009) New hydrogels based on the interpenetration of physical gels of agarose and chemical gels of polyacrylamide. Eur Polym J 45:932–939. doi:10.1016/j.eurpolymj.2008.11.041
22. Bossard F, Aubry T, Gotzamanis G, Tsitsilianis C (2006) *PH*-Tunable rheological properties of a telechelic cationic polyelectrolyte reversible hydrogel. Soft Matter 2:510. doi:10.1039/b601435f

Calculating Structural Properties of Reversibly Crosslinked Polymer Systems Using Self-Consistent Field Theory

Thomas Gruhn, Daming Li, and Heike Emmerich

Abstract

We study the influence of reversible crosslinks on a polymer blend with the help of an extended self-consistent mean field theory. The systems consist of homopolymers of type A and B and copolymers of type AB. Copolymers AB are reversibly crosslinked with a crosslink strength z. The links include monomers of type A and B with weights ω_A and ω_B, respectively. Crosslinking of A and B polymers is prohibited. Without crosslinks the system shows a homogeneous phase, a lamellar phase, a hexagonal phase, and a fully demixed state. Setting $\omega_A + \omega_B = 1$, we find that the total crosslink strength z and the crosslink asymmetry $\Delta\omega \equiv \omega_A - \omega_B$ has a distinct influence on the structure of the system. We show that the microstructure can be switched from a hexagonal to a lamellar structure by increasing z or $\Delta\omega$.

Keywords

Self-consistent field theory • Polymer networks • Crosslinks • Polymer networks • Phase transition • Hydrogels • Self-assembly

Introduction

During the last years, the interest in tailored soft matter systems has strongly increased. New powerful techniques for synthesizing structured macromolecules, manipulating nanosized objects and creating nanoscopic structures in soft matter systems have been developed and methods for analyzing nanostructured soft matter systems have been refined and enhanced, which provide a more detailed insight into the system of interest. The new techniques allow the development of materials with very specific properties. Of special relevance are materials that are soft and easily deformable by external forces, while they preserve some internal structure that brings the material back to its preferred shape as the pressure is released. These materials are important, because they can adapt their shape reversibly to variations of the external conditions like changes in temperature, pH value, molecular composition of the solvent or external fields. Flexible polymer networks like hydrogels are perfectly suited for this purpose [1–3]. Due to the flexibility of the chains and the cavities in between they can be easily deformed, while due to the connectivity of the crosslinked chains, the polymer network preserves its underlying structure after large deformations. Hydrogels and other polymer networks are of great relevance for medical applications like drug delivery systems or tissue engineering, since they can be adapted to the cell biological environment [4–6].

Of great relevance are special hydrogels that are usually called reversible hydrogels. Their connectivity can be changed by external parameters, like the temperature [7, 8], illumination [9], or a change of the molecular composition in the solvent [10]. For example, thermosensitive crosslinks can

T. Gruhn (✉) • H. Emmerich
Materials and Process Simulation (MPS), University of Bayreuth, Bayreuth, Germany
e-mail: thomas.gruhn@uni-bayreuth.de; heike.emmerich@uni-bayreuth.de

D. Li
Department of Mathematics, Shanghai Jiao Tong University, 200240 Shanghai, China
e-mail: lidaming@sjtu.edu.cn

G. Sadowski and W. Richtering (eds.), *Intelligent Hydrogels*, Progress in Colloid and Polymer Science 140, DOI 10.1007/978-3-319-01683-2_18,

be broken by increasing the temperature and form again if the temperature decreased. This cycle can be repeated many times. The effect has various applications. Customized drug delivery in the human body can be achieved by capsules having a shell of reversible hydrogel and active agents inside. The capsules travel through the body until the hydrogel is dissolved, for example by a change of the pH value. Since the hydrogels change their structure and their mechanical properties according to the local environment, they can also be used as sensors. Reversible polymer networks are used for self-healing materials, in which mechanically inferred damages can be healed by dissolving and reconnecting the network, which then fills up and agglutinates the cracks.

Reversible polymer networks are based on crosslinks, whose strength depends sensitively on the environment. For example, at suitably low temperatures, the crosslinks in thermally reversible networks are practically unbreakable, while at higher temperatures the bond strengths are much weaker. One of the most important types of reversibly crosslinking polymers are Diels-Alder-based polymers [8]. At low temperatures (typically, up to about 60 °C), these polymers form a stable crosslinked network with covalent bonds, while at high temperature (typically, at 90 °C or higher) the crosslinks can break. An important state exists at temperatures at which the bond strengths are of the order of the thermal energy $k_B T$ (where k_B is the Boltzmann constant and T is the temperature). Then, crosslinks can form and break during the observation time and the polymer material is an ergodic, equilibrium system. In this state, one speaks of *reversible crosslinks* to distinguish them from *permanent crosslinks* that, practically, do not break in a reasonable time.

In this article, we use an extended version of the self-consistent mean field theory [11] to study polymer networks in a state, in which their crosslinks are reversible. This state is of special interest, since the system can form an equilibrium structure which does not depend on the history of the material. Thus, the equilibrium structure of the system is determined uniquely by the external parameters. If Diels-Alder-based polymers are quenched from such a state to low temperatures, the crosslinks get permanent and the thermal fluctuations of the polymer network are restricted by the crosslinks that were formed in the higher temperature state.

The self-consistent field theory is a well-established method for studying polymer systems. It is based on a concept introduced by Edwards [12] in the 1960s and later continued by Helfand and Yukiko [13]. The main idea is to rewrite the partition function with the help of a Hubbard-Stratonovich transformation. Minimization of the resulting free energy provides the necessary equations for the density distributions and some auxiliary fields that are added by the transformation. The theory has become a standard theory in the polymer community [14–18]. It has been applied on homopolymers [12], diblock copolymers [13, 19–22], polymer blends [23–30], polymer solutions [31–33], polyelectrolytes [34, 35], and systems of lipids [36, 37].

Early theoretical studies on polymer networks have been performed by Flory [38]. Various methods have been used to study the statistical properties of gels [39–41] and gel dynamics under external stimuli [42]. Some works are related to mean field theory. Recently, a mean field theory has been developed for studying the gelation transition in inhomogeneous reversibly bonding systems [43]. The gelation transition due to permanent random crosslinks has been studied by replica field theory [44–49]. The replica field theory is difficult to implement numerically since the fields are defined in the replica space.

An interesting system, which has been studied in detail theoretically [50, 51] and experimentally [52, 53], is a polymer blend of AB copolymers combined with A and B homopolymers. In this article we investigate a modified version of this system, in which reversible crosslinks can form between copolymers AB, while homopolymers consist of monomers A or B, for which the crosslinking groups are disabled by small chemical modification. We assume that the monomers A and B in the copolymer have the same mixing behavior and physical properties as their non-crosslinking counterparts in the homopolymer. In principal, the system should be called an $\mathrm{AB} + \mathrm{A}' + \mathrm{B}'$ system, since monomers A$'$ in the homopolymers are slightly different from monomers A in the copolymers. For the sake of simplicity, we stay with the notation AB + A+ B.

Recently, we have presented an extended self-consistent mean field theory for polymer systems with reversible crosslinks [11]. In this method, the crosslink properties are characterized by an overall crosslink strength z and specific crosslink weights ω_A and ω_B for A and B monomers of the copolymer being involved in crosslinks. Equations have been derived for the case of the AB + A + B model described above and has been tested with one set of parameters. For these parameters, the system showed a lamellar structure, for which the density oscillations were enhanced by enabling the crosslinks. In the current article, the SCMF model is revisited and studied systematically for a large range of crosslink strengths and crosslink asymmetries $\Delta\omega = \omega_A - \omega_B$. We show that the structure of the system can be switched from a hexagonal to a lamellar structure by increasing the crosslink strength or by varying the probability of A and B monomers being involved in crosslinks. The article is organized as follows: We begin with a description of the theoretical method in section "Theoretical Background", starting with the original self-consistent mean field theory in section "Polymer System Without Crosslinks" and adding reversible crosslinks in section "Adding Reversible Crosslinks". Then, numerical details are described in section "Details of the Calculation", before results for polymer blends without crosslinks and with reversible crosslinks are presented in sections "Polymer Blends Without Crosslinks" and "Polymer Blends with Crosslinks", respectively. The dependence of the phase on the crosslink parameters is given in section "Lamellar-Hexagonal Phase Transition" and a summary is given in section "Summary".

Theoretical Background

In this section, we present an extended self-consistent mean field theory that can be used to study polymer networks with reversible crosslinks. The method has been introduced in our previous work [11], which also includes a detailed description of an algorithm. In this article, we describe the theoretical background and refer to the previous work for practical aspects. The paper is dedicated to a mixture of three types of polymers: long AB block copolymers and shorter A and B homopolymers. We start with a description of the self-consistent mean field theory for this system without reversible crosslinks. Afterwards, the extension to systems with reversible crosslinks between the copolymers is described.

Polymer System Without Crosslinks

We consider a polymer mixture that includes three types of polymers, a, b and c with polymerization indices N_a, N_b, and N_c, respectively. Polymers of type a and b are homopolymers and consist of monomers of type A and B, respectively. Polymers of type c are block copolymers, which are divided into two parts $S_A = [0, f]$ and $S_B = (f, N_c]$. The part S_A consists of type A monomers while $S_B = (f, N_c]$ consists of B monomers so that f/N_c is the ratio of A monomers in each polymer c. In the numerical calculations, we restrict ourselves to the case $f/N_c = 1/2$. The system consists of n_j polymers of type $j = a, b, c$. The configuration of the kth polymer of type j is given by $\underline{R}_{j,k}(s)$, parameterized by $s \in [0, N_j]$. We assume that the chains in the system are Gaussian such that the probability distribution of configurations $\underline{R}_{j,k}(s)$ is given by the Wiener measure

$$\mathcal{P}(\underline{R}_{j,k}(s)) = \mathcal{N} \exp\left(-\frac{3}{2\lambda^2}\int_0^{N_j} \mathrm{d}s \left\| \frac{d\underline{R}_{j,k}(s)}{ds}\right\|^2\right), \tag{1}$$

where λ is the statistical length of the monomer segments and $\mathcal{N}$ is a normalization factor. We use $R_j \equiv \left(\underline{R}_{j,k}(s)\right)_{k=1}^{n_j}$ to denote the configuration of all n_j polymers of type j. Then, the microscopic density $\hat{\rho}_A(\underline{x}; R_a, R_c)$ of monomers A at position $\underline{x}$ depends on the configuration R_a of homopolymers of type a and the configuration R_c of copolymers of type c. It can be written as the sum

$$\hat{\rho}_A(\underline{x}; R_a, R_c) = \hat{\rho}_A^{(0)}(\underline{x}; R_a) + \hat{\rho}_A^{(1)}(\underline{x}; R_c). \tag{2}$$

of the microscopic densities

$$\hat{\rho}_A^{(0)}(\underline{x}; R_a) \equiv \sum_{k=1}^{n_a} \int_0^{N_a} \mathrm{d}s \delta(\underline{x} - \underline{R}_{a,k}(s)) \tag{3}$$

$$\hat{\rho}_A^{(1)}(\underline{x}; R_c) \equiv \sum_{k=1}^{n_c} \int_{S_A} \mathrm{d}s \delta(\underline{x} - \underline{R}_{c,k}(s)), \tag{4}$$

of monomers A from chains of type a and c, respectively. Correspondingly, one has $\hat{\rho}_B(\underline{x}; R_b, R_c) = \hat{\rho}_B^{(0)}(\underline{x}; R_b) + \hat{\rho}_B^{(1)}(\underline{x}; R_c)$ with microscopic densities

$$\hat{\rho}_B^{(0)}(\underline{x}; R_b) \equiv \sum_{k=1}^{n_b} \int_0^{N_b} \mathrm{d}s \delta(\underline{x} - \underline{R}_{b,k}(s)) \tag{5}$$

$$\hat{\rho}_B^{(1)}(\underline{x}; R_c) \equiv \sum_{k=1}^{n_c} \int_{S_B} \mathrm{d}s \delta(\underline{x} - \underline{R}_{c,k}(s)) \tag{6}$$

of monomers B from chains of type b and c, respectively. Then, the partition function can be written as

$$Z_{nb} = \prod_{j=a,b,c} \frac{1}{n_j!} \prod_{k=1}^{n_j} \int \hat{\mathcal{D}}\{\underline{R}_{j,k}\} \mathcal{P}(\underline{R}_{j,k}(s))\, \eta_{nb}, \tag{7}$$

where $\int \hat{\mathcal{D}}\{\underline{R}_{j,k}\}$ is the path integral over all configurations of the kth polymer of type j and

$$\eta_{nb} \equiv \delta(\hat{\rho}_A + \hat{\rho}_B - \rho_0) \exp\left(-\frac{1}{\rho_0}\int \mathrm{d}x \chi \hat{\rho}_A \hat{\rho}_B\right). \tag{8}$$

The first term in η_{nb} considers the incompressibility of the system and makes sure that the sum of the local A and B densities is always ρ_0. The second term represents the Flory-Huggins interaction and considers that neighboring monomers of the same type (A-A and B-B) are preferred over unlike neighbors (A-B), which is the case in most of the known polymer blends. The integral expression in the exponent is lower for demixed than for mixed states. Since the Flory-Huggins parameter χ is inverse proportional to the temperature, a phase transition from a mixed to a demixed state can be inferred by decreasing the temperature.

Adding Reversible Crosslinks

Now we assume that reversible crosslinks can form between monomers of the copolymers. A crosslink between copolymers i and j restricts the possible configurations of the polymer system by the fact that the position $\underline{R}_{c,i}(t)$ of copolymer j at $s = t$ must correspond to position $\underline{R}_{c,j}(\tau)$ of copolymer j at $s = \tau$. In a system with M crosslinks, defining a network structure $\mathcal{S}_M = \{i_l, j_l, t_l, \tau_l\}_{l=1}^{M}$, the partition function is reduced to

$$Z[\mathcal{S}_M] = \prod_{j=a,b,c} \frac{1}{n_j!} \prod_{k=1}^{n_j} \int \hat{\mathcal{D}}\{\underline{R}_{j,k}\} \mathcal{P}(\underline{R}_{j,k}(s))\, \eta_{nb}\, \eta_b, \tag{9}$$

with

$$\eta_b \equiv \prod_{l=1}^{M} \delta(\underline{R}_{c,i_l}(t_l) - \underline{R}_{c,j_l}(\tau_l)). \tag{10}$$

So far, Eq. 9 is the partition function of a polymer network with a given network structure and *fixed* crosslinks. We will now consider an ensemble that includes all possible network structures, representing a system with *reversible* crosslinks. First, we write down the partition function $Z^{(M)}$ for M crosslinks that may form between any pair of monomers of any c polymers:

$$Z^{(M)} \equiv \prod_{l=1}^{M} \left[\sum_{i_l=1}^{n_c} \sum_{j_l=1}^{n_c} \int_0^{N_c} \mathrm{d}t_l \int_0^{N_c} \mathrm{d}\tau_l Z[\mathcal{S}_M] \right]. \tag{11}$$

This expression is used to write down the partition function Z for an arbitrary number of crosslinks

$$Z = \sum_{M=0}^{\infty} \frac{z^M}{M!} Z^{(M)}, \tag{12}$$

where each crosslink is weighted with a factor z, which we call the crosslink strength. The partition function can be rewritten as

$$Z = \prod_{j=a,b,c} \frac{1}{n_j!} \int \hat{\mathcal{D}}\{R_j\} \exp\left(-\frac{1}{\rho_0} \int \mathrm{d}x \chi \hat{\rho}_A \hat{\rho}_B \right) \times \delta(\hat{\rho}_A + \hat{\rho}_B - \rho_0) e^{zB(n_c, R_c)}, \tag{13}$$

in which

$$B(n_c, R_c) \equiv \sum_{i_l=1}^{n_c} \sum_{j_l=1}^{n_c} \int_0^{N_c} \mathrm{d}t_l \int_0^{N_c} \mathrm{d}\tau_l \delta(\underline{R}_{c,i_l}(t_l) - \underline{R}_{c,j_l}(\tau_l)) = \int \mathrm{d}x (\hat{\rho}_A^{(1)} + \hat{\rho}_B^{(1)})^2 \tag{14}$$

considers crosslink interactions between any two copolymer chains. The last equality in Eq. 14 is found by inserting Eqs. 4 and 6 on the right-hand side. For a given configuration, the probability for a crosslink is given by $zB(n_c, R_c)$. So far we have assumed that monomers A and B bind with the same probability. This can be changed by considering weighting factors ω_A and ω_B for A and B monomers, respectively. We assume that crosslinks between pairs A+A, A+B, and B+B are weighted by $\omega_A\omega_A$, $\omega_A\omega_B$, and $\omega_B\omega_B$, respectively. The interaction term $B(n_c, R_c)$ is then given by

$$B(n_c, R_c) = \sum_{i_l=1}^{n_c} \sum_{j_l=1}^{n_c} \sum_{X=A,B} \sum_{Y=A,B} \omega_X \omega_Y \times \int_{S_X} \mathrm{d}t_l \int_{S_Y} \mathrm{d}\tau_l \delta(\underline{R}_{c,i_l}(t_l) - \underline{R}_{c,j_l}(\tau_l)) \tag{15}$$

$$= \int \mathrm{d}x (\omega_A \hat{\rho}_A^{(1)} + \omega_B \hat{\rho}_B^{(1)})^2. \tag{16}$$

Then, the probability for a crosslink of a X type monomer of copolymer i with a Y type monomer of polymer j (with $X, Y \in \{A, B\}$) is given by

$$P_{ij}^{XY}(n_c, R_c) \equiv z\omega_X\omega_Y \int_{S_X} dt \int_{S_Y} d\tau \delta\left(\underline{R}_{c,i}(t) - \underline{R}_{c,j}(\tau) \right). \tag{17}$$

In this article, we will use z to control the overall crosslink strength, while we set $\omega_B + \omega_A = 1$. Then, the asymmetry of the crosslink probability is given by $\Delta\omega \equiv \omega_A - \omega_B = 2\omega_A - 1$. In a homogeneously distributed system with $\rho_A^{(1)} = \rho_B^{(1)} = \rho_c/2 = const$, the probability for one crosslink is given by $z\rho_c^2 V$, independent of $\Delta\omega$.

Hubbard-Stratonovich Transformation

The partition function in Eq. 13 can be rewritten with the help of the Hubbard-Stratonovich transformation. First we use

$$\int \mathcal{D}\{\rho_m^{(k)}\} \delta(\rho_m^{(k)} - \hat{\rho}_m^{(k)}) = 1,$$

for the monomer densities $\rho_m^{(k)}$ of monomer type $m = A, B$ in homopolymers ($k = 0$) and copolymers ($k = 1$) to rewrite the partition function Z as

$$Z = \prod_{j=a,b,c} \frac{1}{n_j!} \int \hat{\mathcal{D}}\{R_j\} \prod_{\substack{n=A,B\\k=0,1}} \int \mathcal{D}\{\rho_n^{(k)}\} \prod_{\substack{m=A,B\\ \ell=0,1}} \times \delta(\rho_m^{(\ell)} - \hat{\rho}_m^{(\ell)}) \exp\left[\int \mathrm{d}x \left(-\frac{\chi}{\rho_0} \rho_A \rho_B + z(\omega_A \rho_A^{(1)} + \omega_B \rho_B^{(1)})^2 \right) \right] \delta(\rho_A + \rho_B - \rho_0)$$

where $\rho_m(\underline{x}) = \rho_m^{(0)}(\underline{x}) + \rho_m^{(1)}(\underline{x})$ for $m = A, B$. Then, we use the exponential representations of the δ function

$$\delta(\rho_A + \rho_B - \rho_0) = \int \mathcal{D}\{p\} \exp\left(i \int p(\rho_A + \rho_B - \rho_0)\right), \tag{18}$$

$$\delta(\rho_m^{(k)} - \hat{\rho}_m^{(k)}) = \int \mathcal{D}\{\mathrm{w}_m^{(k)}\} \exp\left(i \int \mathrm{w}_m^{(k)}(\rho_m^{(k)} - \hat{\rho}_m^{(k)})\right) \tag{19}$$

for $m = A, B$ and $k = 0, 1$. The partition function becomes

$$Z = \prod_{\substack{m=A,B\\k=0,1}} \left[\int \mathcal{D}\{\rho_m^{(k)}\} \int \mathcal{D}\{\mathrm{w}_m^{(k)}\}\right] \int \mathcal{D}\{p\}\ \zeta_I \zeta_{II} \tag{20}$$

with

$$\zeta_I \equiv \exp\Bigg[\int \mathrm{d}x \Bigg(-\frac{\chi}{\rho_0}\rho_A\rho_B + z(\omega_A\rho_A^{(1)} + \omega_B\rho_B^{(1)})^2 + \sum_{\substack{m=A,B\\k=0,1}} i(\mathrm{w}_m^{(k)}\rho_m^{(k)}) + ip(\rho_A + \rho_B - \rho_0)\Bigg)\Bigg] \tag{21}$$

and

$$\zeta_{II} = \prod_{j=a,b,c} \frac{1}{n_j!} Q_j \tag{22}$$

with

$$Q_a = Q_a(\mathrm{w}_A^{(0)}) = \int \mathcal{D}\{R_{a,k}\} \mathcal{P}(\underline{R}_{a,k}(s)) \times \exp\left[-i \int_0^{N_a} \mathrm{d}s \mathrm{w}_A^{(0)}(\underline{R}_{a,k}(s))\right],$$

$$Q_b = Q_b(\mathrm{w}_B^{(0)}) = \int \mathcal{D}\{R_{b,k}\} \mathcal{P}(\underline{R}_{b,k}(s)) \times \exp\left[-i \int_0^{N_b} \mathrm{d}s \mathrm{w}_B^{(0)}(\underline{R}_{b,k}(s))\right],$$

$$Q_c = Q_c(\mathrm{w}_A^{(1)}, \mathrm{w}_B^{(1)}) = \int \mathcal{D}\{R_{c,k}\} \mathcal{P}(\underline{R}_{c,k}(s)) \times \exp\left[-i \sum_{m=A,B} \int_{S_m} \mathrm{d}s \mathrm{w}_m^{(1)}(\underline{R}_{c,k}(s))\right].$$

The terms Q_a, Q_b, and Q_c are the partition functions for the single chains of type a, b, and c, respectively. Using Stirling's formula, the partition function Z can be written as

$$Z = \prod_{\substack{m=A,B\\k=0,1}} \left[\int \mathcal{D}\{\rho_m^{(k)}\} \int \mathcal{D}\{\mathrm{w}_m^{(k)}\}\right] \int \mathcal{D}\{p\} \exp(-\beta F)$$

with the free energy in the canonical ensemble

$$\beta F = \int \mathrm{d}x \Bigg[\frac{\chi}{\rho_0}\rho_A\rho_B - z\Bigg(\sum_{m=A,B} \omega_m \rho_m^{(1)}\Bigg)^2 - ip(\rho_A + \rho_B - \rho_0) - \sum_{\substack{m=A,B\\k=0,1}} i(\mathrm{w}_m^{(k)}\rho_m^{(k)})\Bigg] - \sum_{j=a,b,c} n_j \ln \frac{Q_j}{n_j},$$

where β is the inverse temperature.

In the mean field theory we now use the saddle point approximation, which means that we look for the minimum of F as a functional of the polymer fields. With $N \equiv N_c$, we introduce dimensionless variables

$$\tilde{\rho}_m^{(j)} \equiv \rho_m^{(k)}/\rho_0,\ \tilde{\mathrm{w}}_m^{(k)} \equiv iN\mathrm{w}_m^{(k)},\ (m = A, B;\ k = 0, 1) \tag{23}$$

$$\tilde{p}(\underline{x}) \equiv iNp(\underline{x}), \quad \tilde{z} \equiv \rho_0 N z, \tag{24}$$

$$\tilde{Q} \equiv R_g^{-d} Q, \quad \tilde{s} = s/N \tag{25}$$

$$\psi_j \equiv \frac{n_j N}{\rho_0 V}, \quad (j = a, b, c). \tag{26}$$

Here, d is the dimension of the system and

$$R_g = \sqrt{\frac{N\lambda^2}{2d}}. \tag{27}$$

is the gyration radius of a Gaussian chain with the length $N \equiv N_c$ of the copolymers. The gyration radius is also used to define dimensionless lengths $\tilde{x} \equiv x/R_g$.

At the saddle point, $p(\underline{x})$, and $\mathrm{w}_m^{(k)}$ for $m = A, B$ and $k = 0, 1$ are purely imaginary, so that $\tilde{p}(\underline{x})$ and $\tilde{\mathrm{w}}_m^{(k)}$ are real quantities. The variations of the free energy

$$\beta F = \frac{\rho_0}{N} R_g^d \int \mathrm{d}\tilde{x} \Bigg[\chi N \tilde{\rho}_A \tilde{\rho}_B - \tilde{z}\Bigg(\sum_{m=A,B} \omega_m \tilde{\rho}_m^{(1)}\Bigg)^2 - \tilde{p}(\tilde{\rho}_A + \tilde{\rho}_B - 1) - \sum_{\substack{m=A,B\\k=0,1}} (\tilde{\mathrm{w}}_m^{(k)}\tilde{\rho}_m^{(k)})\Bigg] - \sum_{j=a,b,c} \psi_j \ln \frac{R_g^d \tilde{Q}_j}{n_j},$$

over the fields $\tilde{\rho}_m^{(0)}$, $\tilde{\rho}_m^{(1)}$, $\tilde{\mathrm{w}}_m^{(0)}$, $\tilde{\mathrm{w}}_m^{(1)}$, and $\tilde{p}(\underline{x})$ must vanish:

$$0 = \chi N \tilde{\rho}_B - \tilde{\mathrm{w}}_A^{(0)} - \tilde{p}, \tag{28}$$

$$0 = \chi N \tilde{\rho}_A - \tilde{\mathrm{w}}_B^{(0)} - \tilde{p}, \tag{29}$$

$$0 = \chi N \tilde{\rho}_B - 2\tilde{z}\omega_A(\omega_A \tilde{\rho}_A^{(1)} + \omega_B \tilde{\rho}_B^{(1)}) - \tilde{\mathrm{w}}_A^{(1)} - \tilde{p}, \tag{30}$$

$$0 = \chi N \tilde{\rho}_A - 2\tilde{z}\omega_B(\omega_A \tilde{\rho}_A^{(1)} + \omega_B \tilde{\rho}_B^{(1)}) - \tilde{\mathrm{w}}_B^{(1)} - \tilde{p}, \tag{31}$$

$$0 = \tilde{\rho}_A^{(0)} + \frac{\psi_a}{\tilde{Q}_a}\frac{\delta \tilde{Q}_a}{\delta \tilde{\mathrm{w}}_A}, \tag{32}$$

$$0 = \tilde{\rho}_B^{(0)} + \frac{\psi_b}{\tilde{Q}_b}\frac{\delta \tilde{Q}_b}{\delta \tilde{\mathrm{w}}_B}, \tag{33}$$

$$0 = \tilde{\rho}_m^{(1)} + \frac{\psi_c}{\tilde{Q}_c}\frac{\delta \tilde{Q}_c}{\delta \tilde{\mathrm{w}}_m} \qquad (m = A, B), \tag{34}$$

$$0 = \tilde{\rho}_A + \tilde{\rho}_B - 1 \tag{35}$$

Since the polymers are Gaussian, the partition functions can be written as

$$\tilde{Q}_j = \int \mathrm{d}\tilde{x} q_j(\underline{\tilde{x}}, \tilde{s}) q_j(\underline{\tilde{x}}, (N_j/N) - \tilde{s}), \quad (j = a, b), \tag{36}$$

$$\tilde{Q}_c = \int \mathrm{d}\tilde{x} q_c(\underline{\tilde{x}}, \tilde{s}) q_c^{\dagger}(\underline{\tilde{x}}, \tilde{s}). \tag{37}$$

The propagators $q_j(\underline{\tilde{x}}, \tilde{s})$ $(j = a, b, c)$ and $q_c^{\dagger}(\underline{\tilde{x}}, \tilde{s})$ satisfy the modified diffusion equations

$$\left(\frac{\partial}{\partial \tilde{s}} - \Delta + \tilde{\mathrm{w}}_A^{(0)}(\underline{\tilde{x}})\right) q_a(\underline{\tilde{x}}, \tilde{s}) = 0, \quad 0 < \tilde{s} < N_a/N, \tag{38}$$

$$\left(\frac{\partial}{\partial \tilde{s}} - \Delta + \tilde{\mathrm{w}}_B^{(0)}(\underline{\tilde{x}})\right) q_b(\underline{\tilde{x}}, \tilde{s}) = 0, \quad 0 < \tilde{s} < N_b/N, \tag{39}$$

$$\left(\frac{\partial}{\partial \tilde{s}} - \Delta + \tilde{\mathrm{w}}^{(1)}(\underline{\tilde{x}}, \tilde{s})\right) q_c(\underline{\tilde{x}}, \tilde{s}) = 0, \quad (0 < \tilde{s} < 1), \tag{40}$$

$$\left(-\frac{\partial}{\partial \tilde{s}} - \Delta + \tilde{\mathrm{w}}^{(1)}(\underline{\tilde{x}}, \tilde{s})\right) q_c^{\dagger}(\underline{\tilde{x}}, \tilde{s}) = 0, \quad (0 < \tilde{s} < 1), \tag{41}$$

$$\tilde{\mathrm{w}}^{(1)}(\underline{\tilde{x}}, \tilde{s}) \equiv \begin{cases} \tilde{\mathrm{w}}_A^{(1)}(\underline{\tilde{x}}), & \text{if } \tilde{s} \in [0, f/N] \\ \tilde{\mathrm{w}}_B^{(1)}(\underline{\tilde{x}}), & \text{if } \tilde{s} \in [f/N, 1] \end{cases} \tag{42}$$

with boundary conditions $q_j(\underline{\tilde{x}}, 0) = 1$ for $j = a, b, c$ and $q_c^{\dagger}(\underline{\tilde{x}}, N_c) = 1$. Using

$$\frac{\delta \tilde{Q}_a}{\delta \tilde{\mathrm{w}}_A^{(0)}} = \int_0^{N_a} \mathrm{d}\tilde{x} q_a(\underline{\tilde{x}}, s) q_a(\underline{\tilde{x}}, N_a - s), \tag{43}$$

$$\frac{\delta \tilde{Q}_b}{\delta \tilde{\mathrm{w}}_B^{(0)}} = \int_0^{N_b} \mathrm{d}\tilde{x} q_b(\underline{\tilde{x}}, s) q_b(\underline{\tilde{x}}, N_b - s), \tag{44}$$

$$\frac{\delta \tilde{Q}_c}{\delta \tilde{\mathrm{w}}_m^{(0)}} = \int_{S_m} \mathrm{d}\tilde{x} q_c(\underline{\tilde{x}}, s) q_c^{\dagger}(\underline{\tilde{x}}, s). \tag{45}$$

and Eqs. 36, the saddle point equations (28)–(34) can be solved self-consistently.

Details of the Calculation

We investigate a polymer blend with AB copolymers of polymerization index $N \equiv N_c$ and shorter homopolymers with polymerization indices $N_a = N_b = N/5$. The AB copolymers are symmetric in the sense that $f = N/2$. For the sake of a reasonable computational effort, we restrict the calculations to two-dimensional systems. If not mentioned otherwise, systems have box lengths $L = 20R_g$ and the Flory-Huggins parameter is set to $\chi N = 11.3$. In x- and y-direction periodic boundary conditions are used. The overall stoichiometry of the polymer blend is determined by the volume fraction of copolymers

$$\phi_c \equiv \frac{n_c N}{\sum_{k=a,b,c} n_k N_k} \tag{46}$$

and the A ratio of homopolymers

$$\phi_a^{rel} \equiv \frac{\phi_a}{\phi_a + \phi_b} \equiv \frac{n_a N_a}{n_a N_a + n_b N_b}. \tag{47}$$

For $\phi_a^{rel} = 1.0$ ($\phi_a^{rel} = 0.0$), all homopolymers in the system are of type A (type B). If not mentioned otherwise, we use $\phi_c = 0.6$. The crosslink strength is determined by the dimensionless quantity $z_{rel} = \tilde{z}/(\chi N)$.

For a given set of parameters, the equilibrium fields $\tilde{\rho}_m^{(k)}$ for $m = A, B$ and $k = 0, 1$ are obtained iteratively in the following way: First, initial values of $\tilde{\rho}_m^{(k)}$, $\tilde{\mathrm{w}}_m^{(k)}$, and $\tilde{p}$ are chosen. We set $\tilde{\mathrm{w}}_A^{(0)} = 0$ and initialize the other fields with the help of pseudo-random numbers such that Eqs. 28–35 have solutions and the polymer densities are in the range of $0 \leq \tilde{\rho}_m^{(k)} \leq 1$. The procedure is described in detail in [11]. With these initial fields, we solve numerically the modified diffusion equations (32)–(34), with the Crank-Nicolson method [54]. This provides the propagator functions that allow us to calculate $\tilde{Q}_k$ for $k = a, b, c$ and, subsequently, new values of $\tilde{\rho}_m^{(k)}$ for $m = A, B$ and $k = 0, 1$ from Eqs. 32 to 34. For the next iteration, the fields $\tilde{\mathrm{w}}_m^{(k)}$ are updated as

$$\tilde{\mathrm{w}}_m^{(k)} \leftarrow \tilde{\mathrm{w}}_m^{(k)} + \gamma \mu_m^{(k)} + \epsilon(\tilde{\rho} - \tilde{\rho}^{old})$$
$$(m = A, B; \quad k = 0, 1) \tag{48}$$

with numerical parameters $\gamma = 0.1$ and $\epsilon = 0.2$ and the difference $\tilde{\rho} - \tilde{\rho}^{old}$ between the current and the previous total density. The residues $\mu_m^{(k)}$ are defined as

$$\mu_m^{(k)} = \chi N(1 - \tilde{\rho}_m) - \tilde{w}_m^{(k)} - \tilde{\rho} - 2\tilde{z}k\omega_m \sum_{\ell=A,B} \omega_\ell \tilde{\rho}_\ell^{(1)}$$

$$(m = A, B; \quad k = 0, 1). \tag{49}$$

Note that the last part vanishes for $k = 0$. Using Eqs. 28–31, and Eq. 35, we update $\tilde{\rho}$

$$\tilde{\rho} \leftarrow \frac{\chi N}{2} - \frac{\tilde{z}}{2} \sum_{\substack{m=A,B \\ n=A,B}} \omega_m \omega_n \tilde{\rho}_n^{(1)} - \frac{1}{4} \sum_{\substack{m=A,B \\ k=0,1}} \tilde{w}_m^{(k)}. \tag{50}$$

Iterations are performed until the residues $\mu_m^{(k)}$ are smaller than a tolerance $\mu_m^{(k)} < 10^{-9}$.

Results

We will use our extended version of the self-consistent mean field theory to study the influence of crosslinks on the polymer blend consisting of A and B homopolymers and AB copolymers, where the copolymers consist of an A and a B chain of equal length. Initially, we shortly present some structures formed by this system in the absence of crosslinks (which means for $\tilde{z} = 0$).

Polymer Blends Without Crosslinks

We consider a polymer blend without crosslinks that has a copolymer volume fraction of $\phi_c = 0.6$. The system depends on the Flory-Huggins parameter χ and the A ratio of the homopolymers ϕ_a^{rel}. In Fig. 1, equilibrium structures are shown for various parameter pairs $(\chi N, \phi_a^{rel})$. The phase diagram is symmetric in the sense that each system at $(\chi N, \phi_a^{rel})$ has a corresponding system at $(\chi N, 1 - \phi_a^{rel})$ for which the densities of A and B monomers are exchanged. Therefore, in the following, we restrict ourselves to values of $\phi_a^{rel} \geq 0.5$.

For suitably small values of χN, the system must be homogeneous. Homogeneity is still found for $\chi N = 11.3$, if the type A ratio of homopolymers is suitably large, $\phi_a^{rel} \gtrsim 0.7$. For $\phi_a^{rel} = 0.6$ and $\phi_a^{rel} = 0.65$, the system shows a hexagonal structure, while for $\phi_a^{rel} = 0.5$ a lamellar structure is found. Typical density distributions of $\tilde{\rho}_A$ are shown for the lamellar and the hexagonal phase in Fig. 1b, c, respectively. For higher values of χN, only lamellar and hexagonal phases have been observed. In general, hexagonal phases form for large values of ϕ_a^{rel} while lamellar phases form if ϕ_a^{rel} is closer to 0.5 and the ratio of A and B monomers is more symmetric. The hexagonal structure is preferred in more asymmetric blends. Then, the minority component forms round domains that are arranged in a hexagonal structure. If the packing fraction of copolymers is strongly reduced to a value of $\phi_c = 0.1$, and the demixing strengths increased to $\chi N = 20$, the system demixes completely. For $\phi_a^{rel} = 0.6$ a big droplet of type B homopolymer forms which is separated from a surrounding region of A homopolymers by a ring of copolymers. The density $\tilde{\rho}_A$ is shown in Fig. 1d.

Polymer Blends with Crosslinks

Now we consider the case that reversible crosslinks may form between copolymers. Without the loss of generality, we assume that $\omega_A + \omega_B = 1$ and $0 \leq \omega_A \leq 1$ so that $-1 \leq \Delta\omega \leq 1$ determines the asymmetry of the crosslink probability: For $\Delta\omega = 0.0$, monomers of type A and B form crosslinks with the same probability. For $\Delta\omega = 1.0$ only A monomers can form crosslinks and for $\Delta\omega = -1.0$, crosslinks are restricted to B monomers.

For $\phi_a^{rel} = 0.5$, we have studied systematically the structure of systems with crosslink strengths in the range of $0 \leq z_{rel} \leq 0.4$ and $0.05 \leq \Delta\omega \leq 0.4$. In the whole region, the system forms a lamellar phase. Apparently, this phase is strongly favored for systems with a symmetric stoichiometry. As an example, local density distributions are shown in Fig. 2 for a system with $\phi_a^{rel} = 0.5$ and $\Delta\omega = 0.25$. In Fig. 2 the local density $\tilde{\rho}_A$ of the A monomers is shown as a function of the spatial coordinates x and y for a system with $z_{rel} = 0.4$. For a detailed analysis, we investigate monomer densities for a path parameterized by a parameter s along the black arrow in Fig. 2a, which is perpendicular to the stripes. In Fig. 2b, the density of all A monomers $\tilde{\rho}_A(s)$ and the monomer density $\tilde{\rho}_A^{(1)}$ of A monomers being part of a copolymer are shown for two systems with $\phi_a^{rel} = 0.5$ and $\Delta\omega = 0.25$, one with crosslink strengths $z_{rel} = 0.2$ and one with $z_{rel} = 0.4$. The amplitude of the oscillation increases for both curves, $\tilde{\rho}_A(s)$ and $\tilde{\rho}_A^{(1)}(s)$, with increasing crosslink strength $\tilde{z}$. The intensification of the microphase separation has been already shown in our recent paper [11].

The amplitude of the oscillation is also increased if we keep the crosslink strength $\tilde{z}$ constant and increase the crosslink asymmetry $\Delta\omega$. This is shown in Fig. 3a, in which the monomer density $\tilde{\rho}_A$ along the path in Fig. 2a is shown for various $\Delta\omega = 0.05, 0.15, 0.25$ with fixed $\phi_A = 0.5$ and $z_{rel} = 0.4$. If the A part of copolymers is more likely to form crosslinks than the B part, the system profits from a stronger demixing of A and B polymers. The stronger separation reflects the fact that $\omega_A\omega_A + \omega_B\omega_B > 2\omega_A\omega_B$ for $\omega_A \neq \omega_B$ so that the free energy of an A-A and a B-B crosslink is lower than that of two A-B crosslinks.

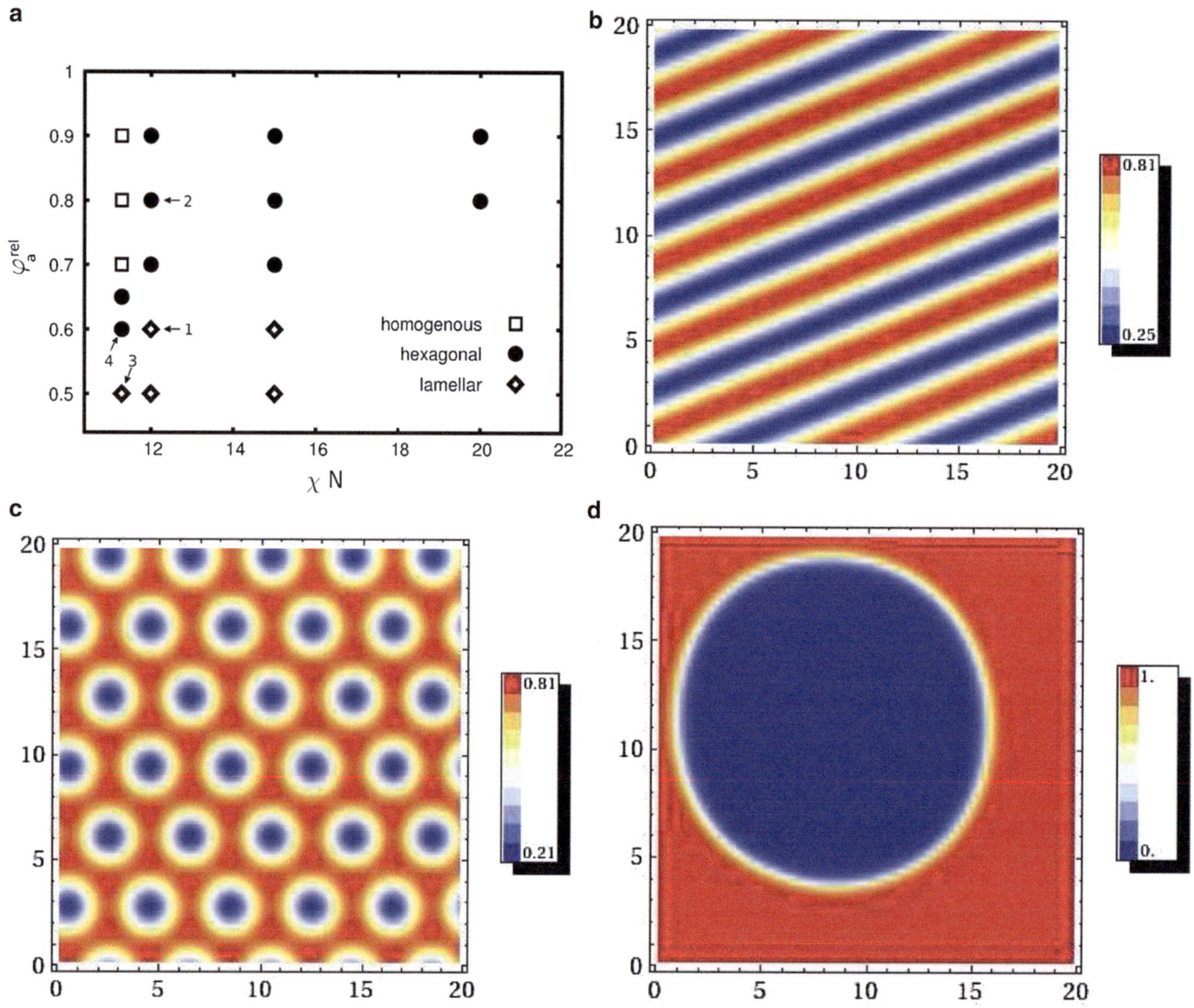

Fig. 1 Phase behavior of a polymer blend including AB copolymers and A and B homopolymers. The copolymers have a volume fraction $\phi_c = 0.6$ and do not form crosslinks ($z = 0$). (**a**) Structures observed for different demixing strengths χN and volume fractions of A polymers ϕ_A^{rel}. (**b**) Density $\tilde{\rho}_A(x, y)$ for $\chi N = 12.0$, $\phi_a^{rel} = 0.6$, (point 1 in 1a). (**c**) Density $\tilde{\rho}_A(x, y)$ for $\chi N = 12.0$, $\phi_a^{rel} = 0.8$, (point 2 in Fig. 1a). (**d**) Density $\tilde{\rho}_A(x, y)$ for a smaller packing fraction of copolymers $\phi_c = 0.1$ and $\chi N = 20.0$, $\phi_a^{rel} = 0.6$

It is interesting to investigate the local monomer density of all copolymers $\tilde{\rho}_c(s)$, independent of their type. This is shown in Fig. 3b. For $\Delta\omega = 0.05$, the copolymer density deviates less than 1 % from the average value $(\tilde{\rho}_c)^{av} = \phi_c = 0.6$. The density shows two maxima at the interface between the A-rich domains and the B-rich domains (at $s \simeq 2$, $s \simeq 6$, and $s \simeq 10$). This reflects the tendency of AB copolymers to be at the interface between A and B rich domains. The density $\tilde{\rho}_c(s)$ shows an absolute minimum at the center of each B rich domain, indicating that the pronounced localization of A monomers in the A rich domains tears parts of the B branch out of the B rich domains. As $\Delta\omega$ is increased, this effect is more pronounced. Obviously, the optimization of the monomer distribution with respect to the ω_A-related part of the free energy becomes more relevant as $\Delta\omega$ increases. At $\Delta\omega = 0.25$ a maximum at the center of the A rich exceeds the values at the interfaces.

Figure 4a shows a hexagonal structure formed by a copolymer blend in the presence of reversible crosslinks with $\phi_A = 0.6$, $z_{rel} = 0.32$, and $\Delta\omega = 0.1$. The structure corresponds qualitatively with the hexagonal phase in the absence of crosslinks shown in Fig. 1a. Almost spherical domains with a high fraction of B monomers are arranged in a regular hexagonal structure. Around these B rich domains, there is a continuous region with a high concentration of A monomers. Within this domain, there are weakly pronounced local maxima of the A monomer density, centered on hexagonal lattice points, which have the largest distance to the B-rich domains. The distance between the maximum of the B density and a neighboring maximum of the A density

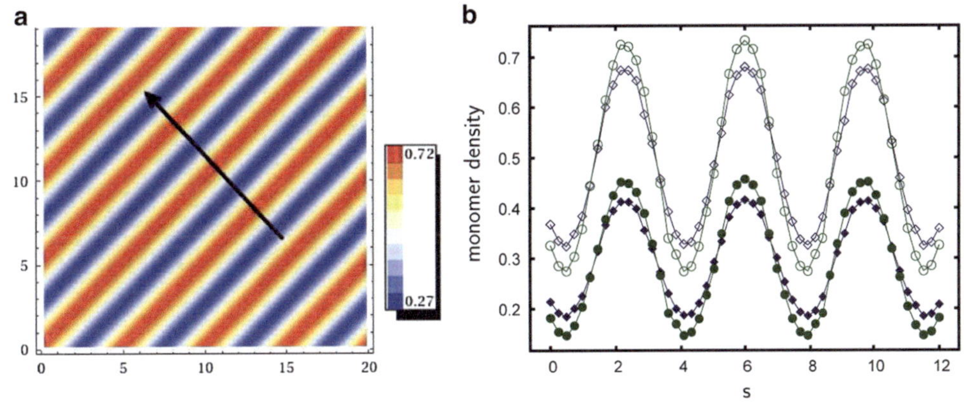

Fig. 2 Influence of reversible crosslinks on the lamellar structure with $\phi_c = 0.6$, $\phi_a^{rel} = 0.5$, and $\Delta\omega = 0.25$. (**a**) Monomer density $\tilde{\rho}_A(x, y)$ for a system with crosslink strength $z_{rel} = 0.4$. The *arrow* indicates the path for the density profiles in Figs. 2b and 3. (**b**) Total A monomer density $\tilde{\rho}_A$ for $z_{rel} = 0.2$ (◇), $z_{rel} = 0.4$ (○) and $\tilde{\rho}_A^{(1)}$ for $z_{rel} = 0.2$ (u), $z_{rel} = 0.4$ (•). Lines are intended to guide the eye

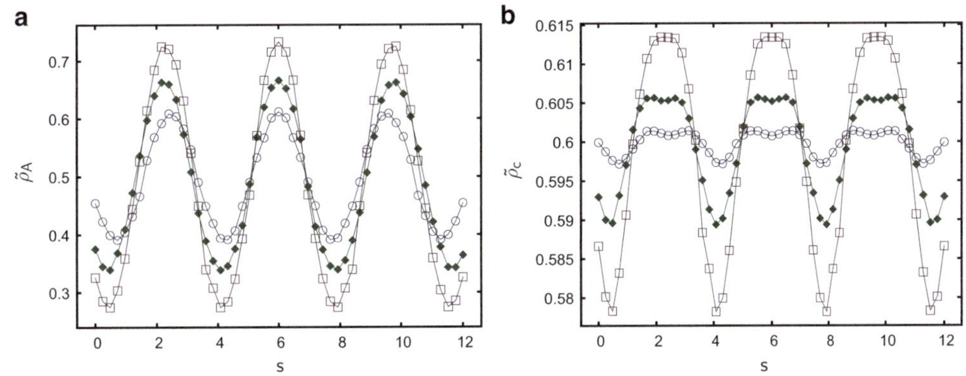

Fig. 3 Influence of the crosslink asymmetry $\Delta\omega = \omega_A - \omega_B$ on the lamellar structure with $\phi_c = 0.6$, $\phi_a^{rel} = 0.5$, and $z_{rel} = 0.4$. (**a**) Monomer density $\tilde{\rho}_A(s)$ and (**b**) monomer density of copolymers $\tilde{\rho}_c(s)$ along the path shown in Fig. 2a. Results are presented for $\Delta\omega = 0.05$ (○), $\Delta\omega = 0.15$ (u), and $\Delta\omega = 0.25$ (□). Lines are intended to guide the eye

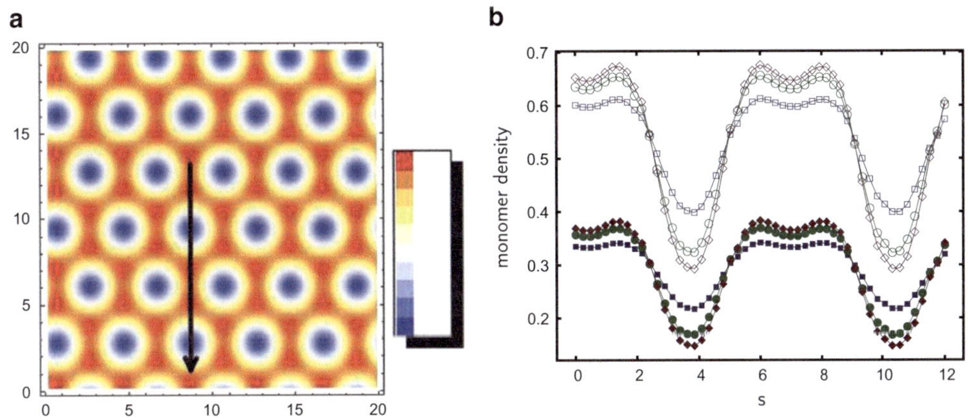

Fig. 4 Influence of reversible crosslinks on the hexagonal structure with $\phi_c = 0.6$ and $\phi_a^{rel} = 0.6$. (**a**) Monomer density $\tilde{\rho}_A(x, y)$ (shown for $z_{rel} = 0.04$, $\Delta\omega = 0.25$). The *black arrow* indicates the path for the density profile in Figs. 4b and 5. (**b**) Densities $\tilde{\rho}_A$ for $\Delta\omega = 0.25$ and $z_{rel} = 0.0$ (□), $z_{rel} = 0.2$ (○), $z_{rel} = 0.32$ (◇) and $\tilde{\rho}_A^{(1)}$ for $\Delta\omega = 0.25$, and $z_{rel} = 0.0$ (■), $z_{rel} = 0.2$ (•), $z_{rel} = 0.32$ (u). Lines are intended to guide the eye

is approximately $2.3R_g \simeq 0.94R_e$, where R_e is the average end-to-end distance of a Gaussian chain with the length of the copolymer. Obviously, only a small fraction of copolymers contribute to the maximum of $\tilde{\rho}_A$ and the maximum of $\tilde{\rho}_B$ at the same time. The local densities are studied in more detail along the path indicated by the black arrow in Fig. 4a.

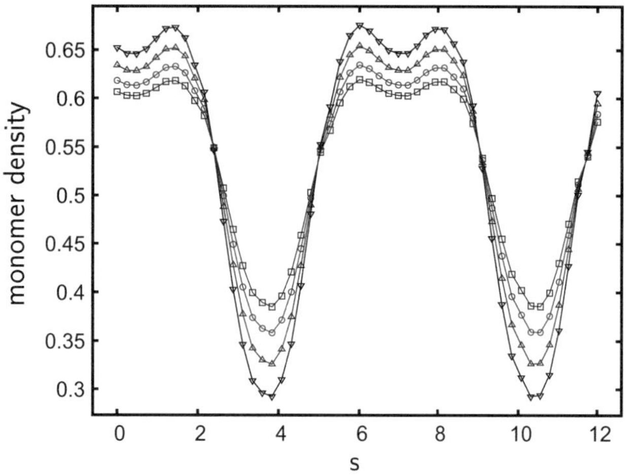

Fig. 5 Influence of the crosslink asymmetry $\Delta\omega = \omega_A - \omega_B$ on the hexagonal structure with $\phi_c = 0.6$, $\phi_a^{rel} = 0.6$, and $z_{rel} = 0.32$. The monomer density $\tilde{\rho}_A(s)$ (along the path in Fig. 4a) is shown for $\Delta\omega = 0.1$ (□), $\Delta\omega = 0.15$ (○), $\Delta\omega = 0.2$ (∇), and $\Delta\omega = 0.25$ (Δ)

The path crosses a minimum of $\tilde{\rho}_A$ than two maxima of $\tilde{\rho}_A$ and then another minimum of $\tilde{\rho}_A$. In Fig. 4b, results for $\tilde{\rho}_A(s)$ and $\tilde{\rho}_A^{(1)}(s)$ are shown for $z_{rel} = 0.0$, $z_{rel} = 0.2$, $z_{rel} = 0.32$. The small two neighboring maxima and the pronounced minima of $\tilde{\rho}_A(s)$ are clearly visible. Like in the case of the lamellar phase, the oscillations are amplified by an increase of the crosslink strength $\tilde{z}$. A corresponding behavior is found for the A monomer density of the copolymers $\tilde{\rho}_A^{(1)}$. Microphase separation is also increased by increasing the $\Delta\omega$ as demonstrated by plots of $\tilde{\rho}_A(s)$ in Fig. 5. Within the studied parameter range for hexagonal structures, the overall copolymer density $\tilde{\rho}_z$ does not depend significantly on $\Delta\omega$ or $\tilde{z}$ (not shown).

Lamellar-Hexagonal Phase Transition

The crosslink parameters $\tilde{z}$ and $\Delta\omega$ can be used to switch between the hexagonal and the lamellar phase. This is important since it can be used to change the structure of the system qualitatively by altering the crosslink probability, for example, by changing the pH value or starting/stopping light irradiation. We investigate the effect for our model system with $\phi_c = 0.6$, $\phi_A^{(rel)} = 0.6$ and $\chi N = 11.3$. The crosslink strength and the asymmetry are studied in the range of $0 \leq z_{rel} \leq 0.8$ and $0 \leq \Delta\omega \leq 0.4$. With systems of box length $L_x = L_y = 20$ we find that for small z_{rel} and small $\Delta\omega$ the system forms a hexagonal state while for large z_{rel} and large $\Delta\omega$ the system is in a lamellar state. However, especially at intermediate values, the final state of the system depends sensitively on the initial configuration that we choose randomly. This is expected for a system close to a first order phase transition, where it shows hysteresis. However, the obtained structures provide reasonable information on the dimension of the unit cells of the phases. For symmetry reasons the smallest rectangular unit cell of the hexagonal phase has a box length ratio of $L_y/L_x = \sqrt{3}$. From the results, found with the large system cells we find that $L_x \simeq 4.0$. For each parameter pair of $\tilde{z}$ and $\Delta\omega$, we perform a systematic study, in which we have applied the self-consistent mean field method on systems with box length L_x from $L_x = 3.4$ to $L_x = 4.8$ in steps of $\Delta L_x = 0.025$. For all L_x we choose $L_y = \sqrt{3}L_x$. A selection of resulting configurations for different system sizes are shown Fig. 6 for $z_{rel} = 0.48$ and $\Delta\omega = 0.3$. Depending on the box size, the system forms different phases, caused by the small box and the periodic boundary conditions. In Fig. 7a, we see the free energy densities for each box size. The symbols in the plot denote the structure of the system. The hexagonal phase is energetically favorable in the range of $3.875 \leq L_x \leq 4.7$ but the lowest energy density exists for a lamellar phase with $L_x = 3.55$. By comparing the free energy densities for the different box sizes one gets the stable phase for each parameter pair $\tilde{z}$ and $\Delta\omega$. The resulting phase diagram is shown in Fig. 7b. It shows a phase transition from hexagonal phase at small values of $\tilde{z}$ and $\Delta\omega$ to a lamellar phase at large $\tilde{z}$ and $\Delta\omega$. The transition value of $\Delta\omega$ decreases almost linearly with increasing crosslink strength. The phase diagram shows that the phase transition can be achieved by either increasing the total crosslink strength $\tilde{z}$ or the asymmetry $\Delta\omega$.

Summary

We have used self-consistent mean field theory for studying the phase behavior of a polymer blend with reversible crosslinks. We have focussed on a system with A and B homopolymers and symmetric AB copolymers in a region in which the system forms hexagonal or lamellar structures. Crosslinks are permitted for copolymers only. The system forms two types of domains, one favored by the A part and the other favored by the B part of the copolymer. We find that, in general, the demixing into microphases can be enhanced by increasing the crosslink strength $\tilde{z}$ or the crosslink asymmetry $\Delta\omega = \omega_A - \omega_B$ of the weights ω_A and ω_B for involving A or B monomers in a crosslink. Large values of $|\Delta\omega|$ favor large values of $(\rho_A^{(1)})^2$ and $(\rho_B^{(1)})^2$ rather than regions in which $\rho_A^{(1)}\rho_B^{(1)}$ is high. This follows from the z-dependent part of F in Eq. 23 and the fact that

$$\omega_A\omega_A + \omega_B\omega_B - 2\omega_A\omega_B = \Delta\omega^2. \tag{51}$$

Thus, if $\Delta\omega^2 > 0$ then a region of size V with large $(\rho_A^{(1)})^2$ and another region of size V with large $(\rho_B^{(1)})^2$ is favored over a region of size $2V$ with large $\rho_A^{(1)}\rho_B^{(1)}$. For finite $\Delta\omega^2$ this effect is increased with increasing z.

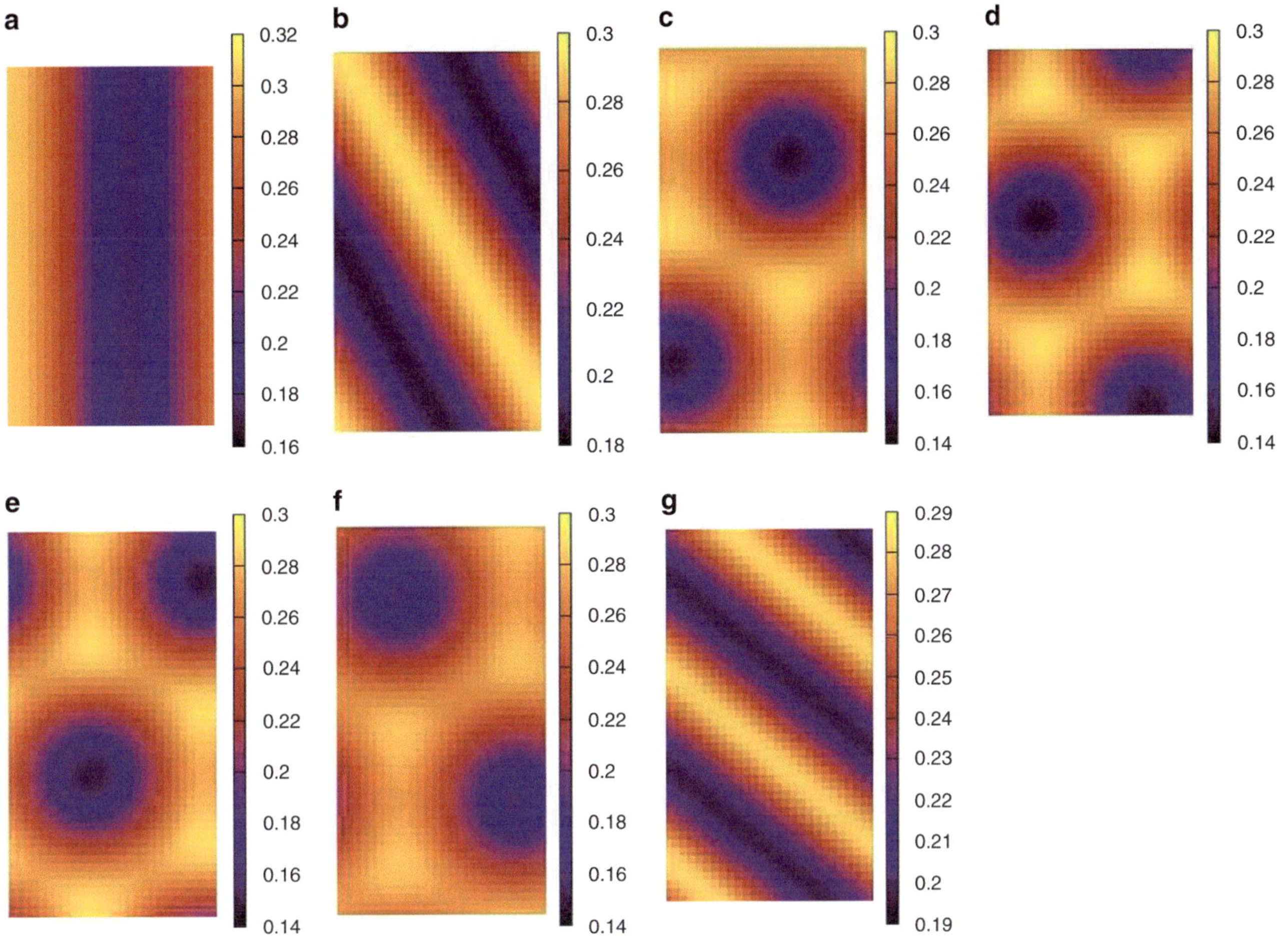

Fig. 6 Unit cells of a polymer blend with reversible crosslinks and parameters $\phi_c = 0.6$, $\phi_A^{rel} = 0.6$, $z_{rel} = 0.48$, and $\Delta\omega = 0.3$. Results are shown for box lengths $L_y = \sqrt{3}L_x$ and (**a**) $L_x = 3.6$, (**b**) $L_x = 3.8$, (**c**) $L_x = 4.0$, (**d**) $L_x = 4.2$, (**e**) $L_x = 4.4$, (**f**) $L_x = 4.6$, and (**g**) $L_x = 4.8$

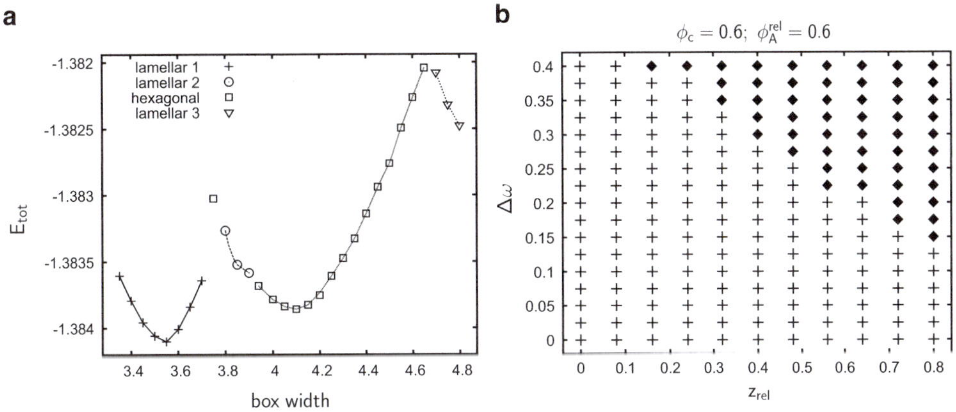

Fig. 7 (**a**) Free energy density of the polymer blend with parameters as given in Fig. 6 as a function of the box width L_x. The symbols indicate different phases (compare with Fig. 6). (**b**) Phase diagram, showing the stabile structure of the system as a function of the crosslink strength z_{rel} and the crosslink asymmetry $\Delta\omega$. *Crosses* (+) denote hexagonal states, *diamonds* (◇) denote lamellar states. Results are shown for $\phi_c = 0.6$ and $\phi_a^{rel} = 0.6$. For suitably low crosslink strength z_{rel} and crosslink asymmetry $\Delta\omega$, the system shows a hexagonal structure, while for suitably high z_{rel} and $\Delta\omega$ the lamellar structure is preferred

Our calculations provide a phase diagram for fixed $\chi N = 11.3$, $\phi_c = 0.6$, and $\phi_a^{rel} = 0.6$, in which the stable phase depends on the choice of $\tilde{z}$ and $\Delta\omega$ so that phase transitions between a hexagonal and a lamellar phase can be inferred by increasing $\tilde{z}$, $\Delta\omega$ or both. This shows a way, how to switch a macroscopically extended phase by changing the crosslink behavior, which in many systems can be altered by changing external conditions such as illumination or temperature.

It may be surprising that increasing the linking asymmetry $\Delta\omega$ switches the system from the hexagonal phase, which is asymmetric with respect to A and B, to the symmetric lamellar phase. As mentioned above, an increase of $(\Delta\omega)^2$ increases the demixing of the phases and favors regions of very large $\rho_A^{(1)}$ or $\rho_B^{(1)}$. We calculated $\int(\rho_A^{(1)})^2 dV$ and $\int(\rho_B^{(1)})^2 dV$ for several hexagonal structures and corresponding lamellar structures with similar free energy and found that these integrals are larger for the lamellar structures. Thus, increasing $|\Delta\omega|$ or z favors strong demixing of A and B of the copolymers. Since the lamellar structure allows a stronger demixing the system switches at suitably large $|\Delta\omega|$ and z from the hexagonal phase to the lamellar phase.

One should note that in the hexagonal phase, the B rich domains are enclosed by the A rich phase such that molecules that prefer the B phase are localized. In the lamellar phase, the same molecules can move over wide distances along the B rich stripes. Thus, our method allows the release of active agents from encapsuled B-rich domains into the surrounding. Alternatively, one could insert molecules that prefer B monomers into a lamellar structure and then switch to the hexagonal phase. At this point, crosslinks may be permanentized (for example, by cooling down the system) and one ends up with a structure in which the molecules of interest are localized in small portions with a defined distance to each other.

In general, the studied behavior of the AB + A + B system with reversibly crosslinked copolymers should also be found in an AB + C + D system with suitable asymmetric AC, AD, BC, BD and CD interactions. This is so, because breakable crosslinks should behave very similar to an effective interaction potential with a finite energy depth and potential width.

The results that we have found show that we can use self-consistent mean field theory to study important aspects inferred by reversible crosslinks. So far, the studies have been restricted to two dimensions. An extension to three dimensions would be of great interest with respect to applications and fundamental aspects of reversibly crosslinked polymer networks.

Acknowledgements This work was supported by DFG SPP 1259: "Intelligent Hydrogels. Modelling and simulation of hydrogel swelling under strong non-equilibrium conditions using the phase-field and phase-field crystal methods" and DFG SFB 840: "Von partikulären Nanosystemen zur Mesotechnologie".

References

1. Liu J (2011) Soft Matter 7:6757
2. Gawel K, Barriet D, Sletmoen M, Stokke BT (2010) Sensors 10:4381
3. Richter A, Paschew G, Klatt S, Lienig J, Arndt K-F, Adler H-P (2008) Sensors 8:561
4. Amin S, Rajabnezhad S, Kohli K (2009) Sci Res Essays 3:1175
5. Deligkaris K, Tadele TS, Olthuis W, van den Berg A (2010) Sens Actuator B-Chem 147:765
6. Shi J, Ouyang J, Li QT, Wang LY, Wu J, Zhong W, Xing MMQ (2012) J Mater Chem 22:23962
7. Liu Y-L, Hsieh C-Y, Chen Y-W (2006) Polymer 47:2581
8. Wei H-L, Yang J, Chu H-J, Yang Z, Ma C-C, Yao K (2011) J Appl Polym Sci 120:974
9. Jiang J, Qi B, Lepage M, Zhao Y (2007) Macromolecules 40:790
10. OReilly RK, Hawkerb CJ, Wooleyc KL (2006) Chem Soc Rev 35:1068
11. Li D, Gruhn T, Emmerich H (2012) J Chem Phys 137:024906
12. Edwards SF (1965) Proc Phys Soc 85:613
13. Helfand E, Yukiko T (1972) J Chem Phys 56:3592
14. Fredrickson G (2006) The equilibrium theory of inhomogeneous polymers. Oxford University Press, New York
15. Fredrickson GH, Ganesan V, Drolet F (2002) Macromolecules 35:16
16. Schmid F (1998) J Phys Condens Matter 10:8108
17. Müller M, Katsov K, Schick M (2006) Phys Rep 434:113
18. Müller M, Schmid F (2005) Adv Polym Sci 185:1
19. Leibler L (1980) Macromolecules 13:1602
20. Matsen MW, Schick M (1994) Phys Rev Lett 72:2660
21. Matsen MW, Bates FS (1996) Macromolecules 29:1091
22. Shull KR (1992) Macromolecules 25:2122
23. Matsen MW, Schick M (1996) Curr Opin Colloid Interface Sci 1:329
24. Matsen MW (1998) Curr Opin Colloid Interface Sci 3:40
25. Philipp KJ, Schick M (1997) Macromolecules 30:137
26. Philipp KJ, Schick M (1997) Macromolecules 30:3916
27. Noolandi J, Hong KM (1982) Macromolecules 15:482
28. Israels R, Jasnow D, Balazs AC, Guo L, Krausch G, Sokolov J, Rafailovich M (1995) J Chem Phys 102:8149
29. Shull KR, Kramer EJ (1990) Macromolecules 23:4769
30. Shull KR (1993) Macromolecules 26:2346
31. Hong KM, Noolandi J (1981) Macromolecules 14:727
32. Noolandi J, Hong KM (1984) Macromolecules 17:1531
33. Noolandi J, Shi A-C (1996) Macromolecules 29:5907
34. Vilgis TA, Borsali R (1991) Phys Rev A 43:6857
35. Wang Q, Taniguchi T, Fredrickson GH (2004) J Phys Chem B 108:6733
36. Müller M, Schick M (1998) Phys Rev E 57:6973
37. Elliott R, Katsov K, Schick M, Szleifer I (2004) J Chem Phys 122:44904
38. Flory PJ (1953) Principles of polymer chemistry. Cornell University Press, Ithaca
39. Edwards SF (1988) J Phys Fr 49:1673
40. Panyukov S, Rabin Y (1996) Phys Rep 269:1
41. Schulz M (2000) J Chem Phys 113:10793
42. Li D, Yang H, Emmerich H (2011) Colloid Polym Sci 289:513
43. Mohan A, Elliott R, Fredrickson GH (2010) J Chem Phys 133:174903
44. Goldbart PM, Castillo H, Zippelius A (1996) Adv Phys 45:393
45. Ulrich S, Mao X, Goldbart M, Zippelius A (2006) Europhys Lett 76:677
46. Xing X, Pfahl S, Mukhopadhyay S, Goldbart M, Zippelius A (2008) Phys Rev E 77:051802
47. Goldbart PM, Zippelius A (1994) Europhys Lett 27:599
48. Benetatos P, Zippelius A (2007) Phys Rev Lett 99:198301
49. Ulrich S, Zippelius A, Benetatos P (2010) Phys Rev E 81:021802
50. Helfand E (1975) J Chem Phys 62:999

51. Düchs D, Ganesan V, Fredrickson GH, Schmid F (2003) Macromolecules 36:9237
52. Bates FS, Maurer WW, Lipic PM, Hillmyer MA, Almdal K, Mortensen K, Fredrickson GH, Lodge TP (1997) Phys Rev Lett 79:849
53. Hillmyer MA, Maurer WW, Lodge TP, Bates FS, Almdal KJ (1999) Phys Chem 103:4814
54. Crank J, Nicolson P (1947) Proc Camb Philos Soc 43:50

Seawater Desalination via Hydrogels: Practical Realisation and First Coarse Grained Simulations

Johannes Höpfner, Tobias Richter, Peter Košovan, Christian Holm, and Manfred Wilhelm

Abstract

We investigated and described a novel approach for water desalination using charged hydrogels under externally applied mechanical forces. The desalination mechanism is based on the unequal distribution of an added salt between gel and surrounding solution phase. We synthesised acrylic acid-based hydrogels of various compositions and investigated their desalination properties with a specifically designed experimental press setup that allowed us to control online the force respective pressure exerted on the gel and to measure the water elution from the gel bed as well as the salt concentration of the eluate. A reference sodium chloride solution was used as a model for desalination applications. The experiments were augmented with a theoretical analysis within a mean-field Donnan model that can semi-quantitatively explain the salt distribution and the desalination process. In addition we performed coarse-grained simulations with explicit ions and charged bead-spring polymers. The simulations provided reference data on well defined systems which could be directly compared with the theoretical predictions. This comparison provided valuable insights into the weak points of the mean-field theory and guidelines for its further development.

Keywords

Hydrogels • Donnan equilibrium • Desalination • Polyelectrolytes • Coarse grained simulations • Absorption under load

Introduction

Hydrogels are a class of materials with a high affinity for water. On a microscopic scale, they are composed of cross-linked polymer chains that incorporate a large amount of water or aqueous solution. Usually, this is achieved using polyelectrolyte gels for which the term "superabsorber" was coined [1]. The material is used in a number of high volume applications which ranks them among the more important polymeric substances. Practical applications of this material today include utilisation in agricultural soil to retain water [2] and in underwater cable insulations [3]. However, the largest application market is personal hygiene products [4]. For this application alone, more than 8×10^6 t of absorbent material were produced in 2011 worldwide [5].

J. Höpfner • M. Wilhelm (✉)
Karlsruhe Institute of Technology (KIT), Institute for Chemical Technology and Polymer Chemistry, Engesserstr. 18, 76131 Karlsruhe, Germany
e-mail: Manfred.Wilhelm@kit.edu

T. Richter • C. Holm
Institute for Computational Physics, University of Stuttgart, Allmandring 3, 70569 Stuttgart, Germany

P. Košovan
Institute for Computational Physics, University of Stuttgart, Allmandring 3, 70569 Stuttgart, Germany

Department of Physical and Macromolecular Chemistry, Faculty of Science, Charles University in Prague, Hlavova 8, 120 00 Praha, Czech Republic

G. Sadowski and W. Richtering (eds.), *Intelligent Hydrogels*, Progress in Colloid and Polymer Science 140,
DOI 10.1007/978-3-319-01683-2_19,

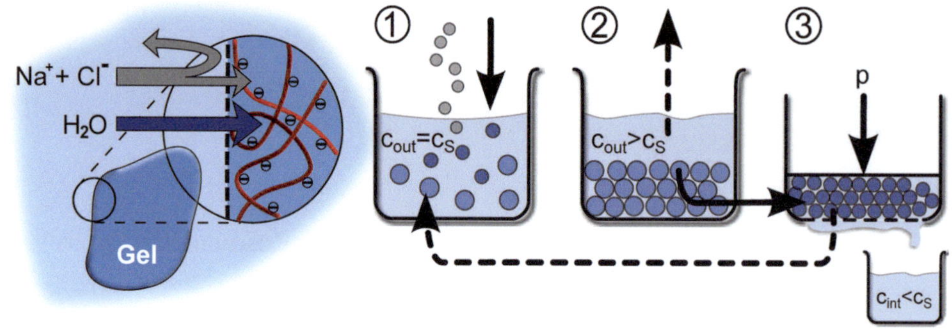

Fig. 1 Schema of the proposed desalination process via polyelectrolyte gels: the dry polymer beads are placed in a brine solution with concentration c_s and brine water is drawn into the polyelectrolyte gel (*1*). The mobile ions of the brine solution are partly rejected from the gel (*left side*). Then, the swollen gel is removed from the brine (*2*). The polymer gel is deswollen by a stimulus e.g. pressure in (*3*) and the desalinated solution with $c_{\mathrm{int}} < c_{\mathrm{out}}$ is obtained. The gel can be swollen and used again

The large water uptake of such networks, often 100 times more than their own weight, is due to the large osmotic potential caused primarily by the ionic groups [6]. However, polyelectrolyte gels also interact with dissolved ions in the swelling agent [7]. Little use has been made of this interaction in terms of practical applications. In this article, we outline a potential large scale application based on this feature. Recently, we were able to show for the first time that hydrogels can be used as a reusable separation agent in the desalination of water [8].

The proposed desalination process can be understood in terms of forward osmosis (FO) [9, 10] in contrast to the routinely employed reverse osmosis (RO) in seawater desalination [11, 12]. In both cases, the salt water feed is separated by a membrane from the filtrate side. A nanoporous membrane allows water to pass, but rejects ions due to the diameter of their hydrate shell [11]. In the RO process, the osmotic pressure of the seawater, about 30 bar, is outweighed by the hydrostatic pressure exerted on the brine phase. In the case of FO, a drawing agent is instead employed to draw water to the filtrate side of the membrane. This agent must have a higher osmotic potential than the feed and should be easily removed by some method. An example of this technique is based on ammonium bicarbonate [13, 14] which decomposes energetically favourably to gaseous products around 60 °C. In a more recent example, hydrogels were used as the drawing agent, which was afterwards deswollen using stimuli such as pressure or temperature [15, 16]. Yet, a nano-filtration membrane was needed for the separation in the above-mentioned examples.

We propose here instead to use the hydrogel as both the drawing agent and the "membrane" simultaneously. When the polymer phase is placed in contact with the solution phase, water is drawn into the polymer due to the osmotic potential, while the added electrolyte is partly rejected and remains in the supernatant solution (Fig. 1, left). This effect has been described already by Katchalsky and Michaeli [7] but raised little subsequent interest. Only a few studies are known that examined the distribution of the charged species e.g. [17–19] between the phases.

The process which we envision here is comprised of three steps as depicted in Fig. 1 on the right. The dry gel beads are first mixed with brine with concentration c_s (step 1). The volume of added solution must be greater than the capacity of the gel, so that a liquid phase remains. Once the gel is swollen to equilibrium, as much as possible of the supernatant water phase is removed (step 2). The polymer is then subjected to an external pressure p_{ext} to remove the stored aqueous solution (step 3). The eluate obtained at the end has a lower salt concentration than initially ($c_{\mathrm{int}} < c_{\mathrm{out}}$), and thus constitutes the first successful desalination step in the envisaged sequence. In step 3 not all water can be removed from the gel with reasonable pressure. Yet, a partly swollen gel can be used in a new cycle of the process.

The main advantage of the proposed process is that it does not depend on a membrane since the interface is the gel surface itself. This has several potential advantages including increased flux and efficient use of space. Membranes are high added value products and therefore rather expensive. They are also prone to bio-fouling and clogging [11]. These problems could be minimised by using a hydrogel instead. Due to these reasons, we consider it worthwhile to pursue the proposed desalination approach, to test its viability and to present different possibilities for optimisation.

In the work presented here, the underlying separation mechanism for water and dissolved electrolytes is investigated for hydrogels prepared from acrylic acid which have been synthesised with varying composition. In the next section we first describe a simple mean-field model which we use to qualitatively describe the desalination process, and which has the nice feature that it is analytically tractable. Following this in section "Simulation Model and Method" we introduce our particle-based simulation model for the hydrogels, and describe the numerical methods for obtaining meaningful simulation data. In section "Experimental" we then describe the synthesis of the used hydrogels and their characterisation without applied external pressure. In section "Design and Use of the Press Setup" we describe in detail our in-house design of the press set-up and the experimental data acquisition. In section "Results and Discussion" we display and critically discuss our findings, and put the obtained results in the perspective to the analytical Donnan theory and the experimental data. The last section summarises our main findings and provides an outlook about future prospects to investigate the desalination process and the optimal hydrogel composition.

Theoretical Treatment

In this section we outline how the proposed desalination process can be described by an analytical theory. We restrict ourselves to a simple description and deliberately neglect some issues which would require a more advanced treatment. Discussion of more complex models and their comparison with simulation data is planned for a forthcoming publication.

Phenomenologically, hydrogel swelling in an aqueous solution can be described as a two-phase system consisting of the gel phase (g) and solution phase (s). The cross-linking of the polymer network prevents polymer segments from entering the solution phase. On the contrary, mobile species (free ions and water molecules) can freely pass through the interface and redistribute between the two phases.

The condition of thermodynamic equilibrium requires that the chemical potential of each mobile species is the same in both phases

$$\mu_{\mathrm{i}}^{\mathrm{g}} = \mu_{\mathrm{i}}^{\mathrm{s}}, \tag{1}$$

where μ_{i} is the chemical potential of species i and the superscripts denote the different phases. Consider a hydrogel placed in a salt solution bath of a 1:1 electrolyte with initial concentration c_{s}. After being swollen to equilibrium the concentration of the bath is c_{out}, which, for an infinite bath, equals still c_{s}. In the bulk solution, the concentrations of individual ions are equal. The concentration of positive and negative ions in the gel is then [20]

$$c_{\pm} = c_{\mathrm{out}} e^{\pm\Psi}, \tag{2}$$

where c_- and c_+ are the concentrations of mobile anions and cations in the gel phase, respectively. $\Psi = e\psi/k_{\mathrm{B}}T$ where ψ is the electric potential difference across the interface (the Donnan potential). The latter arises as a result of differences in the activity coefficients of the ions in the two phases. From Eq. (2) it follows that

$$c_+ c_- = c_{\mathrm{out}}^2, \tag{3}$$

which provides a link between the salt concentration in the solution phase and the concentrations of anions and cations in the gel phase. Electroneutrality has to be satisfied in both phases, leading to

$$c_+ = c_- + f c_{\mathrm{p}}, \tag{4}$$

on the polymer side of the interface where c_{p} is the total concentration of monomer segments in the swollen gel (both charged and uncharged) and f is the fraction of charged units at the given conditions. In the last equation we assumed that the gel-forming polymer is a polyanion and, without loss of generality, we will restrict further derivations to anionic polymers. Analogous relations for polycations can be obtained by symmetrically exchanging c_+ and c_- in all formulas. The amount of salt taken up by the gel is given by the concentration of anions, $c_{\mathrm{int}} = c_-$, while c_+ in addition includes the counterions which compensate the polymer charge, fc_{p}. Combining Eqs. (3) and (4) we obtain the relation between the external salt concentration and the concentration of free ions in the gel

$$c_{\pm} = \left(\left(\frac{fc_{\mathrm{p}}}{2}\right)^2 + c_{\mathrm{out}}^2\right)^{1/2} \pm \frac{fc_{\mathrm{p}}}{2}. \tag{5}$$

From Eq. (5) it follows that always the total concentration of mobile anions in the gel, c_-, is lower than c_{out} while the concentration of mobile cations in the gel, c_+, is higher. This is the key observation for the proposed desalination mechanism. For later discussion we note here that c_{p} is related to the end-to-end distance, R_{e}, of the polymer strands in the gel

$$c_{\mathrm{p}} = \frac{AN}{R_{\mathrm{e}}^3 N_{\mathrm{A}}}, \tag{6}$$

where N_{A} is the Avogadro number, N is the number of monomers per chain and A is a geometrical prefactor, typically on the order of unity, which depends on network structure and topology.

Equation (5) is useful to predict c_{int} in a swollen gel with known (experimentally measured) c_{p}. However, swelling of a hydrogel depends (among other things) on the salt concentration. More specifically, the excess of mobile ions entrapped in the polymer exerts osmotic pressure, which is counteracted by the entropic elasticity of the polymer chains. In addition, contributions coming from interactions between all involved species (polymer, ions, solvent) add to the pressure balance.

The condition of mechanical equilibrium then requires that pressures on both sides of the interface are balanced, i.e.

$$p_{\text{inter}} + p_{\text{elastic}} + p^{\text{g}}_{\text{osm}} = p_{\text{ext}} + p^{\text{s}}_{\text{osm}}\,, \tag{7}$$

where p_{inter} is the contribution to pressure due to intermolecular interactions, p_{elastic} is the elastic contribution of the polymer, $p^{\text{s}}_{\text{osm}}$, $p^{\text{g}}_{\text{osm}}$ are the osmotic contributions of the mobile ions on the solution and gel side, respectively, and p_{ext} is the applied external pressure.

To predict the equilibrium swelling of a hydrogel, a model is needed to link each of the contributions to pressure to structural parameter of the polymer and polymer-solvent interactions. We adopt the classical theory of Flory and Rehner [21] and Flory [6] which assumes all contributions to the total pressure to be independent. A discussion of its validity can be found elsewhere [19,22,23]. The contribution from chain elasticity can be expressed as

$$p_{\text{elastic}} = -\frac{k_{\text{B}}T}{Nb^2 R_{\text{e}}} = -\frac{k_{\text{B}}T}{Nb^2}\left(\frac{c_{\text{p}}N_{\text{A}}}{AN}\right)^{1/3}, \tag{8}$$

where b is the bond length. This is valid for ideal Gaussian chains and is used here as simple approximation which has been shown to work well at swelling equilibrium [22, 23].

The osmotic pressure contribution across the interface can be expressed as

$$p^{\text{g}}_{\text{osm}} - p^{\text{s}}_{\text{osm}} = k_{\text{B}}T(c_+ + c_- - 2c_{\text{out}}), \tag{9}$$

where $c_\pm$ can be obtained from Eq. (5).

The interaction term p_{int} includes the contribution from the non-ideal solvent-polymer mixing, excluded volume and electrostatic interactions. In case of athermal polymer-solvent interactions the mixing enthalpy contribution can be neglected which is a reasonable assumption for hydrogels under good solvent conditions. The excluded volume interactions in a swollen state are negligible but may become important as the hydrogel is compressed. A quantitative discussion of the magnitude of different interaction contributions for simulations of model hydrogels can be found in Ref. [22,23].

Substituting from Eqs. (8) and (9) into Eq. (7) and assuming that the interaction contributions can be neglected, we finally obtain

$$\frac{p_{\text{ext}}}{k_{\text{B}}T} = 2\left(\left(\frac{fc_{\text{p}}}{2}\right)^2 + c^2_{\text{out}}\right)^{1/2} - 2c_{\text{out}} - \frac{1}{Nb^2}\left(\frac{c_{\text{p}}N_{\text{A}}}{AN}\right)^{1/3}, \tag{10}$$

which is an equation of state of the polymer swollen in an infinite salt bath with salt concentration c_{out}. For a finite bath

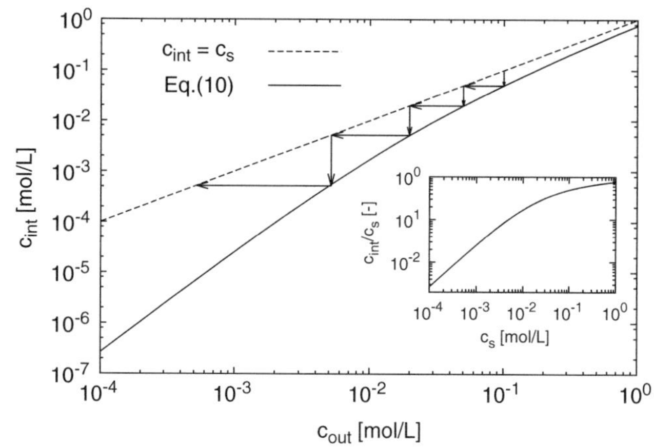

Fig. 2 The salt concentration inside the hydrogel, c_{int}, as a function of external salt concentration, c_{out} computed by numerically solving for R_{e} in Eq. (10) with for a model hydrogel with $N = 100$ and $f = 0.5$ (*solid line*). The *dashed line* shows $c_{\text{int}} = c_{\text{out}}$ as a reference. The *arrows* describe the progress of an idealised desalination experiment as described in the text. Inset: same data plotted as c_{int} / c_{out} to illustrate the relative efficiency of partitioning

we would have to take into account that its concentration also depends on the swelling of the hydrogel: $c_{\text{out}}(c_{\text{p}})$. It does not allow to express explicitly c_+ and c_- as a function of c_{out}, N, f and the prefactor A but it can easily be solved iteratively to obtain c_{p}, which on a common computer takes a fraction of a second. From the equation of state, all thermodynamic variables can be obtained. Specifically, once c_{p} is known, $c_\pm$ can be obtained from Eq. (5). Since it requires little effort as compared to synthesising the gel and performing the experiments, it could be exploited in the initial stages of process optimisation. However, before this can be done, the model must be validated against simulations and experiments on model systems with known structure which will reveal its range of validity and show where it needs to be improved to enable practical use.

In the current simple form, we can use the model to semi-quantitatively illustrate how the desalination experiment works in an idealised realisation. Figure 2 shows the dependence of c_{int} on c_{out} computed for a polymer network with $N = 100$, $f = 0.5$ and $b = 0.5\,\text{nm}$. The concentration inside the hydrogel predicted by the theory is significantly lower than the corresponding external concentration, in accordance with qualitative expectations based on Eq. (5). From the inset it can be seen that the relative difference can range from about 10 % with c_{out} around 1 mol/L to a factor of 50 around 10^{-3} mol/L. Note that the values of c_{int} are sensitive to the geometrical prefactor A but the general shape of the dependence is preserved, independent of A. Arrows in the main plot indicate the progress of an idealised desalination experiment: starting from the bath at $c_{\text{out}} = 10^{-1}$ mol/L, the salt concentration in the hydrogel becomes about 5×10^{-2} mol/L. The solvent from the hydrogel is squeezed out and used as the new salt bath. In the next step, the concentration drops to 2×10^{-2} mol/L and with four desalination steps the final concentration becomes 5×10^{-3} mol/L. Of course, in the real experiment not all water can be squeezed out of a hydrogel and therefore in practice more steps are necessary to achieve the desired final salt concentration. In this context one step of this idealised process could be viewed in analogy to a theoretical plate in fractional distillation.

Simulation Model and Method

Polymer Model

Due to computational demands, particle simulations of charged hydrogels have been feasible only in the twenty-first century. First investigations can be found in Refs. [24–26], and a recent review of hydrogel simulations is also available [27]. The basic idea is to employ a generic coarse-grained bead-spring polymer model that contains charged and uncharged beads. The simulation treats water as an implicit solvent modelled as a dielectric continuum with a dielectric constant $\epsilon_r = 72$, and it can describe, for example, the structure of polyelectrolytes fairly well [28, 29]. Such a model does not represent a specific chemical structure of the polymer but possesses the essential features considered in simple theories or phenomenological approaches: excluded volume, chain connectivity and electrostatic interactions between charged species.

In our present simulations we use a simulation box that consists of a perfect, defect-free network of 16 polyelectrolyte chains end-linked to 8 tetra-functional nodes (see Fig. 3), as has been used before in Ref. [22, 26, 30, 31]. Charged groups on the polymer chains are spaced regularly with the degree of charging $f = 1/(n+1)$ where n is the number of uncharged monomer units between two charged ones. In addition to the polymer, the simulation box contains an appropriate number of monovalent counterions to neutralise the polymer charge. A variable amount of pairs of monovalent salt ions is added as described in more detail below. The polymer network is connected to itself via the periodic boundary conditions, to form a quasi infinite network. While the box size (and hence the gel swelling) is fixed, the polymer chains including the nodes are free to move. The simulations were performed at a fixed number of polymer segments, N, chemical potential of the mobile ions, μ_{ion}, temperature, T, and volume V as described in more detail below.

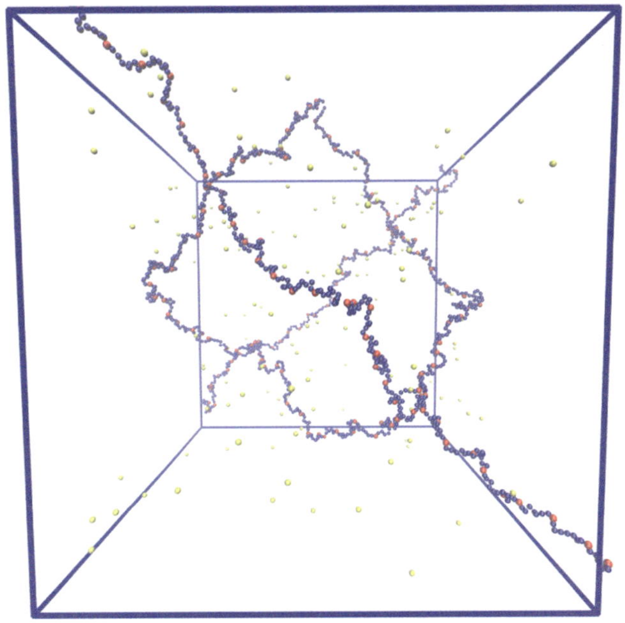

Fig. 3 Simulation snapshot of a tetra-functional hydrogel with 8 nodes, 16 chains with $N = 79$ and $f = 1/8$ (charged beads *red*, neutral ones *blue*) and counterions (*yellow*). Additional salt ion pairs are not shown

Salt Concentration Inside the Gel and Determination of the Swelling Equilibrium

To determine the swelling ratio at equilibrium, we performed a series of grandcanonical simulations at several fixed gel volumes. At equilibrium swelling the total pressure in the box is equal to the (osmotic) pressure of the external salt solution. We determined the pressure for each of the fixed gel volumes mentioned above and interpolating between the obtained values we obtained the box size corresponding to swelling at equilibrium, without applied external pressure. This procedure is shown for one selected system in Fig. 4. In order to determine the swelling ratio, we estimated the volume per chain of the dry hydrogel as the volume of the polymer plus its counterions divided by the maximum packing fraction of hard spheres, $y_{\max}$

$$V_0 = \frac{4\pi N(1+f)\sigma^3}{3y_{\max}}, \tag{11}$$

where σ is the monomer size defined in Eq. (15).

To ensure that the chemical potential of the salt inside the hydrogel is equal to that of the external salt solution, we used a Monte-Carlo scheme to attempt insertion or deletion of salt ions in the box. A Metropolis-like criterion was used to decide if an attempted particle insertion or deletion should

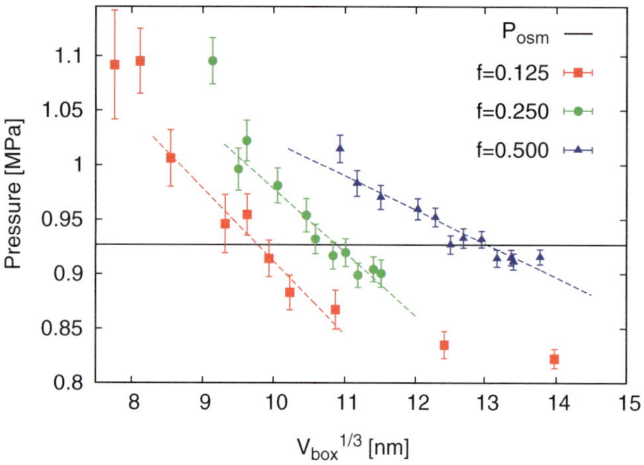

Fig. 4 Determination of swelling equilibrium from simulations at different gel volumes for the hydrogel with $N = 39$, and external salt concentration $c_{\text{out}} = 0.2\,\text{mol/L}$ and various degrees of charging, f, as indicated in the legend. The gel volume at which the pressure equals the osmotic pressure of the salt bath (*horizontal line*) was determined by interpolation (*dashed lines*)

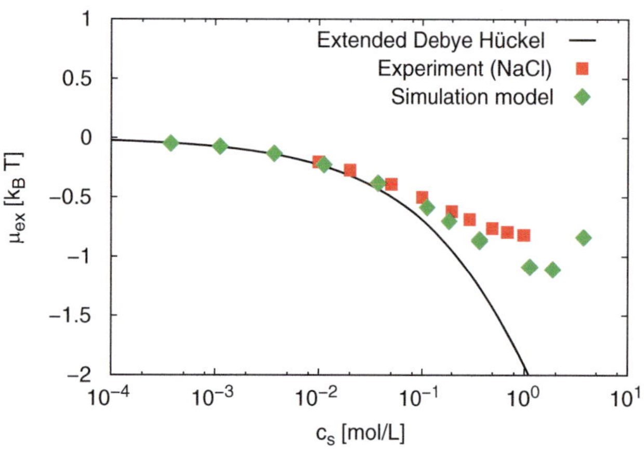

Fig. 5 Dependence of the salt potential chemical (μ_{salt}) of the salt bath on its concentration, c_{out}. Comparison of experimental data on NaCl [33] (*squares*), our simulation model (*diamonds*) and Eq. (14) (*solid line*)

be accepted:

$$P_{\text{acc}}(N \to N+1) = \min\left[1, \frac{1}{\Lambda^3 \varrho_{N+1}} \exp\left[-\beta(-\mu_{\text{salt}} + \Delta E)\right]\right], \quad (12)$$

$$P_{\text{acc}}(N \to N-1) = \min\left[1, \frac{\Lambda^3 \varrho_N}{1} \exp\left[-\beta(\mu_{\text{salt}} + \Delta E)\right]\right], \quad (13)$$

where Λ is the thermal wavelength, ϱ_N is the particle density, $\beta = 1/k_B T$, $\Delta E = E_{\text{new}} - E_{\text{old}}$ is the interaction energy change upon the attempted insertion and μ_{salt} is the chemical potential of the external salt solution.

The dependence of μ_{salt} on c_{out} and the osmotic pressure of the free salt solution without any gel for our particular model have been determined from a separate series of simulations, employing the Widom particle insertion method [32]. To check consistency of our data, we compare in Fig. 5, the c_{out}-dependence of $\mu_{\text{salt}} = \mu_{\text{ideal}} + \mu_{\text{ex}}$ in our model with the experimental data on NaCl [33] and with the extended Debye-Hückel formula

$$\mu_{\text{ex}}^{\text{DH}} = -\frac{\sqrt{8\pi l_B^3 c_{\text{out}} N_A}}{1 + \sqrt{8\pi c_{\text{out}} N_A}\, a / l_B}, \quad (14)$$

with a as the diameter of the respective ion. In contrast to the extended Debye-Hückel formula, our simulation data captures the trend from the experiment well. Considering the coarseness of the model and the fact that no special effort has been devoted to directly match the real data, we take this as an a posteriori validation of the appropriateness of our model for the relevant concentration range.

Interaction Potentials and Simulation Parameters

The excluded-volume interactions between the different particles (monomers, salt ions) were modelled using the Weeks-Chandler-Andersen (WCA) potential [34], which is based on the repulsive part of the commonly used Lennard-Jones (LJ) potential:

$$U_{\text{WCA}}(r) = \begin{cases} 4\epsilon\left(\left(\frac{\sigma}{r}\right)^{12} - \left(\frac{\sigma}{r}\right)^{6} + \frac{1}{4}\right) & \text{for } r \le 2^{1/6}\sigma \\ 0 & \text{for } r \ge 2^{1/6}\sigma \end{cases}, \quad (15)$$

where we used $\epsilon = 1.0\,k_B T$. We used $\sigma = 0.35\,\text{nm}$ for both the monomer size and size of the ions including first hydration shell. With WCA interaction potential, the simulated polymer is purely athermal.

The bonds between the monomer units were modelled using the finite-extensible non-linear elastic (FENE) potential:

$$U_{\text{FENE}}(r) = -\frac{1}{2} k_F r_F^2 \ln\left[1 - \left(\frac{r}{r_F}\right)^2\right], \quad (16)$$

for $r < r_F$ with $r_F = 0.53\,\text{nm}$ and $k_F = 40\,k_B T/\text{nm}^2$.

The ions and ionised units of the polymer carry charges of $q_i = \pm 1$ and interact via the full Coulomb potential

$$U_{\text{Coul}}(r_{ij}) = k_B T l_B q_i q_j / r_{ij}, \quad (17)$$

where l_B is the Bjerrum length defined as

$$l_B = e^2/(4\pi\varepsilon_0\varepsilon_r k_B T), \tag{18}$$

where e is the elementary charge, ε_0 is the permittivity of vacuum, and ε_r is the static dielectric constant. For the current system composed of an aqueous solution of salt, the Bjerrum length was taken to be that of pure water at 25 °C, $l_B = 0.71$ nm. The calculations were performed using the ESPResSo software [35] in version 3.1 [36, 37], employing the P3M algorithm [38] for the electrostatic interactions. The integration of the equations of motion was performed with the standard velocity-Verlet algorithm coupled to a heat bath using a Langevin thermostat with coupling constant $1/\Gamma = \tau$, where τ is the internal simulation time unit, and temperature $T = 298$ K. The integration time step was set to $\Delta t = 0.01\tau$.

Experimental

Synthesis of Hydrogels

The polyelectrolyte gel samples were synthesised specifically for this project. They were made by free radical polymerisation of acrylic acid in water. The synthetic procedure is shown in Fig. 6 and was adapted from similar procedures published recently [39–41]. For a typical gel 1.16 g of the cross-linker *N,N'*-methylenebisacrylamide (MBA, 0.01 eq, 7.5 mmol, 99+ %, Acros, Geel, Belgium) was mixed with 215 mL deionised water in a round-bottom flask. 54 g of freshly distilled acrylic acid (AAc, 1 eq, 0.75 mol, 99 %, Sigma-Aldrich, Steinheim, Germany) was added and the flask was sealed. The mixture was stirred and cooled in an ice bath to $T \approx 5$ °C. Neutralisation was achieved by adding drop-wise 50.1 mL sodium hydroxide in water (33 wt-%, 0.75 eq, extra pure, Acros) so that the temperature did not exceed 12 °C during this exothermic process. 360 mg of the initiator sodium peroxodisulfate (SPS, 0.002 eq, 1.5 mmol, 98+ %, Acros) was dissolved in 2 mL of deionised water and added to the mixture. The solution was further cooled and purged for 30 min with nitrogen gas. To start the reaction, 1.74 g of the accelerator *N,N,N',N'*-tetramethylethylenediamine (TEMED, 0.02 eq, 15 mmol, 99 %, Acros) was added and the cooling unit was removed. The cross-linking reaction was left to proceed for 12 h at room temperature. The resulting clear gel was cut, dried in vacuo at 70 °C, and ground with a mill to a dry bead size of 350–650 μm. All components were used as received, except for AAc which was distilled at 23 mbar and 43 °C prior to use.

The reaction scheme includes several adjustable parameters to obtain gels with different structures. All parameters are defined with respect to the amount of AAc in the mixture.

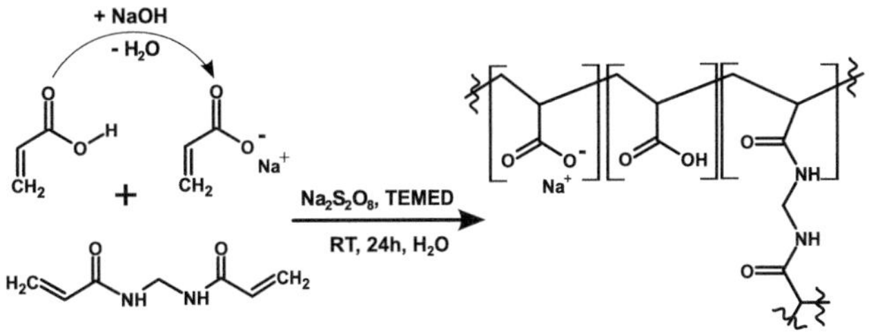

Fig. 6 The reaction scheme for the synthesis of poly(acrylic acid) hydrogels is shown. Free radical polymerisation of partly neutralised acrylic acid in water was in initiated by the redox system sodium peroxodisulfate and *N,N,N',N'*-tetramethylethylenediamine (TEMED). The tetra-functional monomer *N,N'*-methylenebisacrylamide served as the cross-linker

First, the degree of cross-linking (DC) is defined by the amount of cross-linker added to the monomer.

$$\mathrm{DC} = \frac{n(\mathrm{MBA})}{n(\mathrm{AAc})} \times 100\,\%. \tag{19}$$

The fraction of monomer converted to its ionic form is given by the degree of neutralisation (DN).

$$\mathrm{DN} = \frac{n(\mathrm{NaOH})}{n(\mathrm{AAc})} \times 100\,\%. \tag{20}$$

The initial state of the polymeric network is defined at the time of formation by the amount of solvent present in the network structure at the time of synthesis. Here, it is given as the swelling ratio based on mass (Q_{syn}).

$$Q_{syn} = \frac{m(\mathrm{H_2O})}{m(\mathrm{AAc}) + m(\mathrm{MBA})}. \tag{21}$$

The salt solution used to simulate the brine in the desalination and swelling experiments was made from sodium chloride (99.5 %, Acros) with the addition of deionised water to reach the desired concentration of w/v. Before use, sodium chloride was pre-dried at 120 °C. The deionised water exhibited a conductivity of less than 0.5 μS/cm (or >2 MΩ/cm).

Measurement of the Swelling Capacity

For the determination of the equilibrium swelling capacity Q_{eq} of a gel, the polymer was placed in contact with a salt solution and the weight gain was recorded after waiting a sufficiently long time [1, 42]. The dried polymer of mass m(dry) was placed on a sieve with mass m(sieve). The sieve was lowered to be in contact with an excess of the solution. The polymer was then left to swell at room temperature while evaporation was prevented by sealing the chamber. Finally, the sieve was removed and pressed onto a paper towel sheet for 30 s to remove excess solution. It was finally weighed to obtain $m(\text{wet})$. The equilibrium swelling degree was calculated by

$$Q_{\text{eq}} = \frac{m(\text{wet}) - m(\text{sieve}) - m(\text{dry})}{m(\text{dry})}. \quad (22)$$

For each gel the procedure was repeated three times and the average of the swelling capacities was used. The standard deviation of Q_{eq} was found to be below 1 %.

Electrolyte Separation Experiments Without External Pressure

The ion distribution in free swelling (without pressure) was obtained by simply mixing the dry polymer with a brine solution. An excess of solution was used so that two phases were formed. To obtain comparable distribution data, the volume ratio of the phases must be constant when gels of different swelling capacities are considered. It is therefore beneficial to define the relative water content (RWC) of the whole sample (including possible excess supernatant) normalised to the swelling capacity of the same gel

$$\text{RWC} = \frac{m(\text{solvent})}{m_{\text{eq}}(\text{solvent})}. \quad (23)$$

With the above definition, RWC > 1 indicates the presence of excess supernatant in the sample, RWC $= 1$ means the gel has the same water content as it would contain in swelling equilibrium and RWC < 1 occurs when the gel is swollen less than its equilibrium capacity.

The separation experiments were carried out at different values of RWC with a salt solution of the initial concentration c_{s}. The mixture was stirred for at least 12 h at room temperature. The salt concentration was measured in the supernatant phase via its conductivity (c_{out}). The concentration of added salt inside the gel was obtained by the mass balance of added salt ($n_{\text{total}} = n_{\text{int}} + n_{\text{out}}$) using known quantities from

$$c_{\text{int}} = \left(c_{\text{s}} - c_{\text{out}}\left(1 - \frac{1}{\text{RWC}}\right)\right)\text{RWC}. \quad (24)$$

Design and Use of the Press Setup

The swelling state of a hydrogel is controlled by its structure and the external conditions. In this work, external pressure was used to reduce the swelling of the gel and to quantify the resulting electrolyte distribution between the phases by investigating the discharged fluid. For the practical realisation a press setup was needed which is capable of separating polymer and water (step 3 in Fig. 1). A new setup was constructed for this purpose to tailor the form and size as well as the sensors used for the data acquisition [43]. First, the commercially available parts of the setup are described.

Filter Press Components

The pressure sensor was equipped with a flat top membrane (SD-40, Suchy Messtechnik, Lichtenau, Germany) and has a measurement range of 0–100 bar with a max. 0.5 bar deviation from linearity. The distance data was acquired via a digital dial gauge (MarCator 1086, Mahr, Göttingen, Germany) with a 50 mm measurement range and 1 μm accuracy. The conductivity and temperature of the salt solution were measured with a conductivity meter (SevenMulti, Mettler Toledo, Gießen, Germany) resulting in an accuracy of about 0.2 mS/cm and 0.1 °C, respectively.

For the drainage, standard chromatography components were used. These included valves, a T-connector (both 0.06″ bore) and ETFE tubing with 0.75 mm inner diameter. All parts were assembled with 1/4″-28 connectors and obtained from Vici (Schenkon, Switzerland). The filter material was a stack made from three components. Closest to the gel was a cellulose filter paper with 3–5 μm pore size (Carl Roth, Karlsruhe, Germany). It was supported by two layers of wire mesh, the first one with a 0.025 mm mesh size and the second with a 0.13 mm mesh size both made from stainless steel (ANSI 304) obtained from Spörl KG (Sigmaringendorf, Germany).

Setup Design

The press setup is composed of several parts and schematically shown in Fig. 7. In the middle of the setup is the sealed sample chamber where the gel is placed. The solution is separated from the polymer by two sieve elements. The upper sieve element can be moved down together with the piston to compress the sample. The sieve elements (Fig. 7, bottom right) are composed of the sieve holder and the sieve fixture which can be screwed on to fasten the sieve material. The sieve holder (Fig. 7, top right) has a grooved surface and a hole in the middle for the removal of the eluate. A drainage is used to remove the desalinated water and collect it outside the chamber in fractions.

This setup is a substantial improvement to the setup presented by the authors in an earlier publication [8]. In particular it has a more favourable geometry to enable faster separation and is equipped with high precision sensors which allow for automated data acquisition at a high sampling rate.

The independent channels of distance, pressure, and concentration were measured by the sensors described above. The measurements were time-triggered and the data were directly recorded using a computer with a high sampling rate, typically 1 Hz. The pressure in the sample compartment was measured by an axial sensor measuring the pressure inside the sample without the loss at the interfaces. The sensor voltage was collected at 1,000 Hz, decimated and converted

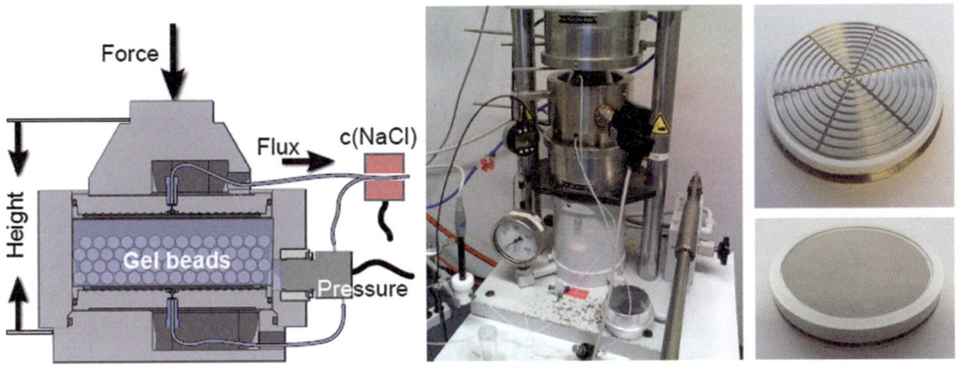

Fig. 7 Schematic of the vertical cross-section and photos of the new test setup used to separate gel from water in the desalination experiment. The assembly and the sensors for pressure, distance, and concentration and the general modus operandi are shown on the *left*. To the *right*, the complete setup is shown in operation with detail photos of the filter elements

online to pressure (see section "Appendix: Experimental Data Analysis"). The volume discharged from the setup was measured indirectly by changes in the distance of the two sieve units as measured by a lateral gauge fastened between the piston and the frame. The change in distance was converted to the change in volume via the area and was subjected to a number of corrections (see section "Appendix: Experimental Data Analysis"). The volume flux and change in RWC were obtained from the same data. The conductivity and temperature of the eluate were measured continuously in a flow cell or by evaluation of the collected fractions. The conductivity was corrected and converted to the salt concentration by a procedure described in the section "Appendix: Experimental Data Analysis".

Operating Procedure

The dry polymer was first mixed with brine (e.g. $c_s = 10$ g/L) at a typical fixed ratio RWC=2 as defined by Eq. (23). The mixture was stirred at room temperature to allow for full equilibration, typically for 12 h (step 1 in Fig. 1). In some experiments part of the supernatant phase with the volume V_{Offset}, was taken directly after equilibration from the sample to measure the conductivity (step 2 in Fig. 1).

Before the sample was loaded into the press, the distance sensor offset was chosen so that, at $d = 0$ mm, the two sieve units were in contact. Then, the pressure measurement was started to acquire the voltage baseline signal U_0 at zero pressure. Next, the sample was added into the chamber, the press was closed and the distance measurement started. The upper sieve unit was lowered slowly to remove most of the trapped air in the setup. The distance d_0 was noted when the first eluate was flowing from the setup.

The pressure was controlled by lowering the piston. Initially, only 1 bar was applied to remove excess solution until the flux decreased considerably (within 20 min). Then, the desired pressure profile was started by raising the pressure in 1 bar steps at the desired time intervals (typically every 5 min at a rate of 12 bar/h). On selected samples we checked by performing compressions at higher and lower rates that the above rate was slow enough not to induce any kinetic effects in the deswelling experiments. Fractions of 10–3 mL were collected depending on the flux. The experiment was stopped either when the maximum applicable pressure was reached, or when the sieve unit has reached its lowest possible position. After the experiment, the obtained fractions were weighed (m_i) and their conductivity σ_i and temperature T_i were measured.

Results and Discussion

The proposed new method for desalination of seawater depends fundamentally on the distribution of low-molecular weight ions between the gel and the solution phase. This distribution is controlled by the type of electrolyte, the gel, and, especially, the swelling state of the gel. The distribution was investigated experimentally for poly(acrylic acid) hydrogels with different values of DC, DN, and Q_{syn} (see Eqs. (19)–(21)) which resulted in considerable differences in their swelling and separation capacities. These results are compared with predictions based on the theoretical Donnan approach and with simulation results on similar systems.

Electrolyte Separation Without an Applied External Pressure

Separation Experiments

The pressure free equilibrium distribution of the added salt was determined by a simple mixing experiment described in section "Experimental". The concentration of mobile ions

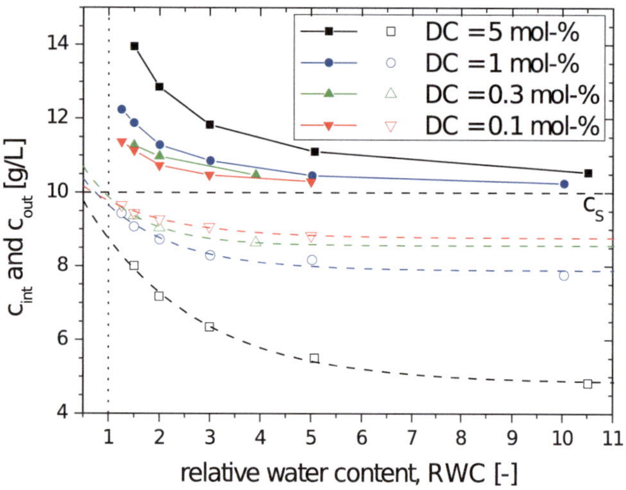

Fig. 8 Hydrogels were immersed in salt solution bath of sizes RWC and the concentrations of added salt inside (c_{int}, *open symbols*) and outside (c_{out}, *filled symbols*) the gel were determined. Gel samples with different degrees of cross-linking (DC) and fixed values of DN = 75 mol-% and Q_{syn} = 4 g/g were used. The initial concentration of NaCl was c_s = 10 g/L. The c_{out} data was measured and the c_{int} values calculated by the mass balance Eq. (24). The c_{int} data was extrapolated to an infinite bath size (*dashed lines*) to obtain data comparable with the Donnan theory and the simulations

in the gel (c_{int}) was calculated from the concentration in the supernatant phase (c_{out}) via a mass balance (Eq. 24). Ideally, one would like to immerse the gel in an infinite bath, so preferential absorption of water by the gel doesn't affect the concentration of the surrounding solution. On the other hand, to be able to compute the salt concentration in the gel, c_{int}, by mass balance, it is desired that there is a well measurable difference between the initial salt concentration, c_s, and the concentration in the solution phase after swelling, c_{out}. To overcome this problem, we determined c_{out} and subsequently also c_{int} using Eq. (24) for several bath sizes. c_{int} as a function of the bath size was extrapolated to an infinite bath (large RWC values). An example of this procedure is given by the dashed lines in Fig. 8 for gels of different cross-linking degrees (DC = 0.1, 0.3, 1, 3 mol-%) at fixed values of DN = 75 mol-% and Q_{syn} = 4 g/g. A values of RWC = 10 is in its effect already close to an infinite bath. As assumed, the external concentrations increase with smaller bath sizes (smaller RWC values).

For comparison with the Donnan theory (next section), the extrapolated values of salt inside the gel were plotted as a function of fc_p in Fig. 9 on the left. It becomes clear that concentration inside the gel was lower at high values of DC.

The same data were measured for hydrogel samples with different values for DN or Q_{syn}. The extrapolated values are shown in Fig. 9 on the right. For Q_{syn} four different samples were investigated at a fixed DC = 1 mol-% and DN = 75 mol- %. The values of Q_{syn} were 1.5, 2, 4, 8 g/g and spanned roughly the entire synthetically accessible range. In the case of DN samples with values of 0, 25, 50, 75, and 95 mol-% were investigated while DC = 1 mol-% and Q_{syn} = 4 g/g were fixed. However, it must be noted that because the DN influences the kinetics of the polymerisation, the obtained gels might not be exactly comparable in terms of their structure.

Comparison with the Donnan Theory

The distribution of an added electrolyte between the gel and surrounding solution phase can be calculated from the Donnan theory using Eq. (5). In Fig. 9 we compare this prediction with the experimental data described above. The three data series given in Fig. 9 with varying DN, DC, and Q_{syn} show similar trends. The decreasing of c_{int} with increasing fc_p predicted by the theory is observed unambiguously in two of the three series (varying DC and varying Q_{syn}). The irregular trend with increasing DN should be discussed in more detail. The apparently outlying data point at very low fc_p corresponds to DN = 0 mol-% and $Q_{eq} \approx 5$. In the uncharged gel at such low swelling, the interactions are dominated by the excluded volume, i.e. a completely different mechanism than considered in the Donnan theory. The other points correspond to higher values of DN. With the exception of the one data point discussed above, in all other cases Eq. (5) predicts significantly lower salt content inside the gel than found experimentally. It is noteworthy that qualitatively very similar discrepancy was observed between the prediction of Eq. (5) and simulation data in Fig. 15. This discrepancy appears to be related to assumptions involved in the derivation of Eq. (5) and its origin is discussed in detail in section "Simulation Results".

Another view of the above results would be that at fixed DN the charge density inside the gel structure is controlled by the swelling capacity. The lower the swelling state the larger is the density of bound ions from the polymer. This results in strong separation of mobile ions between the phases and a low internal salt concentration. A low swelling state and, therefore, a good separation is either introduced by high DC or low Q_{syn}.

Salt Distribution by Forced Deswelling

The separation without pressure, as described above, was mainly a function of the swelling state of the gel, which can be altered by a change in the surroundings (e.g. pressure, temperature). Here, the distribution of electrolyte under load applied with the filter press setup described in section "Design and Use of the Press Setup" was assessed.

Influence of the Degree of Cross-Linking (DC)

The desalination experiments under pressure were carried out for gels with three different DC values of 0.1, 1, and

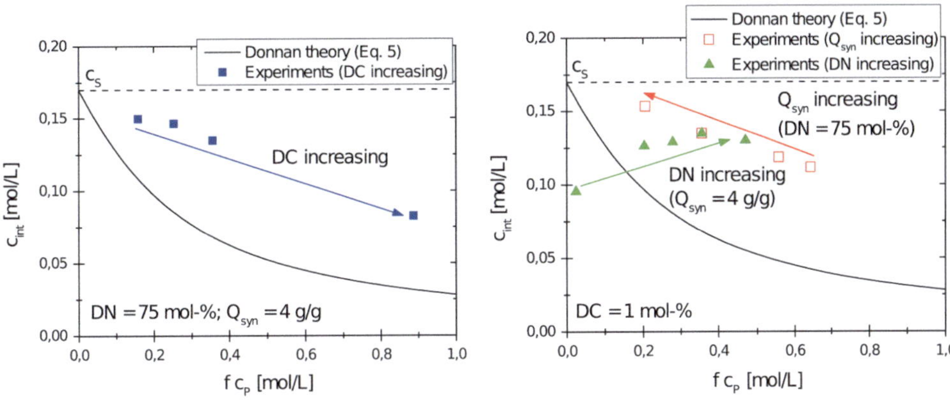

Fig. 9 The concentration of salt taken up by different gels, c_{int}, swollen in excess salt solution with $c_s = 10$ g/L NaCl, as a function of the concentration of bound charges within the network, fc_p, is given. The Donnan theory Eq. (5) (*solid line*) is compared with three data series: (i) *full squares*: fixed DN = 75 mol-% and $Q_{\text{syn}} = 4$ g/g with DC increasing from *left* to *right*; (ii) *open squares*: fixed DC = 1 mol-% and $Q_{\text{syn}} = 4$ g/g with Q_{syn} decreasing from *left* to *right*; (iii) *full triangles*: fixed DC = 1 mol-% and DN = 75 mol-% with DN increasing from *left* to *right*, note $f \approx DN/100$

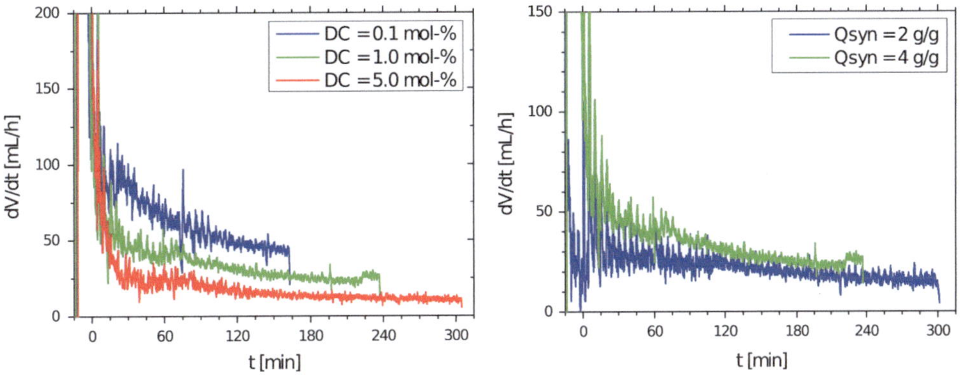

Fig. 10 The results for the flux from hydrogels deswollen by pressure as a function of time is shown. On the *left*, the flux is given for gels with different degrees of cross-linking (DC) with fixed $Q_{\text{syn}} = 4$ g/g and on the *right* the flux is shown for different values of Q_{syn} and a fixed DC of 1 mol-%

5 mol-% of cross-linker. The other synthetic values were fixed at DN = 75 mol-% and $Q_{\text{syn}} = 4$ g/g. The experiments were conducted until maximum pressures of 60, 50, and 34 bar, respectively, were reached. For the DC = 0.1 and 1 mol-% samples a high amount of eluate was pressed out ($V_e \approx 370$ mL) while significantly less was recovered for the highest cross-linked sample (V_e about 300 mL) even though the highest pressure was applied.

The eluate flux from the gel produced by the pressure increase was strongly dependent on DC. For lower values of DC, the flux was higher as shown in Fig. 10 on the left. This result was expected as a higher DC corresponds to mechanically stronger gels that need a higher pressure to deform and deswell. However, in the softer gels the flux decreased faster.

The resulting salt concentration was measured in the obtained fractions and is shown in Fig. 11 as a function of the applied pressure and swelling of the gel relative to the free equilibrium state (RWC). In the beginning of the desalination experiment, the same concentration values were found without applying pressure as described above. As the experiment progressed, the concentration of the eluate was reduced and dropped below c_s over the pressure range between 10 and 20 bar. The reduction of the salt concentration was stronger for higher values of DC in the samples under investigation. Roughly speaking, the lower the concentration is at a given pressure the more efficient is the desalination. The result of the sample with DC − 5.0 mol-% intersected both curves at 23 and 37 bar, respectively. Which makes it more efficient at high applied pressures.

With respect to the relative water content, the drop in concentration only occurred when RWC was smaller than unity. Then, all supernatant solution was removed from the sample. We observe for all experiments that the concentration of the eluate drops below $c_s = 10$ g/L at the same deswelling ratio ($\approx$0.7). At this point we are unable to say if this is a universal feature. In general, a lower swelling state of the gel, results also in a lower salt concentration in the produced eluate.

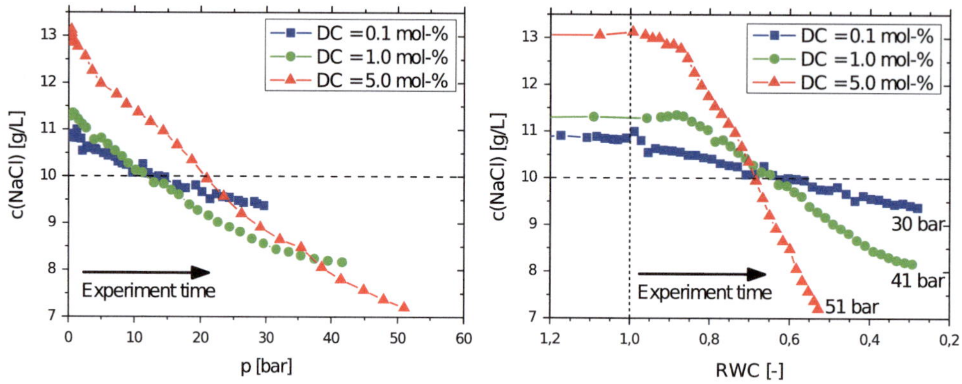

Fig. 11 Pressure driven deswelling results for gels with different cross-linking degrees (DC) as a function of pressure (*left*) and relative water content RWC (*right*) are shown. The concentration of the obtained eluate decreases with increasing applied pressure and decreasing RWC

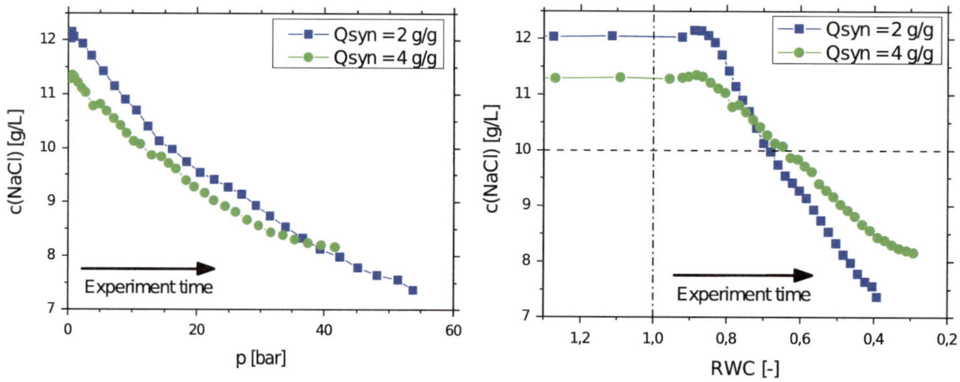

Fig. 12 The results of pressure driven deswelling of gels with different values of Q_{syn} and fixed DC = 1 mol-% are shown. The concentration of the obtained eluate is given as a function of the applied pressure (*left*) and the swelling of the gel relative to the free equilibrium state (relative water content, RWC, *right*)

A high DC resulted, at the end of the experiment, in the lowest salt concentration in the eluate but also the lowest overall flux. This led to the conclusion that the first part of the collected eluate was actually part of the supernatant phase. After the external solution had been removed, water with a lower salt content was released which originated from the inside of the gel particles. The transition between these two cases was gradual because water from both phases mixed at the outlet of the press. Also, the more the gel was compressed, the lower was the salt concentration inside the gel.

Influence of the Swelling State at Synthesis (Q_{syn})

The network structure is fixed by the cross-linking reaction at the time of synthesis and does not change later. Networks with the same values of DC and DN but different values of the swelling state at the time of synthesis, Q_{syn}, may differ in the aspects of topology such as the amount of dangling chain ends and chain length distribution [6]. Therefore, Q_{syn} is an important parameter which determines the swelling capacity of the gel. The higher Q_{syn} the higher is the swelling capacity. Of the four samples for which the salt distribution was investigated without the application of pressure, only two gels were investigated under pressure (Q_{syn} = 2 and 4 g/g).

For the sample Q_{syn}= 2 g/g, the experiment was carried out up to a pressure of 59 bar. The results were compared with the previously shown sample (Q_{syn} = 4 g/g). V_e was lower for the 2 g/g sample, as only 320 mL were retained at the maximum pressure. The flux shown in Fig. 10 on the right gives a similar picture. However, the flux for Q_{syn} = 2 g/g was more stable over time as compared to the 4 g/g. As a result, the rates were almost similar after 4 h. Although, the salt concentration decreased in a qualitatively similar way for both samples as seen in Fig. 12, the extent was different. The drop in the concentration was steeper for the lower Q_{syn} sample but the starting point was about 0.8 g/L higher. As a result, the intersection of the two curves was around 40 bar. Only above this pressure was the desalination performance better for Q_{syn}= 2 g/g. On the right of Fig. 12 it is shown that the water coming from the gel contained much less salt for the same degree of deswelling at the lower Q_{syn} value, proving that the salt rejection was higher while samples with a lower Q_{syn} were better at rejecting the salt, the reduced flux

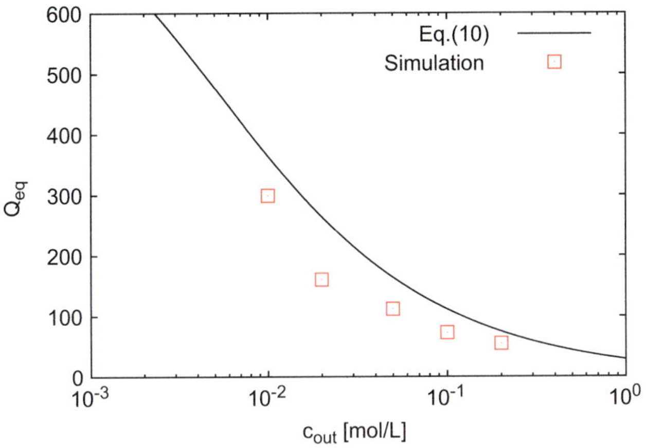

Fig. 13 Equilibrium swelling ratio of a simulated hydrogel with $N = 59$ and $f = 0.33$ compared to the numerical solution of Eq. (10) with $p_{ext} = 0$ and using $Q_{eq} = (16N + 8)/(c_P V_0)$

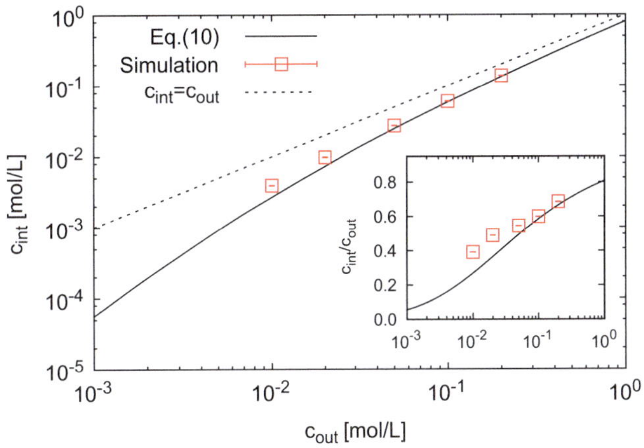

Fig. 14 The salt content in the simulated hydrogel with $N = 59$ and $f = 0.33$ at swelling equilibrium at several outside salt concentrations, compared to the numerical solution of Eq. (10). The inset shows same data plotted as c_{int}/c_{out} as a function of c_{out}

meant that lower salt concentrations were only found at high pressure.

Simulation Results

We performed a series of simulations based on the idealised hydrogel model described in section "Simulation Model and Method". The purpose of the simulations was to investigate the ability of the simple theory from section "Theoretical Treatment", Eq. (10) to describe real (model) data with known structural parameters. The simulations presented here involve the degrees of charging $f = \{0.125, 0.25, 0.33, 0.50\}$ and polymer chain lengths of $N = \{39, 59, 79\}$ monomer units which correspond to cross-link densities (DC) of 2.5, 1.7, and 1.25 mol-%, respectively. The model parameters map on the experimental parameters as $2f \approx \mathrm{DN}/100$ and $N \approx 1/\mathrm{DC}$. It should be noted that the mapping is not exact since the model assumes regular charge distribution on the chain and regular topology with all chains of the same length. For this reason there is no counterpart in the simulation for Q_{syn} because it mainly affects the topology. The experimental system inevitably contains polydisperse chains, irregularities in topology due to incomplete reactions and inhomogeneously distributed charges on the chain. The values of N used in the simulations correspond to the upper range of the experimentally studied DC, while f in the simulations corresponds to the lower range of the experimental DN. In this context it should be noted that in the experiments no significant differences were observed among polymers with DN > 50 mol-%.

We begin the discussion of simulation results by investigation of the free swelling equilibria (without applied external pressure). Figure 13 presents the swelling ratio of the simulated hydrogel with $N = 59$ and $f = 0.33$ as a function of the salt concentration of the infinite bath in which the hydrogel is immersed. The simulation data are compared to Eq. (10) (solid line). The comparison reveals good semi-quantitative agreement. The theoretical prediction captures well the trend in the simulation data but in quantitative terms it overestimates the swelling as measured in the simulation by up to 30 %. This is a surprisingly good agreement if we take into account the rough approximations used in deriving Eq. (10). This success might be attributed to cancellation of various non-ideal contributions which has been discussed in earlier simulation studies [22, 23]. The higher disagreement at low c_{out} could come from the fact that the Gaussian approximation underestimates the elastic force at high swelling when the limit of finite extensibility of the chains is approached.

Figure 14 shows the relation between the internal salt concentration inside a swollen hydrogel and the salt concentration of the bath for the same system as in Fig. 13. Comparison with the prediction of Eq. (10) reveals that the internal salt concentration from the simulation is somewhat higher than the theory predicts. The inset in Fig. 14 shows the same data plotted as c_{int}/c_{out}. Here, the disagreement between the simulation data and the theory becomes more apparent. Relative deviations observed in Fig. 13 are comparable to those in Fig. 14. At a first glance, this might suggest that the theory quite well predicts the gel swelling. However, the deviations in Fig. 14 are in the opposite direction than would be expected from overestimation of Q_{eq} as revealed by Fig. 13. It should lead to an underestimation of fc_p which in turn should translate to an overestimation of c_{int}.

The above discussion suggests that there are some more subtle effects that are not correctly captured by the

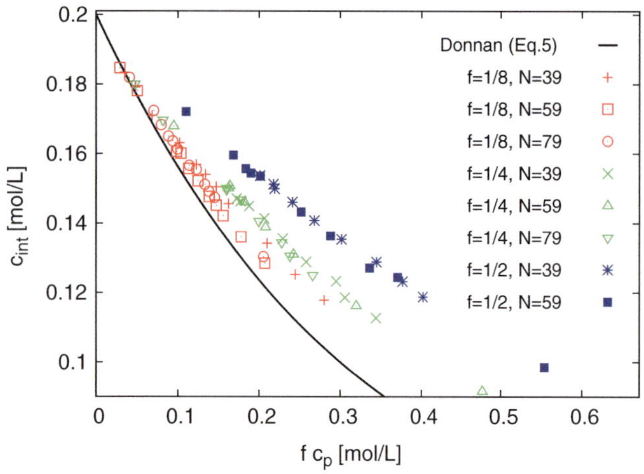

Fig. 15 The salt concentration inside the hydrogel, c_{int}, as a function of concentration of charged groups on the polymer fc_{p} for a series of gels with different parameters as indicated in the legend. The external salt concentration was $c_{\text{out}} = 0.2\,\text{mol/L}$. Simulations were performed at gel volumes smaller or equal to the equilibrium swelling which corresponds to an applied external compression force. The *solid curve* is the prediction of the Donnan theory (Eq. (5))

simple Donnan mean-field theory, beyond the Gaussian chain stretching approximation. Several contributions to the non-ideal behaviour of gel swelling have been thoroughly examined in earlier simulations [22, 23, 26]. Cancellation of various contributions has been shown to result in nearly ideal swelling response even in systems with significant counterion condensation. To our knowledge, only one simulation study [19] has examined PE gel swelling in a salt solution. However, they have not studied the salt partitioning between solution and the gel. In the following, we make the first attempt to do so.

In the previous figures, each data point was obtained from several simulations at varying gel volumes in order to obtain equilibrium swelling. To simplify the approach and reduce computational demands, we took a different approach to examination of the ion partitioning. We made use of all simulation runs at fixed gel volumes which were smaller or (approximately) equal to the a posteriori determined equilibrium volume. These correspond to gels under applied external pressure. We simulated all these gels in equilibrium with a salt bath at a fixed concentration, $c_{\text{out}} = 0.2\,\text{mol/L}$. We note that this deviates from the real experimental protocol, where the supernatant solution was being removed continuously, and consequently at every instant the gel was equilibrated against a solution with a slightly different salt concentration. Our protocol, however, allows for an easier comparison to the Donnan theory.

In Fig. 15 we show the internal salt concentration, c_{int} in gels under compression and in equilibrium with a salt bath at $c_{\text{out}} = 0.2\,\text{mol/L}$. Figure 15 contains a series of different gels with varying $N = \{39, 59, 79\}$ and $f = \{0.125, 0.250, 0.50\}$ as indicated in the legend. The data are plotted as a function of total concentration of charges bound to the hydrogel, fc_{p}, which is the only parameter through which the polymer structure enters Eq. (5). Clearly, deviations from the prediction of Eq. (5) (solid line) are observed. Furthermore, data sets for the same f (same colour) seem to collapse on the same universal curves, with no significant dependence on N. For any f, the simulation data are quite close to the Donnan prediction at small values of fc_{p}, but they deviate more and more as fc_{p} increases. Furthermore, increasing f leads to stronger deviations from the Donnan prediction. We also note, that at equilibrium swelling, which for all studied gels and the given c_{out} occurs at $fc_{\text{p}} \in [0.05 : 0.20]\,\text{mol/L}$, the Donnan theory provides a reasonable approximation for all investigated f, and an excellent approximation for $f = 1/8$. Under compression (high fc_{p}), the theory is still quite good for the gels with low degree of charging while it significantly deviates for gels with a high degree of charging. A similar behaviour was observed experimentally in Fig. 9, where the Donnan theory also strongly underestimates c_{int} and the agreement becomes worse for higher fc_{p}. The Donnan theory should be expected to fail at high ionic strength, since in the current form of Eq. (5) it uses concentrations rather than activities. The observation that the data for the same f collapse on a single curve independent of N suggests that it is the linear charge density on the chains, f, rather than the total ionic strength in the gel, $2fc_{\text{p}} + c_{\text{int}}$, which causes the deviations from ideality. Therefore, in order to augment the Donnan theory and make it applicable to gels under compression, it should not suffice to consider the dependence of activity coefficients on concentration within the framework of the Debye-Hückel theory (Eq. (14)) or a similar more advanced approximation. Instead, one would need to consider inhomogeneity of charge distribution around the polymer, e.g. within the framework of the Poisson-Boltzmann cell model [18]. This approach will be pursued in further work on this problem.

Conclusion

To conclude, we demonstrated how hydrogels can be used for desalination purposes, and presented an exploratory experimental setup with tailored hydrogels. Our current theoretical understanding allows for a qualitative description of this process, but is not yet able to predict the experimental curves quantitatively.

In the present work we have shown that the desalination capacity of a hydrogel depends severely on its composition. Namely, the effective charge density per volume inside the hydrogel (fc_{p}) controls how much salt can enter the hydrogel from the surrounding bath.

The charge density can be easily tuned by the swelling state of the gel. We find that a low swelling state is favourable for good separation. If no pressure is applied, this state is solely a function of the chemical composition and can be tuned by the synthesis parameters. Consequently, a high degree of cross-linking (e.g. DC = 5 mol-%) and a low water content during the synthesis (e.g. Q_{syn}= 2 g/g) are favourable. The degree of neutralisation had relatively little influence on the results if it was greater than 25 mol-%. The separation experiments without applied pressure already gave a good indication of the desalination capacity.

The gel can be compressed by applying pressure with a specialised experimental setup introduced here. This allows to recover most of the salt-depleted water from the gel phase. The lowest concentrations in this eluate, which correspond to the best desalination under pressure was found for gels with high salt rejection without load.

From the experiments and simulations it follows that a higher backbone charge density in the network leads to a smaller salt concentration in the gel phase. The mean-field Donnan theory is able to qualitatively describe this separation process. However, it becomes worse for increasing linear charge density f, and, in fact, systematic deviations between the simulation and the theory can be observed for increasing values of f. This trend is also present in the experimental data that show increasing deviations for increasing charge densities. The simulations suggest different dependencies of salt partitioning for different charge densities on the polymer chain, which appear to be independent of chain length. This behaviour is not captured by the Donnan theory which predicts a single universal dependence of salt partitioning as a function of the concentration of bound ions only. Quantitatively the theory overestimates the equilibrium swelling by up to 30 % and overestimates the efficiency of partitioning of small ions by roughly the same factor. Donnan theory works well at low concentration of bound ions in the gel. A more involved theory should incorporate the spatial inhomogeneity of the charge distribution.

Future work on this subject will focus on further experimental investigation of process parameters and on improving the theoretical description in order to better reproduce the simulation data. Ultimately, a better understanding of the fundamental processes could enable an evaluation of the energy balance of the process which is important for the anticipated industrial applications.

Acknowledgements The authors gratefully acknowledge the financial support of the DFG priority program 1259 "Smart Hydrogels". J.H. acknowledges additional financial support from the Heinrich Böll Stiftung and would like to thank the workshop team as well as Ms. Schlag, Mr. Merger, and Dr. Kübel for help on the project and this article. M.W. thanks his children for this idea while diapering them. P.K. acknowledges support from grant "LK21302 Návrat" from the Czech Ministry of Education.

Appendix: Experimental Data Analysis

Volume and Flux Data

The distance between the sieve units $d(t)$ was measured by a distance gauge. The data points were directly stored on a computer using the program `MarCom` (version 2.0, Mahr GmbH) at typically 1 Hz. The cumulated eluate volume V_e was computed by Eq. (25) using the starting point of the eluate discharge d_0 and the press cross-section $A = 11,300\,\text{mm}^2$

$$V_e(t) = \frac{(d_0 - d(t))A}{1000} + V_{\text{Offset}}. \quad (25)$$

The offset V_{Offset} was added to account for the amount of solution removed from the sample before loading into the press (step 2 in Fig. 1). However, air bubbles are also trapped in the setup at the beginning of the experiment. When the pressure was raised, these bubbles were eventually released with the eluate at up to 10 bar. This caused a decrease in the distance but no volume discharge. The eluate volume was therefore overestimated by the distance data. It was corrected by the the mass of the fractions (Eq. 26), which was not affected by this problem

$$V_i(t) = \sum_i \frac{m_i}{\rho_E}, \quad (26)$$

where m_i are the masses and ρ_E the density of the fractions. The fraction data contained fewer data points than the distance data. Therefore, the fractions were only used to renormalise the volume data while retaining the local fluctuations. An example for the two obtained datasets is shown in Fig. 16.

The flux from the press was calculated from the cumulated volume using a numerical derivative with time (27). However, the values of $d(t)$ showed fluctuations and direct differentiation was not advised. This was overcome by smoothing the distance data using a simple moving average (SMA) algorithm [44]. A 13 pt window width was found to be sufficient but for improved visual inspection an increased smoothing, using 61 pt was found to be beneficial (e.g. as shown in Fig. 10)

$$\dot{V}_n(t) = \frac{\mathrm{d}V_n(t)}{\mathrm{d}t} \approx \frac{\Delta d(t)A3,600}{1000 t_{\text{step}}}. \quad (27)$$

The degree of swelling of the gel at a given time can be calculated from the volume data (28) if the initial solution in the system m(Eluate) and the polymer mass m(Polymer) are known

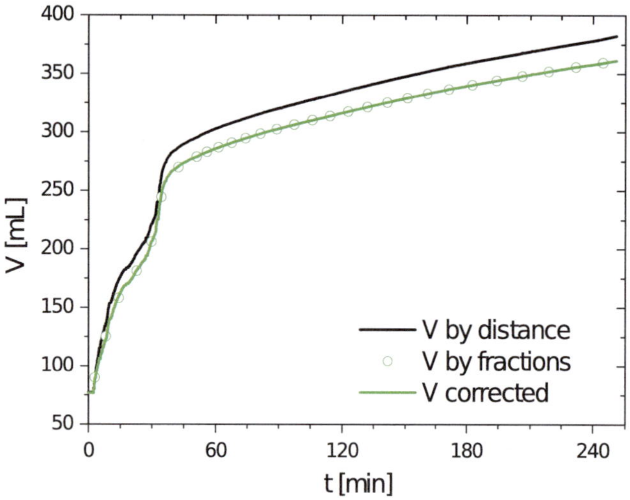

Fig. 16 Typical volume data obtained in a desalination experiment are shown. The cumulative eluate volume (V by distance) as calculated by Eq. (25) and V by fractions calculated by Eq. (26) are given. The corrected data are shown as a function of time by the *green line*

$$\mathrm{RWC}(t) = \frac{Q(t)}{Q_{\mathrm{eq}}} = \frac{m(\mathrm{eluate}) - V(t)\rho_E}{m(\mathrm{polymer})}\frac{1}{Q_{\mathrm{eq}}}. \tag{28}$$

Pressure Data

The pressure sensor was connected and powered by a home-made circuit. The voltage data $U(t)$ was sampled by an ADC (USB-9215, National Instruments, Austin, USA) and collected on a computer by a self-written `LabVIEW`-program (National Instruments). The sampling rate was typically 1,000 Hz and the measured points were decimated and increased in resolution by online oversampling over 250 pt to 4 Hz [45]. The obtained voltage was linear to the applied pressure. The data was baseline corrected with the sensor's signal at zero pressure (U_0) and converted to excess pressure $p(t)$ by Eq. (29). The respective proportionality factor f depends on the connection circuit used

$$p(t) = \frac{U(t) - U_0}{f}. \tag{29}$$

Concentration Data

The conductivity $\sigma(t)$ in mS/cm and the temperature T in °C of the eluate were acquired. The acquisition was done once for each fraction obtained or every 3 s if the flow cell was used. The data was recorded on a computer with the program `LabXpress` (version 2.1, Mettler Toledo). The conductivity data was corrected for thermal effects at the acquired temperature. The corrected value at the reference temperature, $T_{\mathrm{ref}} = 20\,°\mathrm{C}$, was computed using a linear relationship (30). The coefficient α is in the order of 2 %/°C and must be chosen appropriately for each concentration and temperature range. Values can be found in the literature [46]

$$\sigma(T_{\mathrm{ref}}) = \sigma(T)\left(1 + \frac{\alpha}{100\%}(T - T_{\mathrm{ref}})\right). \tag{30}$$

The concentration c of salt in the solution can be computed in g/L (NaCl) using an empirical relation (31) when only one particular electrolyte is present. The coefficients of the third order polynomial were obtained by adjustment to literature data [47]

$$c = A + B\sigma(T_{\mathrm{ref}}) + C\sigma(T_{\mathrm{ref}})^2 + D\sigma(T_{\mathrm{ref}})^3. \tag{31}$$

Typical coefficients for this conversion for $T_{\mathrm{ref}} = 20\,°\mathrm{C}$ and c in the range of 1–20 g/L were as follows

$$A = 0.0343\,\mathrm{g/L} \tag{32}$$

$$B = 0.579\,\mathrm{g\,cm/L\,mS} \tag{33}$$

$$C = 3.23 \times 10^{-3}\,\mathrm{g\,cm^2/L\,mS^2} \tag{34}$$

$$D = -5.87 \times 10^{-6}\,\mathrm{g\,cm^3/L\,mS^3}. \tag{35}$$

Error Estimation and Reproducibility

The desalination experiment is a complex procedure where the overall error as a result of the combination of many points and channels is difficult to assess. As an alternative, four experiments were carried out with the same parameters and polymeric hydrogel to obtain information on the general reproducibility of a complete experiment (not shown).

The experiments show qualitatively the same results but with some quantitative differences. The deviation between the single runs was quite small in the beginning of the experiments and increased over longer times. It was seen that deviations in the pressure especially had a considerable influence on the results. The pressure increase $\mathrm{d}p/\mathrm{d}t$ could usually be controlled within ± 1 % of the desired slope. For the produced volume, a deviation of ± 3, ± 5, and ± 10 mL was found at 3, 18, and 42 bar, respectively. For the eluate concentration, a deviation at the same points of ± 0.1, ± 0.25, and ± 0.4 g/L were found. The uncertainty of the data in $\mathrm{d}c/\mathrm{d}p$ was found to be within 10 %.

References

1. Buchholz FL, Graham AT (1998) Modern superabsorbent polymer technology, 1st edn. Wiley VCH, New York
2. Kazanskii KS, Dubrovskii SA (1992) Adv Polym Sci 104:97
3. Hogari K, Ashiya F (1994) In: Buchholz FL, Peppas NA (eds) Superabsorbent polymers science and technology. ACS symposium series, vol 573, 1st edn. American Chemical Society, Washington, DC, pp 128–140

4. Masuda F (1994) In: Buchholz FL, Peppas NA (eds) Superabsorbent polymers science and technology. ACS symposium series, vol 573, 1st edn. American Chemical Society, Washington, DC, pp 88–98
5. Wiertz P (2011) Sustainability report, 3rd edn. EDANA, Brussels
6. Flory PJ (1953) Principles of polymer chemistry. Cornell University Press, Ithaca
7. Katchalsky A, Michaeli I (1955) J Polym Sci 15:69
8. Höpfner J, Klein C, Wilhelm M (2010) Macromol Rapid Commun 31:1337
9. Cath TY, Childress AE, Elimelech M (2006) J Membr Sci 281:70
10. Kravath RE, Davis JA (1975) Desalination 16:151
11. Fritzmann C, Löwenberg J, Wintgensand T, Melin T (2007) Desalination 216:1
12. Khawaji AD, Kutubkhanah IK, Wie JM (2008) Desalination 221:47
13. McCutcheon JR, McGinnis RL, Elimelech M (2005) Desalination 174:1
14. McGinnis RL, Elimelech M (2007) Desalination 207:370
15. Li D, Zhang X, Yao J, Simon GP, Wang H (2011) Chem Commun 47:1710
16. Li D, Zhang X, Yao J, Zeng Y, Simon GP, Wang H (2011) Soft Matter 7:10048–10056
17. Victorov A, Radke C, Prausnitz J (2005) Phys Chem Chem Phys 8:264
18. Claudio GC, Kremer K, Holm C (2009) J Chem Phys 131:094903
19. Edgecombe S, Schneider S, Linse P (2004) Macromolecules 37(26):10089
20. Victorov AI (2006) Fluid Phase Equilib 241:334
21. Flory JP, Rehner J (1943) J Chem Phys 11:512
22. Mann BA, Holm C, Kremer K (2005) J Chem Phys 122(15):154903
23. Mann BA (2005) The swelling behaviour of polyelectrolyte networks. Ph.D. thesis, Johannes Gutenberg-University, Mainz
24. Schneider S, Linse P (2002) Eur Phys J E 8:457
25. Yan Q, de Pablo JJ (2003) Phys Rev Lett 91(1):018301
26. Mann BA, Everaers R, Holm C, Kremer K (2004) Europhys Lett 67(5):786
27. Košovan P, Richter T, Holm C (2013) Molecular simulations of hydrogels. In: Sadowski G, Richtering W (eds) Intelligent hydrogels. Progress in colloid and polymer science, vol 140. Springer, Switzerland, pp 45–51
28. Stevens MJ, Kremer K (1993) Macromolecules 26:4717
29. Holm C, Hofmann T, Joanny JF, Kremer K, Netz RR, Reineker P, Seidel C, Vilgis TA, Winkler RG (2004) Adv Polym Sci 166:67
30. Mann BA, Holm C, Kremer K (2006) Macromol Symp 237:90
31. Mann BA, Lenz O, Kremer K, Holm C (2011) Macromol Theory Simul 20:721. Cover issue
32. Widom B (1963) J Chem Phys 39:2802
33. Truesdell AH (1968) Science 161(3844):884
34. Weeks JD, Chandler D, Andersen HC (1971) J Chem Phys 54:5237
35. Limbach HJ, Arnold A, Mann BA, Holm C (2006) Comput Phys Commun 174(9):704
36. Arnold A, Lenz O, Kesselheim S, Weeber R, Fahrenberger F, Roehm D, Košovan P, Holm C (2013) In: Griebel M, Schweitzer MA (eds) Meshfree methods for partial differential equations VI, vol 89. Springer, Heidelberg/New York, pp 1–23
37. Espresso homepage, http://www.espressomd.org
38. Deserno M, Holm C (1998) J Chem Phys 109:7678
39. Oppermann W (1992) Swelling behavior and elastic properties of ionic hydrogels. In: Harland RS, Prud'homme RK (eds) Polyelectrolyte gels – properties, preparation, and applications, 1st edn. American Chemical Society, Washington, DC, p 159
40. Umendra D, Metha SK, Choudhary MS, Jain R (1999) Rev Macromol Chem Phys C39:507
41. Vervoort S, Patlazhan S, Weyts J, Budtova T (2005) Polymer 46:121
42. Brendel U (1999) Quellungsverhalten und mechanische eigenschaften von polyelektrolytnetzwerken und technischen superabsorbern. Ph.D. thesis, Johannes Gutenberg-Universität, Mainz
43. Höpfner J (2013) A new method of seawater desalination via acrylic acid based hydrogels: synthesis, characterisation, and experimental realisation. Ph.D. thesis, Karlsruher Institut für Technologie (KIT)
44. Skoog DA, Holler FJ, Crouch SR (2007) Principles of instrumental analysis, 6th edn. Thomson, Belmont
45. van Dusschoten D, Wilhelm M (2001) Rheol Acta 40:395
46. Horvath AL (1985) Handbook of aqueous electrolyte solutions, 1st edn. Ellis Horwood Ltd., Chichester
47. Wolf AV, Brown MG, Prentiss PG (1986) In: Weast RC (ed) CRC handbook of chemistry and physics, 67th edn. CRC Press, Boca Raton, pp D–253 ()

Swelling Behaviour of Functionalized Hydrogels for Application in Chemical Sensors

Margarita Guenther, Thomas Wallmersperger, Karsten Keller, and Gerald Gerlach

Abstract

Thin films of metabolite-specific biocompatible hydrogels were combined with microfabricated piezoresistive pressure transducers to obtain chemomechanical sensors that can serve as selective biochemical sensors for a continuous monitoring of metabolites. The gel swelling pressure has been monitored in simulated physiological solutions by means of the output signal of piezoresistive sensors. Gel response time and accuracy with which hydrogels can track gradual changes in glucose, fructose, pH and CO_2, respectively, were estimated. A significant reduction of the sensor response time has been achieved due to hygroscopic fibres incorporated in the hydrogel layer.

Keywords

Polyelectrolyte hydrogel • Analyte-specific swelling behavior • Piezoresistive microsensor • Chemical sensor

Introduction

Hydrogels are three-dimensionally cross-linked hydrophilic polymer networks swollen in water which are able to undergo a volume phase transition under the influence of environmental changes. The amount of solvent uptake depends on the polymer structure and composition, and can be made particularly responsive to environmental factors, such as temperature, pH-value, ion concentration, salt concentration, analyte molecule concentration in solutions, and solvent composition. The ability to customize the hydrogels' chemical and mechanical characteristics allows for the synthesis of hydrogels optimized for detection of particular target species in a physiological concentration range. In the presence of an analyte, these polymers alter their swelling properties, either by ionization or by formation of analyte-mediated reversible crosslinks.

Hydrogel-based sensors generally consist of two main components, the polymeric hydrogel used for the biochemical detection and a transducer. Transducers involving mechanical [1–6], electrochemical (e.g., capacitive [7], conductometric, amperometric, potentiometric [8]) and optical [9] mechanisms are used to monitor changes in hydrogel volume and structure. The pH-, ionic strength- and glucose-sensitive hydrogels undergo reversible volume transitions with environmental changes. These volume transitions produce swelling pressures which are the basis for the chemical–mechanical transduction of piezoresistive hydrogel-based sensors [1–5]. Hydrogels can readily be miniaturized and directly integrated into piezoresistive pressure sensors. Moreover, hydrogels with various moieties can be used simultaneously in sensor arrays to detect a number of physiological analytes (e.g. glucose, pH, CO_2, ionic strength) [10].

M. Guenther (✉) • G. Gerlach
Solid-State Electronics Lab, Technische Universität Dresden, Helmholtzstr. 10, 01069 Dresden, Germany
e-mail: mguenthe@mail.zih.tu-dresden.de

T. Wallmersperger
Institute of Solid Mechanics, Technische Universität Dresden, George-Bähr-Str. 3c, 01062 Dresden, Germany

K. Keller
Institute for Statics and Dynamics of Aerospace Structures, Universität Stuttgart, Pfaffenwaldring 27, 70569 Stuttgart, Germany

G. Sadowski and W. Richtering (eds.), *Intelligent Hydrogels*, Progress in Colloid and Polymer Science 140,
DOI 10.1007/978-3-319-01683-2_20,

Development of a sensor array that would continuously measure glucose, pH, and CO_2 with high long-term stability and biocompatibility, could considerably reduce the risk of blood glucose excursions outside of the desired range, including life-threatening hypoglycemic events. Furthermore, the early detection of evolving diabetic ketoacidosis would allow patients to seek help early and correct the situation, thus preventing hospitalization for this life-threatening condition. Ketoacidosis is an acute complication in type 1 diabetes, where the blood is overacidified with free fatty acids and ketone bodies, which leads to a fall of blood pH (i.e. arterial blood pH < 7.25) and thereby to metabolic decompensation. A device that would sense CO_2 concentrations would also have benefits to patients with chronic lung disease, giving them early warning of CO_2 retention and seeking treatment before the condition advanced to one that would require hospitalization.

It is well reported that fluctuations in oxygen concentration impact glucose oxidase-based sensor performances, both for amperometric enzyme electrodes [11–13] and for microfabricated sensors with incorporated glucose-sensitive polymeric hydrogels containing glucose oxidase enzyme [14].

In this context, the use of a phenylboronic acid component as a glucose-sensing moiety offers a new prospect for achieving more advanced systems. Phenylboronic acid compounds in water exist in equilibrium between the uncharged, and thus relatively hydrophobic, form and the charged, and thus relatively hydrophilic, form [15]. Upon the addition of glucose, the charged form of phenylborate can form a stable complex with glucose through reversible covalent bonding, whereas the complex between the uncharged form of phenylborate and glucose is quite unstable with a high susceptibility to hydrolysis. Because the complex formed between the charged phenylborate and glucose itself is also anionically charged, the further addition of glucose induces a shift in the above-described equilibrium in the direction of the increasing charged (hydrophilic) forms of phenylborate and vice versa. Consequently, by introducing this property of phenylborates into a polymer gel structure, one obtains a functional polymer gel that undergoes a reversible volume change (volume phase transition) driven by the change in the counterions' osmotic pressure synchronized with the change in glucose concentration [16].

The osmotic pressure due to the concentration variation of ionized polymer groups and due to the distribution of mobile ions as well as the locally different gel swelling and shrinking were numerically investigated in [17, 18] by applying the coupled chemo-electromechanical multi-field formulation for polyelectrolyte gels using the finite element method. The gel swelling behaviour was considered in dependence on the acid dissociation constant value K_a for ionisable polymer groups and on the change of the pH value or salt concentration in solution [19, 20]. In [21], an influence of the change of ambient conditions was simulated for temperature-sensitive polyelectrolyte hydrogels.

In this paper, we use glucose-sensitive enzyme-free copolymers of acrylamide and 3-acrylamidophenylboronic acid (AAm/3-AAmPBA) as well as pH-responsive hydrogels based on *N,N*-dimethylaminoethyl methacrylate (DMAEMA) as chemo-mechanical transducers in piezoresistive biochemical sensors. The glucose and CO_2 response of these hydrogels is due to the increasing fraction of charged side polymer chains in the presence of glucose and CO_2, respectively. The ionization of polymer groups increases the concentration of counter ions in the gel what in its turn leads to the increasing osmotic pressure within the gel and to the gel swelling. The gel swelling pressure has been monitored in simulated physiological solutions by means of the output signal of piezoresistive sensors and accuracy with which sensors can track gradual changes in analytes was estimated.

Experimental

Hydrogel Material Preparation

The monomers used for preparation of the gels were obtained as follows: acrylamide (AAm, Fisher Scientific), 3-acrylamidophenylboronic acid (3-AAmPBA, Frontier Scientific, Logan, UT), *N*-(3-dimethylaminopropyl acrylamide) (DMAPAAm), and *N,N*-methylenebisacrylamide (BIS, Sigma-Aldrich). Hydroxypropylmethacrylate (HPMA), N,N-dimethylaminoethyl methacrylate (DMAEMA), and tetra-ethyleneglycol dimethacrylate (TEGMA) were purchased from Polysciences, Inc. (Warrington, PA). The monomers were used as received. 2-hydroxy-4'-(2-hydroxyethoxy)-2-methyl-propiophenone (HHMP, Sigma-Aldrich), 1-vinyl-2-pyrrolidinone (v-pyrol, Sigma-Aldrich), dimethyl sulfoxide (DMSO, Sigma-Aldrich), 4-(2-Hydroxyethyl)piperazine-1-ethanesulfonic acid (HEPES, Sigma-Aldrich), ammonium peroxydisulfate (APS, Sigma-Aldrich), *N,N,N',N'*-tetramethylethylenediamine (TEMED, Sigma-Aldrich), D(+)-glucose (Roth), and Dulbecco's phosphate-buffered saline solution (1X PBS, Sigma- Aldrich) were also used as received.

Glucose-Sensitive AAm/3-AAmPBA/BIS Hydrogels

A cross-linked glucose-sensitive copolymer of acrylamide and 3-acrylamidophenylboronic acid (AAm/3-AAmPBA) containing AAm/3-AAmPBA/BIS at a nominal mole ratio of 80/20/0.25 was prepared by free-radical crosslinking copolymerization as described in [1]. In brief, stock solutions of AAm and BIS were prepared in distilled water. Appropriate amounts of the two stock solutions were mixed in a vial with TEMED. In order to dissolve 3-AAmPBA into the pre-gel solution, 1 M NaOH was added into the vial. The free-radical initiator APS was introduced after purging the vial with N_2 gas for 10 min, after which the pregel solution was rapidly injected into a cavity (thickness 400 μm) between two square plates (polycarbonate and poly(methyl methacrylate)) of surface area 60 cm^2. The total monomer concentration in the pre-gel solution was 30.2 wt%. After approximately 16 h of reaction at room temperature, the hydrogel slab was removed from the mold and washed for at least 2 days with deionized water and PBS buffer (pH 7.4, ionic strength 0.15 M) before testing. In sugar-free PBS buffer at physiological pH and ionic strength, the hydrogel contains 58 wt% water. The dried hydrogel foils (thickness 330 μm) were prepared by evaporation of water at room temperature and then cut into pieces. The pieces of 3 mm × 3 mm were used for the sensor.

Glucose-Sensitive AAm/3-AAmPBA/DMAPAAm/BIS Hydrogels

Glucose-sensitive hydrogels containing AAm/3-AAmPBA/DMAPAAm/BIS at a nominal mole ratio of 80/8/10/2 were prepared by free radical cross-linking copolymerization as described in [22]. The polymer concentration in the pre-gel solution prior to curing was 20 wt%. In brief, stock solutions were prepared of AAm and BIS in 1 mM HEPES buffer. Appropriate amounts of the two stock solutions were mixed in a vial with DMAPAAm. In order to dissolve 3-AAmPBA into the pre-gel solution, 10 vol.% of DMSO was added into the vial. A mixture of two UV photoinitators (v-pyrol/HHMP = 10/1) was introduced after purging the vial with N_2 gas for 10 minutes. The weight fraction of photoinitator mixture in the pre-gel solution was 2.5 w%. The pre-gel solution was injected into the cavities of the sensor chips with the help of a micropipette (3 μl per cavity) and exposed to UV light (365 nm) in the nitrogen atmosphere.

For the preparation of the hybrid gel material, 10 ml pre-gel solution was mixed with 145 mg hygroscopic cotton fibres (width 7 μm, length 170 μm) before *in situ* polymerization. This mixture was dispensed into the chip cavities and UV-cured.

pH-Sensitive HPMA/DMAEMA/TEGDMA Hydrogels

The pH-responsive biocompatible hydrogel is based on three monomers: hydroxypropyl methacrylate (HPMA), *N,N*-dimethylaminoethyl methacrylate (DMAEMA) and tetra-ethylenglycol dimethacrylate (TEGMA). The DMAEMA monomer contains a pH-sensitive tertiary amine, HPMA was included to obtain a transition pH close to the physiological range [2] and tetra-ethyleneglycol dimethacrylate (TEGMA) acts as crosslinker.

A detailed description of the synthesis procedure of the cationic polyelectrolyte hydrogel containing HPMA/DMAEMA/TEGMA at a nominal mole ratio 70/30/2 is described elsewhere [2, 14]. In brief, appropriate amounts of monomers HPMA and DMAEMA, crosslinker TEGMA as well as accelerator TEMED were mixed in a vial to obtain pregel solution, which was then purged with N_2 gas for about 10 min. Shortly thereafter, the initiator APS was added to the pregel solution and the mixture was vortexed for about 5 s before rapidly injected into a cavity (thickness 400 ± 10 μm) between two square glass plates of surface area 64 cm^2. The hydrogel slab was removed from the glass plate after approximately 4 h of reaction, then washed with PBS solution for at least 2 days to remove unreacted chemicals prior to testing.

Solution Characterization

D(+)-glucose (Roth) and d(−)-fructose (Acros Organics) were dissolved in phosphate buffered saline (PBS, pH7.4) solutions of ionic strength $I = 0.15$ M. The pH value of PBS was adjusted with HCl or with NaOH and was measured using a Knick pH meter with pH/Pt-1000 probe.

Sensor Array Design

The silicon-based microsensor array (2 × 2) is capable of simultaneously measuring the swelling pressure from four stimuli-responsive hydrogels, and hence the concentrations of the target analytes as shown in Fig. 1. Four pressure sensors with a flexible thin silicon flexure plate and with an integrated piezoresistive Wheatstone bridge at the plate surface have been employed as mechano-electrical transducer for the transformation of the plate deflection into an appropriate electrical output signal V_{out}. The polymeric hydrogels were deposited into cavities at the backside of the silicon chip closed with a porous biocompatible and hydrophilic Anopore™ Al_2O_3 membrane (thickness 60 μm, pore size 200 nm). This membrane provides a low protein binding for permanent implants. The solution and analyte molecules diffuse into the chip cavity through the membrane, induce the swelling or shrinking processes of the hydrogel, and lead to the change in the silicon plate deflection.

The plate deflection due to the swelling pressure change ΔP causes a stress state change inside the plate and therefore a change of the resistivity of the piezoresistors affecting proportionally the output voltage V_{out} of the sensor. An increasing value of V_{out} corresponds to a hydrogel swelling, whereas a decreasing one corresponds to a deswelling. The calibration procedure of the sensor has been performed

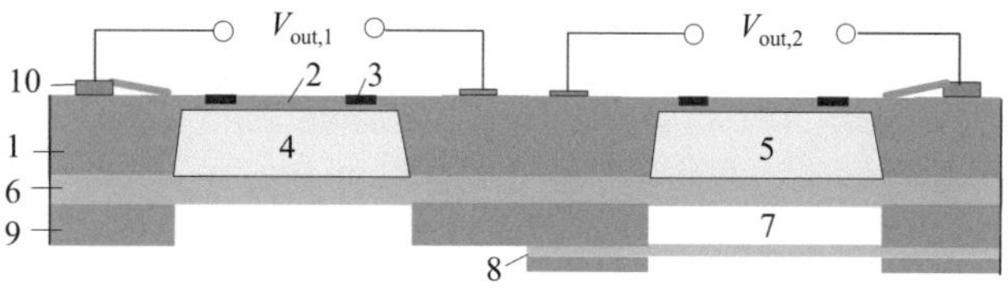

Fig. 1 Sectional view of the 2 × 2 Si chiparray (8 mm × 8 mm × 0.4 mm) with glucose (*left*) and CO_2 (*right*) sensors: *1* Si chip; *2* bending plate; *3* mechano-electrical transducer (piezoresistive bridge); *4* and *5* glucose- and pH-sensitive swellable hydrogels, respectively; *6* porous hydrophilic Anopore™ Al_2O_3 membrane; *7* reservoir with $NaHCO_3$ solution; *8* hydrophobic Severinghaus-type membrane; *9* substrate; *10* interconnect

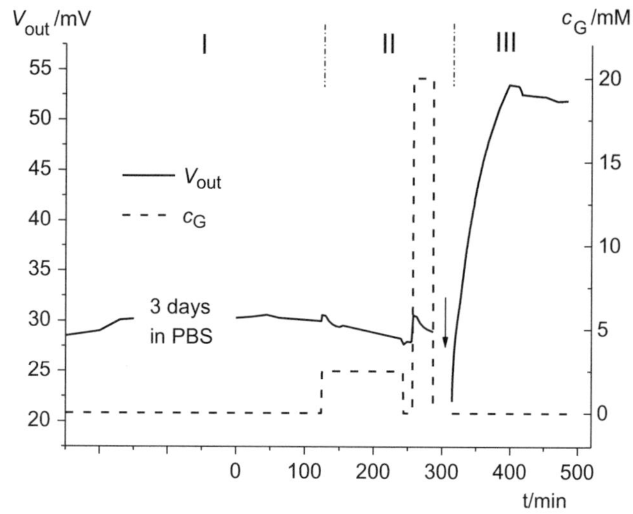

Fig. 2 Time-dependent response of the sensor completely coated with a 3 μm thick parylene C layer during 3 days in PBS solution (*I*), during the cycling between different glucose concentration c_G steps in PBS solutions (*II*) and after removing the parylene C layer from the membrane surface on the sensor's back side (*III*)

by using a pressure controller and the calibration curve $\Delta P = f(\Delta V_{out})$ of the sensor has been obtained.

The CO_2 sensor (Fig. 1, right) contains an additional hydrophobic Severinghaus-type membrane (11 μm thick Parafilm "M"). This outer membrane is permeable to CO_2 and H_2CO_3 but impermeable to ions and water molecules. A 24 mM sodium bicarbonate solution is contained in a reservoir between the outer Severinghaus-type membrane and the rigid inner porous Anopore™ membrane. Thus, the pH value in the reservoir is independent of the pH value in the surrounding fluid. When CO_2 or H_2CO_3 diffuses in or out of the reservoir, the reservoir pH value changes and induces the swelling pressure generation of a confined pH-sensitive hydrogel [5].

The assembly was wire-bonded to a printed circuit board and coated with a 3 μm thick hermetic biocompatible Parylene-C layer using chemical vapor deposition (CVD) to protect the sensor and the electrical connections during wet testing [2]. This layer was removed from membrane surfaces on the array's back side in order to allow the diffusion of analytes and solution into the chip cavities.

After the microsystem preparation, an initial gel swelling in phosphate-buffered saline (PBS, pH 7.4, ionic strength 0.15 M) was performed at room temperature for 24 h. Finally, the array was connected to an appropriate measuring set-up, which is able to monitor the output signals from multiple sensor arrays simultaneously. The measurements were performed under variable, tightly controlled ambient conditions in a temperature chamber. The temperature was controlled with a Vaisala HMP 230 humidity and temperature sensor with an uncertainty of $\pm 0.1\,K$. The uncertainty of the temperature setting was $\pm 0.5\,K$.

Since the piezoresistive pressure sensor chips showed excellent stable and dynamic properties with a response time $t < 1$ s, the long-term stability of the sensor and its response time were solely determined by the stability of the hydrogel characteristics and by the gel swelling/deswelling kinetics, respectively.

Hermeticity Test of the Parylene Encapsulation

The sensor containing the glucose-sensitive hydrogel and completely coated with a 3 μm thick parylene C layer was dipped in PBS solution for 3 days. The sensor's output voltage was monitored using an appropriate measuring set-up. No device failure and no gel swelling were observed during this time (Fig. 2, phase I) as well as during the cycling between the glucose-free and glucose-containing PBS solutions with the glucose concentration c_G (Fig. 2, phase II) confirming the hermeticity of the parylene layer.

Afterwards, this layer was removed from the membrane surface on the sensor's back side in order to allow the diffusion of analytes and solution into the chip cavity. After this, the first hydrogel swelling occurred (Fig. 2, phase III).

Only the back side of the chip comes in contact with the measuring solution, whereas the front side with the electronic components is strictly protected from it. The operational and storage stability of dip sensors coated with a parylene layer had been tested for 10 months in physiological solutions. Parylene showed excellently stable properties as a barrier layer [2].

Results and Discussion

Glucose-Sensitive Hydrogels

The cross-linked enzyme-free glucose-sensitive copolymers of acrylamide and 3-acrylamidophenylboronic acid (AAm/3-AAmPBA) were used in the glucose sensor. The gel sensitivity with regard to glucose concentration in solution was

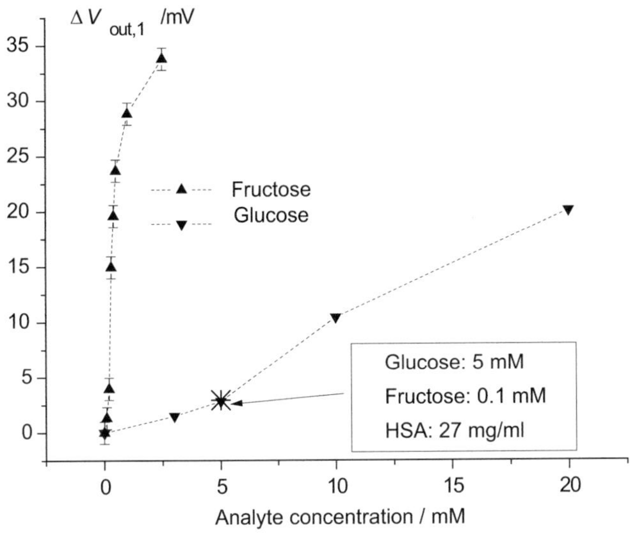

Fig. 3 Response of the sensor with AAm/3-AAmPBA/BIS hydrogel in dependence on glucose (▼) and fructose (▲) concentration, respectively, in PBS solutions (pH 7.4, ionic strength 0.15 M) at 36.5 °C and in PBS solution with $c_G = 5$ mM, $c_F = 0.1$ mM, and $c_{HSA} = 27$ mg/ml (☼)

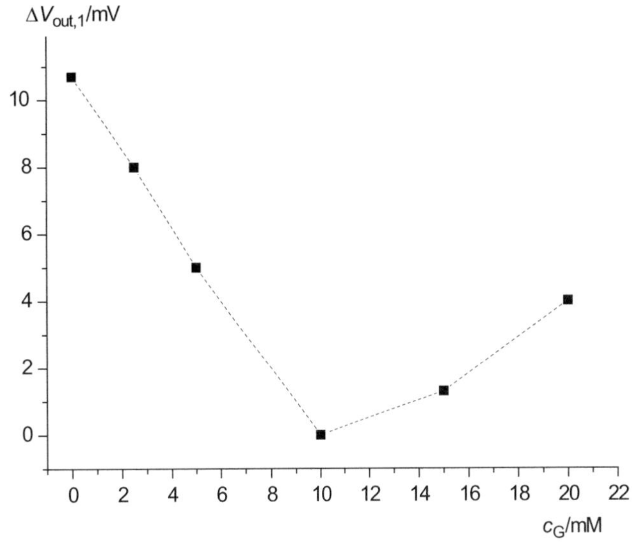

Fig. 4 Response of the sensor with AAm/3-AAmPBA/DMAPAAm/BIS hydrogel in dependence on glucose concentration in PBS solutions (pH 7.4, ionic strength 0.15 M)

investigated within a range from 0 mM to 20 mM of glucose. The swelling process of the hydrogel due to the change of the glucose concentration c_G in PBS solution was monitored by the corresponding change $\Delta V_{out,1}$ in the output voltage of the piezoresistive pressure sensor (Fig. 1, left).

AAm/3-AAmPBA/BIS Hydrogels

The PBA-containing AAm/3-AAmPBA hydrogels show a volume increase with increasing glucose concentration in PBS solution (Fig. 3). This increase occurs because glucose binding favours the charged form of pH-sensitive boronic acid. Hence, when the environmental glucose concentration increases, the fraction of charged boronic acid groups increases too, thereby increasing the osmotic swelling pressure within the gel, which subsequently leads to the gel swelling.

An apparent problem with this glucose sensing mechanism is that PBA cannot distinguish glucose from other cis-diol molecules and may give rise to errors when the glucose measurements are carried out in presence of other sugar molecules, for instance, fructose, which is also found in blood. Although the normal physiological level of fructose is approximately 500 times smaller than that of glucose (8 μM vs. 5.5 mM, respectively), the binding affinity of PBA for fructose exceeds that for glucose by a factor of 40 [1, 7, 22].

In order to investigate this potential interference, the sensor signal was measured in PBS solutions containing fructose within a range from 0 mM to 2.5 mM. Figure 3 shows a greater sensor response to fructose than to glucose. But at small fructose concentrations $c_F < 0.1$ mM no influence of fructose on the sensor signal was observed. As a step to progressively move towards in vivo experiments, an additional test was performed in PBS solution containing 5 mM glucose, 0.1 mM fructose and 27 mg/ml human serum albumin (HSA) in order to mimic body fluids (Fig. 3). No changes were observed in the gel swelling degree due to adding the protein within the physiological concentration range.

AAm/3-AAmPBA/DMAPAAm/BIS Hydrogels

Zwitterionic AAm/3-AAmPBA/DMAPAAm/BIS hydrogels containing anionic phenylboronic acid groups and cationic *N*-(3-dimethylaminopropyl acrylamide) (DMAPAAm) groups with incorporated pendant tertiary amines demonstrate a non-monotonic behaviour with the increasing glucose concentration: a gel shrinking at $c_G < 15$ mM and a gel swelling at $c_G \geq 15$ mM (Fig. 4).

The chosen tertiary amines are Lewis bases that accept a proton to become cations at physiological pH. In addition, the incorporation of tertiary amines facilitates the ionization of nearby boronic acid groups [23] and can form reversible cross-links with them.

The strongly hydrated glucose molecules exhibit strong interactions with water molecules, destroy the water-polymer hydrogen bonding and lead in this way to an enhancement of the gel hydrophobicity and to increasing cross-links between boronic acid groups and tertiary amine groups. This can explain the gel shrinking at small glucose concentrations $c_G < 15$ mM. At glucose concentrations $c_G \geq 15$ mM, the competitive formation of complexes between the charged phenylborate and glucose predominates over hydrophobic interactions and leads to the gel swelling.

The swelling state of hydrogels containing PBA and tertiary amine groups depends on the balance of hydrophilic and hydrophobic interactions in the three-component system of polymer-water-analyte, which determine the polymer swelling/shrinking with changing temperature. The addition of a small amount of glucose to the water-swollen gel perturbs the water-polymer hydrogen bonding and leads to strong polymer-polymer interactions and therefore to a gel shrinking. It was found in [22] that the AAM/3-APB/DMAPAA/BIS polymer network containing adsorbed glucose molecules starts to reswell with increase in temperature at the small glucose concentration $c_G = 5$ mM. In contrast, this hydrogel swells in 5 mM fructose solution at room temperature and shrinks with increasing temperature. A small temperature rise can be used in order to exactly detect the analyte and to enhance the gel selectivity for glucose relative to fructose.

Generally, the additives change the structure of water and thus the free energy of the interaction between polymer and water. The additives influence the solution viscosity and are classified into structure making (cosmotropes) and structure breaking (chaotropes). The viscosity η of an aqueous solution (up to about 0.1 M), relative to the viscosity η_0 of water at the same temperature, varies with the additive concentration c_α according to the Jones-Dole expression

$$\eta/\eta_0 = 1 + A\sqrt{c_\alpha} + Bc_\alpha, \quad (1)$$

where the coefficient A is related to the mobilities and interactions between the additives. The coefficient B reflects the interactions between the additive and the water molecules and results from the degree of water structuring. At higher concentrations, a further term $(Dc_\alpha{}^2)$ may be added to the right hand side of Eq. 1. It has been accepted that the sign of the coefficient B determines whether a certain additive is a water structure breaker (negative B) or a water structure maker (positive B). Adding an analyte-cosmotrope as well as increasing the temperature leads to an enhancement of the polymer chain stiffness, which affects the chain mobility. In contrast, adding an analyte-chaotrope leads to the opposite effect.

Both glucose and fructose molecules have a positive coefficient B and can be classified as water structure maker. The value of B-coefficient for glucose is higher than that for fructose [24]. Fructose exhibits a weaker hydration ability in water and a higher binding affinity to PBA, tends to disperse the polymer chains, preventing them from intramolecular interactions, and, as a result, increases the gel swelling. With the rise of temperature, fructose shows a smaller decrease in the B-coefficient value than glucose [24].

A complex information containing the value of the swelling degree of the functionalised polymer in the presence of the additive as well as the value and direction of temperature-induced changes might allow to exactly detect the analyte as well as its concentration range.

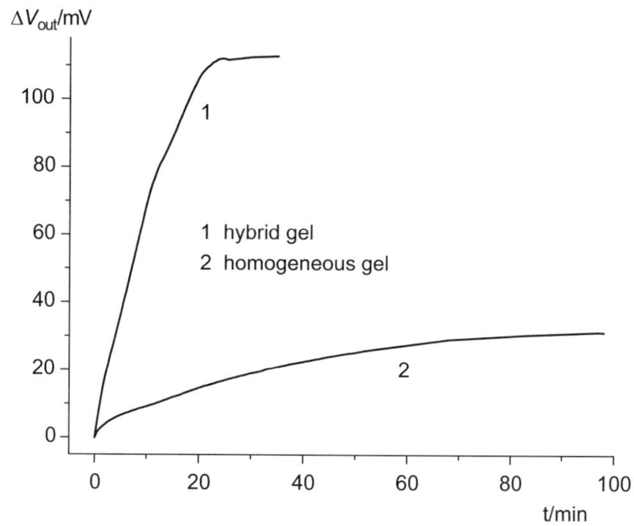

Fig. 5 Sensor output voltage V_{out} during the first swelling of a hybrid AAm/3-AAmPBA/DMAPAAm/BIS gel (*1*) and of a homogeneous AAm/3-AAmPBA/DMAPAAm/BIS gel (*2*) in PBS solution

Improving Gel Response Time

The gel response time depends on the cooperative diffusion coefficient and on the square of the sample dimension. Scaling to micro-dimensions enhances the time response. Consequently, a reduction of the sample size improves the sensor performance. However, a reduction of the gel thickness is limited by the necessity to obtain a sufficiently high sensor signal and, consequently, a sufficient sensitivity. It should be feasible to achieve an optimum between the sensor signal amplitude and the sensor response time by using a gel with an inhomogeneous structure which contains hydrophilic porous fibres and shows a fast solution uptake.

In order to improve the sensor response time, a hybrid hydrogel with hygroscopic fibres was used in the sensors. Figure 5 compares the responses of a hybrid AAm/3-AAmPBA/DMAPAAm/BIS gel and a homogeneous AAm/3-AAmPBA/DMAPAAm/BIS gel. The incorporated hydrophilic porous fibres led to a faster, and at the same time, increased solution uptake. The latter resulted in a higher swelling degree of the hydrogel, and consequently, in a higher sensor signal amplitude.

The hybrid hydrogel shows a rapid volume decrease with increasing glucose concentration in PBS solution (Fig. 6a) compared to the homogeneous hydrogel (Fig. 6b). Figure 6a demonstrates a significant reduction of the sensor response time for the hybrid hydrogel with incorporated hygroscopic fibres which accelerate the diffusion of the solution in the gel and, consequently, the gel swelling/deswelling.

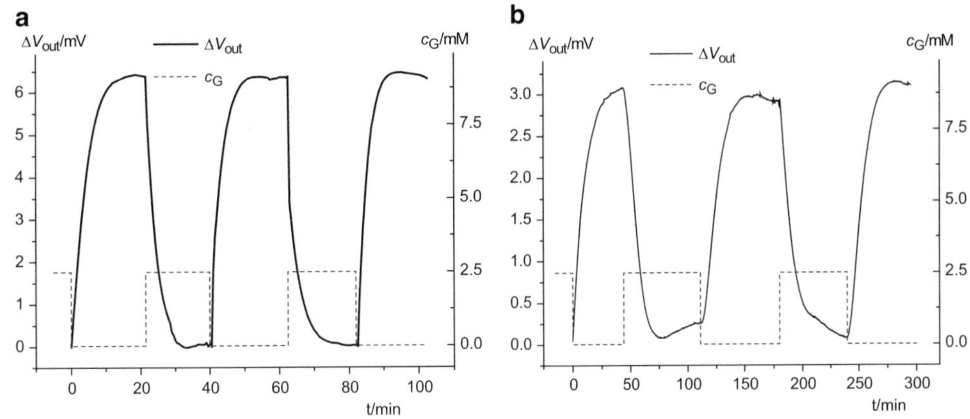

Fig. 6 Change ΔV_{out} of the sensor output voltage during the swelling/deswelling of a hybrid AAm/3-AAmPBA/DMAPAAm/BIS gel (**a**) and of a homogeneous AAm/3-AAmPBA/DMAPAAm/BIS gel (**b**) due to the cycling between glucose-free and glucose-containing PBS solutions

pH- and CO_2 – Sensitive Hydrogels

The pH-responsive biocompatible HPMA/DMAEMA/TEGMA hydrogel was used in the CO_2 sensor (Fig. 1, right). The DMAEMA monomer contains a pH-sensitive tertiary amines, HPMA was included to obtain a transition pH close to the physiological range near pH 7.4 [2, 25]. The sensing mechanism is based on the protonation of the tertiary amines on the DMAEMA backbone. At pH values lower than pH 8, elevated backbone protonation temporarily increases the osmotic swelling pressure within the hydrogel.

In order to evaluate the gel sensitivity to CO_2, various ratios of CO_2 and N_2 (as a reference gas) were bubbled through the 24 mM $NaHCO_3$ solution. The pH value of the solution was controlled using a Knick pH meter with pH/Pt-1000 probe. Figure 7 demonstrates a rapid pH decrease induced by CO_2 and as a consequence an increasing gel swelling. At N_2 bubbling, the pH value increased and a reversible gel deswelling was observed by means of the sensor output voltage.

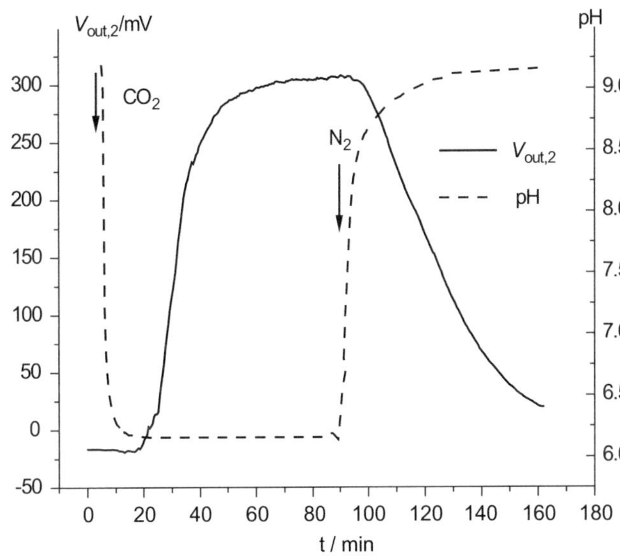

Fig. 7 Response of the sensor with HPMA/DMAEMA/TEGDMA hydrogel on the change of pH value of solution induced by alternating CO_2 and N_2 bubbling

Conclusions

Layers of metabolically sensitive hydrogels have been incorporated in piezoresistive microsensors for a monitoring of metabolite concentration. The response of enzyme-free copolymers with glucose-sensitive phenylboronic acid groups was studied in dependence on glucose concentration in simulated physiological solutions. The accuracy with which a sensor can track gradual changes in glucose was estimated. In order to investigate a potential interference with other components, which are also found in blood and in body fluids, the sensor signal was measured in PBS solutions containing fructose and human serum albumin. No changes were observed in the gel swelling degree due to adding fructose and the protein within the physiological concentration range.

Zwitterionic copolymers with phenylboronic acid and tertiary amine polymer groups demonstrated a non-monotonic behaviour with the increasing glucose concentration due to the competing binding of phenylborates to tertiary amines at small glucose concentrations. The addition of a small amount of glucose to the swollen gel perturbs the water-polymer hydrogen bonding and leads to strong polymer-polymer interactions. These are reflected as desorption of water by the gel. At high glucose concentration, when phenylborate-glucose interactions overtake phenylborate-tertiary amine interactions, the gel reswells. Competing weak interactions between glucose-water and glucose-polymer result in a complex "reentrant" swelling behavior of the hydrogel in glucose solutions.

Zwitterionic copolymers can be used for the enhancement of the gel selectivity for glucose relative to fructose. Fructose exhibits a weaker hydration ability in water and a higher binding affinity to PBA, tends to disperse the polymer chains, preventing them from intramolecular interactions, and as a result, monotonically increases the gel swelling degree with the increasing glucose concentration.

The response of glucose sensors with a hybrid gel and with a homogeneous gel has been investigated using simulated physiological solutions. A significant reduction of the sensor response time has been achieved for the hybrid hydrogel compared to the homogeneous hydrogel.

The sensitivity of *N,N*-dimethylaminoethyl methacrylate-based hydrogels to CO_2 was evaluated for their use in CO_2 sensors.

Combining a smart hydrogel and a micro fabricated pressure sensor chip in piezoresistive biochemical sensors allows to continuously monitor the analyte-dependent swelling of a hydrogel and hence the analyte concentration in ambient aqueous solutions.

Acknowledgments M. N. Avula, F. Solzbacher, P. Tathireddy, S. H. Cho, and J. J. Magda are acknowledged for the help with the parylene C coating as well as for the preparation of pre-gel solutions. The authors gratefully acknowledge support of this work from the Deutsche Forschungsgemeinschaft (SPP 1259, grant Ge 779/14-3).

References

1. Lin G, Chang S, Hao H, Tathireddy P, Orthner M, Magda J, Solzbacher F (2010) Osmotic swelling pressure response of smart hydrogels suitable for chronically implantable glucose sensors. Sens Actuators B-Chem 144:332–336
2. Orthner MP, Lin G, Avula M, Buetefisch S, Magda J, Rieth LW, Solzbacher F (2010) Hydrogel based sensor arrays (2 × 2) with perforated piezoresistive diaphragms for metabolic monitoring (in-vitro). Sens Actuators B-Chem 145(2):807–816. doi:10.1016/j.snb.2010.01.063
3. Guenther M, Gerlach G (2009) Hydrogels for chemical sensors. In: Gerlach G, Arndt K-F (eds) Hydrogel sensors and actuators, Springer series on chemical sensors and biosensors, vol 6. Springer, Berlin, pp 165–195
4. Guenther M, Gerlach G, Wallmersperger T, Solzbacher F, Magda JJ, Lin G, Tathireddy T, Orthner MP (2011) Biochemical microsensors on the basis of metabolically sensitive hydrogels. Proceedings of SPIE, 7976:D1–D9. doi: 10.1117/12.880460
5. Herber S, Bomer J, Olthuis W, Bergveld P, van den Berg A (2005) Miniaturized carbon dioxide gas sensor based on sensing of pH-sensitive hydrogel swelling with a pressure sensor. Biomed Microdevices7(3):197–204
6. Hilt JZ, Gupta AK, Bashir R, Peppas NA (2003) Ultrasensitive biomems sensors based on microcantilevers patterned with environmentally responsive hydrogels. Biomed Microdevices 5(3):177–184
7. Siegel RA, Gu Y, Lei M, Baldi A, Nuxoll E, Ziaie B (2010) Hard and soft micro- and nanofabrication: an integrated approach to hydrogel-based biosensing and drug delivery. J Control Release 141:303–313
8. Buenger D, Topuz F, Groll J (2012) Hydrogels in sensing applications. Prog Polym Sci 37(12):1678–1719
9. Muscatello MMW, Stunja LE, Asher SA (2009) Polymerized crystalline colloidal array sensing of high glucose concentrations. Anal Chem 81:4978–4986
10. Guenther M, Gerlach G, Wallmersperger T, Avula MN, Cho SH, Xie X, Devener BV, Solzbacher F, Tathireddy P, MagdaJJ, Scholz C, Obeid R, Armstrong T (2013) Smart hydrogel-based biochemical microsensor array for medical diagnostics. Adv Sci Technol 85:47–52. doi:10.4028/www.scientific.net/AST.85.47
11. Koschwanez HE, Reichert WM (2007) In vitro, in vivo and post explantation testing of glucose-detecting biosensors: current methods and recommendations. Biomaterials 28:3687–3703
12. Wang J (2008) Electrochemical glucose biosensors. Chem Rev 108:814–825
13. Koudelka M, Rohner-Jeanrenaud F, Terrettaz J, Bobbioni-Harsch E, de Rooij NF, Jeanrenaud B (1991) In-vivo behaviour of hypodermically implanted microfabricated glucose sensors. Biosens Bioelectron 6:31–36
14. Jung DY, Magda JJ, Han IS (2000) Catalase effects on glucose-sensitive hydrogels. Macromolecules 33:3332–3336
15. Matsumoto A, Kurata T, Shiino D, Kataoka K (2004) Swelling and shrinking kinetics of totally synthetic, glucose-responsive polymer gel bearing phenylborate derivative as a glucose-sensing moiety. Macromolecules 37:1502–1510
16. Guenther M, Gerlach G, Wallmersperger T, Solzbacher F, Tathireddy P, Magda JJ, Lin G, Orthner MP (2011) Hydrogel-Based biochemical sensors. Proceedings of 15th international conference on sensors and measurement technology "Sensor + Test 2011", Nuremberg, pp 211–215 (on CD-ROM, ISBN 978-3-9810993-9-3)
17. Wallmersperger T (2009) Modelling and simulation of the chemo-electro-mechanical behaviour. In: Gerlach G, Arndt K-F (eds) Hydrogel sensors and actuators, Springer series on chemical sensors and biosensors, vol 6. Springer, Berlin, pp 137–163
18. Wallmersperger T, Ballhause D, Kröplin B, Günther M, Gerlach G (2009) Coupled multi-field formulation in space and time for the simulation of intelligent hydrogels. J Intel Mat Syst Str 20(12):1483–1492
19. Wallmersperger T, Keller K, Kröplin B, Guenther M, Gerlach G (2011) Chemo-electro-mechanical modeling of pH-sensitive hydrogels. Proc SPIE 7976:101–109
20. Wallmersperger T, Keller K, Kröplin B, Guenther M, Gerlach G (2011) Modeling and simulation of pH-sensitive hydrogels. Coll Pol Sci 289:535–544
21. Keller K, Wallmersperger T, Kröplin B, Guenther M, Gerlach G (2011) Modelling of temperature-sensitive polyelectrolyte gels by the use of the coupled chemo-electro-mechanical formulation. Mech Adv Mater Struct 18:511–523
22. Horkay F, Cho SH, Tathireddy P, Rieth L, Solzbacher F, Magda J (2011) Thermodynamic analysis of the selectivity enhancement obtained by using smart hydrogels that are zwitterionic when detecting glucose with boronic acid moieties. Sens Actuators B-Chem 160:1363–1371.doi:10.1016/j.snb.2011.09.079

23. Hisamitsu I, Kataoka K, Okano T, Sakurai Y (1997) Glucose-responsive gel from phenylborate polymer and poly(vinyl alcohol): prompt response at physiological pH through the interaction of borate with amino group in the gel. Pharm Res 14:289–293
24. Dey PC, Motin MA, Biswas TK, Huque EM (2003) Apparent molar volume and viscosity studies on some carbohydrates in solutions. Chem Mon 134:797–809. doi:10.1007/s00706-002-0530-7
25. Schulz V, Gerlach G, Günther M, Magda JJ, Solzbacher F (2010) Piezoresistive pH microsensors based on stimuli-sensitive polyelectrolyte hydrogels. Tech Mess 77(3):179–186. doi:10.1524/teme.2010.0045

Index

G. Sadowski and W. Richtering (eds.), *Intelligent Hydrogels*, Progress in Colloid and Polymer Science 140,
DOI 10.1007/978-3-319-01683-2, © Springer International Publishing Switzerland 2013

MIX
Papier aus verantwortungsvollen Quellen
Paper from responsible sources
FSC® C105338

If you have any concerns about our products,
you can contact us on
ProductSafety@springernature.com

In case Publisher is established outside the EU,
the EU authorized representative is:
Springer Nature Customer Service Center GmbH
Europaplatz 3, 69115 Heidelberg, Germany

Printed by Libri Plureos GmbH
in Hamburg, Germany